DEGREES AND RADIANS (1.1, 2.1)

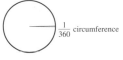

1°

1 radian

$$\frac{\theta_d}{180°} = \frac{\theta_r}{\pi \text{ rad}}$$

$$\text{Degrees} \times \frac{\pi}{180°} = \text{Radians}$$

$$\text{Radians} \times \frac{180°}{\pi} = \text{Degrees}$$

TRIGONOMETRIC FUNCTIONS (2.3)

$$\sin x = \frac{b}{r} \qquad \csc x = \frac{r}{b} \ (b \neq 0)$$

$$\cos x = \frac{a}{r} \qquad \sec x = \frac{r}{a} \ (a \neq 0)$$

$$\tan x = \frac{b}{a} \ (a \neq 0) \qquad \cot x = \frac{a}{b} \ (b \neq 0)$$

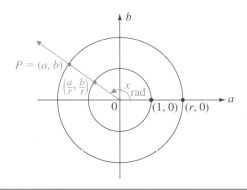

SPECIAL VALUES (2.5)

θ	$\sin\theta$	$\csc\theta$	$\cos\theta$	$\sec\theta$	$\tan\theta$	$\cot\theta$
0° or 0	0	N.D.	1	1	0	N.D.
30° or $\pi/6$	1/2	2	$\sqrt{3}/2$	$2/\sqrt{3}$	$1/\sqrt{3}$	$\sqrt{3}$
45° or $\pi/4$	$1/\sqrt{2}$	$\sqrt{2}$	$1/\sqrt{2}$	$\sqrt{2}$	1	1
60° or $\pi/3$	$\sqrt{3}/2$	$2/\sqrt{3}$	1/2	2	$\sqrt{3}$	$1/\sqrt{3}$
90° or $\pi/2$	1	1	0	N.D.	N.D.	0

N.D. = Not defined

PYTHAGOREAN THEOREM (1.3, C.2)

$$a^2 + b^2 = c^2$$

SPECIAL TRIANGLES (2.5)

30°–60°–90° triangle   45°–45°–90° triangle

NUMBER OF SIGNIFICANT DIGITS (A.3)

Form of Number	Number of Significant Digits
x, with no decimal point	Count the digits of x from left to right, starting with the first digit and ending with the last nonzero digit.
x, with a decimal point	Count the digits of x from left to right, starting with the first nonzero digit and ending with the last digit (which may be zero).

ACCURACY OF CALCULATED VALUES (A.3)

The number of significant digits in a calculation involving multiplication, division, powers, and/or roots is the same as the number of significant digits in the number in the calculation with the smallest number of significant digits.

ACCURACY FOR TRIANGLES (1.3)

Angle to Nearest	Significant Digits for Side Measure
1°	2
10′ or 0.1°	3
1′ or 0.01°	4
10″ or 0.001°	5

(Continued inside back cover)

QUICK REFERENCE CARD

Raymond A. Barnett, Michael R. Ziegler,
Karl E. Byleen, Dave Sobecki,
Analytic Trigonometry with Applications,
Tenth Edition

Also available to help you succeed in this course:

*Student Solutions Manual to Accompany
Analytic Trigonometry with Applications,
Tenth Edition*

A student manual, containing solutions to all
chapter review and cumulative review exercises
and all odd-numbered problems in the book.
(ISBN: 978-0470-28076-8).

Wiley Plus

A powerful on-line assessment system, containing
a large bank of skill-building problems, homework
problems, and solutions. Allows students
unlimited practice. Visit wiley.com/college/wileyplus
for more details.

*For more information about this book or any
John Wiley & Sons, Inc., product, please visit our
web site:* **www.wiley.com/barnett**

John Wiley & Sons, Inc.

TRIGONOMETRIC IDENTITIES

Reciprocal Identities

$$\csc x = \frac{1}{\sin x} \qquad \sec x = \frac{1}{\cos x} \qquad \cot x = \frac{1}{\tan x}$$

Quotient Identities

$$\tan x = \frac{\sin x}{\cos x} \qquad \cot x = \frac{\cos x}{\sin x}$$

Identities for Negatives

$$\sin(-x) = -\sin x \qquad \cos(-x) = \cos x$$
$$\tan(-x) = -\tan x$$

Pythagorean Identities

$$\sin^2 x + \cos^2 x = 1 \qquad \tan^2 x + 1 = \sec^2 x$$
$$1 + \cot^2 x = \csc^2 x$$

Sum Identities

$$\sin(x + y) = \sin x \cos y + \cos x \sin y$$
$$\cos(x + y) = \cos x \cos y - \sin x \sin y$$
$$\tan(x + y) = \frac{\tan x + \tan y}{1 - \tan x \tan y}$$

Difference Identities

$$\sin(x - y) = \sin x \cos y - \cos x \sin y$$
$$\cos(x - y) = \cos x \cos y + \sin x \sin y$$
$$\tan(x - y) = \frac{\tan x - \tan y}{1 + \tan x \tan y}$$

Cofunction Identities

(Replace $\pi/2$ with 90° if x is in degree measure.)

$$\sin\left(\frac{\pi}{2} - x\right) = \cos x \qquad \cos\left(\frac{\pi}{2} - x\right) = \sin x$$
$$\tan\left(\frac{\pi}{2} - x\right) = \cot x \qquad \cot\left(\frac{\pi}{2} - x\right) = \tan x$$
$$\sec\left(\frac{\pi}{2} - x\right) = \csc x \qquad \csc\left(\frac{\pi}{2} - x\right) = \sec x$$

Product–Sum Identities

$$\sin x \cos y = \frac{1}{2}[\sin(x + y) + \sin(x - y)]$$
$$\cos x \sin y = \frac{1}{2}[\sin(x + y) - \sin(x - y)]$$
$$\sin x \sin y = \frac{1}{2}[\cos(x - y) - \cos(x + y)]$$
$$\cos x \cos y = \frac{1}{2}[\cos(x + y) + \cos(x - y)]$$

TRIGONOMETRIC IDENTITIES (cont'd)
LAWS OF SINES AND COSINES

Sum–Product Identities

$$\sin x + \sin y = 2 \sin \frac{x + y}{2} \cos \frac{x - y}{2}$$
$$\sin x - \sin y = 2 \cos \frac{x + y}{2} \sin \frac{x - y}{2}$$
$$\cos x + \cos y = 2 \cos \frac{x + y}{2} \cos \frac{x - y}{2}$$
$$\cos x - \cos y = -2 \sin \frac{x + y}{2} \sin \frac{x - y}{2}$$

Double-Angle Identities

$$\sin 2x = 2 \sin x \cos x \qquad \cos 2x = \begin{cases} \cos^2 x - \sin^2 x \\ 1 - 2 \sin^2 x \\ 2 \cos^2 x - 1 \end{cases}$$

$$\tan 2x = \frac{2 \tan x}{1 - \tan^2 x} = \frac{2 \cot x}{\cot^2 x - 1} = \frac{2}{\cot x - \tan x}$$

Half-Angle Identities

$$\sin \frac{x}{2} = \pm\sqrt{\frac{1 - \cos x}{2}}$$

$$\cos \frac{x}{2} = \pm\sqrt{\frac{1 + \cos x}{2}}$$

$$\tan \frac{x}{2} = \frac{1 - \cos x}{\sin x} = \frac{\sin x}{1 + \cos x} = \pm\sqrt{\frac{1 - \cos x}{1 + \cos x}}$$

Sign is determined by quadrant in which x/2 lies.

$$\sin^2 x = \frac{1 - \cos 2x}{2} \qquad \cos^2 x = \frac{1 + \cos 2x}{2}$$

$$\tan^2 x = \frac{1 - \cos 2x}{1 + \cos 2x}$$

Law of Sines

$$\frac{\sin \alpha}{a} = \frac{\sin \beta}{b} = \frac{\sin \gamma}{c}$$

Law of Cosines

$$a^2 = b^2 + c^2 - 2bc \cos \alpha$$
$$b^2 = a^2 + c^2 - 2ac \cos \beta$$
$$c^2 = a^2 + b^2 - 2ab \cos \gamma$$

Degrees and Radians

$$\frac{1}{360}\text{ circumference}$$

1°

1 radian

$$\frac{\theta_d}{180°} = \frac{\theta_r}{\pi \text{ rad}}$$

$$\text{Degrees} \times \frac{\pi}{180°} = \text{Radians}$$

$$\text{Radians} \times \frac{180°}{\pi} = \text{Degrees}$$

Angles and Arcs

θ in Degrees

$$\frac{\theta}{360°} = \frac{s}{C}$$

θ in Radians

$$\theta = \frac{s}{r}$$

$$s = r\theta$$

C (circumference)

Pythagorean Theorem

$$a^2 + b^2 = c^2$$

N.D. = Not defined

Similar Triangles

$$\frac{a}{a'} = \frac{b}{b'} = \frac{c}{c'}$$

Trigonometric Functions

$$\sin x = \frac{b}{r} \qquad \csc x = \frac{r}{b} \ (b \neq 0)$$

$$\cos x = \frac{a}{r} \qquad \sec x = \frac{r}{a} \ (a \neq 0)$$

$$\tan x = \frac{b}{a} \ (a \neq 0) \qquad \cot x = \frac{a}{b} \ (b \neq 0)$$

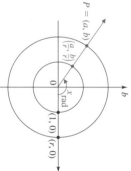

$P = (a, b)$, $\left(\frac{a}{r}, \frac{b}{r}\right)$, 0, x rad, $(1, 0)$, $(r, 0)$

Special Triangles

30°–60°–90° triangle   45°–45°–90° triangle

Special Values

θ	$\sin\theta$	$\csc\theta$	$\cos\theta$	$\sec\theta$	$\tan\theta$	$\cot\theta$
0° or 0	0	N.D.	1	1	0	N.D.
30° or π/6	1/2	2	$\sqrt{3}$/2	2/$\sqrt{3}$	1/$\sqrt{3}$	$\sqrt{3}$
45° or π/4	1/$\sqrt{2}$	$\sqrt{2}$	1/$\sqrt{2}$	$\sqrt{2}$	1	1
60° or π/3	$\sqrt{3}$/2	2/$\sqrt{3}$	1/2	2	$\sqrt{3}$	1/$\sqrt{3}$
90° or π/2	1	1	0	N.D.	N.D.	0

Accuracy for Triangles

Angle to Nearest	Significant Digits for Side Measure
1°	2
10′ or 0.1°	3
1′ or 0.01°	4
10″ or 0.001°	5

Graphing Trigonometric Functions

$$y = A\sin(Bx + C) \qquad y = A\cos(Bx + C)$$

$$\text{Amplitude} = |A| \quad \text{Period} = \frac{2\pi}{B} \quad \text{Frequency} = \frac{B}{2\pi}$$

$$\text{Phase shift} = -\frac{C}{B} \begin{cases} \text{left} & \text{if} -C/B < 0 \\ \text{right} & \text{if} -C/B > 0 \end{cases}$$

$y = \sin x$

$y = \cos x$

$$y = A\tan(Bx + C) \qquad y = A\cot(Bx + C)$$

$$\text{Period} = \frac{\pi}{B} \quad \begin{array}{l}\text{Phase} \\ \text{shift}\end{array} = -\frac{C}{B} \begin{cases} \text{left} & \text{if} -C/B < 0 \\ \text{right} & \text{if} -C/B > 0 \end{cases}$$

$y = \tan x$

$y = \cot x$

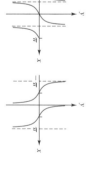

Inverse Trigonometric Functions

$y = \sin^{-1} x$ means $x = \sin y$

where $-\dfrac{\pi}{2} \leq y \leq \dfrac{\pi}{2}$ and $-1 \leq x \leq 1$

$y = \cos^{-1} x$ means $x = \cos y$

where $0 \leq y \leq \pi$ and $-1 \leq x \leq 1$

$y = \tan^{-1} x$ means $x = \tan y$

where $-\dfrac{\pi}{2} < y < \dfrac{\pi}{2}$ and x is any real number

$y = \cot^{-1} x$ means $x = \cot y$

where $0 < y < \pi$ and x is any real number

$y = \sec^{-1} x$ means $x = \sec y$

where $0 \leq y \leq \pi$, $y \neq \dfrac{\pi}{2}$, and $x \leq -1$ or $x \geq 1$

$y = \csc^{-1} x$ means $x = \csc y$

where $-\dfrac{\pi}{2} \leq y \leq \dfrac{\pi}{2}$, $y \neq 0$, and $x \leq -1$ or $x \geq 1$

Applications Index

Acoustics:
  frequency determinations, 155
  musical instruments, 194–196, 200
  noise control, 200–201
  sonic booms, 93–94, 227, 305–306
  sound frequency, 149–150
  sound waves, 221–222
  static, 200
Advertising, 343
Aeronautics:
  aircraft design, 365–366
  air-to-air refueling, 34–35
  altitude determinations, 37
  distance determinations, 37, 38, 49
  navigation, 347, 373, 376, 429, 479
  rocket flight, 222
  slingshot effect, 236–237
  sonic booms, 93–94
  spacecraft orbits, 236–237
Aircraft design, 365–366
Air-to-air refueling, 34–35
Alternating current generators, 175–177,
  187, 327
Angular velocity, 66, 68–69
Apparent diameter, 13
Apparent velocity, 388
Architecture:
  building design, 284–285, 344
  doorway design, 336
  ramp construction, 48
  roof overhang, 38
Astronomy:
  aphelion, 453
  apparent diameter, 13
  Earth's orbit, 64–65, 71
  Earth-Venus distance, 366
  Hubble space telescope, 69
  Jupiter rotation and its moons, 72
  Jupiter's orbit, 71
  light-year, 489
  lunar craters, 20
  lunar geography, 38
  lunar mountains, 20
  lunar phases, 166–168, 334

  mean sidereal day, 68
  mean solar day, 68
  Mercury's orbit, 328
  Neptune's velocity, 71
  perihelion, 453
  planetary orbits, 469–471
  planetary speed, 453
  sun-Mercury distance, 36, 452–453
  sun's diameter, 64
  sun's rotation, 71
  sun-Venus distance, 35
  total solar eclipse, 10
  velocity due to rotation, 71
Auto-racing, 392–393

Balloon flight, 49
Baseball, 337
Beacon lights, 213
Beat (music), 273–274
Bicycling, 65, 70, 72, 349
Billiards, 226
Bioengineering:
  limb movement, 157, 222
  range of motion, 63–64
Biology:
  biorhythms, 174
  bird populations, 308
  eye, 366
  of the eye, 12, 336–337
  modeling body temperature, 189
  perception of light, 177–178
  predator-prey analysis, 213–215
Biomechanics, traction in, 396, 410, 430
Biorhythms, 174
Bird populations, 308
Boating, 92–93, 122
  lightning protection in, 38–39
  navigation in, 343, 365, 376, 388–389,
    395, 478
  velocities in, 395
Boat safety, 37–38, 348
Body temperatures, 189
Bow waves, 92–93, 98, 122
Bungee jumping, 189, 344

Buoy, mass of a, 151, 221
Business cycles, 200, 222, 228, 336

Camera lenses, 20–21
Carbon dioxide, 201
Centrifugal force, 392–393, 395
Climate, 337
Coastal patrol, 364
Coastal piloting, 36–37, 365
Communications, 91
Construction:
  area calculations for, 21
  cost determinations in, 385
  crane positioning, 225
  of fire lookouts, 377
  ladder placement, 37
  length determinations in, 268
  retaining walls, 98
  volume determinations in, 268
Cranes, 225

Damped harmonic motion, 183–186
Daylight, duration of, 328, 349
DeMoivre's theorem, 478
Diagonal parking, 39
Diameter, apparent, 13
Diastolic pressure, 156
Distance, between two points, 248
Diving, 97
Dogsledding, 416
Drive wheels and shafts, 65

Earth, 35–36
  axis of rotation, 83
  circumference, 7–8
  diameter, 8
  distance to Venus, 366
  gravitational constant of, 42
  polar equation of orbit, 471
  radius, 8, 39
  seasons, 83
  speed of orbit, 71
  sunrise, 124
  sunset, 139–140

APPLICATIONS INDEX

Earth science:
  carbon dioxide in atmosphere, 201
  climate, 337
  daylight duration, 349
  hours of daylight, 328
  lake temperatures, 199
  pollution, 221
  twilight duration, 480
Electricity:
  circuits, 22, 155, 171, 227, 348, 478
  current, 84, 122, 175–177, 187, 327,
    343
  wind generators, 67, 79–80
Electromagnetic spectrum, 178
Electromagnetic waves, 177–181,
    188–189, 222, 227
  frequency, 177–178
  gamma rays, 178
  infrared, 178
  microwaves, 178
  radar, 178
  radio, 178
  short waves, 178
  television, 178
  ultraviolet, 178
  visible light, 178
  wavelength, 178–179
  x rays, 178
Emergency vehicles, 228
Energy consumption, 171–172
Engineering:
  bicycle chains, 349
  bicycle movement, 70
  cable tension, 405–406
  compression force, 430
  cross-sectional area, 480
  donkey pumps, 63
  fire lookout, 377
  freeway design, 395
  fuel tanks, 307
  gears, 121, 226–227
  identity use, 268–269, 284, 348
  piston motion, 84, 366–367
  propellers, 158
  pulley systems, 307–308
  race track design, 392–393
  railroad track, 479–480
  rotary and linear motion, 156–157
  rotation in water, 157–158
  shaft motion, 65
  shaft rotation, 70
  static equilibrium, 404–407, 409
  tunnel design, 344, 376–377, 429
  velocity of automobile wheels, 121
Eye, 12, 59, 336–337, 366

Ferris wheels, 156–157, 327
Fiber optics, 91

Field (in physics), 175
Firefighting, 226
Fire spotting, 364
Flight safety, 38
Floating objects, 150–151, 156, 221
Force(s):
  centrifugal, 392–393, 395
  resolution of, 395–396
  resultant, 390–391, 395, 429, 462
Forest fires, 226
Fourier series, 196–197
Freeway design, 395
Frequency, 149–150

Galileo's experiment, 42
Gamma rays, 178
Gardening, 385
Gas storage, 158–159
Gears, 121
Generators, alternating current, 175–177,
    187, 327
Geography:
  depth of canyons, 17–18
  distances between cities, 12, 49, 225
  nautical miles, 12
  parallel of latitude, 40
Geology, 257
Gnomon, 356
Golf, 97, 337
Gravitational constant, 42
Great circle, 306

Harmonics, 159–160, 165–166, 175
Harmonic motion, 183–186
  damped, 183–186
  simple, 159
Hubble space telescope, 69
Huygens probe, 13

Illusions, 95–96
Index of refraction, 86–87, 89, 97, 98
Infrared waves, 178

Javelin throw, 267
Jupiter:
  angular velocity, 71
  linear velocity, point on surface, 71
  rotation of, and its moons, 72

Lawn mowing, 416–417
Lifeguard problem, 40–41
Light cones, 98
Lightning protection, 38–39
Light waves, 177–179, 227
  cone, 98
  fiber optics, 91
  intensity of, 83
  mirage, 88

  polarizing filters, 327
  reflected, 89–91
  refraction, 86–91, 122, 258
  Snell's law, 87, 114
  speed of, 114–115
  telescope, 88
Light-year, 489
Logistics problem, 44, 349

Mach number, 94
Marathon, 5
Mass, determinations for floating objects,
    221
Medicine:
  eye surgery, 336–337
  fiber optics in, 91
  stress tests, 48–49
  traction in, 430
  use of traction, 396, 410
Mercury, distance to sun, 36, 328,
    452–453
Microwaves, 178, 189
Mirage, 88
Mollweide's equation, 364
Moon:
  craters, 20
  diameter of, 59
  in eclipse, 10
  geography, 38
  mountains, 20
  phases, 166–168, 334
Moving wave equation, 182
Music:
  beats in, 273–274, 276, 285
  tones in, 279
Musical instruments:
  modeling sounds produced by,
    194–196, 200
  tuning, 273

Nautical miles, 12
Navigation:
  by airplane, 395
  by airplanes, 347, 373, 376, 429, 479
  beacon rotation, 72
  boat, 39–40
  by boat, 343, 365, 376, 388–389, 395,
    478
  coastal piloting, 36–37
  determining distances, 37–38, 361–362
  on Earth's surface, 225
  and sun, 49
Neptune, linear velocity, 71
Newton's second law of motion, 404
Noise control, 200–201
nth-root theorem, 464–467

Oceanography:

modeling sea temperatures, 158
tidal current curve, 448
Optics:
    modeling light waves, 86–91, 177–179
    refraction, 86–91, 122
    refraction of light, 227
    speed of light in water, 114–115
Optical illusions, 95–96
Optimization, 50
Orbiting Astronomical Observatory, 12

Paddle wheels, 157
Parametric equations, 236
Particle energy, 94–95
Pendulums, 63
Perception (psychology), 95–96
Perspective illusions, 95–96
Phase shift, 172–173
Photography:
    field of view, 12, 64
    lens focusing, 20–21
    polarizing filters, 327
    viewing angle, 305
Physics. See also Acoustics; Optics
    angular velocity, 70
    atomic particle orbit, 70–71
    bow wave modeling, 92–93
    compression determinations, 406–407
    electric current modeling, 175–177, 187
    electromagnetic waves modeling, 177–181, 188–189, 222
    field, 175
    floating objects, 150–151
    Galileo's experiment, 42
    harmonic motion, 183–186
    harmonic motion modeling, 183–186
    light refraction, 258
    mass determinations, 156
    particle energy modeling, 94–95
    pendulums, 63
    spring—mass systems, 155, 189, 221, 222, 227, 228, 279, 285, 324–325, 327, 344, 348
    static equilibrium, 479
    tension determinations, 405–407, 409
    water waves modeling, 181–183, 187–188, 222
    wave movement in, 171
    wind farms, 308
Physiology:
    blood pressure, 156
    of breathing, 155, 199
Pipelines, 41
Pistons, 84
Planetary orbits, 469–471
Pollution, 221
Predator-prey relationships, 213–215

Projectile distance, 267
Propellers, 158
Psychology, of perception, 95–96, 98
Pulleys, 65, 121, 307–308
Pumps, 63

Racetrack design, 393
Radar, 178
Radio towers, 348
Radio waves, 124, 178, 180, 189
Railroads, 479–480
    grades, 226
    tunnels for, 344
Range of motion, 63–64
Refraction, 86–88, 97, 114–115, 227, 258
    angle, 86
    index of, 86
Resolution of forces, 395–396
Resonance, 183–185
Resultant force, 390–391, 395, 429, 462
Resultant velocity, 388
Retaining walls, 98
Rockets, 222
Roof overhang, 38
Rotary motion, 327

Sailboat racing, 452
Sales cycles, 200, 228, 336
Satellites, 70, 121–122, 236, 366, 377, 429
    geostationary, 71
    spy, 64
Satellite telescopes, 12
Saturn, 13, 471
Saws, 226–227
Sawtooth waves, 197
Scales (weighing), 155
Search and rescue, 376
Seasonal business cycles, 200, 222, 228, 336
Seasons, 83–84
Shadow problem, 48, 307
Shafts, 65, 70
Short waves, 178
Simple harmonic motion, 159
Soccer, 306
Solar eclipse, 10
Solar energy, 49–50, 83–84, 479
Sonic booms, 93–94, 227, 305–306
Sound barrier, 70
Sound waves, 93–94, 124, 149–150, 155, 193–196, 200, 221–222. See also Acoustics
    cone, 93–94, 98
    equation, 149–150
    fundamental tone, 194
    overtones, 194
    partial tones, 194

tuning forks, 149
Space flight:
    angular velocity, 70
    communication, 377
    geostationary orbit, 377
    linear velocity, 69
    orbits in, 39, 306, 429
    reduced-gravity simulation, 430
    re-entry of shuttles, 38
    satellites, see Satellites
    satellite telescopes, 12
    shuttle movement relative to Earth, 71–72
Sports:
    auto-racing, 392–393
    baseball, 337
    bicycling, 65, 70, 72, 349
    billiards, 226
    bungee jumping, 344
    golfing, 97, 337
    javelin throwing, 267
    marathon, 5
    sailboat racing, 452
    soccer, 306
    tennis, 19
Spring—mass systems, 155, 183, 189, 221, 222, 227, 228, 279, 285, 324–325, 327, 344, 348
Square wave, 197
Static, modeling sound and, 200
Static equilibrium, 404–407, 409, 479
Statistics, 308
Stress tests, 48–49
Sun, 13
    diameter, 9
    distance to Mercury, 452–453
    eclipse of, 10
    Mercury, distance to, 36
    rotation of, 71
    solar energy, 49–50
Sundials, 356–357
Sunrise times, 124, 139–140, 173
Sunset times, 139–140, 223
Surveying:
    angle measurement, 371–372
    distance measurement, 41, 42, 50, 378, 429
    gas tank height, 64
    Grand Canyon width, 364
    height measurement, 37, 41, 42, 48, 226, 257
    lake length, 375–376, 478
    land cost determinations, 385
    mine shaft, 16, 34
    mountain height, 365
    optimization, 349
    river width, 364, 365
    tree height, 15–16, 365, 478

underground cable, 364–365
volcanic cinder cone, 375
walkways, 227–228
Suspension systems, 183–185
Systolic pressure, 156

Television, 178
Temperature, 158, 173, 228
Temperatures, 199
Tennis, 19
Tidal current curve, 448
Tightrope walking, 409
Trigonometric substitution, 236, 284, 348
Tsunami, 188
Tuning forks, 149, 221–222
Twilight, duration of, 480

Ultraviolet waves, 178, 181

Uranus, 13

Velocity:
  actual, 388
  angular, 66, 68–69
  apparent, 388
  linear, 66–69
  resultant, 388
Venus:
  distance to Earth, 366
  polar equation of orbit, 471
  sun, distance to, 35
Viewing angle, 305, 343
Volume, of a cone, 225
Voyager I, 236
Voyager II, 236

Water:
  moving wave equation, 182
  speed, 182
  tsunami, 188
  wavelength, 181–182
  waves, 124, 171, 181–183, 187–188, 222, 479
Wavelength, 181–182
Wheels:
  angular velocity of, 121
  and shafts, 65
Wind farms, 308
Wind generators, 67, 79–80
Work, 416–417, 419

X rays, 178, 188–189

Achieve Positive Learning Outcomes

www.wileyplus.com

WileyPLUS combines robust course management tools with interactive teaching and learning resources all in one easy-to-use system. It has helped over half a million students and instructors achieve positive learning outcomes in their courses.

WileyPLUS contains everything you and your students need— and nothing more, including:

- ➕ The entire textbook online—with dynamic links from homework to relevant sections. Students can use the online text and save up to half the cost of buying a new printed book.
- ➕ Automated assigning & grading of homework & quizzes.
- ➕ An interactive variety of ways to teach and learn the material.
- ➕ Instant feedback and help for students…available 24/7.

**See and try
WileyPLUS** *in action!*
Details and Demo:
www.wileyplus.com

WILEY PLUS

www.wileyplus.com

Wiley is committed to making your entire *WileyPLUS* experience productive & enjoyable by providing the help, resources, and persona support you & your students need, when you need it.
It's all here: www.wileyplus.com –

TECHNICAL SUPPORT:

- A fully searchable knowledge base of FAQs and help documentation, available 24/7
- Live chat with a trained member of our support staff during business hours
- A form to fill out and submit online to ask any question and get a quick response
- **Instructor-only** phone line during business hours: 1.877.586.0192

FACULTY-LED TRAINING THROUGH THE WILEY FACULTY NETWORK:
Register online: www.wherefacultyconnect.com

Connect with your colleagues in a complimentary virtual seminar, with a personal mentor in your field, or at a live workshop to share best practices for teaching with technology.

1ST DAY OF CLASS...AND BEYOND!

Resources You & Your Students Need to Get Started & Use *WileyPLUS* from the first day forward.

1st DAY OF CLASS ...AND BEYOND!

- 2-Minute Tutorials on how to set up & maintain your *WileyPLUS* course
- User guides, links to technical support & training options
- *WileyPLUS for Dummies*: Instructors' quick reference guide to using *WileyPLUS*
- Student tutorials & instruction on how to register, buy, and use *WileyPLUS*

YOUR *WileyPLUS* ACCOUNT MANAGER:

Your personal *WileyPLUS* connection for any assistance you need!

WILEY PLUS QuickStart

SET UP YOUR *WileyPLUS* COURSE IN MINUTES!

Selected *WileyPLUS* courses with QuickStart contain pre-loaded assignments & presentations created by subject matter experts who are also experienced *WileyPLUS* users.

Interested? See and try WileyPLUS *in action!*
Details and Demo: *www.wileyplus.com*

Analytic Trigonometry

with Applications

10th
Edition

Raymond A. Barnett
Merritt College

Michael R. Ziegler
Marquette University

Karl E. Byleen
Marquette University

Dave Sobecki
Miami University Hamilton

WILEY

John Wiley & Sons, Inc.

PUBLISHER	Laurie Rosatone
ACQUISITIONS EDITOR	Jessica Jacobs
DEVELOPMENT EDITOR	Anne Scanlan-Rohrer
PRODUCTION MANAGER	Dorothy Sinclair
MARKETING MANAGER	Jaclyn Elkins
SENIOR MARKETING ASSISTANT	Chelsee Pengal
SENIOR EDITORIAL ASSISTANT	Jeffrey Benson
SENIOR PRODUCTION EDITOR	Sandra Dumas
SENIOR DESIGNER	Madelyn Lesure
MEDIA EDITOR	Melissa Edwards
PRODUCTION MANAGEMENT SERVICES	Suzanne Ingrao
COVER/CHAPTER OPENER IMAGE	Courtesy of Peter Bennetts and ARM Architecture Urban Design Masterplanning

This book was typeset in 10/12 Times Roman by Prepare and printed and bound by R. R. Donnelley (Jefferson City). The cover was printed by R. R. Donnelley (Jefferson City).

The paper in this book was manufactured by a mill whose forest management programs include sustained yield harvesting of its timberlands. Sustained yield harvesting principles ensure that the number of trees cut each year does not exceed the amount of new growth.

This book is printed on acid-free paper.

ISBN 13   978-0470-28076-8

Printed in the United States of America

10 9 8 7 6 5 4 3 2 1

Preface

The 10th edition of *Analytic Trigonometry with Applications* is designed for a one-term course in trigonometry and for students who have had $1\frac{1}{2}$–2 years of high school algebra or the equivalent. The choice and independence of topics make the text readily adaptable to a variety of courses (see the starred sections in the table of contents). Fundamental to a book's growth and effectiveness is classroom use and feedback. Now in its 10th edition, *Analytic Trigonometry with Applications* has had the benefit of having a substantial amount of both.

■ Emphasis and Style

The text is **written for student comprehension.** Great care has been taken to write a book that is mathematically correct and accessible to students. An **informal style** is used for exposition, definitions, and theorems. Precision, however, is not compromised. General concepts and results are usually presented only after particular cases have been discussed. To gain reader interest quickly, the text moves directly into trigonometric concepts and applications in the first chapter. **Review material** from prerequisite courses is either integrated in certain developments (particularly in Chapters 4 and 5) or can be found in the appendixes. This material can be reviewed as needed by the student or taught in class by an instructor.

■ Examples and Matched Problems

Over 180 completely worked numbered examples are used to introduce concepts and to demonstrate problem-solving techniques. Many examples have lettered parts, significantly increasing the total number of worked examples. Each example is followed by a similar **matched problem for the student to work** while reading the material. This actively involves the student in the learning process. The answers to these matched problems are included at the end of each section for easy reference.

■ Exploration and Discussion

Every section contains **Explore-Discuss** problems interspersed at appropriate places to encourage a student to think about a relationship or process before a result is stated or to investigate additional consequences of a development in the text. **Verbalization** of mathematical concepts, results, and processes is encouraged in these Explore-Discuss boxes, as well as in some matched problems, and in particular problems in almost every exercise set. The Explore-Discuss material can also be used as an in-class or out-of-class **group activity.** In addition, at the end of every chapter, before the chapter review, is a special **Chapter Group Activity** that involves a number of the concepts discussed in the chapter. The Explore-Discuss boxes and the group activities are highlighted with a light screen to emphasize their importance. Problems involving these elements are indicated by color problem numbers.

■ Exercise Sets

The book contains over 3,000 numbered problems. Many problems have lettered parts, significantly increasing the total number of problems. Each exercise set is designed so that an average or below-average student will experience success and a very capable student will be challenged. Exercise sets are mostly divided into A (routine, easy mechanics), B (more difficult mechanics), and C (difficult mechanics and some theory) levels.

■ Applications

A major objective of this book is to give the student substantial experience in **modeling and solving real world problems.** The 10th edition has added many new applications featuring real data where appropriate. Enough applications are included to convince even the most skeptical student that mathematics is really useful (see the Applications Index). Most of the applications are simplified versions of actual real-world problems taken from professional journals and books. No specialized experience is required to solve any of the applications.

■ Technology

The term **calculator** refers to any scientific calculator that can evaluate trigonometric and inverse trigonometric functions. As no tables are included in the text, access to a calculator is assumed. Access to a more sophisticated **graphing calculator is not assumed,** although it is likely that many students will want to make use of one of these devices. To assist these students, **optional graphing calculator activities** are included in appropriate places in the book. The 10th edition features optional applications including using graphing calculators for regression. The activities include brief discussions in the text, examples or portions of examples solved on a graphing calculator, and problems for the student to solve. All of the optional graphing calculator material is clearly identified by ▦ and can be omitted without loss of continuity, if desired.

■ Graphs

All graphs are computer generated to ensure mathematical accuracy. Graphing calculator screens displayed in the text are actual output from a graphing calculator.

■ Changes in the 10th Edition

We have annotated the solution steps of the completely worked examples, making the book even more accessible to students. Hundreds of new exercises have been added, including many that require a verbal explanation (marked by red problem numbers). New applications involving real data have been added where appropriate. Optional applications involving graphing calculators and regression have been added for those who wish to cover this method of curve fitting. The previous system of making the difficult application problems with one or two stars has been dropped.

■ Student Aids

Functional use of color improves the clarity of many illustrations, graphs, and developments and guides students through certain critical steps (see Sections 1.3, 4.2, and 5.3). **Think boxes** (dashed boxes) are used to enclose steps that are usually performed mentally (see Sections 1.1, 2.4, and 6.5). **Screened boxes** are used to highlight important definitions, theorems, results, and step-by-step processes (see Sections 1.3, 4.2, and 6.5). **Caution statements** appear throughout the text where student errors often occur (see Sections 1.3, 2.3, and 4.2). **Boldface type** is used to introduce new terms and highlight important comments. **Chapter review sections** include a review of all important terms and symbols and a comprehensive review exercise. **Cumulative review sections** follow Chapters 3, 5, and 7. Answers to most review exercises, **keyed to appropriate sections,** are included in the back of the book. Additionally, answers to all odd-numbered problems are also in the back of the book. **Formulas and symbols** (keyed to sections in which they are first introduced) and the metric system are summarized on the front and back end papers of the book for convenient reference.

■ Ancillaries for Students

A **Student Solutions Manual** (0-471-74656-8) contains detailed solutions to all chapter review, cumulative review, and odd-numbered problems in the book.

■ Ancillaries for Instructors

An **Instructor's Resource Manual** (0-471-78850-3) contains **answers** to the even-numbered problems not found in the back of the text, plus hints and solutions for the chapter group activities and Explore/Discuss material.

An **Instructor's Solutions Manual** (0-471-78849-X) contains solutions to the even-numbered problems.

A **printed test bank** contains quizzes and exams, plus answers.

■ Error Check

This book has been carefully checked by a number of mathematicians. If any errors are detected, the authors would be grateful if they were sent to kbyleen@wi.rr.com.

▓ Acknowledgments

In addition to the authors, the publication of a book requires the effort and skills of many people. We would like to extend thanks to:

All the people at John Wiley & Sons, who contributed their efforts to the production of this book, especially Jessica Jacobs, Jeff Benson, Anne Scanlan-Rohrer, Suzanne Ingrao, Madelyn Lesure, and Laurie Rosatone.

Fred Safier and Hossein Hamedani, for preparing the student's and instructor's solutions manuals, respectively, and Jeanne Wallace, for producing the manuals.

Hossein Hamedani and Caroline Woods, for checking the mathematical accuracy of the book.

All the people at Scientific Illustrators for their effective illustrations and accurate graphs.

We also wish to thank the many users and reviewers of the 10th and earlier editions:

Doug Aichele,
Oklahoma State University

George Alexander
University of Wisconsin

John Alford,
Sam Houston State University

Charlene Beckman
Grand Valley State University

Steve Butcher
University of Central Arkansas

Ronald Bzoch
University of North Dakota

Anne Cavagnaro,
Yosemite Community College

Robert A. Chaffer
Central Michigan University

Rodney Chase
Oakland Community College

Barbara Chudilowsky
De Anza College

Thomas F. Davis
Sam Houston State University

Jody DeVoe
Valencia Community College

Jorge Dioses
Oakhoma State University

Gay Ellis
Southeast Missouri State University

Richard L. Francis
Southeast Missouri State University

Al Giambrone
Sinclair Community College

Frank Glaser
California State Polytechnic University—Pomona

Jack Goebel
Montana College of Mineral Science and Technology

Deborah Gunter
Langston University

Shellie Gutierrez
Butler County Community College

James Hart
Middle Tennessee State University

Pamala Heard
Langston University

Steven Heath
Southern Utah State College

Robert Henkel
Pierce College

John G. Henson
Mesa College

Fred Hickling
University of Central Arkansas

Quin Higgins
Louisiana State University

JoAnnah Hill
Montana State University

Norma F. James
New Mexico State University— Las Cruces

Judith M. Jones
Valencia Community College

Lisa Juliano
Collin County Community College

Sally Keely
Clark College

Stephanie H. Kurtz
Louisiana State University

Ted Laetsch
University of Arizona

Vernon H. Leach
Pierce College

Lauri Lindberg
Pierce College

Christopher Lippi
Peralta College

Babbette Lowe
Victoria College

Van Ma
Nicholls State University

LaVerne McFadden
Parkland College

John Martin
Santa Rosa Junior College

Susan Meriwether
University of Mississippi

Eldon Miller
University of Mississippi

Georgia Miller
Millisaps College

Margaret Ullmann Miller
Skyline College

Kathy Monaghan
American River College

Michael Montano
Riverside Community College

Charles Mullins
University of Central Arkansas

Karen Murany
Oakland Community College

Karla Neal
Louisiana State University

Harald M. Ness
University of Wisconsin

Nagendra N. Pandey
Montana College of Mineral Science and Technology

Shahrokh Parvini
San Diego Mesa College

Howard Penn
El Centro College

Tammy Potter
Gadsden State University

Gladys Rockin
Oakland Community College

Jean-Pierre Rosay
University of Wisconsin

Miguel San Miguel,
Texas A & M International University

C. Robert Secrist
Kellogg Community College

Michael Seery
Scottsdale Community College

Angelo Segalla
California State University Long Beach

Charles Seifert
University of Central Arkansas

Robyn Elaine Serven
University of Central Arkansas

M. Vali Siadat
Richard J. Daley College

Thomas Spradley
American River College

Lowell Stultz
Kalamazoo Valley Community College

Donna Szott
Community College of Allegheny County

Richard Tebbs
Southern Utah State College

Roberta Tugenberg
Southwestern College

Susan Walker
Montana Tech

Steve Waters
Pacific Union College

Laura Watkins
Maricopa Community College

Randall Wills
Southeastern Louisiana University

Bobby Winters
Pittsburgh State University

Thomas Worthington
Grand Rapids Junior College

W. Xie
California State Polytechnic University—Pomona

Producing this new edition with the help of all these extremely competent people has been a most satisfying experience.

Raymond A. Barnett
Michael R. Ziegler
Karl E. Byleen
Dave Sobecki

CONTENTS

A Note on Calculators   xv

1 RIGHT TRIANGLE RATIOS   1

1.1   Angles, Degrees, and Arcs   2

1.2   Similar Triangles   13

1.3   Trigonometric Ratios and Right Triangles   22

1.4   Right Triangle Applications   33

Chapter 1 Group Activity: A Logistics Problem   44

Chapter 1 Review   45

2 TRIGONOMETRIC FUNCTIONS   51

2.1   Degrees and Radians   52

☆ **2.2**   Linear and Angular Velocity   66

2.3   Trigonometric Functions: Unit Circle Approach   73

☆ **2.4**   Additional Applications   85

2.5   Exact Values and Properties of Trigonometric Functions   99

Chapter 2 Group Activity: Speed of Light in Water   114

Chapter 2 Review   116

3 GRAPHING TRIGONOMETRIC FUNCTIONS   123

3.1   Basic Graphs   125

3.2   Graphing $y = k + A \sin Bx$ and $y = k + A \cos Bx$ 140

3.3   Graphing $y = k + A \sin(Bx + C)$ and $y = k + A \cos(Bx + C)$ 159

☆ Sections marked with a star may be omitted without loss of continuity.

☆ **3.4** Additional Applications 174

3.5 Graphing Combined Forms 190

3.6 Tangent, Cotangent, Secant, and Cosecant Functions Revisited 202

Chapter 3 Group Activity: Predator–Prey Analysis Involving
Coyotes and Rabbits 213

Chapter 3 Review 215

Cumulative Review Exercise, Chapters 1–3 223

4 IDENTITIES 229

4.1 Fundamental Identities and Their Use 230

4.2 Verifying Trigonometric Identities 237

4.3 Sum, Difference, and Cofunction Identities 247

4.4 Double-Angle and Half-Angle Identities 258

☆ **4.5** Product–Sum and Sum–Product Identities 269

Chapter 4 Group Activity: From $M \sin Bt + N \cos Bt$ to $A \sin(Bt + C)$ 277

Chapter 4 Review 279

5 INVERSE TRIGONOMETRIC FUNCTIONS; TRIGONOMETRIC EQUATIONS AND INEQUALITIES 287

5.1 Inverse Sine, Cosine, and Tangent Functions 288

☆ **5.2** Inverse Cotangent, Secant, and Cosecant Functions 309

5.3 Trigonometric Equations: An Algebraic Approach 315

5.4 Trigonometric Equations and Inequalities:
A Graphing Calculator Approach 329

Sections marked with the calculator icon require the use of a graphing calculator. Inclusion of this material will enrich the course, but its omission will not affect the continuity of the course.

Chapter 5 Group Activity: $\sin \dfrac{1}{x} = 0$ *and* $\sin^{-1} \dfrac{1}{x} = 0$ *338*

Chapter 5 Review   339

Cumulative Review Exercise, Chapters 1–5   345

6   # ADDITIONAL TOPICS: TRIANGLES AND VECTORS   351

6.1   Law of Sines   352

6.2   Law of Cosines   367

☆ **6.3**   Areas of Triangles   378

6.4   Vectors: Geometrically Defined   385

6.5   Vectors: Algebraically Defined   396

☆ **6.6**   The Dot Product   410

Chapter 6 Group Activity: The SSA Case and the Law of Cosines   420

Chapter 6 Review   421

7   # POLAR COORDINATES; COMPLEX NUMBERS   431

7.1   Polar and Rectangular Coordinates   432

7.2   Sketching Polar Equations   440

7.3   The Complex Plane   453

7.4   De Moivre's Theorem and the *n*th-Root Theorem   462

Chapter 7 Group Activity: Orbits of Planets   469

Chapter 7 Review   471

Cumulative Review Exercise, Chapters 1–7   475

Appendix A   # COMMENTS ON NUMBERS   481

A.1   Real Numbers   482

A.2   Complex Numbers   485

A.3   Significant Digits   489

Appendix B FUNCTIONS AND INVERSE FUNCTIONS 497

B.1 Functions 498

B.2 Graphs and Transformations 505

B.3 Inverse Functions 516

Appendix C PLANE GEOMETRY: SOME USEFUL FACTS 523

C.1 Lines and Angles 524

C.2 Triangles 525

C.3 Quadrilaterals 527

C.4 Circles 528

Selected Answers *A-1*

Index *I-1*

A Note on Calculators

Use of calculators is emphasized throughout this book. Many brands and types of scientific calculators are available and can be found starting at about $10. Graphing calculators are more expensive; but, in addition to having most of the capabilities of a scientific calculator, they have very powerful graphing capabilities. Your instructor should help you decide on the type and model best suited to this course and the emphasis on calculator use he or she desires.

Whichever calculator you use, it is essential that you read the user's manual for your calculator. A large variety of calculators are on the market, and each is slightly different from the others. Therefore, take the time to read the manual. The first time through do not try to read and understand everything the calculator can do — that will tend to overwhelm and confuse you. Read only those sections pertaining to the operations you are or will be using; then return to the manual as necessary when you encounter new operations.

It is important to remember that a calculator is not a substitute for thinking. It can save you a great deal of time in certain types of problems, but you still must understand basic concepts so that you can interpret results obtained through the use of a calculator.

Right Triangle Ratios

1

1.1 Angles, Degrees, and Arcs

1.2 Similar Triangles

1.3 Trigonometric Ratios and Right Triangles

1.4 Right Triangle Applications

Chapter 1 Group Activity: A Logistics Problem

Chapter 1 Review

$$A = ab$$

b

a

If you were asked to find your height, you would no doubt take a ruler or tape measure and measure it directly. But if you were asked to find the area of your bedroom floor in square feet, you would not be likely to measure the area directly by laying 1 ft squares over the entire floor and counting them. Instead, you would probably find the area **indirectly** by using the formula $A = ab$ from plane geometry, where A represents the area of the room and a and b are the lengths of its sides, as indicated in the figure in the margin.

In general, **indirect measurement** is a process of determining unknown measurements from known measurements by a reasoning process. How do we measure quantities such as the volumes of containers, the distance to the center of the earth, the area of the surface of the earth, and the distances to the sun and the stars? All these measurements are accomplished indirectly by the use of special formulas and deductive reasoning.

The Greeks in Alexandria, during the period 300 BC–AD 200, contributed substantially to the art of indirect measurement by developing formulas for finding areas, volumes, and lengths. Using these formulas, they were able to determine the circumference of the earth (with an error of only about 2%) and to estimate the distance to the moon. We will examine these measurements as well as others in the sections that follow.

It was during the early part of this Greek period that trigonometry, the study of triangles, was born. Hipparchus (160–127 BC), one of the greatest astronomers of the ancient world, is credited with making the first systematic study of the indirect measurement of triangles.

1.1   Angles, Degrees, and Arcs

- Angles
- Degree Measure of Angles
- Angles and Arcs
- Approximation of Earth's Circumference
- Approximation of the Diameters of the Sun and Moon; Total Solar Eclipse

Angles and the degree measure of an angle are the first concepts introduced in this section. Then a very useful relationship between angles and arcs of circles is developed. This simple relationship between an angle and an arc will enable us to measure indirectly many useful quantities such as the circumference of the earth and the diameter of the sun. (If you are rusty on certain geometric relationships and facts, refer to Appendix C as needed.)

■ Angles

Central to the study of trigonometry is the concept of *angle*. An **angle** is formed by rotating a half-line, called a **ray,** around its end point. One ray k, called the **initial side** of the angle, remains fixed; a second ray l, called the **terminal side**

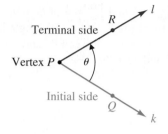

FIGURE 1
Angle θ: A ray rotated around its end point

of the angle, starts in the initial side position and rotates around the common end point *P* in a plane until it reaches its terminal position. The common end point *P* is called the **vertex** (see Fig. 1).

We may refer to the angle in Figure 1 in any of the following ways:

Angle θ $\angle\theta$ Angle *QPR* $\angle QPR$

Angle *P* $\angle P$

The symbol $\angle$ denotes *angle*. The Greek letters theta (θ), alpha (α), beta (β), and gamma (γ) are often used to name angles.

There is no restriction on the amount or direction of rotation in a given plane. When the terminal side is rotated counterclockwise, the angle formed is **positive** (see Figs. 1 and 2a); when it is rotated clockwise, the angle formed is **negative** (see Fig. 2b). Two different angles may have the same initial and terminal sides, as shown in Figure 2c. Such angles are said to be **coterminal.** In this chapter we will concentrate on positive angles; we will consider more general angles in detail in subsequent chapters.

FIGURE 2

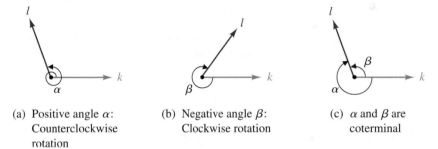

(a) Positive angle α:
     Counterclockwise
     rotation

(b) Negative angle β:
     Clockwise rotation

(c) α and β are
     coterminal

■ Degree Measure of Angles

To compare angles of different sizes, a standard unit of measure is necessary. Just as a line segment can be measured in inches, meters, or miles, an angle is measured in *degrees* or *radians*. (We will postpone our discussion of radian measure of angles until Section 2.1.)

DEGREE MEASURE OF ANGLES

An angle formed by one complete revolution of the terminal side in a counterclockwise direction has **measure 360 degrees**, written **360°.** An angle of **1 degree measure**, written **1°**, is formed by $\frac{1}{360}$ of one complete revolution in a counterclockwise direction. (The symbol ° denotes degrees.)

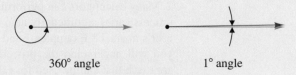

360° angle   1° angle

Angles of measure 90° and 180° represent $\frac{90}{360} = \frac{1}{4}$ and $\frac{180}{360} = \frac{1}{2}$ of complete revolutions, respectively. A 90° angle is called a **right angle** and a 180° angle is called a **straight angle.** An **acute angle** has angle measure between 0° and 90°. An **obtuse angle** has angle measure between 90° and 180° (see Fig. 3).

FIGURE 3
Special angles

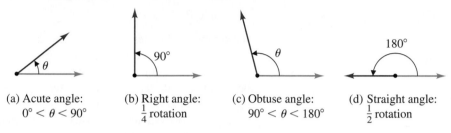

(a) Acute angle:
$0° < \theta < 90°$

(b) Right angle:
$\frac{1}{4}$ rotation

(c) Obtuse angle:
$90° < \theta < 180°$

(d) Straight angle:
$\frac{1}{2}$ rotation

Remark   In Figure 3 we used θ in two different ways: to name an angle and to represent the measure of an angle. This usage is common; the context will dictate the interpretation.   □

Two positive angles are **complementary** if the sum of their measures is 90°; they are **supplementary** if the sum of their measures is 180°(see Fig. 4).

FIGURE 4

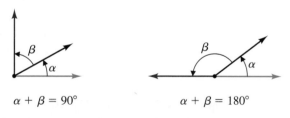

$\alpha + \beta = 90°$

$\alpha + \beta = 180°$

(a) Complementary angles

(b) Supplementary angles

A degree can be divided using decimal notation. For example, 36.25° represents an angle of degree measure 36 plus one-fourth of 1 degree. A degree can also be divided into minutes and seconds just as an hour is divided into minutes and seconds. Each degree is divided into 60 equal parts called minutes (′), and each minute is divided into 60 equal parts called seconds (″). Thus,

5°12′32″

is a concise way of writing 5 degrees, 12 minutes, and 32 seconds.

Degree measure in **decimal degree (DD)** form is useful in some instances, and degree measure in **degree-minute-second (DMS)** form is useful in others. You should be able to go from one form to the other as illustrated in Examples 1 and 2. Many calculators can perform the conversion either way automatically, but the process varies significantly among the various types of calculators—consult your user's manual.* Examples 1 and 2 first illustrate a nonautomatic approach so that you will understand the process. This is followed by an automatic calculator approach that can be used for efficiency.

———————
* User's manuals for many calculators are readily available on the Internet.

Conversion Accuracy   If an angle is measured to the nearest second, the converted decimal form should not go beyond three decimal places, and vice versa. □

EXAMPLE 1   **From DMS to DD**

Convert $12°6'23''$ to decimal degree form correct to three decimal places.

Solution   **Method I**   *Multistep conversion.*   Since

$$6' = \left(\frac{6}{60}\right)^{\circ} \quad \text{and} \quad 23'' = \left(\frac{23}{3,600}\right)^{\circ}$$

then

$$12°6'23'' = \left(12 + \frac{6}{60} + \frac{23}{3,600}\right)^{\circ*}$$
$$= 12.106°$$

To three decimal places

Method II   *Single-step calculator conversion.*   Consult the user's manual for your particular calculator. The conversion shown in Figure 5 is from a graphing calculator.

```
12°6'23"
          12.106
```

FIGURE 5
From DMS to DD   ■

Matched Problem 1†   Convert $128°42'8''$ to decimal degree form correct to three decimal places. ■

EXPLORE/DISCUSS 1

In the 2003 Berlin Marathon, Paul Tergat of Kenya ran the world's fastest recorded time for a marathon: 42.2 km in 2 hr, 4 min, 55 sec. Convert the hour-minute-second form to decimal hours. Explain the similarity of this conversion to the conversion of a degree-minute-second form to a decimal degree form.

---

* Dashed "think boxes" indicate steps that can be performed mentally once a concept or procedure is understood.

† Answers to matched problems are located at the end of each section just before the exercise set.

EXAMPLE 2

From DD to DMS

Convert 35.413° to a degree-minute-second form.

Solution   **Method I**   *Multistep conversion.*

$$35.413° = 35° \; (0.413 \cdot 60)'$$
$$= 35°24.78'$$
$$= 35°24' \; (0.78 \cdot 60)''$$
$$\approx 35°24'47''$$

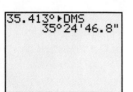

FIGURE 6
From DD to DMS

Method II   *Single-step calculator conversion.*   Consult the user's manual for your particular calculator. The conversion shown in Figure 6 is from a graphing calculator. Rounded to the nearest second, we get the same result as in method I: 35°24'47''.   ■

Matched Problem 2   Convert 72.103° to degree-minute-second form.   ■

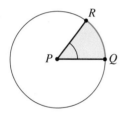

FIGURE 7
Central angle *RPQ* subtended
by arc *RQ*

Angles and Arcs

Given an arc *RQ* of a circle with center *P*, the positive angle *RPQ* is said to be the **central angle** that is **subtended** by the arc *RQ*. We also say that the arc *RQ* is subtended by the angle *RPQ;* see Figure 7.

It follows from the definition of degree that a central angle subtended by an arc $\frac{1}{4}$ the circumference of a circle has degree measure 90; $\frac{1}{2}$ the circumference of a circle, degree measure 180, and the whole circumference of a circle, degree measure 360. In general, to determine the degree measure of an angle θ subtended by an arc of *s* units for a circle with circumference *C* units, use the following proportion:

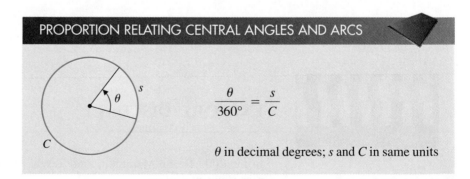

PROPORTION RELATING CENTRAL ANGLES AND ARCS

$$\frac{\theta}{360°} = \frac{s}{C}$$

θ in decimal degrees; *s* and *C* in same units

If we know any two of the three quantities, *s, C,* or θ, we can find the third by using simple algebra. For example, in a circle with circumference 72 in., the degree measure of a central angle θ subtended by an arc of length 12 in. is given by

$$\theta = \frac{12}{72} \cdot 360° = 60°$$

Remark   Note that "an angle of 6°" means "an angle of degree measure 6," and " $\theta = 72°$" means "the degree measure of angle θ is 72." ☐

■ Approximation of Earth's Circumference

The early Greeks were aware of the proportion relating central angles and arcs, which Eratosthenes (240 BC) used in his famous calculation of the circumference of the earth. He reasoned as follows: It was well known that at Syene (now Aswan), during the summer solstice, the noon sun was reflected on the water in a deep well (this meant the sun shone straight down the well and must be directly overhead). Eratosthenes reasoned that if the sun rays entering the well were continued down into the earth, they would pass through its center (see Fig. 8). On the same day at the same time, 5,000 stadia (approx. 500 mi) due north, in Alexandria, sun rays crossed a vertical pole at an angle of 7.5° as indicated in Figure 8. Since sun rays are very nearly parallel when they reach the earth, Eratosthenes concluded that ∠*ACS* was also 7.5°. (Why?)*

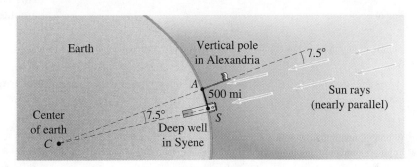

FIGURE 8
Estimating the earth's circumference

Even though Eratosthenes' reasoning was profound, his final calculation of the circumference of the earth requires only elementary algebra:

$$\frac{\theta}{360°} = \frac{s}{C}$$ Substitute $s = 500$ mi and $\theta = 7.5°$.

$$\frac{7.5°}{360°} = \frac{500 \text{ mi}}{C}$$ Multiply both sides by C.

$$C\frac{7.5°}{360°} = 500 \text{ mi}$$ Multiply both sides by $\dfrac{360}{7.5}$.

$$C = \frac{360}{7.5}(500 \text{ mi}) = 24{,}000 \text{ mi}$$

The value calculated today is 24,875 mi.

──────────

* If line p crosses parallel lines m and n, then angles α and β have the same measure.

The diameter of the earth (d) and the radius (r) can be found from this value using the formulas $C = 2\pi r$ and $d = 2r$ from plane geometry. [*Note:* $\pi = C/d$ for all circles.] The constant π has a long and interesting history; a few important dates follow:

1650 BC	Rhind Papyrus	$\pi \approx \frac{256}{81} = 3.16049...$
240 BC	Archimedes	$3\frac{10}{71} < \pi < 3\frac{1}{7}$
		$(3.1408... < \pi < 3.1428...)$
AD 264	Liu Hui	$\pi \approx 3.14159$
AD 470	Tsu Ch'ung-chih	$\pi \approx \frac{355}{113} = 3.1415929...$
AD 1674	Leibniz	$\pi = 4(1 - \frac{1}{3} + \frac{1}{5} - \frac{1}{7} + \frac{1}{9} - \frac{1}{11} + \cdots)$
		$\approx 3.1415926535897932384626$
		(This and other series can be used to compute π to any decimal accuracy desired.)
AD 1761	Johann Lambert	Showed π to be irrational (π as a decimal is nonrepeating and nonterminating)

EXAMPLE 3   **Arc Length**

How large an arc is subtended by a central angle of 6.23° on a circle with radius 10 cm? (Compute the answer to two decimal places.)

Solution   Use the proportion relating angles and arcs.

$$\frac{s}{C} = \frac{\theta}{360°}$$ Substitute $C = 2\pi r$

$$\frac{s}{2\pi r} = \frac{\theta}{360°}$$ Substitute r = 10 cm and θ = 6.23°

$$\frac{s}{2(\pi)(10 \text{ cm})} = \frac{6.23°}{360°}$$ Multiply both sides by $2(\pi)(10$ cm$)$

$$s = \frac{2(\pi)(10 \text{ cm})(6.23)}{360} = 1.09 \text{ cm}$$ ■

Matched Problem 3   How large an arc is subtended by a central angle of 50.73° on a circle with radius 5 m? (Compute the answer to two decimal places.)   ■

■ Approximation of the Diameters of the Sun and Moon; Total Solar Eclipse

EXAMPLE 4

Sun's Diameter

If the distance from the earth to the sun is 93,000,000 mi, find the diameter of the sun (to the nearest thousand miles) if it subtends an angle of 0°31′55″ on the surface of the earth.

Solution For small central angles in circles with very large radii, the **intercepted arc** (arc opposite the central angle) and its **chord** (the straight line joining the end points of the arc) are approximately the same length (see Fig. 9). We thus use the intercepted arc to approximate its chord in many practical problems, particularly when the length of the intercepted arc is easier to compute. We apply these ideas to finding the diameter of the sun as follows:

$$\theta = 0°31′55″ = 0.532°$$

$$\frac{s}{2\pi r} = \frac{\theta}{360°} \qquad \text{Multiply both sides by } 2\pi r$$

$$s = \frac{2\pi r\theta}{360} \qquad \text{Substitute } r = 93,000,000 \text{ mi and } \theta = 0.532°$$

$$= \frac{2(\pi)(93,000,000 \text{ mi})(0.532)}{360} = 864,000 \text{ mi}$$

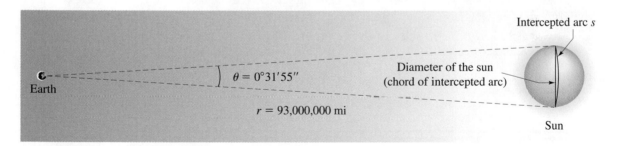

Intercepted arc *s*

$\theta = 0°31′55″$

Diameter of the sun (chord of intercepted arc)

Earth

$r = 93,000,000$ mi

Sun

FIGURE 9

Matched Problem 4 If the moon subtends an angle of about 0°31′5″ on the surface of the earth when it is 239,000 mi from the earth, estimate its diameter to the nearest 10 mi. ■

Answers to Matched Problems

1. 128.702°
2. 72°6′11″
3. 4.43 m
4. 2,160 mi

EXPLORE/DISCUSS 2

A total solar eclipse will occur when the moon passes between the earth and the sun and the angle subtended by the diameter of the moon is at least as large as the angle subtended by the diameter of the sun (see Fig. 10).

(A) Since the distances from the sun and moon to the earth vary with time, explain what happens to the angles subtended by the diameters of the sun and moon as their distances from the earth increase and decrease.

(B) For a total solar eclipse to occur when the moon passes between the sun and the earth, would it be better for the sun to be as far away as possible and the moon to be as close as possible, or vice versa? Explain.

(C) The diameters of the sun and moon are, respectively, 864,000 mi and 2,160 mi. Find the maximum distance that the moon can be from the earth for a total solar eclipse when the sun is at its maximum distance from the earth, 94,500,000 mi.

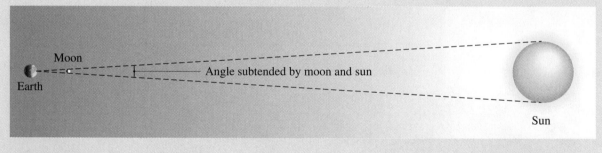

FIGURE 10
Total solar eclipse

EXERCISE 1.1

Throughout the text the problems in most exercise sets are divided into A, B, and C groupings: The A problems are basic and routine, the B problems are more challenging but still emphasize mechanics, and the C problems are a mixture of difficult mechanics and theory.

A *In Problems 1 and 2, find the number of degrees in the angle formed by rotating the terminal side counterclockwise through the indicated fraction of a revolution.*

1. $\frac{1}{2}, \frac{1}{6}, \frac{3}{8}, \frac{7}{12}$ **2.** $\frac{1}{4}, \frac{1}{3}, \frac{7}{8}, \frac{11}{12}$

In Problems 3 and 4, find the fraction of a counterclockwise revolution that will form an angle with the indicated number of degrees.

3. 45°, 150°, 270°   **4.** 30°, 225°, 240°

In Problems 5–12, identify the following angles as acute, right, obtuse, or straight. If the angle is none of these, say so.

5. 50°   **6.** 150°   **7.** 90°

8. 180°   **9.** 135°   **10.** 185°

11. 250°   **12.** 89°

***13.** Discuss the meaning of an angle of 1 degree.

14. Discuss the meaning of minutes and seconds in angle measure.

B *In Problems 15–20, change to decimal degrees accurate to three decimal places:*

15. $25°21'54''$ **16.** $71°43'30''$ **17.** $11°8'5''$

18. $9°3'1''$ **19.** $195°28'10''$ **20.** $267°11'25''$

In Problems 21–26, change to degree-minute-second form:

21. $15.125°$ **22.** $35.425°$ **23.** $79.201°$

24. $52.927°$ **25.** $159.639°$ **26.** $235.253°$

27. Which of the angle measures, $47°33'41''$ or $47.556°$, is larger? Explain how you obtained your answer.

28. Runner A ran a marathon in 3 hr, 43 min, 24 sec, and runner B in 3.732 hr. Which runner is faster? Explain how you obtained your answer.

In Problems 29–34, write $\alpha < \beta, \alpha = \beta,$ or $\alpha > \beta$, as appropriate.

29. $\alpha = 47°23'31''$ **30.** $\alpha = 32°51'54''$
    $\beta = 47.386°$ $\beta = 32.865°$

31. $\alpha = 125°27'18''$ **32.** $\alpha = 80.668°$
    $\beta = 125.455°$ $\beta = 80°40'20''$

33. $\alpha = 20.512°$ **34.** $\alpha = 242.311°$
    $\beta = 20°30'50''$ $\beta = 242°18'32''$

In Problems 35–38, perform the indicated operations directly on a calculator. Express the answers in DMS form. For example, a particular graphing calculator performs the following calculation $(67°13'4'' - 45°27'32'')$ as follows:

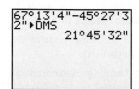

35. $47°37'49'' + 62°40'15''$

36. $105°53'22'' + 26°38'55''$

37. $90° - 67°37'29''$

38. $180° - 121°51'22''$

In Problems 39 and 40, θ is a central angle in a circle with radius 1.

39. What is the length of the arc subtended by $\theta = 0°$?

40. What is the length of the arc subtended by $\theta = 360°$?

In Problems 41–48, find C, θ, s, or r as indicated. Refer to the figure:

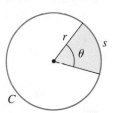

41. $C = 1{,}000$ cm, $\theta = 36°$, $s = ?$ (exact)

42. $s = 12$ m, $C = 108$ m, $\theta = ?$ (exact)

43. $s = 25$ km, $\theta = 20°$, $C = ?$ (exact)

44. $C = 740$ mi, $\theta = 72°$, $s = ?$ (exact)

C **45.** $r = 5{,}400{,}000$ mi, $\theta = 2.6°$, $s = ?$
    (to the nearest 10,000 mi)

46. $s = 38{,}000$ cm, $\theta = 45.3°$, $r = ?$
    (to the nearest 1,000 cm)

47. $\theta = 12°31'4''$, $s = 50.2$ cm, $C = ?$
    (to the nearest 10 cm)

48. $\theta = 24°16'34''$, $s = 14.23$ m, $C = ?$
    (to one decimal place)

In Problems 49–52, find A or θ as indicated. Refer to the figure. The ratio of the area of a sector of a circle (A) to the total area of the circle (πr^2) is the same as the ratio of the central angle of the sector (θ) to 360°:

$$\frac{A}{\pi r^2} = \frac{\theta}{360°}$$

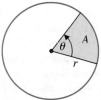

49. $r = 25.2$ cm, $\theta = 47.3°$, $A = ?$ (to the nearest unit)

50. $r = 7.38$ ft, $\theta = 24.6°$, $A = ?$ (to one decimal place)

51. $r = 12.6$ m, $A = 98.4$ m^2, $\theta = ?$ (to one decimal place)

52. $r = 32.4$ in., $A = 347$ in.^{2}, $\theta = ?$ (to one decimal place)

---

* A red problem number indicates that a problem may involve critical thinking and verbalization of concepts.

Applications

Biology:Eye In Problems 53–56, find r, θ, or s as indicated. Refer to the following: The eye is roughly spherical with a spherical bulge in front called the cornea. (Based on the article, "The Surgical Correction of Astigmatism" by Sheldon Rothman and Helen Strassberg in The UMAP Journal, *Vol. V, No. 2, 1984.*)

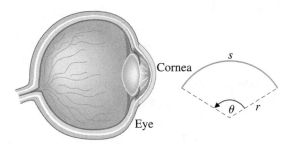

Figure for 53–56

53. $s = 11.5$ mm, $\theta = 118.2°$, $r = ?$
    (to two decimal places)

54. $s = 12.1$ mm, $r = 5.26$ mm, $\theta = ?$
    (to one decimal place)

55. $\theta = 119.7°$, $r = 5.49$ mm, $s = ?$
    (to one decimal place)

56. $\theta = 117.9°$, $s = 11.8$ mm, $r = ?$
    (to two decimal places)

Geography/Navigation In Problems 57–60, find the distance on the surface of the earth between each pair of cities (to the nearest mile), given their respective latitudes. Latitudes are given to the nearest 10'. Note that each chosen pair of cities has approximately the same longitude (i.e., lies on the same north–south line). Use $r = 3,960$ mi for the earth's radius. The figure shows the situation for San Francisco and Seattle. [Hint: $C = 2\pi r$]

57. San Francisco, CA, 37°50′N;   Seattle, WA, 47°40′N

58. Phoenix, AZ, 33°30′N;   Salt Lake City, UT, 40°40′N

59. Dallas, TX, 32°50′N;   Lincoln, NE, 40°50′N

60. Buffalo, NY, 42°50′N;   Durham, NC, 36°0′N

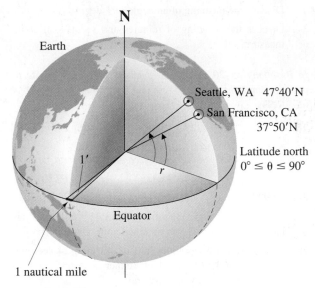

Figure for 57–64

Geography/Navigation In Problems 61–64, find the distance on the earth's surface, to the nearest nautical mile, between each pair of cities. A *nautical mile* is the length of 1' of arc on the equator or any other circle on the surface of the earth having the same center as the equator. See the figure. Since there are $60 \times 360 = 21,600'$ in 360°, the length of the equator is 21,600 nautical miles.

61. San Francisco, CA, 37°50′N;   Seattle, WA, 47°40′N

62. Phoenix, AZ, 33°30′N;   Salt Lake City, UT, 40°40′N

63. Dallas, TX, 32°50′N;   Lincoln, NE, 40°50′N

64. Buffalo, NY, 42°50′N;   Durham, NC, 36°0′N

65. **Photography**

    (A) The angle of view of a 300 mm lens is 8°. Approximate the width of the field of view to the nearest foot when the camera is at a distance of 500 ft.

    (B) Explain the assumptions that are being made in the approximation calculation in part (A).

66. **Satellite Telescopes** In the Orbiting Astronomical Observatory launched in 1968, ground personnel could direct telescopes in the vehicle with an accuracy of 1' of arc. They claimed this corresponds to hitting a 25¢ coin at a distance of 100 yd (see the figure on the next page). Show that this claim is approximately correct; that is, show that the length of an arc subtended by 1' at 100 yd is approximately the diameter of a quarter, 0.94 in.

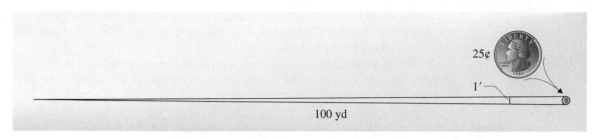

Figure for 66

The *apparent diameter* of an object is the measure (in degrees) of the arc intercepted by the object as seen by an observer. For example, on earth the apparent diameter of the sun is 0.532° and the apparent diameter of the moon is 0.518° (see Example 4 and Matched Problem 4). If r is the distance from the observer to the object and s is the diameter of the object, then the apparent diameter θ is given by

$$\theta = \frac{360s}{2\pi r} \quad \text{r and s are expressed in the same units of measurement}$$

In Problems 67–72, round the answers to three significant digits.

67. Apparent Diameter The planet Saturn is 886,000,000 miles from the sun and the sun has a diameter of 864,000 miles. What is the apparent diameter of the sun for an observer on Saturn? Express your answer in decimal degrees to three decimal places.

68. Apparent Diameter The planet Uranus is 1,780 million miles from the sun and the sun has a diameter of 864,000 miles. To an observer on Uranus, what is the apparent diameter of the sun? Express your answer in decimal degrees to three decimal places.

69. Apparent Diameter In January 2005 the European Space Agency's Huygens landed on the surface of Titan, one of Saturn's moons. Titan has a diameter of 3,200 miles. When viewed from Saturn, Titan has an apparent diameter of 0.242°. How far is Titan from Saturn? Round your answer to the nearest thousand miles.

70. Apparent Diameter Juliet, one of Uranus's moons, has a diameter of 52 miles. When viewed from Uranus, Juliet has an apparent diameter of 0.075°. How far is Juliet from Uranus? Round your answer to the nearest thousand miles.

71. Apparent Diameter Epimetheus, another of Saturn's moons, is 94,100 miles from Saturn. When viewed from Saturn, Epimetheus has an apparent diameter of 0.053°. What is the diameter of Epimetheus? Round your answer to the nearest mile.

72. Apparent Diameter Cordelia, another of Uranus's moons, is 31,000 miles from Uranus. When viewed from Uranus, Cordelia has an apparent diameter of 0.03°. What is the diameter of Cordelia? Round your answer to the nearest mile.

1.2 Similar Triangles

- Euclid's Theorem and Similar Triangles
- Applications

Properties of similar triangles, stated in Euclid's theorem that follows, are central to this section and form a cornerstone for the development of trigonometry. Euclid (300 BC), a Greek mathematician who taught in Alexandria, was one of the most influential mathematicians of all time. He is most famous for writing the *Elements*, a collection of thirteen books (or chapters) on geometry, geometric algebra, and number theory. In the Western world, next to the Bible, the *Elements* is probably the most studied text of all time.

The ideas on indirect measurement presented here are included to help you understand basic concepts. More efficient methods will be developed in the next section.

Remark   Since calculators routinely compute to eight or ten digits, one could easily believe that a computed result is far more accurate than warranted. **Generally, a final result cannot be any more accurate than the least accurate number used in the calculation. Regarding calculation accuracy, we will be guided by Appendix A.3 on significant digits throughout the text.**   □

◼ Euclid's Theorem and Similar Triangles

In Section 1.1, problems were included that required knowledge of the distances from the earth to the moon and the sun. How can inaccessible distances of this type be determined? Surprisingly, the ancient Greeks made fairly accurate calculations of these distances as well as many others. The basis for their methods is the following elementary theorem of Euclid:

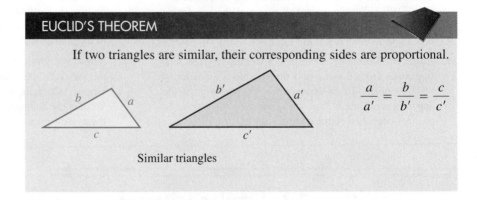

EUCLID'S THEOREM

If two triangles are similar, their corresponding sides are proportional.

$$\frac{a}{a'} = \frac{b}{b'} = \frac{c}{c'}$$

Similar triangles

Remark   Recall from plane geometry (see Appendix C.2) that the interior angles of any polygon are always positive. For triangles, the sum of the measures of the three angles is always 180°. Two triangles are similar if two angles of one triangle have the same measure as two angles of the other. If the two triangles happen to be right triangles, then they are similar if an acute angle in one has the same measure as an acute angle in the other. Take a moment to draw a few figures and to think about these statements.   □

EXPLORE/DISCUSS 1

Identify pairs of triangles from the following that are similar and explain why.

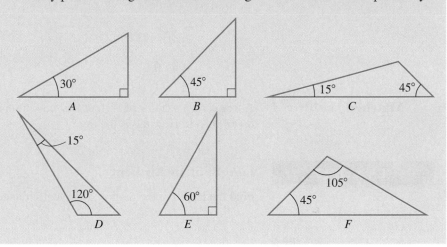

■ Applications

We now solve some elementary problems using Euclid's theorem.

EXAMPLE 1

Height of a Tree

A tree casts a shadow of 32 ft at the same time a vertical yardstick (3.0 ft) casts a shadow of 2.2 ft (see Fig. 1). How tall is the tree?

FIGURE 1

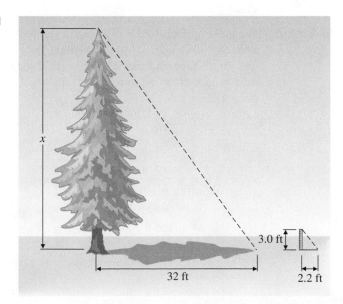

Solution   The parallel sun rays make the same angle with the tree and the yardstick. Since both triangles are right triangles and have an acute angle of the same measure, the triangles are similar. The corresponding sides are proportional, and we can write

$$\frac{x}{3.0 \text{ ft}} = \frac{32 \text{ ft}}{2.2 \text{ ft}}$$

$$x = \frac{3.0}{2.2}(32 \text{ ft})$$

$$= 44 \text{ ft} \qquad \text{To two significant digits} \qquad \blacksquare$$

Matched Problem 1   A tree casts a shadow of 31 ft at the same time a 5.0 ft vertical pole casts a shadow of 0.56 ft. How tall is the tree?   ■

   **EXAMPLE 2**   **Length of an Air Vent**

Find the length of the proposed air vent indicated in Figure 2.

FIGURE 2
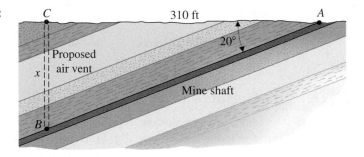

Solution   We make a careful scale drawing of the mine shaft relative to the proposed air vent as follows: Pick any convenient length, say 2.0 in., for $A'C'$; copy the 20° angle CAB and the 90° angle ACB using a protractor (see Fig. 3 on the next page). Now measure $B'C'$ (approx. 0.7 in.), and set up an appropriate proportion:

$$\frac{x}{0.7 \text{ in.}} = \frac{310 \text{ ft}}{2.0 \text{ in.}} \qquad \text{Multiply both sides by 0.7 in.}$$

$$x = \frac{0.7 \text{ in.}}{2.0 \text{ in.}}(310 \text{ ft}) \qquad (0.7 \text{ in.})/(2.0 \text{ in.}) = 0.35 \text{ is a dimensionless* number.}$$

$$= (0.35)(310 \text{ ft}) \qquad \text{Multiply}$$

$$= 100 \text{ ft} \qquad \text{To one significant digit}$$

---

* If both numbers in a ratio are expressed in the same unit of measurement, then the result is a **dimensionless** number.

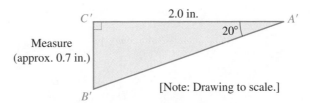

FIGURE 3

[*Note:* The use of scale drawings for finding indirect measurements is included here only to demonstrate basic ideas. Scale drawings can introduce considerable error in a calculation. A more efficient and more accurate method will be developed in Section 1.3.] ▪

Matched Problem 2 Suppose in Example 2 that $AC = 550$ ft and $\angle A = 30°$. If in a scale drawing $A'C'$ is chosen to be 3.0 in. and $B'C'$ is measured as 1.76 in., find BC, the length of the proposed mine shaft. ▪

EXPLORE/DISCUSS 2

We want to measure the depth of a canyon from a point on its rim by using similar triangles and no scale drawings. The process illustrated in Figure 4 on page 18 was used in medieval times.* A vertical pole of height a is moved back from the canyon rim so that the line of sight from the top of the pole passes the rim at D to a point G at the bottom of the canyon. The setback DC is then measured to be b. The same vertical pole is moved to the canyon rim at D, and a horizontal pole is moved out from the rim through D until the line of sight from B to G includes the end of the pole at E. The pole overhang DE is measured to be c. We now have enough information to find y, the depth of the canyon.

(A) Explain why triangles ACD and GFD are similar.

(B) Explain why triangles BDE and GFE are similar.

(C) Set up appropriate proportions, and with the use of a little algebra, show that

$$x = \frac{bc}{b - c} \qquad y = \frac{ac}{b - c}$$

Continued

---

* This process is mentioned in an excellent article by Victor J. Katz (University of the District of Columbia) titled *The Curious History of Trigonometry, in The UMAP Journal,* Winter 1990.

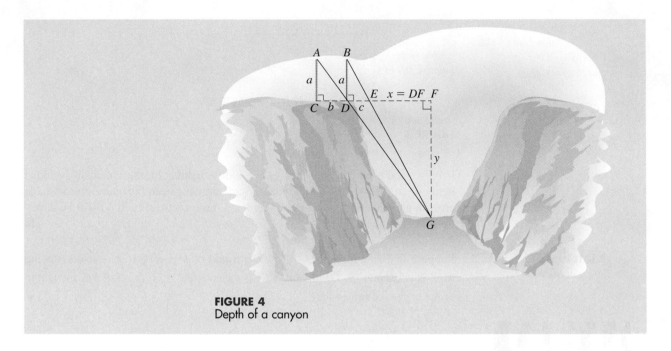

FIGURE 4
Depth of a canyon

<table>
<tr><td rowspan="2">**Answers to**
Matched Problems</td><td>**1.** 280 ft (to two significant digits)</td></tr>
<tr><td>**2.** 320 ft (to two significant digits)</td></tr>
</table>

EXERCISE 1.2

Round each answer to an appropriate accuracy; use the guidelines from Appendix A.3.

A **1.** If two angles of one triangle have the same measure as two angles of another triangle, what can you say about the measure of the third angle of each triangle? Why?

2. If an acute angle of one right triangle has the same measure as the acute angle of another right triangle, what can you say about the measure of the second acute angle of each triangle? Why?

3. The calculator value of (1.7) (2.8) is 4.76. If 1.7 and 2.8 are measurements, what is the correct rounded value according to Appendix A. 3?

4. The calculator value of (10) (0.33) is 3.3. If 10 and 0.33 are measurements, what is the correct rounded value according to Appendix A.3?

The triangles in the figure are similar. In Problems 5–10, find the indicated length.

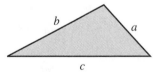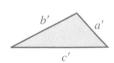

Figure for 5–10

5. $a = 5$, $b = 15$, $a' = 7$, $b' = ?$

6. $b = 3$, $c = 24$, $b' = 1$, $c' = ?$

7. $c = ?$, $a = 12$, $c' = 18$, $a' = 2.4$

8. $a = 51$, $b = ?$, $a' = 17$, $b' = 8.0$

9. $b = 52,000$; $c = 18,000$, $b' = 8.5$, $c' = ?$

10. $a = 640,000$, $b = ?$, $a' = 15$, $b' = 0.75$

B **11.** Can two similar triangles have equal sides? Explain.

12. If two triangles are similar and a side of one triangle is equal to the corresponding side of the other, are the remaining sides of the two triangles equal? Explain.

The triangles in the figure are similar. In Problems 13–20, find the indicated length.

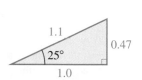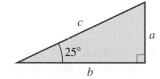

Figure for 13–20

13. $b = 51$ in., $a = ?$, $c = ?$

14. $b = 32$ cm, $a = ?$, $c = ?$

15. $a = 23.4$ m, $b = ?$, $c = ?$

16. $a = 63.19$ cm, $b = ?$, $c = ?$

17. $b = 2.489 \times 10^9$ yd, $a = ?$, $c = ?$

18. $b = 1.037 \times 10^{13}$ m, $a = ?$, $c = ?$

19. $c = 8.39 \times 10^{-5}$ mm, $a = ?$, $b = ?$

20. $c = 2.86 \times 10^{-8}$ cm, $a = ?$, $b = ?$

C *In Problems 21 and 22, find the unknown quantities. (If you have a protractor, make a scale drawing and complete the problem using your own measurements and calculations. If you do not have a protractor, use the quantities given in the problem.)*

21. Suppose in the figure that $\angle A = 70°$, $\angle C = 90°$, and $a = 101$ ft. If a scale drawing is made of the triangle by choosing a' to be 2.00 in. and c' is then measured to be 2.13 in., estimate c in the original triangle.

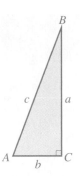

Figure for 21 and 22

22. Repeat Problem 21, choosing $a' = 5.00$ in. and c', a measured quantity, to be 5.28 in.

 Applications

23. **Tennis** A ball is served from the center of the baseline into the deuce court. If the ball is hit 9 ft above the ground and travels in a straight line down the middle of the court, and the net is 3 ft high, how far from the base of the net will the ball land if it just clears the top of the net? (See the figure.) Assume all figures are exact and compute the answer to one decimal place.

24. **Tennis** If the ball in Problem 23 is hit 8.5 ft above the ground, how far away from the base of the net will the ball land? Assume all figures are exact and compute the answer to two decimal places. (Now you can see why tennis players try to spin the ball on a serve so that it curves downward.)

25. **Indirect Measurement** Find the height of the tree in the figure on the next page, given that $AC = 24$ ft, $CD = 2.1$ ft, and $DE = 5.5$ ft.

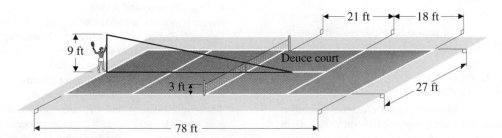

Figure for 23 and 24

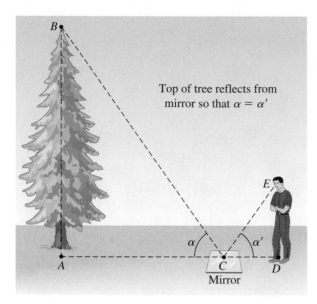

Top of tree reflects from
mirror so that $\alpha = \alpha'$

Figure for 25 and 26

26. Indirect Measurement   Find the height of the tree in
the figure, given that $AC = 25$ ft, $CD = 2$ ft 3 in., and
$DE = 5$ ft 9 in.

27. Indirect Measurement   A flagpole that is 20 ft high
casts a 32 ft shadow. At the same time, a second flagpole
casts a 44 ft shadow. How tall is the second flagpole?

28. Indirect Measurement   Jack is 6 ft tall and casts an
8 ft shadow. At the same time, Jill casts a 7 ft shadow.
How tall is Jill?

29. Indirect Measurement   A surveyor's assistant positions an 8.0 ft vertical pole so that the top of the pole and
the top of a distant tree are aligned in the surveyor's line
of sight. The surveyor's eye level is 5.7 ft above the
ground, the distance between the surveyor and the pole is
16 ft, and the distance between the surveyor and the tree
is 200 ft. How tall is the tree?

30. Indirect Measurement   A streetlight is at the top of a
12 ft pole. A 5.0 ft tall person is standing 25 ft from the
bottom of the pole. How long is the person's shadow?

*Problems 31 and 32 are optional for those who have protrac-
tors and can make scale drawings.*

31. Astronomy   The following figure illustrates a method
that is used to determine depths of moon craters from
observatories on earth. If sun rays strike the surface of
the moon so that $\angle BAC = 15°$ and AC is measured to be
4.0 km, how high is the rim of the crater above its floor? ,

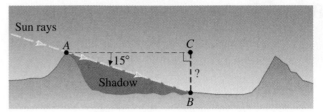

Depth of moon crater

Figure for 31

32. Astronomy   The figure illustrates how the height of a
mountain on the moon can be determined from earth. If
sun rays strike the surface of the moon so that
$\angle BAC = 28°$ and AC is measured to be 4.0×10^3m,
how high is the mountain?

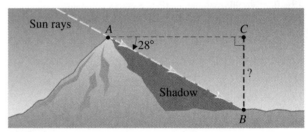

Height of moon mountain

Figure for 32

33. Fundamental Lens Equation   Parallel light rays (light
rays from infinity) approach a thin convex lens and are
focused at a point on the other side of the lens. The distance f from the lens to the **focal point F** is called the
focal length of the lens [see part (a) of the figure on the
next page]. A standard lens on many 35 mm cameras has
a focal length of 50 mm (about 2 in.), a 28 mm lens is a
wide-angle lens, and a 200 mm lens is telephoto. How
does a lens focus an image of an object on the film in a
camera? Part (b) of the figure shows the geometry
involved for a thin convex lens. Point P at the top of the
object is selected for illustration (any point on the object
would do). Light rays from point P travel in all directions. Those that go through the lens are focused at P'.
Since light rays PA and CP' are parallel, AP' and CP
pass through focal points F' and F, respectively, which
are equidistant from the lens; that is, $FB = BF' = f$.
Also note that $AB = h$, the height of the object, and
$BC = h'$, the height of the image.

(A) Explain why triangles *PAC* and *FBC* are similar and why triangles *ACP′* and *ABF′* are similar.

(B) From the properties of similar triangles, show that

$$\frac{h + h'}{u} = \frac{h'}{f} \quad \text{and} \quad \frac{h + h'}{v} = \frac{h}{f}$$

(C) Combining the results in part (B), derive the important lens equation

$$\frac{1}{u} + \frac{1}{v} = \frac{1}{f}$$

where *u* is the distance between the object and the lens, *v* is the distance between the focused image on the film and the lens, and *f* is the focal length of the lens.

(D) How far (to three decimal places) must a 50 mm lens be from the film if the lens is focused on an object 3 m away?

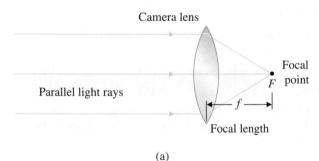

Camera lens

Focal point *F*

Parallel light rays

f

Focal length

(a)

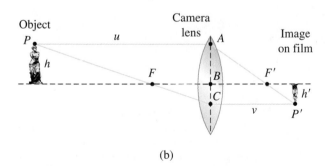

Object
P

h

u

Camera lens

A

F

B

C

F′

v

P′

h′

Image on film

(b)

Figure for 33 and 34

34. Fundamental Lens Equation (Refer to Problem 33.) For a 50 mm lens it is said that if the object is more than 20 m from the lens, then the rays coming from the object are very close to being parallel, and *v* will be very close to *f*.

(A) Use the lens equation given in part (C) of Problem 33 to complete the following table (to two decimal places) for a 50 mm lens. (Convert meters to millimeters before using the lens equation.)

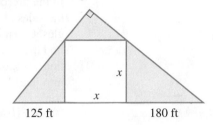

u (m)	10	20	30	40	50	60
v (m)						

(B) Referring to part (A), how does *v* compare with *f* as *u* increases beyond 20? (Note that 0.1 mm is less than the diameter of a period on this page.)

35. Construction A contractor wants to locate a square building on a triangular lot (see the figure). Find the width of the building.

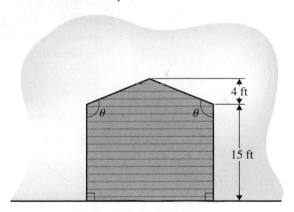

125 ft   180 ft

x

x

36. Construction A carpenter is building a deck in the shape of a pentagon (see the figure). If the total area of the deck is 340 sq ft, find the width of the deck.

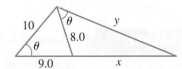

4 ft

15 ft

θ θ

37. Geometry Find *x* and *y* in the figure.

10

θ

θ

8.0

9.0

y

x

38. Geometry Find *x* and *y* in the figure.

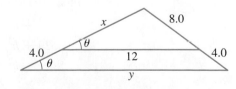

x

8.0

4.0

θ

θ

12

4.0

y

1.3   Trigonometric Ratios And Right Triangles

- Pythagorean Theorem
- Trigonometric Ratios
- Calculator Evaluation
- Solving Right Triangles

In the preceding section we used Euclid's theorem to find relationships between the sides of similar triangles. In this section, we turn our attention to right triangles. Given information about the sides and acute angles in a triangle, we want to find the remaining sides and angles. This is called **solving a triangle**. In this section we show how this can be done without using scale drawings. The concepts introduced here will be generalized extensively as we progress through the book.

■ Pythagorean Theorem

We start our discussion of right triangles with the familiar Pythagorean theorem.

PYTHAGOREAN THEOREM

In a right triangle, the side opposite the right angle is called the **hypotenuse** and the other two sides are called **legs**. If a, b, and c are the lengths of the legs and hypotenuse, respectively, then $a^2 + b^2 = c^2$.

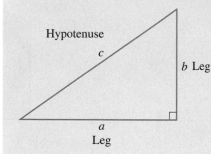

$$a^2 + b^2 = c^2$$

EXAMPLE 1

Area of a Triangle

Find the area of the triangle in Figure 1 on page 23 (see Appendix C.2).

Solution   We need both the base (23) and the altitude (h) to find the area of the triangle. First, we use the Pythagorean theorem to find h.

$$23^2 + h^2 = 27^2 \qquad \text{Evaluate squared terms}$$
$$529 + h^2 = 729 \qquad \text{Subtract 529 from both sides}$$
$$h^2 = 729 - 529 \qquad \text{Simplify}$$
$$= 200 \qquad \text{Take the square root of both sides}$$

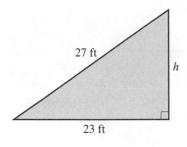

FIGURE 1

$$h = \sqrt{200}$$ 	Discard negative square root. (Why?)

$$= 14 \text{ ft}$$ 	To two significant digits

The area of the triangle is

$$A = \frac{1}{2}(\text{base})(\text{altitude})$$ 	Substitute base = 23 and altitude = 14

$$= \frac{1}{2}(23)(14)$$ 	Evaluate

$$= 160 \text{ sq ft}$$ 	To two significant digits

Matched Problem 1 Find the area of the triangle in Figure 1 if the altitude is 17 feet and the base is unknown. ∎

■ Trigonometric Ratios

If θ is an acute angle ($0° < \theta < 90°$) in a right triangle, then one of the legs of the triangle is referred to as the **side opposite** angle θ and the other leg is referred to as the **side adjacent** to angle θ (see Fig. 2). As before, the hypotenuse is the side opposite the right angle. There are six possible ratios of the lengths of these three sides.

FIGURE 2

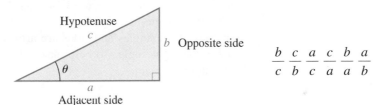

$$\frac{b}{c} \quad \frac{c}{b} \quad \frac{a}{c} \quad \frac{c}{a} \quad \frac{b}{a} \quad \frac{a}{b}$$

These ratios are referred to as trigonometric ratios, and because of their importance, each is given a name: sine (sin), cosine (cos), tangent (tan), cosecant (csc), secant (sec), and cotangent (cot). And each is written in abbreviated form as follows:

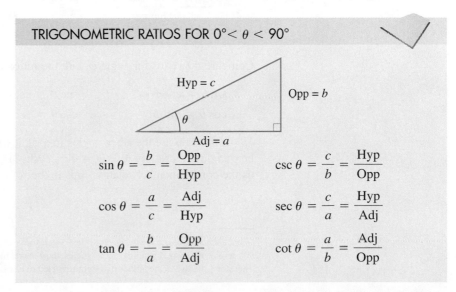

TRIGONOMETRIC RATIOS FOR $0° < \theta < 90°$

$$\sin \theta = \frac{b}{c} = \frac{\text{Opp}}{\text{Hyp}} \qquad \csc \theta = \frac{c}{b} = \frac{\text{Hyp}}{\text{Opp}}$$

$$\cos \theta = \frac{a}{c} = \frac{\text{Adj}}{\text{Hyp}} \qquad \sec \theta = \frac{c}{a} = \frac{\text{Hyp}}{\text{Adj}}$$

$$\tan \theta = \frac{b}{a} = \frac{\text{Opp}}{\text{Adj}} \qquad \cot \theta = \frac{a}{b} = \frac{\text{Adj}}{\text{Opp}}$$

EXPLORE/DISCUSS 1

For a given acute angle θ in a right triangle, use Euclid's theorem (from Section 1.2) to explain why the value of any of the six trigonometric ratios for that angle is independent of the size of the triangle.

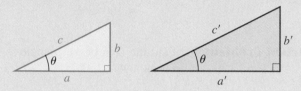

The trigonometric ratios should be learned. They will be used extensively in the work that follows. We recommend that you remember the descriptive forms (Opp/Hyp, etc.), instead of the formulas (b/c, etc.). The descriptive forms are independent of the labels used for the sides or angles. It is important to note that the right angle in a right triangle can be oriented in any position and that the names of the angles and sides are arbitrary, but the hypotenuse is always opposite the right angle. To help make all of this clear, look at the ratios for this triangle:

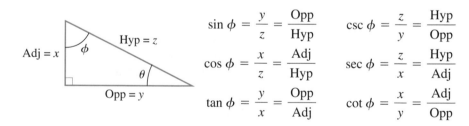

$$\sin \phi = \frac{y}{z} = \frac{\text{Opp}}{\text{Hyp}} \qquad \csc \phi = \frac{z}{y} = \frac{\text{Hyp}}{\text{Opp}}$$

$$\cos \phi = \frac{x}{z} = \frac{\text{Adj}}{\text{Hyp}} \qquad \sec \phi = \frac{z}{x} = \frac{\text{Hyp}}{\text{Adj}}$$

$$\tan \phi = \frac{y}{x} = \frac{\text{Opp}}{\text{Adj}} \qquad \cot \phi = \frac{x}{y} = \frac{\text{Adj}}{\text{Opp}}$$

Compare the ratios for angle ϕ with the ratios for angle θ in the definition:

$$\sin \theta = \cos \phi \qquad \sec \theta = \csc \phi \qquad \tan \theta = \cot \phi$$

$$\cos \theta = \sin \phi \qquad \csc \theta = \sec \phi \qquad \cot \theta = \tan \phi$$

If θ and ϕ are the acute angles in a right triangle, then they are complementary angles and satisfy $\theta + \phi = 90°$. (Why?)* Substituting $90° - \theta$ for ϕ gives us the complementary relationships in the following box.

---

* Since the sum of the measures of all three angles in a triangle is 180°, and a right triangle has one 90° angle, the two remaining acute angles must have measures that sum to $180° - 90° = 90°$. Therefore, the two acute angles in a right triangle are always complementary.

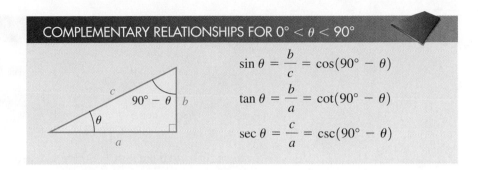

COMPLEMENTARY RELATIONSHIPS FOR $0° < \theta < 90°$

$$\sin \theta = \frac{b}{c} = \cos(90° - \theta)$$

$$\tan \theta = \frac{b}{a} = \cot(90° - \theta)$$

$$\sec \theta = \frac{c}{a} = \csc(90° - \theta)$$

Note that the sine of θ is the same as the cosine of the complement of θ (which is $90° - \theta$ in the triangle shown), the tangent of θ is the cotangent of the complement of θ, and the secant of θ is the cosecant of the complement of θ. The trigonometric ratios cosine, cotangent, and cosecant are sometimes referred to as the **cofunctions** of sine, tangent, and secant, respectively.

■ Calculator Evaluation

For the trigonometric ratios to be useful in solving right triangle problems, we must be able to find each for any acute angle. Scientific and graphing calculators can approximate (almost instantly) these ratios to eight or ten significant digits. Scientific and graphing calculators generally use different sequences of steps. Consult the user's manual for your particular calculator.

> **The use of a scientific calculator is assumed throughout the book, and a graphing calculator is required for many optional problems. A graphing calculator is a scientific calculator with additional capabilities, including graphing.**

Calculators have two trigonometric modes: degree and radian. Our interest now is in *degree mode*. Later we will discuss radian mode in detail.

 Caution Refer to the user's manual accompanying your calculator to determine how it is to be set in degree mode, and set it that way. *This is an important step and should not be overlooked.* Many errors can be traced to calculators being set in the wrong mode. □

If you look at the function keys on your calculator, you will find three keys labeled

$$\boxed{\sin} \qquad \boxed{\cos} \qquad \boxed{\tan}$$

These keys are used to find sine, cosine, and tangent ratios, respectively. The calculator also can be used to compute cosecant, secant, and cotangent ratios using the reciprocal* relationships, which follow directly from the definition of the six trigonometric ratios.

---

* Recall that two nonzero numbers a and b are **reciprocals** of each other if $ab = 1$; then we may write $a = 1/b$ and $b = 1/a$.

RECIPROCAL RELATIONSHIPS FOR $0° < \theta < 90°$

$$\csc \theta \sin \theta = \frac{c}{b} \cdot \frac{b}{c} = 1 \qquad \text{so} \qquad \csc \theta = \frac{1}{\sin \theta}$$

$$\sec \theta \cos \theta = \frac{c}{a} \cdot \frac{a}{c} = 1 \qquad \text{so} \qquad \sec \theta = \frac{1}{\cos \theta}$$

$$\cot \theta \tan \theta = \frac{a}{b} \cdot \frac{b}{a} = 1 \qquad \text{so} \qquad \cot \theta = \frac{1}{\tan \theta}$$

⚠ Caution When using reciprocal relationships, many students tend to associate cosecant with cosine and secant with sine: Just the opposite is correct. □

 EXAMPLE 2 **Calculator Evaluation**

Evaluate to four significant digits using a calculator:

(A) sin 23.72°   (B) tan 54°37′

(C) sec 49.31°   (D) cot 12.86°

Solution First, set the calculator in degree mode.

(A) sin 23.72° = 0.4023

(B) Some calculators require 54°37′ to be converted to decimal degrees first; others can do the calculation directly—check your user's manual.

$$\tan 54°37′ \boxed{= \tan 54.6166...} = 1.408$$

(C) sec 49.31° = $\dfrac{1}{\cos 49.31°}$ = 1.534

(D) cot 12.86° = $\dfrac{1}{\tan 12.86°}$ = 4.380 ∎

```
sin(23.72)
           .4022673787
tan(54+37/60)
           1.408003909
1/cos(49.31)
           1.533822141
1/tan(12.86)
           4.380279399
```

FIGURE 3

Figure 3 shows the calculations in Example 2 on a TI-84 graphing calculator set in degree mode.

Matched Problem 2 Evaluate to four significant digits using a calculator:

(A) cos 38.27°   (B) sin 37°44′

(C) cot 49.82°   (D) csc 77°53′ ∎

EXPLORE/DISCUSS 2

Experiment with your calculator to determine which of the following two window displays from a graphing calculator is the result of the calculator being set in degree mode and which is the result of the calculator being set in radian mode.

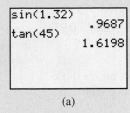

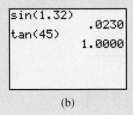

(a)   (b)

Now we reverse the process illustrated in Example 2. Suppose we are given

$$\sin \theta = 0.3174$$

How do we find θ? That is, how do we find the acute angle θ whose sine is 0.3174? The solution to this problem is written symbolically as either

$$\theta = \arcsin 0.3174 \qquad \text{"arcsin" and "sin}^{-1}\text{" both represent the same thing.}$$

or

$$\theta = \sin^{-1} 0.3174$$

Both of these expressions are read "θ is the angle whose sine is 0.3174."

 Caution   It is important to note that $\sin^{-1} 0.3174$ does not mean $1/(\sin 0.3174)$; the -1 "exponent" is a superscript that is part of a *function symbol*. More will be said about this in Chapter 5, where a detailed discussion of these concepts is given.   □

We can find θ directly using a calculator. The function key $\boxed{\sin^{-1}}$ or its equivalent takes us from a trigonometric sine ratio back to the corresponding acute angle in decimal degrees when the calculator is in degree mode. Thus, if $\sin \theta = 0.3174$, then we can write $\theta = \arcsin 0.3174$ or $\theta = \sin^{-1} 0.3174$. We choose the latter and proceed as follows:

$$\theta = \sin^{-1} 0.3174$$
$$= 18.506° \qquad \text{To three decimal places}$$
$$\text{or } 18°30'21'' \qquad \text{To the nearest second}$$

✔ **Check** $\sin 18.506° = 0.3174$ Perform the check on a calculator.

EXAMPLE 3

Finding Inverses

Find each acute angle θ to the accuracy indicated:

(A) $\cos \theta = 0.7335$ (to three decimal places)

(B) $\theta = \tan^{-1} 8.207$ (to the nearest minute)

(C) $\theta = \arcsin 0.0367$ (to the nearest $10'$)

Solution   First, set the calculator in degree mode.

(A) If $\cos \theta = 0.7335$, then
$$\theta = \cos^{-1} 0.7335$$
$$= 42.819° \quad \text{To three decimal places}$$

(B) $\theta = \tan^{-1} 8.207$
$$= 83.053$$
$$= 83°3' \quad \text{To the nearest minute}$$

(C) $\theta = \arcsin 0.0367$
$$= 2.103°$$
$$= 2°10' \quad \text{To the nearest } 10'$$ ∎

```
cos⁻¹(0.7335)
       42.81938042
tan⁻¹(8.207)
       83.05291494
sin⁻¹(0.0367)
       2.103227424
```

FIGURE 4

Figure 4 shows the inverse function calculations in Example 3 on a TI-84 graphing calculator set in degree mode.

Matched Problem 3   Find each acute angle θ to the accuracy indicated:

(A) $\tan \theta = 1.739$ (to two decimal places)

(B) $\theta = \sin^{-1} 0.2571$ (to the nearest $10''$)

(C) $\theta = \arccos 0.0367$ (to the nearest minute) ∎

We postpone discussion of $\cot^{-1}$, $\sec^{-1}$, and $\csc^{-1}$ until Chapter 5. The preceding discussion will handle all our needs at this time.

■ Solving Right Triangles

To solve a right triangle is to find, given the measures of two sides or the measures of one side and an acute angle, the measures of the remaining sides and angles. Solving right triangles is best illustrated through examples. Note at the outset that accuracy of the computations is governed by Table 1 (which is reproduced inside the front cover for easy reference).

TABLE 1	
Angle to nearest	*Significant digits for side measure*
$1°$	2
$10'$ or $0.1°$	3
$1'$ or $0.01°$	4
$10''$ or $0.001°$	5

Remark When we use the equal sign (=) in the following computations, it should be understood that equality holds only to the number of significant digits justified by Table 1. The approximation symbol (≈) is used only when we want to emphasize the approximation. □

EXAMPLE 4

Solving a Right Triangle

Solve the right triangle in Figure 5.

Solution First set your calculator in degree mode.

Solve for the complementary angle.

$$90° - \theta = 90° - 35.7° = 54.3°$$ *Remember, 90° is exact.*

Solve for b. Since $\theta = 35.7°$ and $c = 124$ m, we look for a trigonometric ratio that involves θ and c (the known quantities) and b (an unknown quantity). Referring to the definition of the trigonometric ratios (page 23), we see that both sine and cosecant involve all three quantities. We choose sine and proceed as follows:

$$\sin \theta = \frac{b}{c}$$ *Solve for b.*
$$b = c \sin \theta$$ *Substitute $b = 124$ m and $\theta = 35.7°$.*
$$= (124 \text{ m})(\sin 35.7°)$$ *Evaluate on a calculator.*
$$= 72.4 \text{ m}$$ *To three significant digits*

FIGURE 5

Solve for a. Now that we have b, we can use the tangent, cotangent, cosine, or secant to find a. We choose the cosine:

$$\cos \theta = \frac{a}{c}$$ *Solve for a.*
$$a = c \cos \theta$$ *Substitute $c = 124$ m and $\theta = 35.7°$.*
$$= (124 \text{ m})(\cos 35.7°)$$ *Evaluate on a calculator.*
$$= 101 \text{ m}$$ *To three significant digits* ■

Matched Problem 4 Solve the triangle in Example 4 with $\theta = 28.3°$ and $c = 62.4$ cm ■

EXAMPLE 5

Solving a Right Triangle

Solve the right triangle in Figure 6 for θ and $90° - \theta$ to the nearest 10′ and for a to three significant digits.

Solution *Solve for θ.*

$$\sin \theta = \frac{b}{c} = \frac{42.7 \text{ km}}{51.3 \text{ km}}$$

$$\sin \theta = 0.832$$

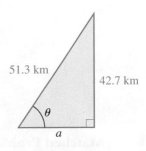

FIGURE 6

Given the sine of θ, how do we find θ? We can find θ directly using a calculator as discussed in Example 3.

The $\boxed{\sin^{-1}}$ key or its equivalent takes us from a trigonometric ratio back to the corresponding angle in decimal degrees (if the calculator is in degree mode).

$$\begin{aligned}\theta &= \sin^{-1} 0.832 &&\text{\textit{Use a calculator}}\\ &= 56.3° &&(0.3)(60) = 18' \approx 20'\\ &= 56°20' &&\text{\textit{To the nearest 10'}}\end{aligned}$$

Solve for the complementary angle.

$$\begin{aligned}90° - \theta &= 90° - 56°20'\\ &= 33°40'\end{aligned}$$

Solve for a. Use cosine, secant, cotangent, or tangent. We will use tangent:

$$\begin{aligned}\tan \theta &= \frac{b}{a} &&\text{\textit{Solve for a.}}\\ a &= \frac{b}{\tan \theta} &&\text{\textit{Substitute a = 42.7 km and }}\theta\text{\textit{ = 56°20'.}}\\ &= \frac{42.7 \text{ km}}{\tan 56°20'} &&\text{\textit{Use a calculator.}}\\ &= 28.4 \text{ km} &&\text{\textit{To three significant digits}}\end{aligned}$$

✔ **Check**   We check by using the Pythagorean theorem (see Fig. 7).

$$28.4^2 + 42.7^2 \overset{?}{=} 51.3^2 \qquad \text{\textit{Compute both sides to three significant digits.}}$$

$$2{,}630 \overset{\checkmark}{=} 2{,}630$$

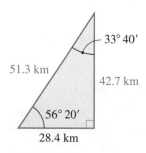

33° 40′

51.3 km

42.7 km

56° 20′

28.4 km

FIGURE 7

Matched Problem 5   Repeat Example 5 with $b = 23.2$ km and $c = 30.4$ km.

In this section we have concentrated on technique. In the next section we will consider a large variety of applications involving the techniques discussed here.

Answers to
Matched Problems

1. 180 sq ft
2. (A) 0.7851   (B) 0.6120   (C) 0.8445   (D) 1.023
3. (A) 60.10°   (B) 14°53′50″   (C) 87°54′
4. $90° - \theta = 61.7°$, $b = 29.6$ cm, $a = 54.9$ cm
5. $\theta = 49°40′$, $90° - \theta = 40°20′$, $a = 19.7$ km

EXERCISE 1.3

A *In Problems 1–6, refer to the figure and identify each of the named trigonometric ratios with one of the following quotients: a/b, b/a, a/c, c/a, b/c, c/b. Try not to look back in the text.*

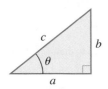

Figure for 1–12

1. $\cos \theta$	**2.** $\sin \theta$	**3.** $\tan \theta$
4. $\cot \theta$	**5.** $\sec \theta$	**6.** $\csc \theta$

In Problems 7–12, refer to the figure and identify each of the quotients with a named trigonometric ratio from the following list: sin θ, cos θ, sec θ, csc θ, tan θ, cot θ. Try not to look back in the text.

7. b/c	**8.** a/c	**9.** b/a
10. a/b	**11.** c/b	**12.** c/a

In Problems 13–24, find each trigonometric ratio to three significant digits.

13. $\sin 34.7°$	**14.** $\cos 18.9°$	**15.** $\tan 29°45′$
16. $\cot 15°35′$	**17.** $\sec 42.2°$	**18.** $\csc 22.5°$
19. $\cos 83.4°$	**20.** $\sin 59.3°$	**21.** $\cot 66.7°$
22. $\tan 72.6°$	**23.** $\csc 81°20′$	**24.** $\sec 48°50′$

B *In Problems 25–32, find each acute angle θ to the accuracy indicated.*

25. $\cos \theta = 0.5$ (to the nearest degree)

26. $\tan \theta = 1$ (to the nearest degree)
27. $\sin \theta = 0.8125$ (to two decimal places)
28. $\tan \theta = 2.25$ (to two decimal places)
29. $\theta = \arcsin 0.4517$ (to the nearest 10′)
30. $\theta = \arccos 0.2557$ (to the nearest 10′)
31. $\theta = \tan^{-1}(2.753)$ (to the nearest minute)
32. $\theta = \cos^{-1}(0.0125)$ (to the nearest minute)

In Problems 33–36, if you are given the indicated measures in a right triangle, explain why you can or cannot solve the triangle.

33. The measures of two adjacent sides a and b
34. The measures of one side and one angle
35. The measures of two acute angles
36. The measure of the hypotenuse c

In Problems 37–46, solve the right triangle (labeled as in the figure at the beginning of the exercises) given the information in each problem.

37. $\theta = 58°40′$, $c = 15.0$ mm
38. $\theta = 62°10′$, $c = 33.0$ cm
39. $\theta = 83.7°$, $b = 3.21$ km
40. $\theta = 32.4°$, $a = 42.3$ m
41. $\theta = 71.5°$, $b = 12.8$ in.
42. $\theta = 44.5°$, $a = 2.30 \times 10^6$ m
43. $b = 63.8$ ft, $c = 134$ ft   (angles to the nearest 10′)
44. $b = 22.0$ km, $a = 46.2$ km   (angles to the nearest 10′)
45. $b = 132$ mi, $a = 108$ mi   (angles to the nearest 0.1°)
46. $a = 134$ m, $c = 182$ m   (angles to the nearest 0.1°)

47. The graphing calculator screen that follows shows the solution of the accompanying triangle for sides a and b. Clearly, something is wrong. Explain what is wrong and find the correct measures for the two sides.

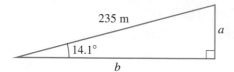

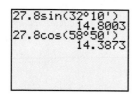

48. In the figure that follows, side a was found two ways: one using α and sine, and the other using β and cosine. The graphing calculator screen shows the two calculations, but there is an error. Explain the error and correct it.

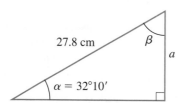

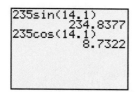

In Problems 49–52, verify the truth of each statement for the indicated values.

49. $(\sin \theta)^2 + (\cos \theta)^2 = 1$
   (A) $\theta = 11°$ (B) $\theta = 6.09°$ (C) $\theta = 43°24'47''$

50. $(\sin \theta)^2 + (\cos \theta)^2 = 1$
   (A) $\theta = 34°$ (B) $\theta = 37.281°$ (C) $\theta = 87°23'41''$

51. $\sin \theta - \cos(90° - \theta) = 0$
   (A) $\theta = 19°$ (B) $\theta = 49.06°$ (C) $\theta = 72°51'12''$

52. $\tan \theta - \cot(90° - \theta) = 0$
   (A) $\theta = 17°$ (B) $\theta = 27.143°$ (C) $\theta = 14°12'33''$

C *In Problems 53–58, solve the right triangles (labeled as in the figure at the beginning of the exercises).*

53. $a = 23.82$ mi, $\theta = 83°12'$

54. $a = 6.482$ m, $\theta = 35°44'$

55. $b = 42.39$ cm, $a = 56.04$ cm
   (angles to the nearest $1'$)

56. $a = 123.4$ ft, $c = 163.8$ ft
   (angles to the nearest $1'$)

57. $b = 35.06$ cm, $c = 50.37$ cm
   (angles to the nearest $0.01°$)

58. $b = 5.207$ mm, $a = 8.030$ mm
   (angles to the nearest $0.01°$)

59. Show that $(\sin \theta)^2 + (\cos \theta)^2 = 1$, using the definition of the trigonometric ratios (page 23) and the Pythagorean theorem.

60. Without looking back in the text, show that for each acute angle θ:
   (A) $\csc \theta = \dfrac{1}{\sin \theta}$ (B) $\cos(90° - \theta) = \sin \theta$

61. Without looking back in the text, show that for each acute angle θ:
   (A) $\cot \theta = \dfrac{1}{\tan \theta}$ (B) $\csc(90° - \theta) = \sec \theta$

62. Without looking back in the text, show that for each acute angle θ:
   (A) $\sec \theta = \dfrac{1}{\cos \theta}$ (B) $\cot(90° - \theta) = \tan \theta$

Geometric Interpretation of Trigonometric Ratios *Problems 63–68 refer to the figure, where O is the center of a circle of radius 1, θ is the acute angle AOD, D is the intersection point of the terminal side of angle θ with the circle, and EC is tangent to the circle at D.*

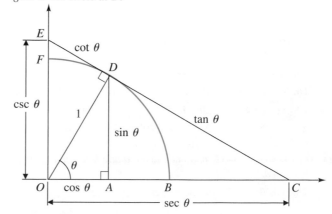

Figure for 63–68

63. Show that:

(A) $\sin \theta = AD$ (B) $\tan \theta = DC$ (C) $\csc \theta = OE$

64. Show that:

(A) $\cos \theta = OA$ (B) $\cot \theta = DE$ (C) $\sec \theta = OC$

65. Explain what happens to each of the following as the acute angle θ approaches 90°:

(A) $\sin \theta$ (B) $\tan \theta$ (C) $\csc \theta$

66. Explain what happens to each of the following as the acute angle θ approaches 90°:

(A) $\cos \theta$ (B) $\cot \theta$ (C) $\sec \theta$

67. Explain what happens to each of the following as the acute angle θ approaches 0°:

(A) $\cos \theta$ (B) $\cot \theta$ (C) $\sec \theta$

68. Explain what happens to each of the following as the acute angle θ approaches 0°:

(A) $\sin \theta$ (B) $\tan \theta$ (C) $\csc \theta$

In Problems 69–72, find the area of the polygon in the figure.

69. Geometry

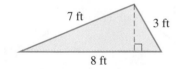

70. Geometry

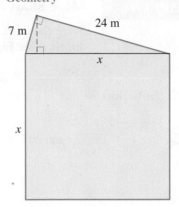

71. Geometry

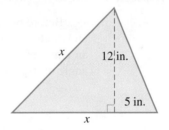

72. Geometry

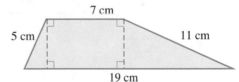

1.4 Right Triangle Applications

Now that you know how to solve right triangles, we can consider a variety of interesting and significant applications.

EXPLORE/DISCUSS 1

Discuss the minimum number of sides and/or angles that must be given in a right triangle in order for you to be able to solve for the remaining angles and sides.

Mine Shaft Application

Solve the mine shaft problem in Example 2, Section 1.2, without using a scale drawing (see Fig. 1).

Solution

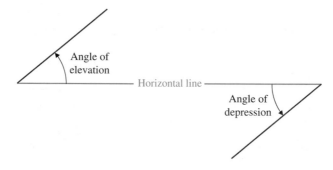

FIGURE 1

$$\tan \theta = \frac{\text{Opp}}{\text{Adj}}$$ Substitute $\theta = 20°$, Opp = x, and Adj = 310 ft.

$$\tan 20° = \frac{x}{310 \text{ ft}}$$ Solve for x.

$$x = (310 \text{ ft})(\tan 20°)$$ Use a calculator.

$$= 110 \text{ ft}$$ To two significant digits   ■

Matched Problem 1   Solve the problem in Example 1 if $AC = 550$ ft and $\angle A = 30°$.   ■

Before proceeding further, we introduce two new terms: **angle of elevation** and **angle of depression.** An angle measured from the horizontal upward is called an angle of elevation; one measured from the horizontal downward is called an angle of depression (see Fig. 2).

FIGURE 2

Length of Air-to-Air Fueling Hose

To save time or because of the lack of landing facilities for large jets in some parts of the world, the military and some civilian companies use air-to-air refueling for some planes (see Fig. 3). If the angle of elevation of the refueling plane's hose is $\theta = 32°$ and b in Figure 3 is 120 ft, how long is the hose?

FIGURE 3
Air-to-air fueling

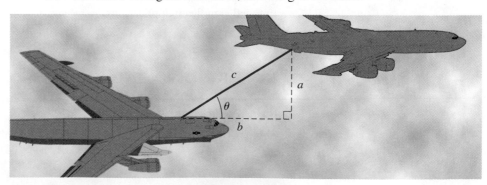

Solution

$$\sec \theta = \frac{c}{b} \qquad \text{\textit{Solve for c.}}$$

$$c = b \sec \theta \qquad \text{\textit{Use } } \sec \theta = \frac{1}{\cos \theta}.$$

$$= \frac{b}{\cos \theta} \qquad \text{\textit{Substitute } } b = 120 \text{ ft and } \theta = 32°.$$

$$= \frac{120 \text{ ft}}{\cos 32°} \qquad \text{\textit{Use a calculator.}}$$

$$= 140 \text{ ft} \qquad \text{\textit{To two significant digits}} \qquad ■$$

Matched Problem 2 The horizontal shadow of a vertical tree is 23.4 m long when the angle of eleva-tion of the sun is 56.3°. How tall is the tree?   ■

 **EXAMPLE 3**   **Astronomy**

If we know that the distance from the earth to the sun is approximately 93,000,000 mi, and we find that the largest angle between the earth–sun line and the earth–Venus line is 46°, how far is Venus from the sun? (Assume that the earth and Venus have circu-lar orbits around the sun—see Fig. 4.)

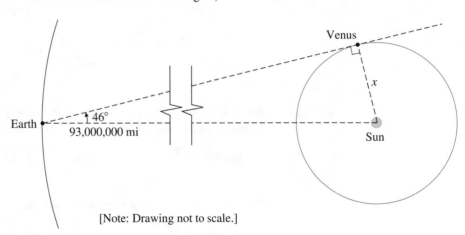

[Note: Drawing not to scale.]

FIGURE 4

Solution The earth–Venus line at its largest angle to the earth–sun line must be tangent to Venus's orbit. Thus, from plane geometry, the Venus–sun line must be at right angles to the earth–Venus line at this time. The sine ratio involves two known quantities and the unknown distance from Venus to the sun. To find x we proceed as follows:

$$\sin 46° = \frac{x}{93,000,000} \qquad \text{\textit{Solve for x.}}$$

$$x = 93,000,000 \sin 46° \qquad \text{\textit{Use a calculator.}}$$

$$= 67,000,000 \text{ mi} \qquad \text{\textit{To two significant digits}} \qquad ■$$

Matched Problem 3   If the largest angle that the earth–Mercury line makes with the earth–sun line is 28°, how far is Mercury from the sun? (Assume circular orbits.)   ■

Coastal Piloting

A boat is cruising along the coast on a straight course. A rocky point is sighted at an angle of 31° from the course. After continuing 4.8 mi, another sighting is taken and the point is found to be 55° from the course (see Fig. 5). How close will the boat come to the point?

FIGURE 5

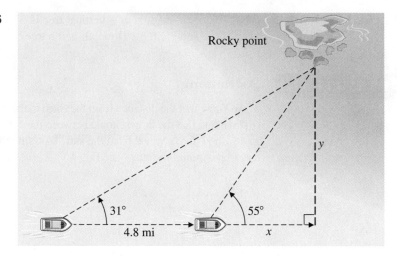

Solution   Referring to Figure 5, y is the closest distance that the boat will be to the point. To find y we proceed as follows: From the small right triangle we note that

$$\cot 55° = \frac{x}{y} \qquad \text{Solve for x.}$$
$$x = y \cot 55° \tag{1}$$

Now, from the large right triangle, we see that

$$\cot 31° = \frac{4.8 + x}{y} \qquad \text{Multiply both sides by y.}$$
$$y \cot 31° = 4.8 + x \tag{2}$$

Substituting equation (1) into (2), we obtain

$$y \cot 31° = 4.8 + y \cot 55° \qquad \text{Subtract y cot 55° from both sides.}$$

$$y \cot 31° - y \cot 55° = 4.8 \qquad \text{Factor.}$$
$$y(\cot 31° - \cot 55°) = 4.8 \qquad \text{Solve for y.}$$
$$y = \frac{4.8}{\cot 31° - \cot 55°} \qquad \text{Use a calculator.}$$
$$y = 5.0 \text{ mi} \qquad \text{To two significant digits*} \qquad ■$$

─────────

* We will continue to use the guidelines in Appendix A.3 for determining the accuracy of a calculation, but we will no longer explicitly state this accuracy.

Matched Problem 4   Repeat Example 4 after replacing 31° with 28°, 55° with 49°, and 4.8 mi with 5.5 mi.   ∎

Answers to Matched Problems

1. 320 ft (to two significant digits)
2. 35.1 m (to three significant digits)
3. 44,000,000 mi (to two significant digits)
4. 5.4 mi (to two significant digits)

EXERCISE 1.4

Applications

A   **1. Construction**   A ladder 8.0 m long is placed against a building. The angle between the ladder and the ground is 61°. How high will the top of the ladder reach up the building?

8.0 m

61°

Figure for 1

2. Construction   An 18 ft ladder is leaning against a house. It touches the bottom of a window that is 14 ft 6 in. above the ground. What is the measure of the angle that the ladder forms with the ground?

3. Surveying   When the angle of elevation of the sun is 58°, the shadow cast by a tree is 28 ft long. How tall is the tree?

4. Surveying   Find the angle of elevation of the sun at the moment when a 21 m flag pole casts a 14 m shadow.

5. Flight   An airplane flying at an altitude of 5,100 ft is 12,000 ft from an airport. What is the angle of elevation of the plane?

6. Flight   The angle of elevation of a kite at the end of a 400.0 ft string is 28.5°. How high is the kite?

7. Surveying   A flagpole stands in the middle of a flat, level parking lot. Twenty-nine ft away the angle of elevation to the top of the flagpole is 38°. How tall is the flagpole?

8. Surveying   The angle of elevation to the top of a dam at a point 38.4 m from the base of the dam is 64°30'. How high is the dam?

9. Surveying   You have walked 21 m away from a tree. At that point the angle of elevation to the top of the tree is 75°. How tall is the tree?

10. Flight   A large airplane (plane A) flying at 31,000 ft sights a smaller plane (plane B) traveling at an altitude of 28,000 ft. The angle of depression is 37°. What is the line-of-sight distance between the two planes?

11. Surveying   When the sun's angle of elevation is 41°, a building casts a shadow of 38 m. How high is the building?

12. Navigation   From the top of a lighthouse 21 m high, a sailboat is sighted at an angle of depression of 6°. How far from the base of the lighthouse is the boat?

13. Boat Safety   Use the information in the figure to find the distance x from the boat to the base of the cliff.

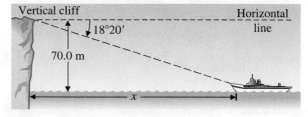

Vertical cliff   Horizontal line

18°20'

70.0 m

x

Figure for 13

14. Boat Safety In Problem 13, how far is the boat from the top of the cliff?

15. Geography on the Moon Find the depth of the moon crater in Problem 31, Exercise 1.2, without using a scale drawing.

16. Geography on the Moon Find the height of the mountain on the moon in Problem 32, Exercise 1.2, without using a scale drawing.

17. Flight Safety A glider is flying at an altitude of 8,240 m. The angle of depression from the glider to the control tower at an airport is 15°40′. What is the horizontal distance (in kilometers) from the glider to a point directly over the tower?

18. Flight Safety The height of a cloud or fog cover over an airport can be measured as indicated in the figure. Find h in meters if $b = 1.00$ km and $\alpha = 23.4°$ ·

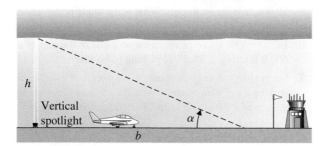

Figure for 18

19. Space Flight The figure shows the reentry flight pattern of a space shuttle. If at the beginning of the final approach the shuttle is at an altitude of 3,300 ft and its ground distance is 8,200 ft from the beginning of the

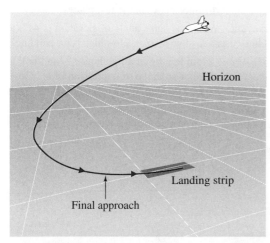

Figure for 19 and 20

landing strip, what glide angle (angle of depression) must be used for the shuttle to touch down at the beginning of the landing strip?

20. Space Flight If at the beginning of the final approach the shuttle in Problem 19 is at an altitude of 3,600 ft and its ground distance is 9,300 ft from the beginning of the landing strip, what glide angle must be used for the shuttle to touch down at the beginning of the landing strip?

B 21. Architecture An architect who is designing a two-story house in a city with a 40°N latitude wishes to control sun exposure on a south-facing wall. Consulting an architectural standards reference book, she finds that at this latitude the noon summer solstice sun has a sun angle of 75° and the noon winter solstice sun has a sun angle of 27° (see the figure).

(A) How much roof overhang should she provide so that at noon on the day of the summer solstice the shadow of the overhang will reach the bottom of the south-facing wall?

(B) How far down the wall will the shadow of the overhang reach at noon on the day of the winter solstice?

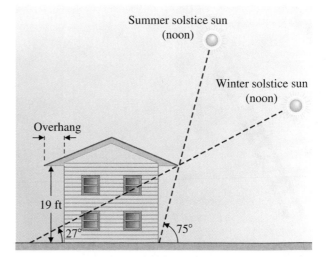

Figure for 21

22. Architecture Repeat Problem 21 for a house located at 32°N latitude, where the summer solstice sun angle is 82° and the winter solstice sun angle is 35°.

23. Lightning Protection A grounded lightning rod on the mast of a sailboat produces a cone of safety as indicated in the following figure. If the top of the rod is 67.0 ft above the water, what is the diameter of the circle of safety on the water?

Figure for 23 and 24

24. Lightning Protection   In Problem 23 how high should the top of the lightning rod be above the water if the diameter of the circle on the water is to be 100 ft?

25. Diagonal Parking   To accommodate cars of most sizes, a parking space needs to contain an 18 ft by 8.0 ft rectangle as shown in the figure. If a diagonal parking space makes an angle of 72° with the horizontal, how long are the sides of the parallelogram that contain the rectangle?

Figure for 25

26. Diagonal Parking   Repeat Problem 25 using 68° instead of 72°.

27. Earth Radius   A person in an orbiting spacecraft (see the figure) h mi above the earth sights the horizon on the earth at an angle of depression of α. (Recall from geometry that a line tangent to a circle is perpendicular to the radius at the point of tangency.) We wish to find an expression for the radius of the earth in terms of h and α.

(A) Express $\cos \alpha$ in terms of r and h.

(B) Solve the answer to part (A) for r in terms of h and α.

(C) Find the radius of the earth if $\alpha = 22°47'$ and $h = 335$ mi.

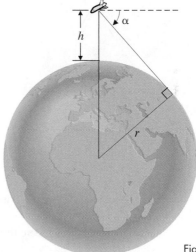

Figure for 27

28. Orbiting Spacecraft Height   A person in an orbiting spacecraft sights the horizon line on earth at an angle of depression α. (Refer to the figure in Problem 27.)

(A) Express $\cos \alpha$ in terms of r and h.

(B) Solve the answer to part (A) for h in terms of r and α.

(C) Find the height of the spacecraft h if the sighted angle of depression $\alpha = 24°14'$ and the known radius of the earth, $r = 3,960$ mi, is used.

29. Navigation   Find the radius of the circle that passes through points $P, A,$ and B in part (a) of the figure on the next page. [*Hint:* The central angle in a circle subtended by an arc is twice any inscribed angle subtended by the same arc—see figure (b).] If A and B are known objects on a maritime navigation chart, then a person on a boat at point P can locate the position of P on a circle on the chart by sighting the angle APB and completing the calculations as suggested. By repeating the procedure with another pair of known points, the position of the boat on the chart will be at an intersection point of the two circles.

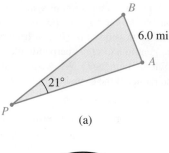

(a)

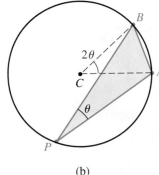

(b)

Figure for 29

30. Navigation Repeat Problem 29 using 33° instead of 21° and 7.5 km instead of 6.0 mi.

31. Geography Assume the earth is a sphere (it is nearly so) and that the circumference of the earth at the equator is 24,900 mi. A **parallel of latitude** is a circle around the earth at a given latitude that is parallel to the equator (see the figure). Approximate the length of a parallel of latitude passing through San Francisco, which is at a latitude of 38°N. See the figure, where θ is the latitude, R is the radius of the earth, and r is the radius of the parallel of latitude. In general, show that if E is the length of the equator and L is the length of a parallel of latitude at a latitude θ, then $L = E \cos \theta$.

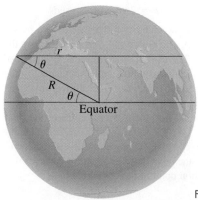

Figure for 31

32. Geography Using the information in Problem 31 and the fact that the circumference of the earth at the equator is 40,100 km, determine the length of the Arctic Circle (66°33′N) in kilometers.

33. Precalculus: Lifeguard Problem A lifeguard sitting in a tower spots a distressed swimmer, as indicated in the figure. To get to the swimmer, the lifeguard must run some distance along the beach at rate p, enter the water, and swim at rate q to the distressed swimmer.

Figure for 33 and 34

(A) To minimize the total time to the swimmer, should the lifeguard enter the water directly or run some distance along the shore and then enter the water? Explain your reasoning.

(B) Express the total time T it takes the lifeguard to reach the swimmer in terms of θ, d, c, p, and q.

(C) Find T (in seconds to two decimal places) if $\theta = 51°$, $d = 380$ m, $c = 76$ m, $p = 5.1$ m/sec, and $q = 1.7$ m/sec.

(D) The following table from a graphing calculator display shows the various times it takes to reach the swimmer for θ from 55° to 85° (X represents θ and Y_1 represents T). Explain the behavior of T relative to θ. For what value of θ in the table is the total time T minimum?

X	Y₁	
55.000	118.65	
60.000	117.53	
65.000	116.89	
70.000	116.66	
75.000	116.80	
80.000	117.28	
85.000	118.08	
X=55		

(E) How far (to the nearest meter) should the lifeguard run along the shore before swimming to achieve the minimal total time estimated in part (D)?

34. Precalculus: Lifeguard Problem   Refer to Problem 33.

(A) Express the total distance D covered by the lifeguard from the tower to the distressed swimmer in terms of d, c, and θ.

(B) Find D (to the nearest meter) for the values of d, c, and θ in Problem 33C.

(C) Using the values for distances and rates in Problem 33C, what is the time (to two decimal places) it takes the lifeguard to get to the swimmer for the shortest distance from the lifeguard tower to the swimmer? Does going the shortest distance take the least time for the lifeguard to get to the swimmer? Explain. (See the graphing calculator table in Problem 33D.)

35. Precalculus: Pipeline   An island is 4 mi offshore in a large bay. A water pipeline is to be run from a water tank on the shore to the island, as indicated in the figure.

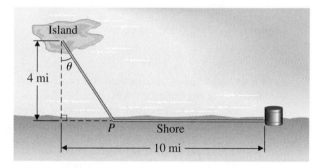

Figure for 35 and 36

The pipeline costs $40,000 per mile in the ocean and $20,000 per mile on the land.

(A) Do you think that the total cost is independent of the angle θ chosen, or does it depend on θ? Explain.

(B) Express the total cost C of the pipeline in terms of θ.

(C) Find C for $\theta = 15°$ (to the nearest hundred dollars).

(D) The following table from a graphing calculator display shows the various costs for θ from 15° to 45° (X represents θ and Y_1 represents C). Explain the behavior of C relative to θ. For what value of θ in the table is the total cost C minimum? What is the minimum cost (to the nearest hundred dollars)?

X	Y_1
15.000	344208
20.000	341151
25.000	339236
30.000	338564
35.000	339307
40.000	341737
45.000	346274

X=15

(E) How many miles of pipe (to two decimal places) should be laid on land and how many miles placed in the water for the total cost to be minimum?

36. Precalculus: Pipeline   Refer to Problem 35.

(A) Express the total length of the pipeline L in terms of θ.

(B) Find L (to two decimal places) for $\theta = 35°$.

(C) What is the cost (to the nearest hundred dollars) of the shortest pipeline from the island to the water tank? Is the shortest pipeline from the island to the water tank the least costly? Explain. (See the graphing calculator table in Problem 35D.)

37. Surveying   Use the information in the figure to find the height y of the mountain.

$$\tan 42° = \frac{y}{x} \qquad \tan 25° = \frac{y}{1.0 + x}$$

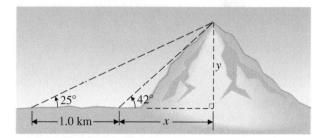

Figure for 37

38. Surveying   Standing on one bank of a river, a surveyor measures the angle to the top of a tree on the opposite bank to be 23°. He backs up 45 ft and remeasures the angle to the top of the tree at 18°. How wide is the river?

39. Surveying   A surveyor wants to determine the height of a tall tree. He stands at some distance from the tree and determines that the angle of elevation to the top of the tree is 45°. He moves 36 ft closer to the tree, and now the angle of elevation is 55°. How tall is the tree?

40. Surveying   Using the figure, show that:

$$h = \frac{d}{\cot \alpha - \cot \beta}$$

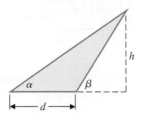

Figure for 40

41. Surveying   From the sunroof of Janet's apartment building, the angle of depression to the base of an office building is 51.4° and the angle of elevation to the top of the office building is 43.2° (see the figure). If the office building is 847 ft high, how far apart are the two buildings and how high is the apartment building?

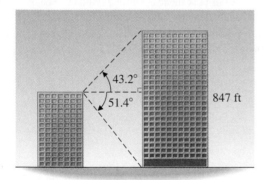

Figure for 41

42. Surveying   At a position across the street 65 ft from a building, the angle to the top of the building is 42°. An antenna sits on the front edge of the roof of the building. The angle to the top of the antenna is 55°. How tall is the building? How tall is the antenna itself, not including the height of the building?

43. Surveying   From the top of a lighthouse 175 ft high, the angles of depression of the top and bottom of a flagpole are 40°20′ and 44°40′, respectively. If the base of the lighthouse and the base of the flagpole lie on the same horizontal plane, how tall is the pole?

44. Surveying   Using the figure, show that

$$h = \frac{d}{\cot \alpha + \cot \beta}$$

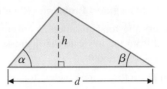

Figure for 44

45. Precalculus: Physics   In physics one can show that the velocity (*v*) of a ball rolling down an inclined plane (neglecting air resistance and friction) is given by

$$v = g(\sin \theta)t$$

where *g* is the gravitational constant and *t* is time (see the figure).

Galileo's experiment

Figure for 45

Galileo (1564–1642) used this equation in the form

$$g = \frac{v}{(\sin \theta)t}$$

so he could determine *g* after measuring *v* experimentally. (There were no timing devices available then that were accurate enough to measure the velocity of a free-falling body. He had to use an inclined plane to slow the motion down, and then he was able to calculate an approximation for *g*.) Find *g* if at the end of 2.00 sec a ball is traveling at 11.1 ft/sec down a plane inclined at 10.0°.

46. Precalculus: Physics   In Problem 45 find *g* if at the end of 1.50 sec a ball is traveling at 12.4 ft/sec down a plane inclined at 15.0°.

47. Geometry   What is the altitude of an equilateral triangle with side 4.0 m? [An equilateral triangle has all sides (and all angles) equal.]

48. Geometry   The altitude of an equilateral triangle is 5.0 cm. What is the length of a side?

49. Geometry   Two sides of an isosceles triangle have length 12 yd and the included angle, opposite the base of the triangle, is 48°. Find the length of the base.

50. Geometry   The perpendicular distance from the center of an eight-sided regular polygon (octagon) to each of the sides is 14 yd. Find the perimeter of the octagon.

51. Geometry Find the length of one side of a nine-sided regular polygon inscribed in a circle with radius 8.32 cm (see the figure).

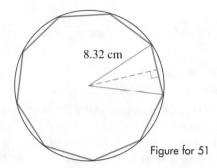

8.32 cm

Figure for 51

52. Geometry What is the radius of a circle inscribed in the polygon in Problem 51? (The circle will be tangent to each side of the polygon, and the radius will be perpendicular to the tangent line at the point of tangency.)

53. Geometry In part (a) of the figure, M and N are midpoints to the sides of a square. Find the exact value of sin θ. [*Hint:* The solution utilizes the Pythagorean theorem, similar triangles, and the definition of sine. Some useful auxiliary lines are drawn in part (b) of the figure.]

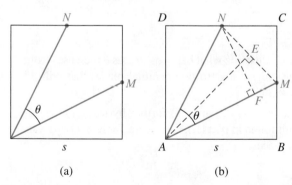

(a) (b)

Figure for 53

54. Geometry Find r in the figure. The circle is tangent to all three sides of the isosceles triangle. (An isosceles triangle has two sides equal.) [*Hint:* The radius of a circle and a tangent line are perpendicular at the point of tangency. Also, the altitude of the isosceles triangle will pass through the center of the circle and will divide the original triangle into two congruent triangles.]

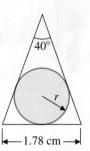

40°

r

|← 1.78 cm →| Figure for 54

55. Geometry A square is located inside an equilateral triangle as shown in the figure. Find the length of a side of the square.

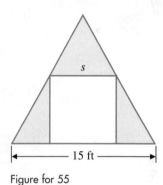

s

|← 15 ft →|

Figure for 55

56. Geometry The length of each side of a six-sided regular polygon (hexagon) is 20.0 cm. Find the distances x and y in the figure.

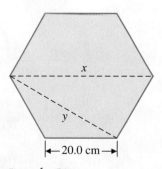

x

y

|← 20.0 cm →|

Figure for 56

CHAPTER 1 GROUP ACTIVITY

A Logistics Problem

A log of length L floats down a canal that has a right angle turn, as indicated in Figure 1.

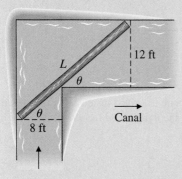

FIGURE 1

(A) Can logs of any length be floated around the corner? Discuss your first thoughts on this problem without any mathematical analysis.

(B) Express the length of the log L in terms of θ and the two widths of the canal (neglect the width of the log), assuming the log is touching the sides and corner of the canal as shown in Figure 1. (In calculus, this equation is used to find the longest log that will float around the corner. Here, we will try to approximate the maximum log length by using noncalculus tools.)

(C) An expression for the length of the log in part (B) is

$$L = 12 \csc \theta + 8 \sec \theta$$

What are the restrictions on θ? Describe what you think happens to the length L as θ increases from values close to $0°$ to values close to $90°$.

(D) Complete Table 1, giving values of L to one decimal place.

TABLE 1							
θ	30°	35°	40°	45°	50°	55°	60°
L	33.2						

(E) Using the values in Table 1, describe what happens to L as θ increases from 30° to 60°. Estimate the longest log (to one decimal place) that will go around the corner in the canal.

(F) We can refine the results in parts (D) and (E) further by using more values of θ close to the result found in part (E). Complete Table 2, giving values of L to two decimal places.

TABLE 2							
θ	44°	46°	48°	50°	52°	54°	56°
L	28.40						

(G) Using the values in Table 2, estimate the longest log (to two decimal places) that will go around the corner in the canal.

(H) Discuss further refinements on this process. (A graphing calculator with table-producing capabilities would be helpful in carrying out such a process.)

CHAPTER 1 REVIEW

1.1
ANGLES, DEGREES, AND ARCS

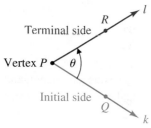

FIGURE 1

An **angle** is formed by rotating a half-line, called a **ray,** around its end point. See Figure 1: One ray k, called the **initial side** of the angle, remains fixed; a second ray l, called the **terminal side** of the angle, starts in the initial side position and is rotated around the common end point P in a plane until it reaches its terminal position. The common end point is called the **vertex.** An angle is **positive** if the terminal side is rotated counterclockwise and **negative** if the terminal side is rotated clockwise. Different angles with the same initial and terminal sides are called **coterminal.** An angle of **1 degree** is $\frac{1}{360}$ of a complete revolution in a counterclockwise direction. Names for special angles are noted in Figure 2.

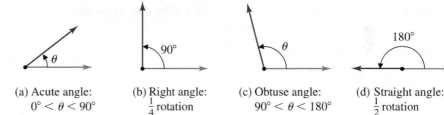

(a) Acute angle:
    $0° < \theta < 90°$

(b) Right angle:
    $\frac{1}{4}$ rotation

(c) Obtuse angle:
    $90° < \theta < 180°$

(d) Straight angle:
    $\frac{1}{2}$ rotation

FIGURE 2

Two positive angles are **complementary** if the sum of their measures is 90°; they are **supplementary** if the sum of their measures is 180°.

Angles can be represented in terms of **decimal degrees** or in terms of **minutes** ($\frac{1}{60}$ of a degree) and **seconds** ($\frac{1}{60}$ of a minute). Calculators can be used to convert from decimal degrees (DD) to degrees-minutes-seconds (DMS), and vice versa. The **arc length** s of an arc **subtended** by a **central angle** θ in a circle of radius r (Fig. 3) satisfies

FIGURE 3

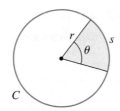

$$\frac{\theta}{360°} = \frac{s}{C} \qquad C = 2\pi r = \pi d$$

θ in decimal degrees; s and C in same units

1.2
SIMILAR TRIANGLES

The properties of similar triangles stated in **Euclid's theorem** are central to the development of trigonometry: If two triangles are similar, their corresponding sides are proportional (see Fig. 4).

FIGURE 4

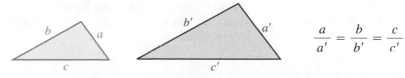

$$\frac{a}{a'} = \frac{b}{b'} = \frac{c}{c'}$$

Similar triangles

1.3
TRIGONOMETRIC RATIOS AND RIGHT TRIANGLES

If a, b, and c are the legs and hypotenuse, respectively, of a right triangle, then the **Pythagorean theorem** states that $a^2 + b^2 = c^2$. The six **trigonometric ratios** for the angle θ in a right triangle (see Fig. 5) with **opposite side b, adjacent side a,** and **hypotenuse c** are:

$$\sin \theta = \frac{b}{c} = \frac{\text{Opp}}{\text{Hyp}} \qquad \csc \theta = \frac{c}{b} = \frac{\text{Hyp}}{\text{Opp}}$$

$$\cos \theta = \frac{a}{c} = \frac{\text{Adj}}{\text{Hyp}} \qquad \sec \theta = \frac{c}{a} = \frac{\text{Hyp}}{\text{Adj}}$$

$$\tan \theta = \frac{b}{a} = \frac{\text{Opp}}{\text{Adj}} \qquad \cot \theta = \frac{a}{b} = \frac{\text{Adj}}{\text{Opp}}$$

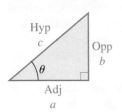

Right triangle

FIGURE 5

The complementary relationships that follow illustrate why cosine, cotangent, and cosecant are called the **cofunctions** of sine, tangent, and secant, respectively (see Fig. 6).

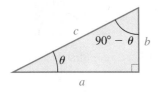

FIGURE 6

Complementary Relationships	Reciprocal Relationships
$\sin \theta = \dfrac{b}{c} = \cos(90° - \theta)$	$\csc \theta = \dfrac{1}{\sin \theta}$
$\tan \theta = \dfrac{b}{a} = \cot(90° - \theta)$	$\sec \theta = \dfrac{1}{\cos \theta}$
$\sec \theta = \dfrac{c}{a} = \csc(90° - \theta)$	$\cot \theta = \dfrac{1}{\tan \theta}$

Solving a right triangle involves finding the measures of the remaining sides and acute angles when given the measure of two sides or the measure of one side and one acute angle. Accuracy of these computations is governed by the following table:

Angle to nearest	Significant digits for side measure
1°	2
10′ or 0.1°	3
1′ or 0.01°	4
10″ or 0.001°	5

1.4
RIGHT TRIANGLE APPLICATIONS

An angle measured upward from the horizontal is called an **angle of elevation,** and one measured downward from the horizontal is called an **angle of depression.**

CHAPTER 1 REVIEW EXERCISE

Work through all the problems in this chapter review and check answers in the back of the book. Answers to all review problems are there, and following each answer is a number in italics indicating the section in which that type of problem is discussed. Where weaknesses show up, review appropriate sections in the text.

A **1.** $2°9'54'' = ?''$

2. An arc of $\frac{1}{8}$ the circumference of a circle subtends a central angle of how many degrees?

3. Given two similar triangles, as shown in the figure, find a if $c = 20{,}000$, $a' = 4$, and $c' = 5$.

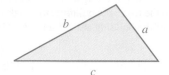

Figure for 3

4. Change $36°20'$ to decimal degrees (to two decimal places).

5. Write a definition of an angle of degree measure 1.

6. Which of the following triangles are similar? Explain why.

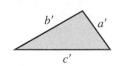

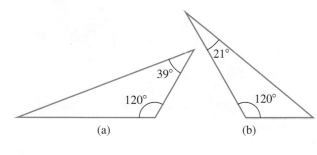

Figure for 6

7. Is it possible in two triangles that are similar to have one with an obtuse angle and the other with no obtuse angle? Explain.

8. Explain why a triangle cannot have more than one obtuse angle.

9. If an office building casts a shadow of 40 ft at the same time a vertical yardstick (36 in.) casts a shadow of 2.0 in., how tall is the building?

10. For the triangle shown here, identify each ratio:

(A) $\sin \theta$ (B) $\sec \theta$ (C) $\tan \theta$
(D) $\csc \theta$ (E) $\cos \theta$ (F) $\cot \theta$

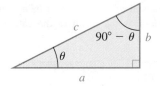

Figure for 10

11. Solve the right triangle in Problem 10, given $c = 20.2$ cm and $\theta = 35.2°$.

B **12.** Find the degree measure of a central angle subtended by an arc of 8.00 cm in a circle with circumference 20.0 cm.

13. If the minute hand of a clock is 2.00 in. long, how far does the tip of the hand travel in exactly 20 min?

14. One angle has a measure of $27°14'$ and another angle has a measure of $27.25°$. Which is larger? Explain how you obtained your answer.

15. Use a calculator to:
(A) Convert $67°42'31''$ to decimal degree form.
(B) Convert $129.317°$ to degree-minute-second form.

16. Perform the following calculations on a calculator and write the results in DMS form:
(A) $82°14'37'' - 16°32'45''$
(B) $3(13°47'18'' + 95°28'51'')$

17. For the triangles in Problem 3, find b to two significant digits if $a = 4.1 \times 10^{-6}$ mm, $a' = 1.5 \times 10^{-4}$ mm, and $b' = 2.6 \times 10^{-4}$ mm.

18. For the triangle in Problem 10, identify by name each of the following ratios relative to angle θ:
(A) a/c (B) b/a (C) b/c
(D) c/a (E) c/b (F) a/b

19. For a given value θ, explain why $\sin\theta$ is independent of the size of a right triangle having θ as an acute angle.

20. Solve the right triangle in Problem 10, given $\theta = 62°20'$ and $a = 4.00 \times 10^{-8}$ m.

21. Find each θ to the accuracy indicated.
(A) $\tan\theta = 2.497$ (to two decimal places)
(B) $\theta = \arccos 0.3721$ (to the nearest $10'$)
(C) $\theta = \sin^{-1} 0.0559$ (to the nearest second)

22. Which of the following window displays from a graphing calculator is the result of the calculator being set in degree mode and which is the result of the calculator being set in radian mode?

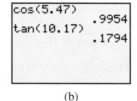

(a)   (b)

23. Solve the right triangle in Problem 10, given $b = 13.3$ mm and $a = 15.7$ mm. (Find angles to the nearest $0.1°$.)

24. Find the angles in Problem 23 to the nearest $10'$.

25. If an equilateral triangle has a side of 10 ft, what is its altitude (to two significant digits)?

C **26.** A curve of a railroad track follows an arc of a circle of radius 1,500 ft. If the arc subtends a central angle of $36°$, how far will a train travel on this arc?

27. Find the area of a sector with central angle $36.5°$ in a circle with radius 18.3 ft. Compute your answer to the nearest unit.

28. Solve the triangle in Problem 10, given $90° - \theta = 23°43'$ and $c = 232.6$ km.

29. Solve the triangle in Problem 10, given $a = 2,421$ m and $c = 4,883$ m. (Find angles to the nearest $0.01°$.)

30. Use a calculator to find csc $67.1357°$ to four decimal places.

Applications

31. Precalculus: Shadow Problem   A person is standing 20 ft away from a lamppost. If the lamp is 18 ft above the ground and the person is 5 ft 6 in. tall, how long is the person's shadow?

32. Surveying   Two ladders leaning against a wall make the same angle with the ground. The 12 ft ladder reaches 9 ft up the wall. How much farther up the wall does the 20 ft ladder reach?

33. Surveying   If a person casts an 11 ft shadow when the elevation of the sun is $27°$, how tall is the person?

34. Surveying   At two points 95 ft apart on a horizontal line perpendicular to the front of a building, the angles of elevation of the top of the building are $25°$ and $16°$. How tall is the building?

35. Construction   The front porch of a house is 4.25 ft high. The angle of elevation of a ramp from the ground to the porch is $10.0°$ (see the figure). How long is the ramp? How far is the end of the ramp from the porch?

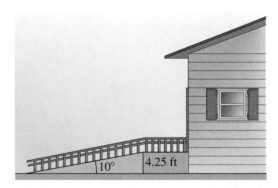

Figure for 35

36. Medicine: Stress Test   Cardiologists give stress tests by having patients walk on a treadmill at various speeds and inclinations. The amount of inclination may be given as an angle or as a percentage (see the figure). Find the angle of inclination if the treadmill is set at a 4% incline. Find the percentage of inclination if the angle of inclination is $4°$.

a   Angle of inclination: θ

Percentage of inclination: $\dfrac{a}{b}$

Figure for 36

37. Geography/Navigation   Find the distance (to the nearest mile) between Green Bay, WI, with latitude 44°31′N, and Mobile, AL, with latitude 30°42′N. (Both cities have approximately the same longitude.) Use $r = 3{,}960$ mi for the earth's radius.

38. Precalculus: Balloon Flight   The angle of elevation from the ground to a hot air balloon at an altitude of 2,800 ft is 64°. What will be the new angle of elevation if the balloon descends straight down 1,400 ft?

39. Surveying   Use the information in the figure to find the length x of the island.

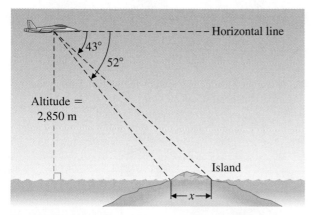

Figure for 39

40. Precalculus: Balloon Flight   Two tracking stations 525 m apart measure angles of elevation of a weather balloon to be 73.5° and 54.2°, as indicated in the

figure. How high is the balloon at the time of the measurements?

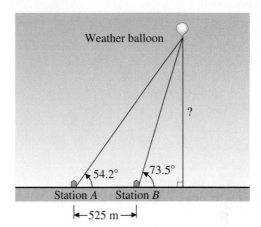

Figure for 40

41. Navigation: Chasing the Sun   Your flight is westward from Buffalo, NY, and you notice the sun just above the horizon. How fast would the plane have to fly to keep the sun in the same position? (The latitude of Buffalo is 42°50′N, the radius of the earth is 3,960 mi, and the earth makes a complete rotation about its axis in 24 hr.)

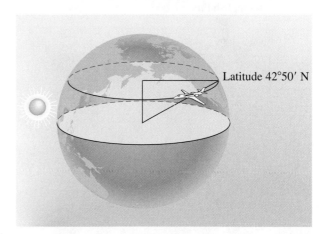

Figure for 41

42. Solar Energy   A truncated conical solar collector is aimed directly at the sun as shown in part (a) of the figure on the next page. An analysis of the amount of solar energy absorbed by the collecting disk requires certain equations relating the quantities shown in part (b) of the figure. Find an equation that expresses

(A) β in terms of α
(B) r in terms of α and h
(C) $H - h$ in terms of r, R, and α

(Based on the article "The Solar Concentrating Properties of a Conical Reflector," by Don Leake in *The UMAP Journal,* Vol. 8, No. 4, 1987.)

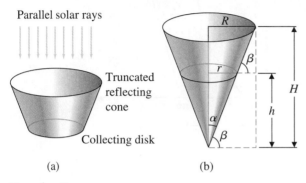

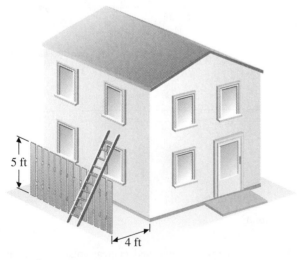

(a)   (b)

Figure for 42

43. **Precalculus: Optimization**   A 5 ft fence is 4 ft away from a building. A ladder is to go from the ground, across the top of the fence to the building. (See the figure.) We want to find the length of the shortest ladder that can accomplish this.

Figure for 43

(A) How is the required length of the ladder affected as the foot of the ladder moves away from the fence?

How is the length affected as the foot moves closer to the fence?

(B) Express the length of the ladder in terms of the distance from the fence to the building, the height of the fence, and θ, the angle of elevation that the ladder makes with the level ground.

(C) Complete the table, giving values of L to two decimal places:

θ	25°	35°	45°	55°	65°	75°	85°
L	16.25						

(D) Explain what happens to L as θ moves from 25° to 85°. What value of θ produces the minimum value of L in the table? What is the minimum value?

(E) Describe how you can continue the process to get a better estimate of the minimum ladder length.

44. **Surveying**   A triangular plot is marked by posts at A, B, and C (see the figure). Using AB as a baseline, a surveyor measures angle CAB. Then she measures the distance from A to B and angle ABC. Find, to the nearest foot, the distance from A to C and the distance from B to C.

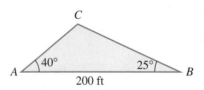

Figure for 44

45. **Surveying**   Find the area of the plot of land in the figure to the nearest square foot.

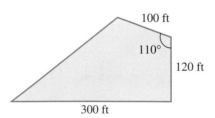

Figure for 45

Trigonometric Functions

2

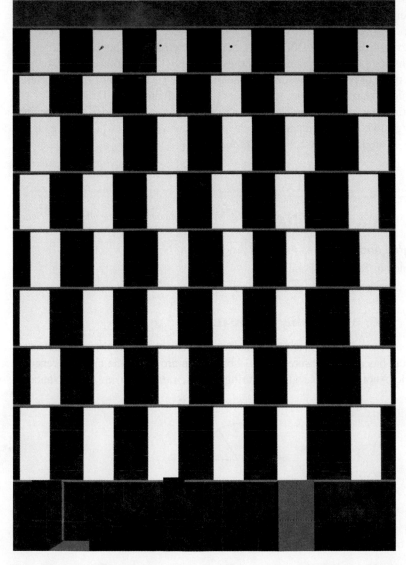

2.1 Degrees and Radians

☆**2.2** Linear and Angular Velocity

2.3 Trigonometric Functions: Unit Circle Approach

☆**2.4** Additional Applications

2.5 Exact Values and Properties of Trigonometric Functions

Chapter 2 Group Activity: Speed of Light in Water

Chapter 2 Review

☆ Sections marked with a star may be omitted without loss of continuity.

T he trigonometric ratios we studied in Chapter 1 provide a powerful tool for indirect measurement. It was for this purpose only that trigonometry was used for nearly 2,000 years. Astronomy, surveying, mapmaking, navigation, construction, and military uses catalyzed the extensive development of trigonometry as a tool for indirect measurement.

A turning point in trigonometry occurred after the development of the rectangular coordinate system (credited mainly to the French philosopher-mathematician René Descartes, 1596–1650). The trigonometric ratios, through the use of this system, were generalized into trigonometric functions. This generalization increased their usefulness far beyond the dreams of those originally responsible for this development. The Swiss mathematician Leonhard Euler (1707–1783), probably the greatest mathematician of his century, made substantial contributions in this area. (In fact, there were very few areas in mathematics in which Euler did not make significant contributions.)

Through the demands of modern science, the *periodic* nature of these new functions soon became apparent, and they were quickly put to use in the study of various types of periodic phenomena. The trigonometric functions began to be used on problems that had nothing whatsoever to do with angles and triangles.

In this chapter we generalize the concept of trigonometric ratios along the lines just suggested. Before we undertake this task, however, we will introduce another form of angle measure called the *radian.*

2.1 Degrees and Radians

- Degree and Radian Measure of Angles
- Angles in Standard Position
- Arc Length and Area of a Sector of a Circle

■ Degree and Radian Measure of Angles

In Chapter 1 we defined an angle and its degree measure. Recall that a central angle in a circle has angle measure $1°$ if it subtends an arc $\frac{1}{360}$ of the circumference of the circle. Another approach to measuring angles, *radian measure,* has advantages

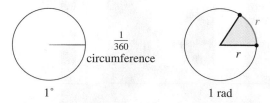

FIGURE 1
Degree and radian measure

in mathematics and the sciences. A central angle subtended by an arc of length equal to the radius of the circle is defined to be an angle of **radian measure 1** (see Fig. 1). When we write $\theta = 2°$, we are referring to an angle of degree measure 2. When we write $\theta = 2$ rad, we are referring to an angle of radian measure 2.

EXPLORE/DISCUSS 1

Discuss why the radian measure and degree measure of an angle are independent of the size of the circle having the angle as a central angle.

It follows from the definition of radian measure that the radian measure of a central angle θ subtended by an arc of length s is found by determining how many times the length of the radius r, used as a unit length, is contained in the arc length s.

RADIAN MEASURE OF CENTRAL ANGLES

The *radian measure* of a central angle of a circle is given by

$$\theta = \frac{s}{r} \text{ radians}$$

where s is the length of the arc opposite θ and r is the radius of the circle.

[*Note:* s and r must be in the same units.]

What is the radian measure of a central angle subtended by an arc of 32 cm in a circle of radius 8 cm?

$$\theta = \frac{32 \text{ cm}}{8 \text{ cm}} = 4 \text{ rad}$$

Remark The word *radian* or its abbreviation *rad* is sometimes omitted when we are dealing with the radian measure of angles. An angle of measure 4 is thus understood to mean 4 radians, not 4 degrees; if degrees are intended, we write 4°. □

What is the radian measure of an angle of 180°? A central angle of 180° is subtended by an arc $\frac{1}{2}$ of the circumference of the circle. If C is the circumference of a circle, then $\frac{1}{2}$ of the circumference is given by

$$s = \frac{C}{2} = \frac{2\pi r}{2} = \pi r \qquad \text{and} \qquad \theta = \frac{s}{r} = \frac{\pi r}{r} = \pi \text{ rad}$$

Therefore, 180° corresponds to π rad. This is important to remember, since the radian measures of many special angles can be obtained from this correspondence. For example, 90° is 180°/2; therefore, 90° corresponds to $\pi/2$ rad. Since 360° is twice 180°, 360° corresponds to 2π rad. Similarly, 60° corresponds to $\pi/3$ rad, 45° to $\pi/4$ rad, and 30° to $\pi/6$ rad. These special angles and their degree and radian measures will be referred to frequently throughout this book. Table 1 summarizes these special correspondences for ease of reference.

TABLE 1						
Radians	$\pi/6$	$\pi/4$	$\pi/3$	$\pi/2$	π	2π
Degrees	30	45	60	90	180	360

The following proportion can be used to convert the degree measure θ_d of angle θ to its radian measure θ_r. It applies to all angles, negative as well as positive, including angles with measures greater than 360° or less than −360°.

RADIAN–DEGREE CONVERSION FORMULAS

$$\frac{\theta_d}{180°} = \frac{\theta_r}{\pi \text{ rad}} \qquad \text{or}$$

$$\theta_d = \frac{180°}{\pi \text{ rad}} \theta_r \qquad \text{Radians to degrees}$$

$$\theta_r = \frac{\pi \text{ rad}}{180°} \theta_d \qquad \text{Degrees to radians}$$

[*Note:* We will omit units in calculations until the final answer.]

EXAMPLE 1

Radian–Degree Conversion

(A) Find the degree measure of −1.5 rad in exact form and in decimal form to four decimal places.

(B) Find the radian measure of 44° in exact form and in decimal form to four decimal places.

(C) Use a calculator with automatic conversion capability to perform the conversions in parts (A) and (B).

Solution (A) $\theta_d = \dfrac{180°}{\pi \text{ rad}}\,\theta_r$

$= \dfrac{180}{\pi}\,(-1.5)$

$= -\dfrac{270°}{\pi}$ *Exact form*

$= -85.9437°$ *To four decimal places*

(B) $\theta_r = \dfrac{\pi \text{ rad}}{180°}\,\theta_d$

$= \dfrac{\pi}{180}\,(44)$

$= \dfrac{11\pi}{45}\text{ rad}$ *Exact form*

$= 0.7679 \text{ rad}$ *To four decimal places*

(C) Check the user's manual for your calculator to find out how to convert parts (A) and (B) with an automatic routine. The following window display is from a graphing calculator with an automatic conversion routine:

```
(-1.5)ʳ
            -85.9437
44°
               .7679
```

Matched Problem 1 (A) Find the degree measure of 1 rad in exact form and in decimal form to four decimal places.

(B) Find the radian measure of $-120°$ in exact form and in decimal form to four decimal places.

(C) Use a calculator with automatic conversion capability to perform the conversions in parts (A) and (B). ∎

■ Angles in Standard Position

To generalize the concept of trigonometric ratios, we first locate an angle in **standard position** in a rectangular coordinate system. To do this, we place the vertex at the origin and the initial side along the positive *x* axis. Recall that when the rotation is counterclockwise, the angle is positive and when the rotation is clockwise, the angle is negative. Figure 2 illustrates several angles in standard position.

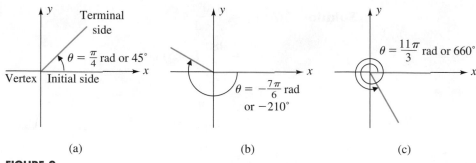

FIGURE 2
Angles in standard position

EXAMPLE 2

Sketching Angles in Standard Position

Sketch the following angles in their standard positions:

(A) $-60°$ (B) $3\pi/2$ rad   (C) -3π rad   (D) $405°$

Solution

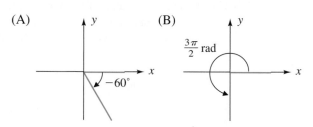

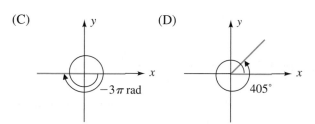

Matched Problem 2   Sketch the following angles in their standard positions:

(A) $120°$ (B) $-\pi/6$ rad   (C) $7\pi/2$ rad   (D) $-495°$

Two angles are said to be **coterminal** if their terminal sides coincide when both angles are placed in their standard positions in the same rectangular coordinate system. Figure 3 shows two pairs of coterminal angles.

Remarks   **1.** The degree measures of two coterminal angles differ by an integer* multiple of $360°$.

*An integer is a positive or negative whole number or 0; that is, the set of integers is $\{\ldots, -4, -3, -2, -1, 0, 1, 2, 3, 4, \ldots\}$.

2. The radian measures of two coterminal angles differ by an integer multiple of 2π. □

FIGURE 3
Coterminal angles

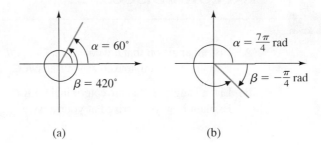

(a)   (b)

 **EXAMPLE 3**

Recognizing Coterminal Angles

Which of the following pairs of angles are coterminal?

(A) $\alpha = -135°$ (B) $\alpha = 120°$
    $\beta = 225°$ $\beta = -420°$

(C) $\alpha = -\pi/3 \text{ rad}$ (D) $\alpha = \pi/3 \text{ rad}$
    $\beta = 2\pi/3 \text{ rad}$ $\beta = 7\pi/3 \text{ rad}$

Solution (A) The angles are coterminal if $\alpha - \beta$ is an integer multiple of 360°.
$$\alpha - \beta = (-135°) - 225° = -360° = -1(360°)$$
Therefore, α and β are coterminal

(B) $\alpha - \beta = 120° - (-420°) = 540°$
The angles are not coterminal, since 540° is not an integer multiple of 360°.

(C) The angles are coterminal if $\alpha - \beta$ is an integer multiple of 2π.
$$\alpha - \beta = \left(-\frac{\pi}{3}\right) - \frac{2\pi}{3} = -\frac{3\pi}{3} = -\pi$$
The angles are not coterminal, since $-\pi$ is not an integer multiple of 2π.

(D) $\alpha - \beta = \dfrac{\pi}{3} - \dfrac{7\pi}{3} = -\dfrac{6\pi}{3} = -2\pi = (-1)(2\pi)$
Therefore, α and β are coterminal. ■

Matched Problem 3 Which of the following pairs of angles are coterminal?

(A) $\alpha = 90°$ (B) $\alpha = 750°$
    $\beta = -90°$ $\beta = 30°$

(C) $\alpha = -\pi/6 \text{ rad}$ (D) $\alpha = 3\pi/4 \text{ rad}$
    $\beta = -25\pi/6 \text{ rad}$ $\beta = 7\pi/4 \text{ rad}$ ■

EXPLORE/DISCUSS 2

(A) List all angles β that are coterminal with $\theta = \pi/3$ rad, $-5\pi \leq \beta \leq 5\pi$. Explain how you arrived at your answer.

(B) List all angles β that are coterminal with $\theta = -45°$, $-900° \leq \beta \leq 900°$. Explain how you arrived at your answer.

■ Arc Length and Area of a Sector of a Circle

FIGURE 4
Sector of a circle

At first it may appear that radian measure of angles is more complicated and less useful than degree measure. However, just the opposite is true. Formulas for arc length and the area of a sector of a circle, which we obtain in this section, should begin to convince you of some of the advantages of radian measure over degree measure. Refer to Figure 4 in the following discussion.

From the definition of radian measure of an angle,

$$\theta = \frac{s}{r} \text{ rad}$$

Solving for s, we obtain a **formula for arc length:**

$$s = r\theta \qquad \theta \text{ in radians} \qquad (1)$$

If θ is in degree measure, we must multiply by $\pi/180$ first (to convert to radians); then formula (1) becomes

$$s = \frac{\pi}{180} r\theta \qquad \theta \text{ in degrees} \qquad (2)$$

We see that the formula for arc length is much simpler when θ is in radian measure.

EXAMPLE 4

Arc Length

In a circle of radius 4.00 cm, find the arc length subtended by a central angle of:

(A) 3.40 rad   (B) 10.0°

Solution   (A) $s = r\theta$ (B) $s = \dfrac{\pi}{180} r\theta$

$= 4.00(3.40) = 13.6$ cm $= \dfrac{\pi}{180}(4.00)(10.0) = 0.698$ cm ■

Matched Problem 4   In a circle of radius 6.00 ft, find the arc length subtended by a central angle of:

(A) 1.70 rad   (B) 40.0°   ■

EXPLORE/DISCUSS 3

When you look at a full moon, the image appears on the retina of your eye as shown in Figure 5.

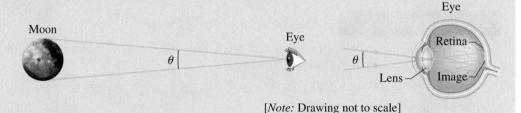

[*Note:* Drawing not to scale]

FIGURE 5
Vision

(A) Discuss how you would estimate the diameter d' of the moon's image on the retina, given the moon's diameter d, the distance u from the moon to the eye lens, and the distance v from the image to the eye lens. [Recall from Section 1.1: For small central angles in circles with very large radii, the intercepted arc and its chord are approximately the same length.]

(B) The moon's image on the retina is smaller than a period on this page! To verify this, find the diameter of the moon's image d' on the retina to two decimal places, given:

Moon's diameter: $d = 2{,}160$ mi

Moon's distance from the eye: $u = 239{,}000$ mi

Distance from eye lens to retina: $v = 17.4$ mm

A **sector of a circle** is the region swept out by segment *CA* as it rotates through the central angle θ (see Fig. 4 on page 58); we require θ to be positive but less than one complete revolution. The formula for the **area A of a sector of a circle** with radius r and central angle θ in radian measure can be found by starting with the following proportion:

$$\frac{A}{\pi r^2} = \frac{\theta}{2\pi}$$

$$A = \frac{1}{2}r^2\theta \qquad \theta \text{ in radians} \tag{3}$$

If θ is in degree measure, we must multiply by $\pi/180$ first (to convert to radians); then formula (3) becomes

$$A = \frac{\pi}{360} r^2 \theta \qquad \theta \text{ in degrees} \tag{4}$$

The formula for the area of a sector of a circle is also simpler when θ is in radian measure.

EXAMPLE 5   Area of a Sector of a Circle

In a circle of radius 3 m, find the area (to three significant digits) of the sector with central angle:

(A) 0.4732 rad   (B) 25°

Solution   (A) $A = \dfrac{1}{2} r^2 \theta$ (B) $A = \dfrac{\pi}{360} r^2 \theta$

               $= \dfrac{1}{2}(3)^2(0.4732) = 2.13 \text{ m}^2$ $= \dfrac{\pi}{360}(3)^2(25) = 1.96 \text{ m}^2$ ■

Matched Problem 5   In a circle of radius 7 in., find the area (to four significant digits) of the sector with central angle:

(A) 0.1332 rad   (B) 110°   ■

It is important to gain experience in the use of radian measure, since the concept will be used extensively in this book. By the time you finish the book you should feel as comfortable with radian measure as you now do with degree measure.

Answers to Matched Problems   **1.** (A) $180°/\pi$; 57.2958°   (C)

                            (B) $-2\pi/3$ rad; -2.0944 rad

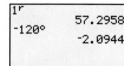

        2. (A)   (B)

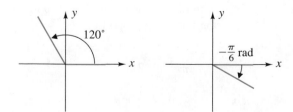

(C)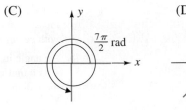

(D)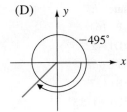

3. (A) Not coterminal   (B) Coterminal   (C) Coterminal   (D) Not coterminal
4. (A) 10.2 ft   (B) 4.19 ft
5. (A) 3.263 in.^{2} (B) 47.04 in.^{2}

EXERCISE 2.1

A

1. Explain what it means for an angle to have radian measure 1.

2. Explain what it means for an angle to have degree measure 1.

3. Mentally convert the following to exact radian measure by starting with 30° and taking multiples (remember that 180° corresponds to π radians):

   30°, 60°, 90°, 120°, 150°, 180°, 210°, 240°, 270°, 300°, 330°, 360°

4. Mentally convert the following to exact radian measure by starting with 45° and taking multiples (remember that 180° corresponds to π radians):

   45°, 90°, 135°, 180°, 225°, 270°, 315°, 360°

5. Explain what it means for an angle to be in standard position.

6. Explain what it means for two angles in standard position to be coterminal.

B *In Problems 7–12, sketch each angle in its standard position and find the degree measure of the two nearest angles (one positive and one negative) that are coterminal with the given angle.*

7. 30°   8. 135°   9. 225°
10. 60°   11. −75°   12. −140°

In Problems 13–18, sketch each angle in its standard position and find the radian measure of the two nearest angles (one positive and one negative) that are coterminal with the given angle.

13. $\pi/8$ rad   14. $\pi/3$ rad
15. $-2\pi/5$ rad   16. $-7\pi/4$ rad
17. $2\pi/3$ rad   18. $6\pi/5$ rad

19. Which is larger: an angle of degree measure 60 or an angle of radian measure 1? Explain.

20. Which is smaller: an angle of radian measure 4 or an angle of degree measure 240°? Explain.

In Problems 21–26, find the radian measure for each angle. Express the answer in exact form and also in approximate form to four significant digits.

21. 3°   22. 5°
23. 110°   24. 190°
25. 36°   26. 108°

In Problems 27–32, find the degree measure for each angle. Express the answer in exact form and also in approximate form (in decimal degrees) rounded to four significant digits.

27. 2 rad   28. 3 rad
29. 0.6 rad   30. 1.4 rad
31. $2\pi/7$ rad   32. $2\pi/9$ rad

In Problems 33–36, use a calculator with an automatic radian–degree conversion routine.

33. Find the degree measure of:
    (A) 0.50 rad   (B) 1.4 rad
    (C) 6.20 rad   (D) −4.59 rad

34. Find the degree measure of:
    (A) 0.750 rad   (B) 1.5 rad
    (C) 3.80 rad   (D) −7.21 rad

35. Find the radian measure of:

(A) 25°   (B) 164°

(C) 648°   (D) −221.7°

36. Find the radian measure of:

(A) 35°   (B) 187°

(C) 437°   (D) −175.3°

37. If the radius of a circle is 5.0 m, find the radian measure and the degree measure of an angle subtended by an arc of length:

(A) 2 m   (B) 6 m

(C) 12.5 m   (D) 20 m

38. If the radius of a circle is 40.0 cm, find the radian measure and the degree measure of an angle subtended by an arc of length:

(A) 15 cm   (B) 65 cm

(C) 123.5 cm   (D) 210 cm

39. In a circle of radius 7 m, find the length of the arc subtended by a central angle of:

(A) 1.35 rad   (B) 42.0°

(C) 0.653 rad   (D) 125°

40. In a circle of radius 5 in., find the length of the arc subtended by a central angle of:

(A) 0.228 rad   (B) 37.0°

(C) 1.537 rad   (D) 175.0°

41. If the radian measure of an angle is doubled, is the degree measure of the same angle also doubled? Explain.

42. If the degree measure of an angle is cut in half, is the radian measure of the same angle also cut in half? Explain.

43. An arc length s on a circle is held constant while the radius of the circle r is doubled. Explain what happens to the central angle subtended by the arc.

44. The radius of a circle r is held constant while an arc length s on the circle is doubled. Explain what happens to the central angle subtended by the arc.

45. In a circle of radius 12.0 cm, find the area of the sector with central angle:

(A) 2.00 rad   (B) 25.0°

(C) 0.650 rad   (D) 105°

46. In a circle of radius 10.5 ft, find the area of the sector with central angle:

(A) 0.150 rad   (B) 15.0°

(C) 1.74 rad   (D) 105°

C **47.** Explain why an angle in standard position and of radian measure m intercepts an arc of length m on a unit circle with center at the origin.

48. An angle in standard position intercepts an arc of length s on a unit circle with center at the origin. Explain why the radian measure of the angle is also s.

In Problems 49–58, indicate the quadrant* in which the terminal side of each angle lies.

49. 495°   **50.** $9\pi/4$ rad

51. $-17\pi/6$ rad   **52.** 696°

53. −937°   **54.** $-11\pi/3$ rad

55. $29\pi/4$ rad   **56.** −672°

57. 10 rad   **58.** −20 rad

In Problems 59–66, find each value to four decimal places. Convert radians to decimal degrees.

59. 56.1225° = ? rad   **60.** 116.9853° = ? rad

61. 0.4638 rad = ?°   **62.** 2.562 rad = ?°

63. 87°39′42″ = ? rad   **64.** 261°15′45″ = ? rad

65. $19\pi/7$ rad = ?°   **66.** $43\pi/11$ rad = ?°

In Problems 67–74, recall from geometry that an angle that is inscribed in a circle (see ∠DEF in the figure) has measure θ/2, where θ is the measure of the central angle that subtends the same arc as the inscribed angle.

67. An angle that is inscribed in a circle of radius 3 m subtends an arc of length 2 m. Find its radian measure.

68. An angle that is inscribed in a circle of radius 4 m subtends an arc of length 5 m. Find its radian measure.

69. An angle that is inscribed in a circle of radius 6 ft subtends an arc of length 13 ft. Find its degree measure to the nearest degree.

70. An angle that is inscribed in a circle of radius 15 ft subtends an arc of length 34 ft. Find its degree measure to the nearest degree.

---

* Recall that a rectangular coordinate system divides a plane into four parts called quadrants. These quadrants are numbered in a counterclockwise direction starting in the upper right-hand corner.

71. An angle of 3° is inscribed in a circle of radius 85 km. Find the length of the arc it subtends.

72. An angle of 27° is inscribed in a circle of radius 52 km. Find the length of the arc it subtends.

73. An angle of radian measure $\pi/8$ is inscribed in a circle and subtends an arc of length 42 cm. Find the radius of the circle.

74. An angle of radian measure 0.25 is inscribed in a circle and subtends an arc of length 26 cm. Find the radius of the circle.

Applications

75. **Radian Measure**   What is the radian measure of the smaller angle made by the hands of a clock at 2:30 (see the figure)? Express the answer in terms of π and as a decimal fraction to two decimal places.

Figure for 75 and 76

76. **Radian Measure**   Repeat Problem 75 for 4:30.

77. **Pendulum**   A clock has a pendulum 22 cm long. If it swings through an angle of 32°, how far does the bottom of the bob travel in one swing?

Figure for 77 and 78

78. **Pendulum**   If the bob on the bottom of the 22 cm pendulum in Problem 77 traces a 9.5 cm arc on each swing, through what angle (in degrees) does the pendulum rotate on each swing?

79. **Engineering**   Oil is pumped from some wells using a donkey pump as shown in the figure. Through how many degrees must an arm with a 72 in. radius rotate to produce a 24 in. vertical stroke at the pump down in the ground? Note that a point at the end of the arm must travel through a 24 in. arc to produce a 24 in. vertical stroke at the pump.

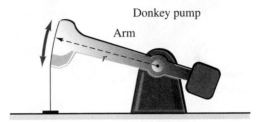

Figure for 79 and 80

80. **Engineering**   In Problem 79, find the arm length r that would produce an 18 in. vertical stroke while rotating through 21°.

81. **Bioengineering**   A particular woman, when standing and facing forward, can swing an arm in a plane perpendicular to her shoulders through an angle of $3\pi/2$ rad (see the figure). Find the length of the arc (to the nearest centimeter) her fingertips trace out for one complete swing of her arm. The length of her arm from the pivot point in her shoulder to the end of her longest finger is 54.3 cm while it is kept straight and her fingers are extended with palm facing in.

Figure for 81

82. Bioengineering A particular man, when standing and facing forward, can swing a leg through an angle of $2\pi/3$ rad (see the figure). Find the length of the arc (to the nearest centimeter) his heel traces out for one complete swing of the leg. The length of his leg from the pivot point in his hip to the bottom of his heel is 102 cm while his leg is kept straight and his foot is kept at right angles to his leg.

Figure for 82

83. Astronomy The sun is about 1.5×10^8 km from the earth. If the angle subtended by the diameter of the sun on the surface of the earth is 9.3×10^{-3} rad, approximately what is the diameter of the sun? [*Hint:* Use the intercepted arc to approximate the diameter.]

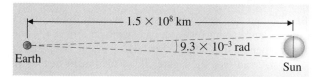

Figure for 83

84. Surveying If a natural gas tank 5.000 km away subtends an angle of $2.44°$, approximate its height to the nearest meter (see Problem 83).

85. Photography The angle of view for a 300 mm telephoto lens is $8°$ (see the figure). At 1,250 ft, what is the approximate width of the field of view? Use an arc length to approximate the chord length to the nearest foot.

86. Photography The angle of view for a 1,000 mm telephoto lens is $2.5°$ (see the figure). At 865 ft, what is the approximate width of the field of view? Use an arc length to approximate the chord length to the nearest foot.

87. Spy Satellites Some spy satellites have cameras that can distinguish objects that subtend angles of as little as 5×10^{-7} rad. If such a satellite passes over a particular

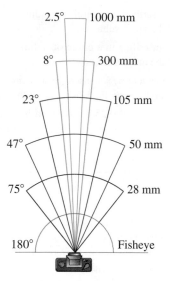

Figure for 85 and 86

country at an altitude of 250 mi, how small an object can the camera distinguish? Give the answer in meters to one decimal place and also in inches to one decimal place. (1 mi = 1,609 m; 1 m = 39.37 in.)

88. Spy Satellites Repeat Problem 87 if the satellite passes over the country at an altitude of 155 mi. Give the answers to three decimal places.

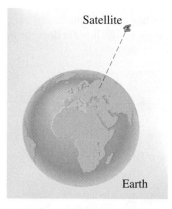

Figure for 87 and 88

89. Astronomy Assume that the earth's orbit is circular. A line from the earth to the sun sweeps out an angle of how many radians in 1 week? Express the answer in terms of π and as a decimal fraction to two decimal places. (Assume exactly 52 weeks in a year.)

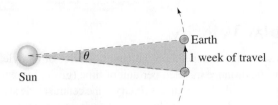

Figure for 89 and 90

90. Astronomy   Repeat Problem 89 for 13 weeks.

91. Astronomy   In measuring time, an error of 1 sec per day may not seem like a lot. But suppose a clock is in error by at most 1 sec per day. Then in 1 year the accumulated error could be as much as 365 sec. If we assume the earth's orbit about the sun is circular, with a radius of 9.3×10^7 mi, what would be the maximum error (in miles) in computing the distance the earth travels in its orbit in 1 year?

92. Astronomy   Using the clock described in Problem 91, what would be the maximum error (in miles) in computing the distance that Venus travels in a "Venus year"? Assume Venus's orbit around the sun is circular, with a radius of 6.7×10^7 mi, and that Venus completes one orbit (a "Venus year") in 224 earth days.

93. Geometry   A sector of a circle has an area of 52.39 ft^2 and a radius of 10.5 ft. Calculate the perimeter of the sector to the nearest foot.

94. Geometry   A sector of a circle has an area of 145.7 cm^2 and a radius of 8.4 cm. Calculate the perimeter of the sector to the nearest centimeter.

95. Revolutions and Radians

(A) Describe how you would find the number of radians generated by the spoke in a bicycle wheel that turns through n revolutions.

(B) Does your answer to part (A) depend on the size of the bicycle wheel? Explain.

(C) Find the number of radians generated for a wheel turning through 5 revolutions; through 3.6 revolutions.

96. Revolutions and Radians

(A) Describe how you would find the number of radians through which a 10 cm diameter pulley turns if u meters of rope are pulled through without slippage.

(B) Does your answer to part (A) depend on the diameter of the pulley? Explain.

(C) Through how many radians does the pulley in part (A) turn when 5.75 m of rope are pulled through without slippage?

97. Engineering   Rotation of a drive wheel causes a shaft to rotate (see the figure). If the drive wheel completes 3 revolutions, how many revolutions will the shaft complete? Through how many radians will the shaft turn? Compute answers to one decimal place.

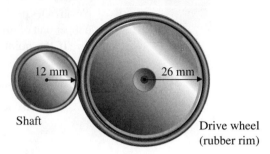

Figure for 97

98. Engineering   In Problem 97, find the radius (to the nearest millimeter) of the drive wheel required for the 12 mm shaft to make 7 revolutions when the drive wheel makes 3 revolutions.

99. Radians and Arc Length   A bicycle wheel of diameter 32 in. travels a distance of 20 ft. Find the angle (to the nearest degree) swept out by one of the spokes.

100. Radians and Arc Length   A bicycle has a front wheel with a diameter of 24 cm and a back wheel with a diameter of 60 cm. Through what angle (in radians) does the front wheel turn if the back wheel turns through 12 rad?

Figure for 100

101. Cycling   A bicycle has 28 in. diameter tires. The largest front gear (at the pedals) has 48 teeth and the smallest has 20 teeth. The largest rear gear has 32 teeth and the smallest has 11 teeth. Determine the maximum distance traveled (to the nearest inch) in one revolution of the pedals.

102. Cycling   Refer to Problem 101. Determine the minimum distance traveled (to the nearest inch) in one revolution of the pedals.

☆2.2   Linear and Angular Velocity

Suppose that a point moves with uniform speed on the circumference of a circle of radius r. Its **linear velocity** is the distance traveled per unit of time (given, for example, in miles per hour or feet per second). Its **angular velocity** is the central angle swept out per unit of time (given, for example, in radians per second or degrees per minute).

If the point moves a distance s on the circumference of a circle in time t (Fig. 1), then

$$V = \text{Linear velocity of point on the circle}$$
$$= \text{Arc length per unit of time}$$
$$= \frac{s}{t}$$

Similarly, if the point sweeps out a central angle θ in time t (Fig. 1), then

$$\omega = \text{Angular velocity of point on the circle}$$
$$= \text{Angular measure per unit of time}$$
$$= \frac{\theta}{t}$$

FIGURE 1

There is a simple connection between linear velocity and angular velocity that follows from the formula for arc length: Recall that $s = r\theta$ when θ is measured in radians (Section 2.1).

$$s = r\theta \qquad \text{Divide both sides by } t.$$
$$\frac{s}{t} = r\frac{\theta}{t} \qquad \text{Substitute } V = \frac{s}{t} \text{ and } \omega = \frac{\theta}{t}.$$
$$V = r\omega$$

In words, the linear velocity is equal to the radius times the angular velocity.

The preceding results are summarized in the box.

LINEAR AND ANGULAR VELOCITY

Suppose that a point moves with uniform speed on the circumference of a circle of radius r. If the point moves an arc length s and sweeps out a central angle of θ radians in time t, then:

Linear velocity: $V = \dfrac{s}{t}$

Angular velocity: $\omega = \dfrac{\theta}{t}$

Furthermore, $V = r\omega$.

☆ Sections marked with a star may be omitted without loss of continuity.

EXPLORE/DISCUSS 1

If the central angle θ is measured in radians, then the arc length formula $s = r\theta$ implies the formula $V = r\omega$. Recall that if θ is measured in degrees, then the arc length formula is more complicated (see Section 2.1). What is the formula connecting linear and angular velocity in that case?

EXAMPLE 1

Electrical Wind Generator

An electrical wind generator (see Fig. 2) has propeller blades that are 5.00 m long. If the blades are rotating at 8π rad/sec, what is the linear velocity (to the nearest meter per second) of a point on the tip of one of the blades?

FIGURE 2
Electrical wind generator

Solution

$$V = r\omega$$
$$= 5.00(8\pi)$$
$$= 126 \text{ m/sec}$$

Matched Problem 1   If a 3.0 ft diameter wheel turns at 12 rad/min, what is the velocity of a point on the wheel (in feet per minute)?

EXAMPLE 2

Angular Velocity

A point on the rim of a 6.0 in. diameter wheel is traveling at 75 ft/sec. What is the angular velocity of the wheel (in radians per second)?

Solution $V = r\omega$, so $\omega = \dfrac{V}{r}$

$$= \frac{75}{0.25} = 300 \text{ rad/sec} \quad [\textit{Note: } 3.0 \text{ in. } = 0.25\text{ft}] \quad \blacksquare$$

Matched Problem 2   A point on the rim of a 4.00 in. diameter wheel is traveling at 88.0 ft/sec. What is the angular velocity of the wheel (in radians per second)?   ■

EXAMPLE 3

Linear Velocity

If a 6 cm diameter drive shaft is rotating at 4,000 rpm (revolutions per minute), what is the speed of a point on its surface (in centimeters per minute, to two significant digits)?

Solution   Since 1 revolution is equivalent to 2π rad, we multiply 4,000 by 2π to obtain the angular velocity of the shaft in radians per minute:

$$\omega = 8,000\pi \text{ rad/min}$$

Now we use $V = r\omega$ to complete the solution:

$$V = 3(8,000\pi) = 75,000 \text{ cm/min} \quad \blacksquare$$

Matched Problem 3   If an 8 cm diameter drive shaft in a boat is rotating at 350 rpm, what is the speed of a point on its surface (in centimeters per second, to three significant digits)?   ■

EXPLORE/DISCUSS 2

The earth follows an elliptical path around the sun while rotating on its axis. Relative to the sun, the earth completes one rotation every 24 hr, which is called the **mean solar day.** However, relative to certain fixed stars used in astronomy, the average time it takes the earth to rotate about its axis, called the **mean sidereal day,** is 23 hr 56 min 4.091 sec of mean solar time. The earth's radius at the equator is 3,963.205 mi. Use a mean sidereal day in the following discussions.

(A) Discuss how you would find the angular velocity ω of the earth's rotation in radians per hour; then find ω to three decimal places.

(B) Explain how you would find how fast a person standing on the equator is moving in miles per hour; then find this linear velocity.

EXAMPLE 4

Hubble Space Telescope

The 25,000 lb Hubble space telescope (see Fig. 3) was launched April 1990 and placed in a 380 mi circular orbit above the earth's surface. It completes one orbit every 97 min, going from a dawn-to-dusk cycle nearly 15 times a day. If the radius of the earth is 3,964 mi, what is the linear velocity of the space telescope in miles per hour (mph)?

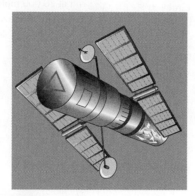

FIGURE 3
Hubble space telescope

Solution   The telescope completes 1 revolution (2π rad) in

$$\frac{97}{60} \text{ hr} = 1.6 \text{ hr}$$

The angular velocity generated by the space telescope relative to the center of the earth is

$$\omega = \frac{\theta}{t} = \frac{2\pi}{1.6} = 3.9 \text{ rad/hr}$$

The linear velocity of the telescope is

$$\begin{aligned} V &= r\omega \\ &= (3{,}964 + 380)(3.9) \\ &= 17{,}000 \text{ mph} \end{aligned}$$ ∎

Matched Problem 4   A space shuttle was placed in a circular orbit 250 mi above the earth's surface. One orbit is completed in 1.51 hr. If the radius of the earth is 3,964 mi, what is the linear velocity of the shuttle in miles per hour (to three significant digits)? ∎

Answers to
Matched Problems

1. 18 ft/min   **2.** 528 rad/sec   **3.** 147 cm/sec
4. 17,500 mph

EXERCISE 2.2

A **1.** Explain the meaning of linear velocity for a point moving on the circumference of a circle.

2. Explain the meaning of angular velocity for a point moving on the circumference of a circle.

3. A point moves on the circumference of a circle at 100 radians per second. Is this a linear velocity or an angular velocity? Explain.

4. A car travels around a circular race track at 185 miles per hour. Is this a linear velocity or an angular velocity? Explain.

In Problems 5–8, use the indicated information to find the velocity V of a point on the rim of a wheel.

5. $r = 12$ cm, $\omega = 0.7$ rad/min

6. $r = 125$ mm, $\omega = 0.07$ rad/sec

7. $r = 1.2$ ft, $\omega = 200$ rad/hr

8. $r = 0.567$ km, $\omega = 145$ rad/hr

In Problems 9–12, use the indicated information to find the angular velocity ω.

9. $r = 250$ mi, $V = 1,950$ mi/hr

10. $r = 8.0$ ft, $V = 90$ ft/sec

11. $r = 98$ m, $V = 210$ m/min

12. $r = 125$ cm, $V = 355$ cm/sec

B *In Problems 13–16, find the angular velocity of a wheel turning through θ radians in time t.*

13. $\theta = 3\pi$ rad, $t = 0.056$ hr

14. $\theta = 5\pi$ rad, $t = 22.0$ sec

15. $\theta = 9.62$ rad, $t = 1.53$ min

16. $\theta = 4.75$ rad, $t = 12.9$ sec

17. Explain the concept of a mean solar day.

18. Explain the concept of a mean sidereal day.

 Applications

19. **Engineering**   A 16 mm diameter shaft rotates at 1,500 rps (revolutions per second). Find the speed of a point on its surface (to the nearest meter per second).

20. **Engineering**   A 6 cm diameter shaft rotates at 500 rps. Find the speed of a point on its surface (to the nearest meter per second).

21. **Space Science**   An earth satellite travels in a circular orbit at 20,000 mph. If the radius of the orbit is 4,300 mi, what angular velocity (in radians per hour, to three significant digits) is generated?

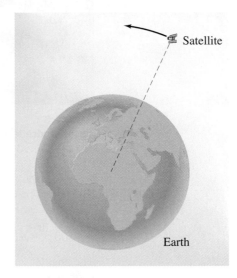

Figure for 21

22. **Engineering**   A bicycle is ridden at a speed of 7.0 m/sec. If the wheel diameter is 64 cm, what is the angular velocity of the wheel in radians per second?

23. **Physics**   The velocity of sound in air is approximately 335.3 m/sec. If an airplane has a 3.000 m diameter propeller, at what angular velocity will its tip pass through the sound barrier?

24. **Physics**   If an electron in an atom travels around the nucleus in a circular orbit (see the figure on the next page) at 8.11×10^6 cm/sec, what angular velocity (in radians per second) does it generate, assuming the radius of the orbit is 5.00×10^{-9} cm?

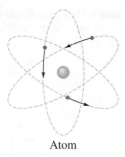

Atom

Figure for 24

25. **Astronomy**   The earth revolves about the sun in an orbit that is approximately circular with a radius of 9.3×10^7 mi (see the figure). The radius of the orbit sweeps out an angle with what exact angular velocity (in radians per hour)? How fast (to the nearest hundred miles per hour) is the earth traveling along its orbit?

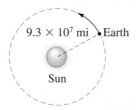

Velocity of the earth

Figure for 25

26. **Astronomy**   Take into consideration only the daily rotation of the earth to find out how fast (in miles per hour) a person halfway between the equator and North Pole would be moving. The radius of the earth is approx. 3,964 mi, and a daily rotation takes 23.93 hr.

27. **Astronomy**   Jupiter makes one full revolution about its axis every 9 hr 55 min. If Jupiter's equatorial diameter is 88,700 mi:

(A) What is its angular velocity relative to its axis of rotation (in radians per hour)?

(B) What is the linear velocity of a point on Jupiter's equator?

28. **Astronomy**   The sun makes one full revolution about its axis every 27.0 days. Assume 1 day = 24 hr. If its equatorial diameter is 865,400 mi:

(A) What is the sun's angular velocity relative to its axis of rotation (in radians per hour)?

(B) What is the linear velocity of a point on the sun's equator?

29. **Space Science**   For an earth satellite to stay in orbit over a given stationary spot on earth, it must be placed in orbit 22,300 mi above the earth's surface (see the figure). It will then take the satellite the same time to complete one orbit as the earth, 23.93 hr. Such satellites are called **geostationary satellites** and are used for communications and tracking space shuttles. If the radius of the earth is 3,964 mi, what is the linear velocity of a geostationary satellite?

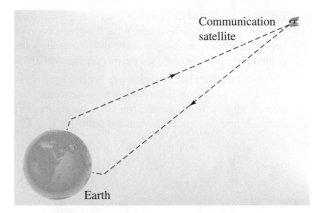

Communication satellite

Earth

Figure for 29

30. **Astronomy**   Until 1999 the planet Neptune was the planet farthest from the sun, 2.795×10^9 mi. (After 1999 and for the next 228 years Pluto will have this honor.) If Neptune takes 164 years to complete one orbit, what is its linear velocity in miles per hour?

31. **Space Science**   The earth rotates on its axis once every 23.93 hr, and a space shuttle revolves around the earth in the plane of the earth's equator once every 1.51 hr. Both are rotating in the same direction (see the figure on the next page). What is the length of time between consecutive passages of the shuttle over a particular point P on the equator? [*Hint:* The shuttle will make one complete revolution (2π rad) and a little more to be over the same point P again. Thus, the central angle generated by point P on earth must equal the central angle generated by the space shuttle minus 2π.]

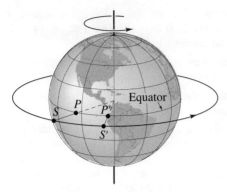

Figure for 31

32. Astronomy One of the moons of Jupiter rotates around the planet in its equatorial plane once every 42 hr 30 min. Jupiter rotates around its axis once every 9 hr 55 min. Both are rotating in the same direction. What is the length of time between consecutive passages of the moon over the same point on Jupiter's equator? [See the hint for Problem 31.]

33. Precalculus: Rotating Beacon A beacon light 15 ft from a wall rotates clockwise at the rate of exactly 1 rps (see the figure). To answer the following questions, start counting time (in seconds) when the light spot is at C.

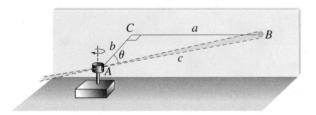

Figure for 33 and 34

(A) Describe how you can represent θ in terms of time t; then represent θ in terms of t.

(B) Describe how you can represent a, the distance the light travels along the wall, in terms of t; then represent a in terms of t.

(C) Complete Table 1, relating a and t, to two decimal places. (If your calculator has a table-producing capability, use it.) What does Table 1 seem to tell you about the motion of the light spot on the wall as t increases from 0.00 to 0.24? What happens to a when $t = 0.25$?

TABLE 1							
t (sec)	0.00	0.04	0.08	0.12	0.16	0.20	0.24
a (ft)		3.85					

34. Precalculus: Rotating Beacon

(A) Referring to the rotating beacon in Problem 33, describe how you can represent c, the length of the light beam, in terms of t; then represent c in terms of t.

(B) Complete Table 2, relating c and t, to two decimal places. (If your calculator has a table-producing capability, use it.) What does Table 2 seem to tell you about the rate of change of the length of the light beam as t increases from 0.00 to 0.24? What happens to c when $t = 0.25$?

TABLE 2							
t (sec)	0.00	0.04	0.08	0.12	0.16	0.20	0.24
c (ft)		15.49					

35. Cycling A racing bicycle has 28 in. diameter tires. The front gears (at the pedals) have 53 teeth and 39 teeth. The largest rear gear has 21 teeth and the smallest has 11 teeth. Determine the maximum speed of the bicycle (to the nearest tenth of a mile per hour) if the cyclist's cadence is 90 revolutions of the pedals per minute.

36. Cycling Refer to Problem 35. When the bicycle is in its highest gear, what cadence (to the nearest rpm) is required to maintain a speed of 26.5 miles per hour? ("Highest gear" refers to the largest ratio of teeth on the front gear to teeth on the rear gear.)

2.3 Trigonometric Functions: Unit Circle Approach

- Definition of Trigonometric Functions
- Calculator Evaluation
- Application
- Summary of Sign Properties

The correspondence that associates each acute angle θ with the trigonometric ratio $\sin \theta$ is a function.* We will remove the requirement that θ be acute in this section's unit circle approach to the sine function. We will define the sine of *any* angle in standard position, positive, negative, or zero. Equivalently, since any angle in standard position has a radian measure, we can define the sine of any real number: positive, negative, or zero. In a similar manner we will define the cosine, tangent, cosecant, secant, and cotangent functions. This generalization of trigonometric ratios frees the trigonometric functions from angles and opens them up to a large variety of significant applications that are not directly related to triangles.

■ Definition of Trigonometric Functions

The **unit circle** in a rectangular coordinate system is the circle of radius 1 with center the origin (Fig. 1). If (a, b) is a point on a circle of radius $r > 0$, then, by the Pythagorean theorem,

$$a^2 + b^2 = r^2 \qquad \text{Divide both sides by } r^2.$$

$$\left(\frac{a}{r}\right)^2 + \left(\frac{b}{r}\right)^2 = 1$$

FIGURE 1

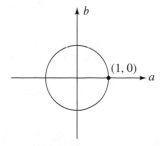

Therefore, $(a/r, b/r)$ is a point on the unit circle that lies on the same ray from the origin as the point (a, b). We define the trigonometric functions in terms of the coordinates of points on the unit circle. There are six trigonometric functions: sine, cosine, tangent, cotangent, secant, and cosecant. The values of these functions at a real number x are denoted by $\sin x$, $\cos x$, $\tan x$, $\cot x$, $\sec x$, and $\csc x$, respectively. The real number x represents the radian measure of an angle in standard position.

* A brief review of Appendix B.1 on functions might prove helpful before you start this section.

Definition 1

TRIGONOMETRIC FUNCTIONS

Remarks

For an arbitrary angle in standard position having radian measure x, let $P = (a, b)$ be the point of intersection of the terminal side of the angle and the circle of radius $r > 0$. Then:

$$\sin x = \frac{b}{r} \qquad\qquad \csc x = \frac{r}{b} \quad (b \neq 0)$$

$$\cos x = \frac{a}{r} \qquad\qquad \sec x = \frac{r}{a} \quad (a \neq 0)$$

$$\tan x = \frac{b}{a} \quad (a \neq 0) \qquad \cot x = \frac{a}{b} \quad (b \neq 0)$$

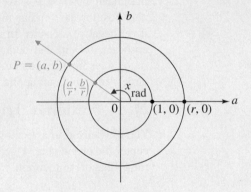

1. Note that $\sin x$ equals the second coordinate of the point $(a/r, b/r)$ on the unit circle, $\cos x$ equals the first coordinate, and $\tan x$ equals the second coordinate divided by the first coordinate.

2. If $r = 1$ then the point (a, b) of Definition 1 is on the unit circle and the values of the six trigonometric functions are simply

$$\sin x = b \qquad\qquad \csc x = \frac{1}{b} \quad (b \neq 0)$$

$$\cos x = a \qquad\qquad \sec x = \frac{1}{a} \quad (a \neq 0)$$

$$\tan x = \frac{b}{a} \quad (a \neq 0) \qquad \cot x = \frac{a}{b} \quad (b \neq 0)$$

3. If we associate angles with their radian measures, we obtain a one-to-one correspondence between angles in standard position and real numbers. This makes it possible to use the symbol x of Definition 1 to denote an angle as well as its radian measure. So when it is convenient we say, for example, "the angle $\pi/2$" rather than "the angle with radian measure $\pi/2$." A trigonometric function of an angle in degree measure is defined to be the trigono-

metric function of the same angle in radian measure; for example, $\sin 270° = \sin(3\pi/2) = -1$ because $(0, -1)$ is the point on the unit circle that lies on the terminal side of the angle $3\pi/2$.   □

Any real number x is the radian measure of some angle in standard position, and the ordinate b of the corresponding point P on the unit circle lies between -1 and 1. Furthermore, any real number b such that $-1 \le b \le 1$ is the ordinate of at least one point on the unit circle. Therefore, the domain of the function $y = \sin x$ is the set of all real numbers, and its range is the set of all real numbers y such that $-1 \le y \le 1$. The function $y = \cos x$, by similar reasoning, has the same domain and range as $y = \sin x$. These results are summarized in the following box.

DOMAIN AND RANGE FOR SINE AND FOR COSINE

Domain: All real numbers

Range: $-1 \le y \le 1$, y a real number

The restrictions in the definitions of the other four trigonometric functions (see Definition 1) imply that their domains do not contain all real numbers. Their domains and ranges will be determined in Chapter 3.

   **EXAMPLE 1**   **Evaluating Trigonometric Functions**

Find the exact values of each of the six trigonometric functions for the angle x with terminal side containing the point $(-4, -3)$.

Solution   The distance* from the point $(-4, -3)$ to the origin is

$$r = \sqrt{(-4)^2 + (-3)^2} = \sqrt{16 + 9} = \sqrt{25} = 5$$

We apply Definition 1 with $a = -4$, $b = -3$, and $r = 5$:

$$\sin x = \frac{b}{r} = -\frac{3}{5} \qquad\qquad \csc x = \frac{r}{b} = -\frac{5}{3}$$

$$\cos x = \frac{a}{r} = -\frac{4}{5} \qquad\qquad \sec x = \frac{r}{a} = -\frac{5}{4}$$

$$\tan x = \frac{b}{a} = \frac{3}{4} \qquad\qquad \cot x = \frac{a}{b} = \frac{4}{3}$$ ■

---

* The Pythagorean theorem implies that the distance between two points $P_1 = (x_1, y_1)$ and $P_2 = (x_2, y_2)$ in a rectangular coordinate system is given by the formula $d(P_1, P_2) = \sqrt{(x_2 - x_1)^2 + (y_2 - y_1)^2}$. Distance is always greater than or equal to zero.

Matched Problem 1   Find the exact values of each of the six trigonometric functions for the angle x with terminal side containing the point $(-6, 8)$.   ■

Remark   In Section 2.5 we discuss a variant of the method of Example 1 using *reference triangles*.   □

EXAMPLE 2

Using Given Information to Evaluate Trigonometric Functions

Find the exact value of each of the other five trigonometric functions for the angle x—without finding x—given that the terminal side of x is in quadrant IV and $\cos x = \frac{3}{5}$.

Solution   Because $\cos x = a/r = \frac{3}{5}$, we let $a = 3$, $r = 5$. We find b so that (a, b) is the point in quadrant IV that lies on the circle of radius $r = 5$ with center the origin.

$$a^2 + b^2 = r^2 \qquad \text{Substitute } a = 3, r = 5.$$
$$9 + b^2 = 25 \qquad \text{Solve for } b^2.$$
$$b^2 = 16 \qquad b \text{ is negative since } (a, b) \text{ lies in quadrant IV.}$$
$$b = -4$$

We now apply Definition 1 with $a = 3$, $b = -4$, and $r = 5$:

$$\sin x = \frac{b}{r} = -\frac{4}{5}$$

$$\tan x = \frac{b}{a} = -\frac{4}{3}$$

$$\csc x = \frac{r}{b} = -\frac{5}{4}$$

$$\sec x = \frac{r}{a} = \frac{5}{3}$$

$$\cot x = \frac{a}{b} = -\frac{3}{4}$$ ■

Matched Problem 2   Find the exact value of each of the other five trigonometric functions for the angle x—without finding x—given that the terminal side of x is in quadrant I and $\sin x = \frac{5}{13}$.   ■

EXAMPLE 3

Using Given Information to Evaluate Trigonometric Functions

Find the exact value of each of the other five trigonometric functions for the angle x—without finding x—given that the terminal side of x is in quadrant III and $\cot x = 2$.

Solution   Because $\cot x = a/b = 2$, we let $a = -2$, $b = -1$ (both coordinates of points in quadrant III are negative). The distance from (a, b) to the origin is

$$r = \sqrt{(-2)^2 + (-1)^2} = \sqrt{4 + 1} = \sqrt{5}$$

We apply Definition 1 with $a = -2$, $b = -1$, and $r = \sqrt{5}$:

$$\sin x = \frac{b}{r} = -\frac{1}{\sqrt{5}}$$

$$\cos x = \frac{a}{r} = -\frac{2}{\sqrt{5}}$$

$$\tan x = \frac{b}{a} = \frac{1}{2}$$

$$\csc x = \frac{r}{b} = -\sqrt{5}$$

$$\sec x = \frac{r}{a} = -\frac{\sqrt{5}}{2}$$ ∎

Matched Problem 3   Find the exact value of each of the other five trigonometric functions for the angle x—without finding x—given that the terminal side of x is in quadrant II and $\tan x = -3$.   ∎

EXPLORE/DISCUSS 1

(A) Find the coordinates of the eight equally spaced points A, B, C, D, E, F, G, H on the unit circle.

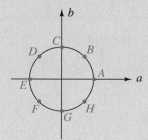

(B) Use the coordinates of the eight points and Definition 1 to complete Table 1 (note that tan x is undefined when $x = \pi/2$ or $3\pi/2$).

TABLE 1								
x	0	$\pi/4$	$\pi/2$	$3\pi/4$	π	$5\pi/4$	$3\pi/2$	2π
sin x	0	$\sqrt{2}/2$	1					
cos x	1	$\sqrt{2}/2$	0					
tan x	0	1	–				–	

■ Calculator Evaluation

We used a calculator in Section 1.3 to approximate trigonometric ratios for acute angles in degree measure. These same calculators are internally programmed to approximate (to eight or ten significant digits) trigonometric functions for *any* angle (however large or small, positive or negative) in degree or radian measure, or for *any* real number. (Remember, most graphing calculators use different sequences of steps than scientific calculators. Consult the owner's manual for your calculator.) In Section 2.5 we will show how to obtain exact values for certain special angles (integer multiples of 30° and 45° or integer multiples of $\pi/6$ and $\pi/4$) without the use of a calculator.

Caution

1. Set your calculator in **degree mode** when evaluating trigonometric functions of angles in degree measure.

2. Set your calculator in **radian mode** when evaluating trigonometric functions of angles in radian measure or trigonometric functions of real numbers.   □

We generalize the reciprocal relationships stated in Section 1.3 to evaluate secant, cosecant, and cotangent.

RECIPROCAL RELATIONSHIPS

For x any real number or angle in degree or radian measure,

$$\csc x = \frac{1}{\sin x} \qquad \sin x \neq 0$$

$$\sec x = \frac{1}{\cos x} \qquad \cos x \neq 0$$

$$\cot x = \frac{1}{\tan x} \qquad \tan x \neq 0$$

EXAMPLE 4

Calculator Evaluation of Trigonometric Functions

With a calculator, evaluate to four significant digits:

(A) $\sin 286.38°$ (B) $\tan(3.472 \text{ rad})$ (C) $\cot 5.063$

(D) $\cos(-107°35')$ (E) $\sec(-4.799)$ (F) $\csc 192°47'22''$

Solution

(A) $\sin 286.38° = -0.9594$ *Degree mode*

(B) $\tan(3.472 \text{ rad}) = 0.3430$ *Radian mode*

(C) $\cot 5.063 \boxed{= \dfrac{1}{\tan 5.063}}$

   $= -0.3657$ *Radian mode*

(D) $\cos(-107°35') \boxed{= \cos(-107.5833\ldots)}$

   $= -0.3021$ *Degree mode*

(E) $\sec(-4.799) \boxed{= \dfrac{1}{\cos(-4.799)}}$

   $= 11.56$ *Radian mode*

(F) $\csc 192°47'22'' \boxed{= \dfrac{1}{\sin 192.7894\ldots}}$

   $= -4.517$ *Degree mode*   ■

Matched Problem 4

With a calculator, evaluate to four significant digits:

(A) $\cos 303.73°$ (B) $\sec(-2.805)$ (C) $\tan(-83°29')$

(D) $\sin(12 \text{ rad})$ (E) $\csc 100°52'43''$ (F) $\cot 9$ ■

■ **Application**

An application of trigonometric functions that does not involve any angles will now be considered.

EXAMPLE 5

Wind Generators

In 2008 the Department of Energy reported that wind energy could produce 20% of the nation's total electrical output by the year 2030. "Wind farms" are springing up in many parts of the United States (see Fig. 3 on page 80). A particular wind generator can generate alternating current given by the equation

$$I = 50 \cos(120\pi t + 45\pi)$$

where t is time in seconds and I is current in amperes. What is the current I (to two decimal places) when $t = 1.09$ sec? (More will be said about alternating current in subsequent sections.)

FIGURE 3
Wind generators

Solution   Set the calculator in radian mode; then evaluate the equation for $t = 1.09$:

$$I = 50 \cos[120\pi(1.09) + 45\pi] = 40.45 \text{ amperes} \qquad ■$$

Matched Problem 5   Repeat Example 5 for $t = 2.17$ sec.   ■

■ **Summary of Sign Properties**

We close this important section by having you summarize the sign properties of
the six trigonometric functions in Table 2 in Explore/Discuss 2. Note that Table 2
does not need to be committed to memory because particular cases are readily determined from the definitions of the functions involved (see Fig. 4).

EXPLORE/DISCUSS 2

Using Figure 4 as an aid, complete Table 2 by determining the sign of each of
the six trigonometric functions in each quadrant. In Table 2, x is associated with
an angle that terminates in the respective quadrant, (a, b) is a point on the terminal side of the angle, $r = \sqrt{a^2 + b^2} > 0$, and $(a/r, b/r)$ is the point on the
terminal side of the angle that lies on the unit circle.

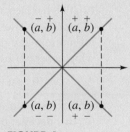

FIGURE 4

TABLE 2

	Quadrant I			Quadrant II			Quadrant III			Quadrant IV		
	a $+$	b $+$	r $+$	a $-$	b $+$	r $+$	a $-$	b $-$	r $+$	a $+$	b $-$	r $+$
$\sin x = b/r$ $\csc x = r/b$										$-$		
$\cos x = a/r$ $\sec x = r/a$												
$\tan x = b/a$ $\cot x = a/b$	$+$											

Answers to Matched Problems

1. $\sin x = \frac{4}{5}, \cos x = -\frac{3}{5}, \tan x = -\frac{4}{3}, \csc x = \frac{5}{4}, \sec x = -\frac{5}{3}, \cot x = -\frac{3}{4}$

2. $\cos x = \frac{12}{13}, \tan x = \frac{5}{12}, \csc x = \frac{13}{5}, \sec x = \frac{13}{12}, \cot x = \frac{12}{5}$

3. $\sin x = 3/\sqrt{10}, \cos x = -1/\sqrt{10}, \csc x = \sqrt{10}/3,$
   $\sec x = -\sqrt{10}, \cot x = -1/3$

4. (A) 0.5553   (B) −1.059   (C) −8.754   (D) −0.5366
   (E) 1.018   (F) −2.211

5. $I = -15.45$ amperes

EXERCISE 2.3

A *In Problems 1–6, verify that the point Q lies on the unit circle and find the exact value of each of the six trigonometric functions if the terminal side of angle x contains Q.*

1. $Q - (\frac{3}{5}, \frac{4}{5})$ 2. $Q = (\frac{5}{13}, \frac{12}{13})$
3. $Q = (1/\sqrt{2}, 1/\sqrt{2})$ 4. $Q = (\sqrt{3}/2, 1/2)$
5. $Q = (-1/2, \sqrt{3}/2)$ 6. $Q = (-1/\sqrt{2}, -1/\sqrt{2})$

In Problems 7–12, find the exact value of each of the six trigonometric functions if the terminal side of angle x contains the point P.

7. $P = (8, 6)$ 8. $P = (-12, 9)$
9. $P - (-7, -24)$ 10. $P = (-5, -12)$
11. $P = (-12, 5)$ 12. $P = (24, 7)$

In Problems 13–24, find the exact value of each of the other five trigonometric functions for the angle x, (without finding x), given the indicated information.

13. $\sin x = \frac{3}{5}$;   x is a quadrant I angle
14. $\cos x = \frac{12}{13}$;   x is a quadrant I angle
15. $\cos x = \frac{5}{13}$;   x is a quadrant IV angle
16. $\sin x = -\frac{4}{5}$;   x is a quadrant III angle
17. $\tan x = \frac{3}{2}$;   x is a quadrant III angle
18. $\cot x = 4$;   x is a quadrant I angle
19. $\cot x = \frac{1}{2}$;   x is a quadrant I angle
20. $\tan x = -\frac{1}{3}$;   x is a quadrant IV angle
21. $\sec x = \sqrt{2}$;   x is a quadrant IV angle
22. $\csc x = -\frac{25}{7}$;   x is a quadrant III angle

23. $\csc x = -\frac{25}{24};$ x is a quadrant III angle

24. $\sec x = -\sqrt{2};$ x is a quadrant II angle

25. Is it possible to find a real number x such that $\cos x$ is positive and $\sec x$ is negative? Explain.

26. Is it possible to find an angle x such that $\tan x$ is negative and $\cot x$ is positive? Explain.

Use a calculator to find the answers to Problems 27–44 to four significant digits. Make sure the calculator is in the correct mode (degree or radian) for each problem.

27. $\sin 62°$ **28.** $\cos(7 \text{ rad})$ **29.** $\tan(6 \text{ rad})$

30. $\cot 5$ **31.** $\sec 4$ **32.** $\csc 129°$

33. $\cos 208°$ **34.** $\sin 198°$ **35.** $\cot 312°$

36. $\tan 483°$ **37.** $\csc 2$ **38.** $\sec 39\pi/5$

39. $\sin 11\pi/7$ **40.** $\csc(-9)$ **41.** $\sec(-1)$

42. $\cos(-55°)$ **43.** $\cot(-22°)$ **44.** $\tan(-138°)$

B *In Problems 45–50, refer to Definition 1 and the remarks that follow the definition.*

45. Explain why the largest possible value of $\sin x$ is 1.

46. Explain why the smallest possible value of $\cos x$ is -1.

47. Explain why there is no largest possible value of $\tan x$.

48. Explain why there is no largest possible value of $\csc x$.

49. For any angle x in radians, explain why $\sin x$ is equal to $\sin (x + 2\pi)$.

50. For any angle x in radians, explain why $\cos x$ is equal to $\cos (x - 2\pi)$.

In Problems 51–54, find the exact value of each of the six trigonometric functions for an angle x that has a terminal side containing the indicated point.

51. $(\sqrt{3}, 1)$ **52.** $(1, 1)$

53. $(1, -\sqrt{3})$ **54.** $(-1, \sqrt{3})$

In which quadrants can the terminal side of an angle x lie in order for each of the following to be true?

55. $\cos x > 0$ **56.** $\sin x > 0$ **57.** $\tan x > 0$

58. $\cot x > 0$ **59.** $\sec x > 0$ **60.** $\csc x > 0$

61. $\sin x < 0$ **62.** $\cos x < 0$ **63.** $\cot x < 0$

64. $\tan x < 0$ **65.** $\csc x < 0$ **66.** $\sec x < 0$

In Problems 67–72, find the exact value of each of the other five trigonometric functions for an angle x (without finding x), given the indicated information.

67. $\sin x = \frac{1}{2}; \tan x < 0$ **68.** $\cos x = -\frac{1}{2}; \cot x > 0$

69. $\sec x = \frac{3}{2}; \sin x < 0$ **70.** $\csc x = \sqrt{5}; \cos x < 0$

71. $\cot x = -\sqrt{3}; \sin x < 0$

72. $\tan x = -\frac{1}{2}; \cos x > 0$

73. If angles α and β, $\alpha \neq \beta$, are in standard position and are coterminal, are $\cos \alpha$ and $\cos \beta$ equal? Explain.

74. If $\cos \alpha = \cos \beta, \alpha \neq \beta$, are α and β coterminal? Explain.

75. From the following display on a graphing calculator, explain how you would find $\cot x$ without finding x. Then find $\cot x$ to five decimal places.

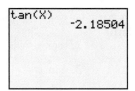

76. From the following display on a graphing calculator, explain how you would find $\sec x$ without finding x. Then find $\sec x$ to five decimal places.

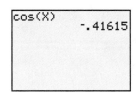

Use a calculator to find the answers to Problems 77–94 to four significant digits.

77. $\cos 1.539$ **78.** $\sin 37.85°$ **79.** $\csc 26°42'18''$

80. $\sec 107.53°$ **81.** $\cot(-3.86°)$ **82.** $\tan 4.738$

83. $\sec(-245.06°)$ **84.** $\csc(-0.408)$ **85.** $\tan 12°38'27''$

86. $\cot 352°5'55''$ **87.** $\sin 12.48$ **88.** $\cos(-432.18°)$

89. $\csc(-595.62°)$ **90.** $\sin 605°42'75''$ **91.** $\cos 6.77$

92. $\sec 3.55$ **93.** $\tan 482°12'52''$ **94.** $\cot(-567.43°)$

C **95.** Which trigonometric functions are not defined when the terminal side of an angle lies along the positive or negative vertical axis? Explain.

96. Which trigonometric functions are not defined when the terminal side of an angle lies along the positive or negative horizontal axis? Explain.

For Problems 97–100, refer to the figure:

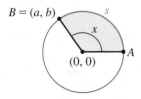

97. In the figure, the coordinates of the center of the circle are $(0, 0)$. If the coordinates of A are $(5, 0)$ and arc length s is exactly 6 units, find:

(A) The exact radian measure of x

(B) The coordinates of B (to three significant digits)

98. In the figure, the coordinates of the center of the circle are $(0, 0)$. If the coordinates of A are $(4, 0)$ and arc length s is exactly 10 units, find:

(A) The exact radian measure of x

(B) The coordinates of B (to three significant digits)

99. In the figure, the coordinates of the center of the circle are $(0, 0)$. If the coordinates of A are $(1, 0)$ and the arc length s is exactly 2 units, find:

(A) The exact radian measure of x

(B) The coordinates of B (to three significant digits)

100. In the figure, the coordinates of the center of the circle are $(0, 0)$. If the coordinates of A are $(1, 0)$ and the arc length s is exactly 4 units, find:

(A) The exact radian measure of x

(B) The coordinates of B (to three significant digits)

101. A circle with its center at the origin in a rectangular coordinate system passes through the point $(4, 3)$. What is the length of the arc on the circle in the first quadrant between the positive horizontal axis and the point $(4, 3)$? Compute the answer to two decimal places.

102. Repeat Problem 101 with the circle passing through $(3, 4)$.

103. An angle x in standard position has its terminal side in quadrant IV. If $\tan x = -\frac{1}{3}$, find the exact values of the other five trigonometric functions of x.

104. An angle x in standard position has its terminal side in quadrant III. If $\cot x = \frac{1}{2}$, find the exact values of the other five trigonometric functions of x.

Applications

105. **Solar Energy**   Light intensity I on a solar cell changes with the angle of the sun and is given by the formula in the following figure. Find the intensity in terms of the constant k for $\theta = 0°, \theta = 20°, \theta = 40°, \theta = 60°$, and $\theta = 80°$. Compute each answer to two decimal places.

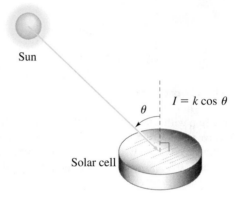

Figure for 105 and 106

106. **Solar Energy**   In Problem 105, at what angle will the light intensity I be 50% of the vertical intensity?

Sun's Energy and Seasons   The reason we have summers and winters is that the earth's axis of rotation tilts 23.5° away from the perpendicular, as indicated in the figure. The formula given in Problem 107 quantifies this phenomenon.

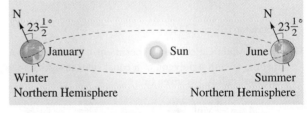

107. The amount of heat energy E from the sun received per square meter per unit of time in a given region on the surface of the earth is approximately proportional to the cosine of the angle θ that the sun makes with the vertical (see the figure on page 84). That is,

$$E = k \cos \theta$$

where k is the constant of proportionality for a given region. For a region with a latitude of 40°N, compare the energy received at the summer solstice ($\theta = 15°$)

with the energy received at the winter solstice ($\theta = 63°$). Express answers in terms of k to two significant digits.

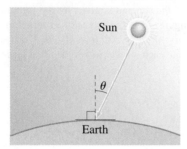

Figure for 107 and 108

108. For a region with a latitude of 32°N, compare the energy received at the summer solstice ($\theta = 8°$) with the energy received at the winter solstice ($\theta = 55°$). Refer to Problem 107, and express answers in terms of k to two significant digits.

For Problems 109 and 110, refer to the figure:

Figure for 109 and 110

109. **Precalculus: Calculator Experiment**   It can be shown that the area of a polygon of n equal sides inscribed in a circle of radius 1 is given by

$$A_n = \frac{n}{2} \sin\left(\frac{360}{n}\right)^{\circ} \qquad \text{Refer to the figure.}$$

(A) Complete the table, giving A_n to five decimal places:

n	6	10	100	1,000	10,000
A_n					

(B) As n gets larger and larger, what number does A_n seem to approach? [*Hint:* What is the area of a circle with radius 1?]

(C) Will an inscribed polygon ever be a circle for any n, however large? Explain.

110. **Precalculus: Calculator Experiment**   It can be shown that the area of a polygon of n equal sides circumscribed around a circle of radius 1 is given by

$$A_n = n \tan\left(\frac{180}{n}\right)^{\circ}$$

(A) Complete the table, giving A_n to five decimal places:

n	6	10	100	1,000	10,000
A_n					

(B) As n gets larger and larger, what number does A_n seem to approach? [*Hint:* What is the area of a circle with radius 1?]

(C) Will a circumscribed polygon ever be a circle for any n, however large? Explain.

111. **Engineering**   The figure shows a piston connected to a wheel that turns at 10 revolutions per second (rps). If P is at (1, 0) when $t = 0$, then $\theta = 20\pi t$, where t is time in seconds. Show that

$$x = a + \sqrt{5^2 - b^2} = \cos 20\pi t + \sqrt{25 - (\sin 20\pi t)^2}$$

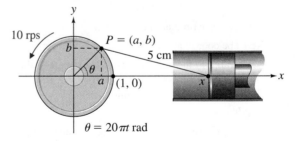

$\theta = 20\pi t$ rad

Figure for 111 and 112

112. **Engineering**   In Problem 111, find the position (to two decimal places) of the piston (the value of x) for $t = 0$ and $t = 0.01$ sec.

113. **Alternating Current**   An alternating current generator produces an electric current (measured in amperes) that is described by the equation

$$I = 35 \sin(48\pi t - 12\pi)$$

where t is time in seconds. (See Example 5 and the figure that follows.) What is the current I when $t = 0.13$ sec?

Figure for 113 and 114

114. Alternating Current What is the current *I* in Problem 113 when *t* = 0.310 sec?

115. Precalculus: Angle of Inclination The slope of a non-vertical line passing through points $P_1 = (x_1, y_1)$ and $P_2 = (x_2, y_2)$ is given by the formula

$$\text{Slope} = m = \frac{y_2 - y_1}{x_2 - x_1}$$

The angle θ that the line *L* makes with the *x* axis, $0° \leq \theta < 180°$, is called the angle of inclination of the line *L* (see the figure). Thus,

$$\text{Slope} = m = \tan \theta \qquad 0° \leq \theta < 180°$$

(A) Compute the slopes (to two decimal places) of the lines with angles of inclination 63.5° and 172°.

(B) Find the equation of a line passing through $(-3, 6)$ with an angle of inclination 143°. [*Hint:* Recall $y - y_1 = m(x - x_1)$.] Write the answer in the form $y = mx + b$, with *m* and *b* to two decimal places.

116. Precalculus: Angle of Inclination Refer to Problem 115.

(A) Compute the slopes (to two decimal places) of the lines with angles of inclination 89.2° and 179°.

(B) Find the equation of a line passing through $(7, -4)$ with an angle of inclination 101°. Write the answer in the form $y = mx + b$, with *m* and *b* to two decimal places.

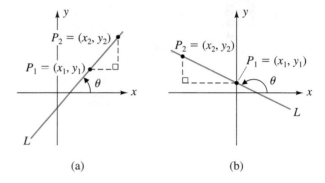

(a)   (b)

Figure for 115 and 116

☆2.4 Additional Applications

- Modeling Light Waves and Refraction
- Modeling Bow Waves
- Modeling Sonic Booms
- High-Energy Physics: Modeling Particle Energy
- Psychology: Modeling Perception

If time permits, the material in this section will provide additional understanding of the use of trigonometry relative to several interesting applications. If the material must be omitted, at least look over the next few pages to gain a better appreciation of some additional applications of trigonometric functions.

☆ Sections marked with a star may be omitted without loss of continuity.

■ Modeling Light Waves and Refraction

Did you ever look at a pencil in a glass of water or a straight pole pushed into a clear pool of water? The objects appear to bend at the surface (see Fig. 1). This bending phenomenon is caused by *refracted light.* The principle behind refracted light is **Fermat's least-time principle.** In 1657, the French mathematician Pierre Fermat proposed the intriguing idea that light, in going from one point to another, travels along the path that takes the least time. In this single simple statement, Fermat captured a benchmark principle in optics. It can be shown, using more advanced mathematics, that the least-time path of light traveling from point A in medium M_1 to point B in medium M_2 is a bent line as shown in Figure 2. For the light reflected off the surface of medium M_2 back into medium M_1, the angle of incidence equals the angle of reflection (see Fig. 2). This principle, called the **law of reflection,** was known to Euclid in the third century BC.

FIGURE 1
One or two pencils? There is actually only one.

FIGURE 2
Refraction

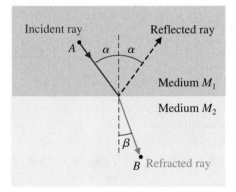

In physics it is shown that

$$\frac{c_1}{c_2} = \frac{\sin \alpha}{\sin \beta} \tag{1}$$

where c_1 is the speed of light in medium M_1, c_2 is the speed of light in medium M_2, and α and β are as indicated in Figure 2.

A more convenient form of equation (1) uses the notion of the **index of refraction,** which is the ratio of the speed of light in a vacuum to the speed of light in a given substance:

$$\text{Index of refraction, } n = \frac{\text{Speed of light in a vacuum}}{\text{Speed of light in a substance}}$$

If we let c represent the speed of light in a vacuum, then

$$\frac{c_1}{c_2} = \frac{c_1/c}{c_2/c} = \frac{1/n_1}{1/n_2} = \frac{n_2}{n_1} \tag{2}$$

where n_1 is the index of refraction in medium M_1, and n_2 is the index of refraction in medium M_2. (The index of refraction of a substance, rather than the velocity of light in that substance, is the property that is generally tabulated.) Substituting (2) into (1), we obtain **Snell's law:**

$$\frac{n_2}{n_1} = \frac{\sin \alpha}{\sin \beta} \qquad \text{Snell's law} \tag{3}$$

[*Note:* n_2 is on the top and n_1 is on the bottom in (3), while c_1 is on the top and c_2 is on the bottom in (1).]

For any two given substances, the ratio n_2/n_1 is a constant. Thus, equation (3) is equivalent to

$$\frac{\sin \alpha}{\sin \beta} = \text{Constant}$$

The fact that the sines of the angles of incidence and refraction are in a constant ratio to each other was discovered experimentally by the Dutch astronomer-mathematician Willebrod Snell (1591–1626) and was deduced later from basic principles by the French mathematician-philosopher René Descartes (1596–1650). The **law of refraction** stated in equation (3), **Snell's law,** is known in France as **Descartes' law.**

EXAMPLE 1

Refracted Light

A spotlight shining on a pond strikes the water so that the angle of incidence α in Figure 3 is 23.5°. Find the refracted angle β. Use Snell's law and the fact that $n = 1.33$ for water and $n = 1.00$ for air. (Ignore the reflected ray.)

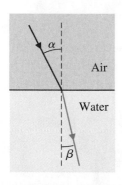

FIGURE 3
Refraction

Solution   Use

$$\frac{n_2}{n_1} = \frac{\sin \alpha}{\sin \beta}$$

where $n_2 = 1.33$, $n_1 = 1.00$, and $\alpha = 23.5°$, and solve for β:

$$\frac{1.33}{1.00} = \frac{\sin 23.5°}{\sin \beta}$$

$$\sin \beta = \frac{\sin 23.5°}{1.33}$$

$$\beta = \sin^{-1}\left(\frac{\sin 23.5°}{1.33}\right) = 17.4° \qquad \blacksquare$$

Matched Problem 1   Repeat Example 1 with $\alpha = 18.4°$. $\blacksquare$

The fact that light bends when passing from one medium into another is what makes telescopes, microscopes, and cameras possible. It is the carefully controlled bending of light rays that produces the useful results in optical instruments. Figures 4–6 illustrate a variety of phenomena connected with refracted light.

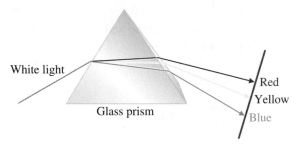

FIGURE 4
Light spectrum — different wavelengths, different indexes of refraction, different colors

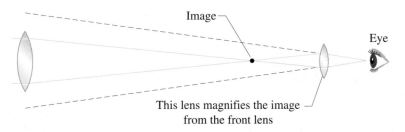

FIGURE 5
Telescope optics

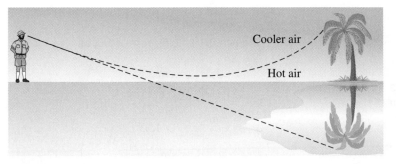

FIGURE 6
Mirage — light is refracted and bent to appear as if it came from water on the ground

Table 1 lists refractive indexes for several common materials.

TABLE 1
Refractive Indexes

Material	Refractive index
Air	1.0003
Crown glass	1.52
Diamond	2.42
Flint glass	1.66
Ice	1.31
Water	1.33

 **EXAMPLE 2**

Reflected Light

If an underwater flashlight is directed toward the surface of a swimming pool, at what angle of incidence α will the light beam be totally reflected?

Solution   The index of refraction for water is $n_1 = 1.33$ and that for air is $n_2 = 1.00$. Find the angle of incidence α in Figure 7 such that the angle of refraction β is 90°.

$$\frac{\sin \alpha}{\sin \beta} = \frac{n_2}{n_1}$$

$$\sin \alpha = \frac{1.00}{1.33} \sin 90°$$

$$\boxed{\sin \alpha = \frac{1.00}{1.33} \ (1)}$$

$$\alpha = \sin^{-1} \frac{1.00}{1.33} = 48.8°$$

FIGURE 7
Underwater light refraction

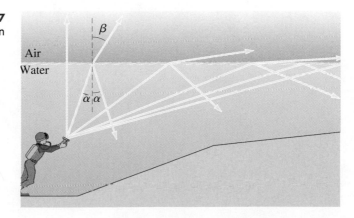

The light will be totally reflected if $\alpha \geq 48.8°$ (no light will be transmitted through the surface).   ■

Matched Problem 2   Show that the light beam passing through the flint glass prism shown in Figure 8 is totally reflected off the slanted surface.

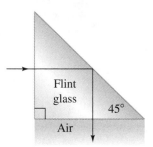

FIGURE 8
Flint glass prism

■

EXPLORE/DISCUSS 1

Figure 9 shows a person spear fishing from a rocky point over clear water. The real fish is the dark one, and the apparent fish (the fish that is actually seen) is the light one. Discuss this phenomenon relative to refraction. Should the person aim high or low relative to the apparent image in order to spear the fish? Explain why.

FIGURE 9
Aim high or low?

In general, the angle of incidence α such that the angle of refraction β is 90° is called the **critical angle.** For any angle of incidence larger than the critical angle, a light ray will be totally reflected. This critical angle (for total reflection) is important in the design of many optical instruments (such as binoculars) and is at the heart of the science of fiber optics. A small-diameter glass fiber bent in a curve (shown in Fig. 10) will "trap" a light ray entering one end, and the ray will be totally reflected and emerge out the other end.

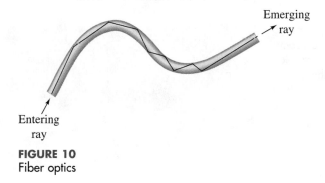

FIGURE 10
Fiber optics

Important uses of fiber optics are found in medicine and communications. Physicians use fiber-optic instruments to see inside functioning organs. Surgeons use fiber optics to perform surgery involving only small incisions and outpatient facilities. A fiber-optic communication cable carries information using high-frequency pulses of laser light that are essentially distortion-free. Using underwater armored cable (Fig. 11), fiber-optic communication networks are now in place worldwide. (The Atlantic cable was completed in 1988, and the Pacific cable was completed in 1989.) One fiber-optic communication cable can carry 40,000 simultaneous telephone conversations or transmit the entire contents of the *Encyclopaedia Britannica* in less than 1 minute.

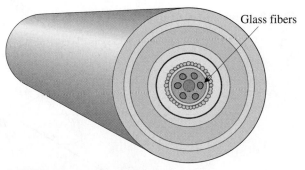

FIGURE 11
Transoceanic fiber-optic armored
cable (about 1 in. diameter)

■ Modeling Bow Waves

A boat moving at a constant rate, faster than the water waves it produces, generates a **bow wave** that extends back from the bow of the boat at a given angle (Fig. 12). If we know the speed of the boat and the speed of the waves produced by the boat, then we can determine the angle of the bow wave. Actually, if we know any two of these quantities, we can always find the third. Surprisingly, the solution to this problem also can be applied to sonic booms and high-energy particle physics, as we will see later in this section.

FIGURE 12
Bow waves of boats, sonic booms, and high-energy physics are related in a curious way; this section explains how

Referring to Figure 13, we reason as follows: When a boat is at P_1, the water wave it produces will radiate out in a circle, and by the time the boat reaches P_2, the wave will have moved a distance of r_1, which is less than the distance between P_1 and P_2 since the boat is assumed to be traveling faster than the wave. By the time the boat reaches the apex position B in Figure 13, the wave motion at P_2 will have moved r_2 units, and the wave motion at P_1 will have continued on out to r_3. Because of the constant speed of the boat and the constant speed of the wave motion, these circles of wave radiation will all have a common tangent that passes through the boat. Of course, the motion of the boat is continuous, and what we have said about P_1 and P_2 applies to all points along the path of the boat. The result of this phenomenon is the clearly visible wave front produced by the bow

FIGURE 13

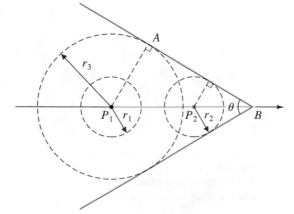

of the boat. Refer again to Figure 13, and you will see that the boat travels from P_1 to B in the same time t that the bow wave travels from P_1 to A; hence, if S_b is the speed of the boat and S_w is the speed of the bow wave, then (using $d = rt$),

$$\text{Distance from } P_1 \text{ to } A = S_w t \qquad \text{Distance from } P_1 \text{ to } B = S_b t$$

and, since triangle P_1BA is a right triangle,

$$\sin\frac{\theta}{2} = \frac{S_w t}{S_b t} = \frac{S_w}{S_b}$$

Thus,

$$\sin\frac{\theta}{2} = \frac{S_w}{S_b}$$

where S_w is the speed of the bow wave, S_b is the speed of the boat, and $S_b > S_w$.

EXAMPLE 3

Bow Wave Speed

If a speedboat travels at 45 km/hr and the angle between the bow waves is 72°, how fast is the bow wave traveling?

Solution   We use

$$\sin\frac{\theta}{2} = \frac{S_w}{S_b}$$

where $\theta = 72°$ and $S_b = 45$ km/hr. Then we solve for S_w:

$$S_w = (45 \text{ km/hr})(\sin 36°) \approx 26 \text{ km/hr} \qquad ∎$$

Matched Problem 3   If a speedboat is traveling at 121 km/hr and the angle between the bow waves is 74.5°, how fast is the bow wave moving? ∎

▮ Modeling Sonic Booms

If we follow exactly the same line of reasoning as in the discussion of bow waves, we see that an aircraft flying faster than the speed of sound produces sound waves that pile up behind the aircraft in the form of a cone (see Fig. 14).

FIGURE 14
Sound cones and sonic booms

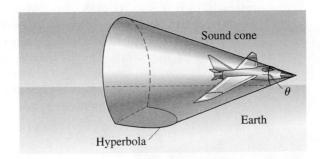

The cone intersects the ground in the form of a hyperbola, and along this curve we experience a phenomenon called a **sonic boom.** As in the bow wave analysis, we have

$$\sin \frac{\theta}{2} = \frac{S_s}{S_a}$$

where S_s is the speed of the sound wave, S_a is the speed of the aircraft, and $S_a > S_s$.

EXPLORE/DISCUSS 2

Chuck Yeager (1923–) was a 24-year-old U.S. Air Force test pilot when he broke the sound barrier in 1947 in a Bell X-1 rocket plane, ushering in the age of supersonic flight. The speed of supersonic aircraft is measured in terms of Mach numbers. [The designation Mach is named for Ernst Mach (1838–1916), an Austrian physicist who made significant contributions to the study of sound.] A **Mach number** is the ratio of the speed of an object to the speed of sound in the surrounding medium. Mach 1 is the speed of an object flying at the speed of sound, approximately 750 mph at sea level.

(A) Write a formula for the Mach number M of a supersonic aircraft, if S_a is the speed of the aircraft and S_s is the speed of sound in the surrounding medium.

(B) Rewrite the formula for the angle θ of the sound cone in Figure 14 in terms of M, assuming $S_a > S_s$.

(C) Find the angle θ of the sound cone of the British-French commercial aircraft Concorde (now retired), flying at Mach 2.

■ High-Energy Physics: Modeling Particle Energy

Nuclear particles can be made to move faster than the velocity of light in certain materials, such as glass. In 1958, three physicists (Cerenkov, Frank, and Tamm) jointly received a Nobel prize for the work they did based on this fact. Interestingly, the bow wave analysis for boats applies equally well here. Instead of a sound cone, as in Figure 14, they obtained a light cone, as shown in Figure 15 on page 95.

By measuring the cone angle θ, Cerenkov, Frank, and Tamm were able to determine the speed of the particle because the speed of light in glass is readily determined. They used the formula

$$\sin \frac{\theta}{2} = \frac{S_\ell}{S_p}$$

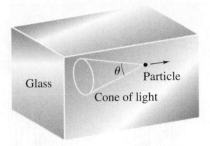

FIGURE 15
Particle energy

where S_ℓ is the speed of light in glass, S_p is the speed of the particle, and $S_p > S_\ell$. By determining the speed of the particle, they were then able to determine its energy by routine procedures.

◼ Psychology: Modeling Perception

An important field of study in psychology concerns sensory perception—hearing, seeing, smelling, feeling, and tasting. It is well known that individuals see certain objects differently in different surroundings. Lines that appear to be parallel in one setting may appear to be curved in another. Lines of the same length may appear to have different lengths in two different settings. Is a square always a square (see Fig. 16)?

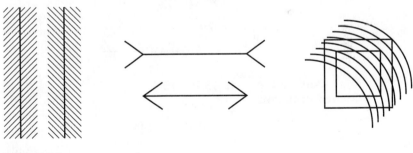

FIGURE 16
Illusions

Figure 17 on page 96 illustrates a perspective illusion in which the people farthest away appear to be larger than those closest—they are actually all the same size.

An interesting experiment in visual perception was conducted by psychologists Berliner and Berliner. A tilted field of parallel lines was presented to several subjects, who were then asked to estimate the position of a horizontal line in the field. Berliner and Berliner (*American Journal of Psychology*, vol. 65, pp. 271–277, 1952) reported that most subjects were consistently off and that the difference in degrees d between their estimates and the actual horizontal could be approximated by the equation

$$d = a + b \sin 4\theta$$

FIGURE 17
Which person is tallest?

where a and b are constants associated with a particular individual and θ is the angle of tilt of the visual field in degrees (see Fig. 18).

FIGURE 18
Visual perception

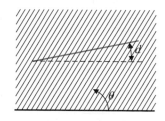

Answers to Matched Problems

1. $13.7°$

2. $\dfrac{\sin 45°}{\sin \beta} = \dfrac{1.00}{1.66}$

   $\sin \beta = 1.66 \sin 45°$

   $\sin \beta = 1.17$

3. 73.2 km/hr

Since $\sin \beta$ cannot exceed 1 (see Definition 1 of trigonometric functions on page 74), the condition $\sin \beta = 1.17$ cannot physically happen! In other words, the light must be totally reflected, as indicated in Figure 8.

EXERCISE 2.4

Applications

1. Explain Fermat's least-time principle.

2. Explain why the index of refraction of a substance is a number that is greater than 1.

3. What is an angle of incidence?

4. What is an angle of reflection?

5. Explain the law of reflection.

6. What is an angle of refraction?

7. Explain the law of refraction, also known as Snell's law.

8. Explain the concept of critical angle.

In Problems 9–16, refer to the following figure and table (repeated from the text for convenience):

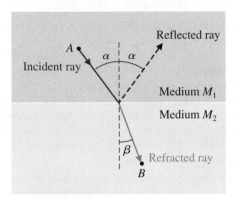

Refractive Indexes	
Material	Refractive index
Air	1.0003
Crown glass	1.52
Diamond	2.42
Flint glass	1.66
Ice	1.31
Water	1.33

9. A light ray passing through air strikes the surface of a pool of water so that the angle of incidence is $\alpha = 40.6°$. Find the angle of refraction β.

10. Repeat Problem 9 with $\alpha = 34.2°$

11. A light ray from an underwater spotlight passes through a porthole (of flint glass) of a sunken ocean liner. If the angle of incidence α is 32.0°, what is the angle of refraction β?

12. Repeat Problem 11 with $\alpha = 45.0°$

13. If light inside a diamond strikes one of its facets (flat surfaces), what is the critical angle of incidence α for total reflection? (The diamond is surrounded by air.) Compute your answer in decimal degrees to three significant digits.

Figure for 13

14. If light inside a triangular flint glass prism strikes one of the flat surfaces, what is the critical angle of incidence α for total reflection? (The prism is surrounded by air.)

15. A golfer hits a ball into a pond. From the side of the pond, the ball is spotted on the bottom. Does the ball appear to be above or below the real ball? Explain.

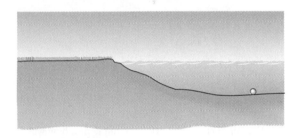

16. The figure shows a diver looking up on a very calm day. Explain why the diver sees a bright circular field surrounded by darkness. What does the diver see within this bright circular field? The bright circular field with the diver's eye at the apex forms a cone. What is the angle θ of the cone?

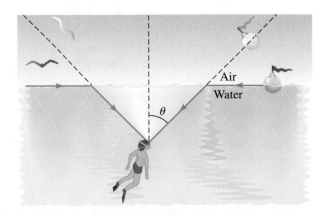

17. Index of Refraction An experiment is set up to find the index of refraction of an unknown liquid. A rectangular container is filled to the top with the liquid and the experimenter moves the container up and down until the far bottom edge of the container is just visible (see the figure). Use the information in the figure to determine the index of refraction of the liquid.

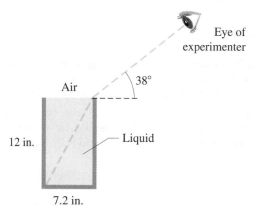

Figure for 17 and 18

18. Index of Refraction Repeat Problem 17 with 38° replaced by 43°.

19. Explain what is meant by the angle of a bow wave.

20. Explain the concept of Mach number.

21. Bow Waves If the bow waves of a boat travel at 20 km/hr and create an angle of 60°, how fast is the boat traveling?

22. Bow Waves A boat traveling at 55 km/hr produces bow waves that separate at an angle of 54°. How fast is a bow wave moving?

23. Sound Cones If a supersonic jet flies at Mach 1.5, what will be the cone angle (to the nearest degree)?

24. Sound Cones If a supersonic jet flies at Mach 3.2, what will be the cone angle (to the nearest degree)?

25. Light Cones In crown glass, light travels at approx. 1.97×10^{10} cm/sec. If a high-energy particle passing through this glass creates a light cone of 92°, how fast is it traveling?

26. Light Cones Repeat Problem 25 using a light cone angle of 126°.

Psychology: Perception *In Problems 27 and 28, use the empirical formula $d = a + b \sin 4\theta$, from Berliner and Berliner's study of perception, and determine d (to the nearest degree) for the given values of a, b, and θ.*

27. $a = -2.2,$ $b = -4.5,$ $\theta = 30°$

28. $a = -1.8,$ $b = -4.2,$ $\theta = 40°$

29. Construction A construction company builds residential and commercial retaining walls from concrete blocks that weigh 87 lb. The tapered sides of the blocks allow for both straight and curved walls. The top of each block is an isosceles trapezoid with dimensions $a = 16''$, $b = 14''$ and $h = 12''$ (see figure). If the indicated distance t between consecutive blocks is $2''$, the wall is straight. If t is less than $2''$, the end points of the faces lie on a circular arc. Determine the radius r of the circular arc (to the nearest inch) if $t = 1''$.

30. Construction Stone Strong casts 5,600 lb concrete blocks with 24 sq ft faces that are used to build massive retaining walls. The dimensions of the top of each block are $a = 96''$, $b = 90.5''$ and $h = 42''$ (see figure). Determine the distance t between consecutive blocks (to the nearest quarter inch) so that the end points of the faces lie on a circular arc of radius $192'$.

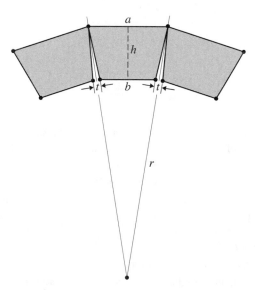

Figure for 29 and 30

2.5 Exact Values and Properties of Trigonometric Functions

- Exact Values of Trigonometric Functions at Special Angles
- Reference Triangles
- Periodic Functions
- Fundamental Identities

In Section 2.3 we defined the trigonometric functions in terms of coordinates of points on the unit circle. Given an arbitrary angle of radian measure x, it may be impossible to give the *exact* values of the trigonometric functions of x; in that case we use a calculator to give approximate values. However, for certain special angles, we can employ some basic geometric facts to find the exact values of the trigonometric functions.

We will also study more general properties of the trigonometric functions that can be quickly deduced from their definitions.

◼ Exact Values of Trigonometric Functions at Special Angles

For certain special angles it is easy to give exact values of the six trigonometric functions without using a calculator.

A **quadrantal angle** is an integer multiple of $\pi/2$ or $90°$. A quadrantal angle has its terminal side on one of the coordinate axes. Because we know the coordinates of the four points on the unit circle that lie on the coordinate axes (Fig. 1), we can easily give the exact values of the trigonometric functions for any quadrantal angle (some values may be undefined).

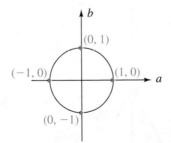

FIGURE 1

EXAMPLE 1

Evaluating Trigonometric Functions of Quadrantal Angles

Find:

(A) $\sin 90°$ (B) $\cos \pi$ (C) $\tan(-2\pi)$ (D) $\cot(-180°)$

Solution For each angle, visualize the point (a, b) on the unit circle that lies on the terminal side of the angle. Then determine the value of the trigonometric function from a and b (see Definition 1 in Section 2.3).

(A) $(a, b) = (0, 1)$; $\sin 90° = b = 1$
(B) $(a, b) = (-1, 0)$; $\cos \pi = a = -1$
(C) $(a, b) = (1, 0)$; $\tan(-2\pi) = b/a = 0/1 = 0$
(D) $(a, b) = (-1, 0)$; $\cot(-180°) = a/b = -1/0$ Not defined ■

Matched Problem 1 Find:

(A) $\sin(3\pi/2)$ (B) $\sec(-\pi)$ (C) $\tan 90°$ (D) $\cot(-270°)$ ■

EXPLORE/DISCUSS 1

In Example 1D, notice that $\cot(-180°)$ is not defined. Discuss other angles in degree measure for which the cotangent is not defined. For what angles in degree measure is the cosecant function not defined?

We can also calculate exact values of the trigonometric functions, without using a calculator, for angles that are multiples of $\pi/4$ (45°) or $\pi/6$ (30°).

Before doing so, it is helpful to recall the multiples of $\pi/4$ (45°), $\pi/6$ (30°), and $\pi/3$ (60°). See Figure 2.

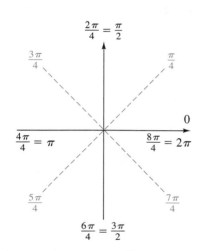

(a) Multiples of $\frac{\pi}{4}$ (45°)

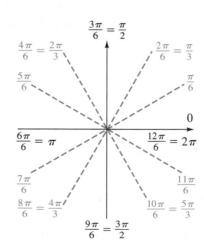

(b) Multiples of $\frac{\pi}{6}$ (30°)

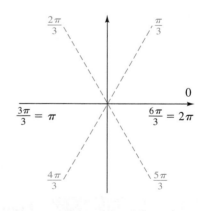

(c) Multiples of $\frac{\pi}{3}$ (60°)

FIGURE 2
Multiples of special angles

To find exact values of the trigonometric functions at multiples of $\pi/4$ (45°), let $F = (a, b)$ be the point on the unit circle that lies on the terminal side of a 45° angle. Then $a = b$ (why?), and we can use the equation of the unit circle to solve for a:

$$a^2 + b^2 = 1 \qquad \text{Substitute } a \text{ for } b.$$
$$a^2 + a^2 = 1$$
$$2a^2 = 1$$
$$a^2 = \tfrac{1}{2} \qquad a \text{ is positive since } F \text{ is in quadrant I.}$$
$$a = \frac{1}{\sqrt{2}}$$

The point F is $(1/\sqrt{2}, 1/\sqrt{2})$. By symmetry and reflection in the coordinate axes we obtain the coordinates of the points on the unit circle at angles of $3\pi/4, 5\pi/4$, and $7\pi/4$ (Fig. 3).

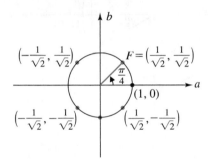

FIGURE 3

EXAMPLE 2

Evaluating Trigonometric Functions of Multiples of $\pi/4$

Find:

(A) $\sec(5\pi/4)$ (B) $\tan(-45°)$ (C) $\sin(135°)$ (D) $\cos(9\pi/4)$

Solution   (A) $(a, b) = (-1/\sqrt{2}, -1/\sqrt{2})$; $\sec(5\pi/4) = 1/a = -\sqrt{2}$

(B) $(a, b) = (1/\sqrt{2}, -1/\sqrt{2})$; $\tan(-45°) = b/a = -1$

(C) $(a, b) = (-1/\sqrt{2}, 1/\sqrt{2})$; $\sin(135°) = b = 1/\sqrt{2}$

(D) $(a, b) = (1/\sqrt{2}, 1/\sqrt{2})$; $\cos(9\pi/4) = a = 1/\sqrt{2}$ ∎

Matched Problem 2   Find:

(A) $\cos(225°)$ (B) $\tan(3\pi/4)$ (C) $\csc(45°)$ (D) $\sec(-\pi/4)$ ∎

To find exact values of the trigonometric functions at angles that are multiples of $\pi/6$ (30°), let $T = (a, b)$ be the point on the unit circle that lies on the terminal side of an angle of $\pi/6$ radians (Fig. 4). Let T' be the point at $-\pi/6$. The triangle TOT' is isosceles and has a 60° angle, so all three angles are 60°, and triangle TOT' is equilateral.

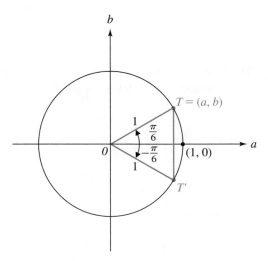

FIGURE 4

Therefore, $b = 1/2$ (why?), and we can use the equation of the unit circle to solve for a:

$$a^2 + b^2 = 1 \qquad \text{Substitute } b = \frac{1}{2}.$$

$$a^2 + \left(\frac{1}{2}\right)^2 = 1$$

$$a^2 = 1 - \left(\frac{1}{2}\right)^2$$

$$a^2 = \frac{3}{4} \qquad \text{a is positive since } T \text{ is in quadrant I.}$$

$$a = \frac{\sqrt{3}}{2}$$

The point T is $(\sqrt{3}/2, 1/2)$, and by symmetry and reflection we obtain the coordinates of the points on the unit circle at angles of $\pi/3$ (60°), $2\pi/3$ (120°), $5\pi/6$ (150°), $7\pi/6$ (210°), $4\pi/3$ (240°), $5\pi/3$ (300°), and $11\pi/6$ (330°) (see Fig. 5 on page 103, which also incorporates the points in Figs. 1 and 3).

FIGURE 5
Multiples of $\pi/6$ and $\pi/4$ on a unit circle

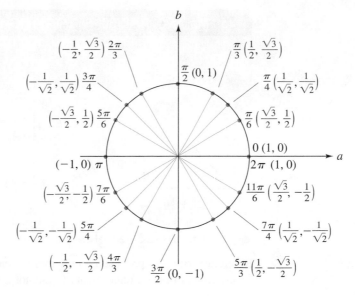

Unit circle $a^2 + b^2 = 1$

EXAMPLE 3

Evaluating Trigonometric Functions Exactly

Find each exactly:

(A) $\sin(7\pi/6)$ (B) $\sec(-60°)$ (C) $\tan(120°)$ (D) $\cos(-11\pi/6)$

Solution   Refer to Figure 5 to find the point (a, b) on the unit circle that lies on the terminal side of the angle.

(A) $(a, b) = (-\sqrt{3}/2, -1/2)$; $\sin(7\pi/6) = b = -1/2$

(B) $(a, b) = (1/2, -\sqrt{3}/2)$; $\sec(-60°) = 1/a = 2$

(C) $(a, b) = (-1/2, \sqrt{3}/2)$; $\tan(120°) = b/a = -\sqrt{3}$

(D) $(u, b) = (\sqrt{3}/2, 1/2)$; $\cos(-11\pi/6) = a = \sqrt{3}/2$ ∎

Matched Problem 3   Find each exactly:

(A) $\cot(5\pi/6)$ (B) $\csc(330°)$ (C) $\sin(315°)$ (D) $\tan(4\pi/3)$ ∎

Because $\sin 45° = \cos 45° = 1/\sqrt{2}$, any right triangle with a 45° angle has sides that are proportional to the sides of the first triangle in Figure 6 on page 104. Because $\sin 30° = 1/2$ and $\cos 30° = \sqrt{3}/2$, any right triangle with a 30° angle has sides that are proportional to the sides of the second triangle in Figure 6. It is helpful to remember the side lengths of these two special right triangles.

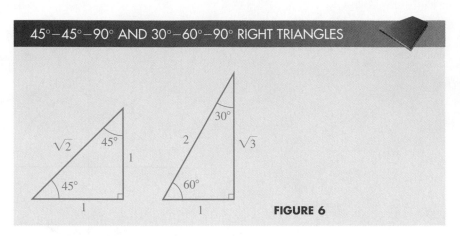

FIGURE 6

■ Reference Triangles

Reference triangles provide a handy tool for calculating values of trigonometric functions and make it unnecessary to memorize Figure 5.

REFERENCE TRIANGLE AND REFERENCE ANGLE

For a nonquadrantal angle θ:

   1. Drop a perpendicular from a point $P = (a, b)$ on the terminal side of θ to the horizontal axis. Let F denote the foot of the perpendicular. The right triangle PFO is a **reference triangle** for θ.

   2. The **reference angle** α is the acute angle (always taken positive) between the terminal side of θ and the horizontal axis (see Fig. 7).

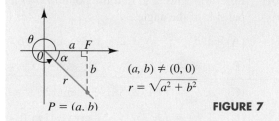

FIGURE 7

    The point $P = (a, b)$ in Definition 1 is allowed to be any point on the terminal side of the angle other than the origin; P is not required to lie on the unit circle.

EXAMPLE 4

Reference Triangles and Angles

Sketch the reference triangle and find the reference angle α for each of the following angles:

(A) $\theta = 330°$ (B) $\theta = -315°$ (C) $\theta = -\pi/4$ (D) $\theta = 4\pi/3$

Solution   (A)

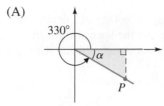

$$\alpha = 360° - 330° = 30°$$

FIGURE 8

(B)

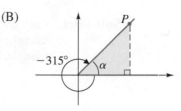

$$\alpha = 360° - 315° = 45°$$

FIGURE 9

(C)

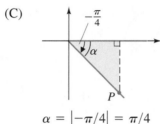

$$\alpha = |-\pi/4| = \pi/4$$

FIGURE 10

(D)

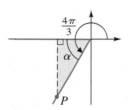

$$\alpha = 4\pi/3 - \pi = \pi/3$$

FIGURE 11   ■

Matched Problem 4   Sketch the reference triangle and find the reference angle α for each of the following angles:

(A) $\theta = -225°$ (B) $\theta = 420°$ (C) $\theta = -5\pi/6$ (D) $\theta = 2\pi/3$ ■

To compute values of the trigonometric functions of a nonquadrantal angle θ, label the legs of the reference triangle with the coordinates of P (note that one or both of these labels may be negative depending on the quadrant in which P lies) and label the hypotenuse with $r = \sqrt{a^2 + b^2}$ (r is always positive) as indicated in Figure 7. Then use the following formulas, interpreting Adj, Opp, and Hyp as the labels a, b, and r, respectively:

$$\sin \theta = \frac{\text{Opp}}{\text{Hyp}} \qquad \csc \theta = \frac{\text{Hyp}}{\text{Opp}}$$

$$\cos \theta = \frac{\text{Adj}}{\text{Hyp}} \qquad \sec \theta = \frac{\text{Hyp}}{\text{Adj}}$$

$$\tan \theta = \frac{\text{Opp}}{\text{Adj}} \qquad \cot \theta = \frac{\text{Adj}}{\text{Opp}}$$

Note that the calculations are similar to finding the trigonometric ratios of the acute reference angle a, except that Adj or Opp (or both) may be negative.

EXAMPLE 5

Evaluation of Trigonometric Functions Using Reference Triangles

Use reference triangles to evaluate exactly:

(A) $\cos(2\pi/3)$ (B) $\tan 315°$ (C) $\sin(-3\pi/4)$ (D) $\cot 300°$

Solution Each angle has a $45°-45°-90°$ or $30°-60°-90°$ right triangle as its reference triangle; we can use the special right triangles in Figure 6. Locate the reference triangle, label its sides, and evaluate the trigonometric function (see Figs. 12–15).

(A) $\cos\dfrac{2\pi}{3} = \dfrac{a}{r} = \dfrac{-1}{2} = -\dfrac{1}{2}$ (B) $\tan 315° = \dfrac{b}{a} = \dfrac{-1}{1} = -1$

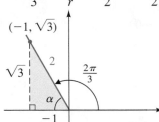

FIGURE 12

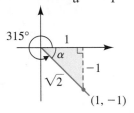

FIGURE 13

(C) $\sin(-3\pi/4) = \dfrac{b}{r} = \dfrac{-1}{\sqrt{2}}$ (D) $\cot 300° = \dfrac{a}{b} = \dfrac{1}{-\sqrt{3}}$

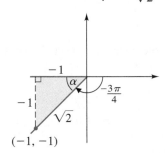

FIGURE 14

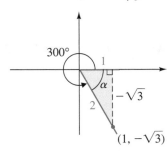

FIGURE 15   ■

Matched Problem 5 Use reference triangles to evaluate exactly:

(A) $\sin(-30°)$ (B) $\cot(7\pi/6)$ (C) $\sec(135°)$ (D) $\tan(7\pi/4)$ ■

EXAMPLE 6 **Finding Special Angles**

Find the least positive θ in degree and radian measure for which each is true.

(A) $\sin\theta = \sqrt{3}/2$ (B) $\cos\theta = -1/\sqrt{2}$

Solution (A) Draw a reference triangle (Fig. 16) in the first quadrant with side opposite reference angle $\sqrt{3}$ and hypotenuse 2. Observe that this is a special $30°-60°-90°$ triangle:

FIGURE 16

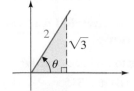

$\theta = 60°$ or $\dfrac{\pi}{3}$

(B) Draw a reference triangle (Fig. 17) in the second quadrant with side adjacent reference angle -1 and hypotenuse $\sqrt{2}$. Observe that this is a special $45°{-}45°{-}90°$ triangle:

FIGURE 17

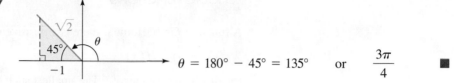

$$\theta = 180° - 45° = 135° \quad \text{or} \quad \frac{3\pi}{4} \qquad \blacksquare$$

Matched Problem 6   Repeat Example 6 for:

(A) $\tan \theta = 1/\sqrt{3}$ (B) $\sec \theta = -\sqrt{2}$ ■

■ Periodic Functions

Let $Q = (a, b)$ be the point on the unit circle that lies on the terminal side of an angle having radian measure x (Fig. 18). Then, since there are 2π radians in one complete rotation, the same point lies on the terminal side of $x + 2\pi$.

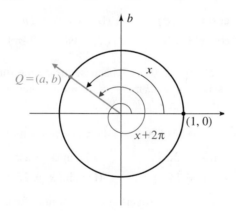

FIGURE 18

Therefore,

$$b = \sin x = \sin(x + 2\pi)$$
$$a = \cos x = \cos(x + 2\pi)$$

More generally, if we add any integer multiple of 2π to x, we will return to the same point Q. So

$$\sin x = \sin(x + 2k\pi)$$
$$\cos x = \cos(x + 2k\pi)$$

for $k = 0, \pm1, \pm2, \pm3, \ldots$

Functions with this kind of repetitive behavior are called **periodic functions.** In general:

PERIODIC FUNCTIONS

A function f is **periodic** if there is a positive real number p such that

$$f(x + p) = f(x)$$

for all x in the domain of f. The smallest such positive p, if it exists, is called **the period of** f.

From the definition of a periodic function, we conclude that:

Both the sine function and cosine function have a period of 2π.

The other four trigonometric functions also have periodic properties, which we discuss in detail in the next chapter.

EXAMPLE 7   **Using Periodic Properties**

If $\cos x = -0.0315$, what is the value of each of the following?

(A) $\cos(x + 2\pi)$ (B) $\cos(x - 2\pi)$

(C) $\cos(x + 18\pi)$ (D) $\cos(x - 34\pi)$

Solution   All are equal to -0.0315, because the cosine function is periodic with period 2π. That is, $\cos(x + 2k\pi) = \cos x$ for *all* integers k. In part (A), $k = 1$; in part (B), $k = -1$; in part (C), $k = 9$; and in part (D), $k = -17$.   ∎

Matched Problem 7   If $\sin x = 0.7714$, what is the value of each of the following?

(A) $\sin(x + 2\pi)$ (B) $\sin(x - 2\pi)$

(C) $\sin(x + 14\pi)$ (D) $\sin(x - 26\pi)$ ∎

The periodic properties of the circular functions are of paramount importance in the development of further mathematics as well as the applications of mathematics. As we will see in the next chapter, the circular functions are made to order for the analysis of real-world periodic phenomena: light, sound, electrical, and water waves; motion in buildings during earthquakes; motion in suspension systems in automobiles; planetary motion; business cycles; and so on.

EXPLORE/DISCUSS 2

The function $y = \tan x$ is a periodic function but its period is smaller than 2π. What is the period of the tangent function? Make a conjecture by considering the values of $y = \tan x$ at angles x that are multiples of $\pi/4$ and $\pi/6$.

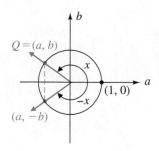

FIGURE 19

Fundamental Identities

Let $Q = (a, b)$ be the point on the unit circle that lies on the terminal side of an angle having radian measure x (Fig. 19). Then

$$b = \sin x \quad \text{and} \quad a = \cos x$$

The definitions of the other four trigonometric functions imply the following useful relationships:

$$\csc x = \frac{1}{b} = \frac{1}{\sin x} \tag{1}$$

$$\sec x = \frac{1}{a} = \frac{1}{\cos x} \tag{2}$$

$$\cot x = \frac{a}{b} = \frac{1}{b/a} = \frac{1}{\tan x} \tag{3}$$

$$\tan x = \frac{b}{a} = \frac{\sin x}{\cos x} \tag{4}$$

$$\cot x = \frac{a}{b} = \frac{\cos x}{\sin x} \tag{5}$$

Because the terminal points of x and $-x$ are symmetric with respect to the horizontal axis (see Fig. 19), we have the following sign properties:

$$\sin(-x) = -b = -\sin x \tag{6}$$

$$\cos(-x) = a = \cos x \tag{7}$$

$$\tan(-x) = \frac{-b}{a} = -\frac{b}{a} = -\tan x \tag{8}$$

Finally, because $(a, b) = (\cos x, \sin x)$ is on the unit circle $a^2 + b^2 = 1$, it follows that

$$(\cos x)^2 + (\sin x)^2 = 1$$

which is usually written in the form

$$\sin^2 x + \cos^2 x = 1 \tag{9}$$

where $\sin^2 x$ and $\cos^2 x$ are concise ways of writing $(\sin x)^2$ and $(\cos x)^2$, respectively.

⚠ **Caution** Note that $\sin^2 x$ is *not* the same as $\sin x^2$: $\sin^2 x = (\sin x)^2$ but $\sin x^2 = \sin (x^2)$. This caution applies to cosine as well. ☐

Equations (1)–(9) are called **fundamental identities**. They hold true for all replacements of x by real numbers (or angles in degree or radian measure) for which both sides of an equation are defined.

EXAMPLE 8

Use of Identities

Simplify each expression using the fundamental identities.

(A) $\dfrac{\sin^2 x + \cos^2 x}{\tan x}$ (B) $\dfrac{\sin(-x)}{\cos(-x)}$

Solution   (A) $\dfrac{\sin^2 x + \cos^2 x}{\tan x}$ Use identity (9).

$= \dfrac{1}{\tan x}$ Use identity (3).

$= \cot x$

(B) $\dfrac{\sin(-x)}{\cos(-x)}$ Use identity (4).

$= \tan(-x)$ Use identity (8).

$= -\tan x$ ∎

Matched Problem 8   Simplify each expression using the fundamental identities.

(A) $\dfrac{1 - \cos^2 x}{\sin^3 x}$ (B) $\tan(-x)\cos(-x)$ ∎

Some of the fundamental identities will be used in Chapter 3 as aids to graphing trigonometric functions. A detailed discussion of identities is found in Chapter 4.

Answers to Matched Problems

1. (A) -1 (B) -1 (C) Not defined   (D) 0
2. (A) $-1/\sqrt{2}$ (B) -1 (C) $\sqrt{2}$ (D) $\sqrt{2}$
3. (A) $-\sqrt{3}$ (B) -2 (C) $-1/\sqrt{2}$ (D) $\sqrt{3}$
4. (A)   (B)

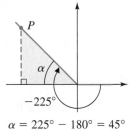

$\alpha = 225° - 180° = 45°$ $\alpha = 420° - 360° = 60°$

(C)   (D)

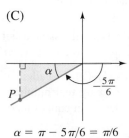

$\alpha = \pi - 5\pi/6 = \pi/6$ $\alpha = \pi - 2\pi/3 = \pi/3$

5. (A) $-1/2$ (B) $\sqrt{3}$ (C) $-\sqrt{2}$ (D) -1

6. (A) $30°$ or $\pi/6$ (B) $135°$ or $3\pi/4$

7. All are 0.7714

8. (A) $\csc x$ (B) $-\sin x$

EXERCISE 2.5

A **1.** What is a quadrantal angle?

2. Which trigonometric functions are undefined at certain of the quadrantal angles? Explain.

3. For a nonquadrantal angle θ, explain how to construct the reference triangle for θ.

4. Explain the concept of a reference angle.

In Problems 5–16, sketch the reference triangle and find the reference angle α.

5. $\theta = 60°$ **6.** $\theta = 45°$ **7.** $\theta = -60°$

8. $\theta = -45°$ **9.** $\theta = \dfrac{-\pi}{3}$ **10.** $\theta = \dfrac{-\pi}{4}$

11. $\theta = \dfrac{3\pi}{4}$ **12.** $\theta = \dfrac{5\pi}{6}$ **13.** $\theta = -210°$

14. $\theta = -150°$ **15.** $\theta = \dfrac{-5\pi}{4}$ **16.** $\theta = \dfrac{-5\pi}{3}$

In Problems 17–34, find the exact value of each trigonometric function.

17. $\sin 0°$ **18.** $\cos 0°$ **19.** $\tan 0$

20. $\cot 0$ **21.** $\cos 60°$ **22.** $\sin 30°$

23. $\cot 45°$ **24.** $\tan 30°$ **25.** $\sec \dfrac{\pi}{6}$

26. $\csc \dfrac{\pi}{3}$ **27.** $\sin \dfrac{-\pi}{2}$ **28.** $\cos \dfrac{-3\pi}{2}$

29. $\cot \pi$ **30.** $\tan \dfrac{\pi}{2}$ **31.** $\cos \dfrac{-3\pi}{4}$

32. $\sin \dfrac{-\pi}{3}$ **33.** $\tan (-60°)$ **34.** $\cot (-30°)$

B *In Problems 35–46, find the exact value of each trigonometric function.*

35. $\sin \dfrac{5\pi}{4}$ **36.** $\cos \dfrac{7\pi}{4}$

37. $\tan \dfrac{5\pi}{3}$ **38.** $\cot \dfrac{11\pi}{6}$

39. $\sec \dfrac{-3\pi}{2}$ **40.** $\csc (-2\pi)$

41. $\cos (-210°)$ **42.** $\sin (-300°)$

43. $\cot 405°$ **44.** $\tan 480°$

45. $\csc(-495°)$ **46.** $\sec(-510°)$

In Problems 47–50, find all angles θ, $0 \le \theta \le 2\pi$, for which the following functions are not defined. Explain why.

47. tangent   **48.** cotangent

49. cosecant   **50.** secant

In each graphing calculator display in Problems 51–54, indicate which values are not exact and find the exact value.

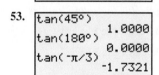

51.
```
sin(π/6)
              .5000
sin(-45°)
             -.7071
cos(0)
             1.0000
```

52.
```
cos(90°)
             0.0000
sin(2π/3)
              .8660
sin(π/6)
              .5000
```

53.
```
tan(45°)
             1.0000
tan(180°)
             0.0000
tan(-π/3)
            -1.7321
```

54.
```
sin(-150°)
             -.5000
tan(-150°)
              .5774
tan(5π/4)
             1.0000
```

In Problems 55–60, find the least positive θ in (A) degree measure (B) radian measure for which each is true.

55. $\sin \theta = \dfrac{1}{2}$ **56.** $\cos \theta = \dfrac{1}{\sqrt{2}}$

57. $\cos \theta = \dfrac{-1}{2}$ **58.** $\sin \theta = \dfrac{-1}{2}$

59. $\tan \theta = -\sqrt{3}$ **60.** $\cot \theta = -1$

61. Is there a difference between $\sin^2 x$ and $\sin x^2$? Explain.

62. Is there a difference between $(\cos x)^2$ and $\cos^2 x$? Explain.

63. If sin $x = 0.9525$, what is the value of each of the following?

(A) $\sin(x + 2\pi)$ (B) $\sin(x - 2\pi)$

(C) $\sin(x + 10\pi)$ (D) $\sin(x - 6\pi)$

64. If cos $x = -0.0379$, what is the value of each of the following?

(A) $\cos(x + 2\pi)$ (B) $\cos(x - 2\pi)$

(C) $\cos(x + 8\pi)$ (D) $\cos(x - 12\pi)$

65. Evaluate tan x and $(\sin x)/(\cos x)$ to two significant digits for:

(A) $x = 1$ (B) $x = 5.3$ (C) $x = -2.376$

66. Evaluate cot x and $(\cos x)/(\sin x)$ to two significant digits for:

(A) $x = -1$ (B) $x = 8.7$ (C) $x = -12.64$

67. Evaluate $\sin(-x)$ and $-\sin x$ to two significant digits for:

(A) $x = 3$ (B) $x = -12.8$ (C) $x = 407$

68. Evaluate $\cos(-x)$ and cos x to two significant digits for:

(A) $x = 5$ (B) $x = -13.4$ (C) $x = -1,003$

69. Evaluate $\sin^2 x + \cos^2 x$ to two significant digits for:

(A) $x = 1$ (B) $x = -8.6$ (C) $x = 263$

70. Evaluate $1 - \sin^2 x$ and $\cos^2 x$ to two significant digits for:

(A) $x = 14$ (B) $x = -16.3$ (C) $x = 766$

Simplify each expression using the fundamental identities.

71. $\sin x \csc x$ **72.** $\cos x \sec x$

73. $\cot x \sec x$ **74.** $\tan x \csc x$

75. $\dfrac{\sin x}{1 - \cos^2 x}$ **76.** $\dfrac{\cos x}{1 - \sin^2 x}$

77. $\cot(-x) \sin(-x)$ **78.** $\tan(-x) \cos(-x)$

C **79.** Find the exact value of all the angles between $0°$ and $360°$ for which $\sin \theta = -\sqrt{3}/2$.

80. Find the exact value of all the angles between $0°$ and $360°$ for which $\tan \theta = -\sqrt{3}$.

81. Find the exact value of all the angles between 0 rad and 2π rad for which $\cot \theta = -\sqrt{3}$

82. Find the exact value of all the angles between 0 rad and 2π rad for which $\cos \theta = -\sqrt{3}/2$.

For each graphing calculator display in Problems 83–86, find the least positive exact X in radian measure that produces the result shown.

83.
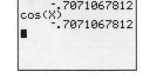

84.

```
1/√(2)
        .7071067812
sin(X)
        .7071067812
■
```

85.

```
1/√(3)
        .5773502692
tan(X)
        .5773502692
■
```

86.
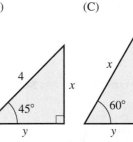

Find the exact values of x and y in Problems 87 and 88.

87. (A)   (B)   (C)

88. (A)   (B)   (C)

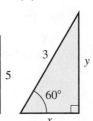

Problems 89 and 90 show the coordinates of a point on a unit circle in a graphing calculator window. Let s be the length of the least positive arc from (1, 0) to the point. Find s to three decimal places.

89.

1.5

Y1=√(1-X2)

−2.27 | 2.27

X=.58064516 Y=.81415674

−1.5

90.

1.5

Y1=√(1-X2)

−2.27 | 2.27

X=.87096774 Y=.4913402

−1.5

For Problems 91 and 92, fill the blanks in the Reason column with the appropriate identity, (1)–(9).

91. Statement

$\tan^2 x + 1 = \left(\dfrac{\sin x}{\cos x}\right)^2 + 1$ (A) _____

$= \dfrac{\sin^2 x}{\cos^2 x} + 1$ Algebra

$= \dfrac{\sin^2 x + \cos^2 x}{\cos^2 x}$ Algebra

$= \dfrac{1}{\cos^2 x}$ (B) _____

$= \left(\dfrac{1}{\cos x}\right)^2$ Algebra

$= \sec^2 x$ (C) _____

Reason

92. Statement

$\cot^2 x + 1 = \left(\dfrac{\cos x}{\sin x}\right)^2 + 1$ (A) _____

$= \dfrac{\cos^2 x}{\sin^2 x} + 1$ Algebra

$= \dfrac{\cos^2 x + \sin^2 x}{\sin^2 x}$ Algebra

Reason

$= \dfrac{1}{\sin^2 x}$ (B) _____

$= \left(\dfrac{1}{\sin x}\right)^2$ Algebra

$= \csc^2 x$ (C) _____

93. What is the period of the cosecant function?

94. What is the period of the secant function?

95. Explain why the function $f(x) = 5\sin(2\pi x)$ is periodic and find its period.

96. Explain why the function $g(x) = 4\cos(x/3)$ is periodic and find its period.

97. Explain why the function $h(x) = (\sin x)/x$ is not periodic.

98. Explain why the function $k(x) = 2x\cos x$ is not periodic.

Applications

99. Precalculus: Pi Estimate   With s_n as shown in the figure, a sequence of numbers is formed as indicated. Compute the first five terms of the sequence to six decimal places and compare the fifth term with the value of $\pi/2$.

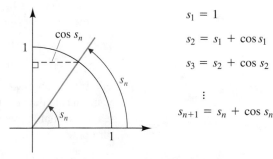

$s_1 = 1$

$s_2 = s_1 + \cos s_1$

$s_3 = s_2 + \cos s_2$

$\vdots$

$s_{n+1} = s_n + \cos s_n$

100. Precalculus: Pi Estimate   Repeat Problem 99 using $s_1 = 0.5$ as the first term of the sequence.

CHAPTER 2 GROUP ACTIVITY

Speed of Light in Water

Given the speed of light in air, how can you determine the speed of light in water by conducting a simple experiment? To start, read "Modeling Light Waves and Refraction" in Section 2.4. Snell's law for refracted light is stated here again for easy reference:

FIGURE 1
Refraction

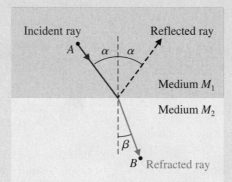

$$\frac{c_1}{c_2} = \frac{\sin \alpha}{\sin \beta} \qquad \text{Snell's law}$$

where c_1 is the speed of light in medium M_1, c_2 is the speed of light in medium M_2, and α and β are as indicated in Figure 1.

The speed of light in clean air is $c_a = 186{,}225$ mi/sec. The problem is to find the speed of light in clear water, c_w. Start by showing that

$$c_w = c_a \frac{\sin \beta}{\sin \alpha}$$

If you can find α and β for light going from a point in air to a point in water, you can calculate c_w using equation (1). The following simple experiment using readily available materials produces fairly good estimates for α and β, which in turn produce a surprisingly good estimate for the speed of light in water.

EXPERIMENT

Materials needed

- A large, straight-sided, light-colored coffee mug—on the inside of the mug, opposite the handle mark, draw a small dark line along the edge where the bottom meets the side. Use a black or red marking pen.

- A centimeter scale, with each centimeter unit divided into tenths (millimeters).

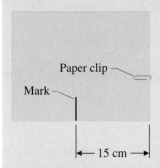

Paper clip

Mark

← 15 cm →

FIGURE 2

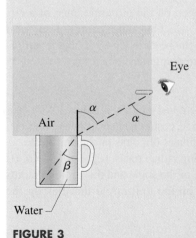

Eye

Air

α

α

β

Water

FIGURE 3

• A piece of light-colored stiff cardboard approximately 8.5 in. by 11 in. (or 28 cm by 22 cm)—cut out of a box, or use a manila file folder. Place the cardboard with the long side down and make a clear mark perpendicular to the bottom edge 15 cm from the lower right corner. Place a paper clip on the right side with the edge parallel to the bottom (see Fig. 2).

Procedure

Perform the experiment working in pairs or as individuals; then compare results with other members of your group and average your estimates for the speed of light in water. Discuss any problems and how you could improve your estimates.

Step 1 Place the mug on a flat surface where there is good overhead light and position the handle toward you. Fill the mug to the top with clear tap water.

Step 2 Place the cardboard vertically on top of the cup with the 15 cm mark at the inside edge of the cup. The cardboard should be lined up over the handle and the mark you made on the inside bottom edge of the cup opposite the handle (see Fig. 3).

Step 3 Holding the vertical cardboard as instructed in step 2, and with your face 2 or 3 in. away from the right edge of the board, move your eye up and down until the near inside top edge of the cup lines up with the dark mark on the opposite bottom edge. Move the paper clip until the bottom right-hand corner is on the line of sight indicated above (see Fig. 3). Remove the cardboard and measure (to the nearest millimeter) the distance from the bottom right corner to the bottom edge of the paper clip.

Step 4 Discuss how you can find α, and find it.

Step 5 Discuss how you can find β, and find it.

Step 6 Use equation (1) to find your estimate of the speed of light in water, and compare your value with the values obtained by others in the group.

Step 7 The speed of light in clear water, determined through more precise measurements, is $c_w = 140{,}061$ miles per second. How does the value you found compare with this value?

Step 8 Discuss why you think the average of all the values found by your group should be more or less precise than your individual value. Calculate the average of all the individual values produced by the group. What is your conclusion?

CHAPTER 2 REVIEW

2.1
DEGREES AND
RADIANS

An angle of **radian measure 1** is a central angle of a circle subtended by an arc having the same length as the radius. The **radian measure** of a central angle subtending an arc of length s in a circle of radius r is $\theta = s/r$ radians (rad). **Radian measure and degree measure** are related by

$$\frac{\theta_d}{180°} = \frac{\theta_r}{\pi \text{ rad}}$$

An angle with its vertex at the origin and initial side along the positive x axis is in **standard position.** Two angles are **coterminal** if their terminal sides coincide when both angles are placed in their standard position in the same coordinate system. The arc length and the area of a **sector of a circle** (see Fig. 1) are given by

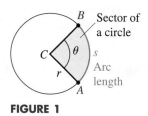

FIGURE 1

	Radian Measure	Degree Measure
Arc length	$s = r\theta$	$s = \dfrac{\pi}{180} r\theta$
Area	$A = \dfrac{1}{2} r^2\theta$	$A = \dfrac{\pi}{360} r^2\theta$

*2.2
LINEAR AND
ANGULAR VELOCITY

Suppose that a point moves with uniform speed on the circumference of a circle of radius r, covering an arc length s and sweeping out a central angle of θ radians, in time t. Then the **linear velocity** of the point is $V = s/t$, and its **angular velocity** is $\omega = \theta/t$. The linear velocity and angular velocity are related by $V = r\omega$.

2.3
TRIGONOMETRIC
FUNCTIONS: UNIT
CIRCLE APPROACH

The **unit circle** in a rectangular coordinate system is the circle of radius 1 with center the origin. The six trigonometric functions—sine, cosine, tangent, cotangent, secant, and cosecant—are defined in terms of the coordinates of points on the unit circle. For an arbitrary angle in standard position having radian measure x, let $P = (a, b)$ be the point of intersection of the terminal side of the angle and the circle of radius $r > 0$ (Fig. 2). Then $(a/r, b/r)$ is the point on the unit circle that lies on the

FIGURE 2

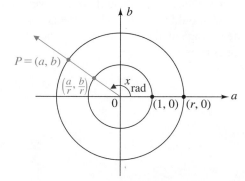

same ray from the origin as point P. The six trigonometric functions of x are defined by:

$$\sin x = \frac{b}{r} \qquad\qquad \csc x = \frac{r}{b} \quad (b \neq 0)$$

$$\cos x = \frac{a}{r} \qquad\qquad \sec x = \frac{r}{a} \quad (a \neq 0)$$

$$\tan x = \frac{b}{a} \quad (a \neq 0) \qquad \cot x = \frac{a}{b} \quad (b \neq 0)$$

The domain of the function $y = \sin x$ is the set of all real numbers, and its range is the set of real numbers y such that $-1 \leq y \leq 1$. The function $y = \cos x$ has the same domain and range as $y = \sin x$. The restrictions in the definitions of the other four trigonometric functions imply that their domains do not contain all real numbers.

If x is any real number or any angle in degree or radian measure, then the following **reciprocal relationships** hold (division by 0 excluded):

$$\csc x = \frac{1}{\sin x} \qquad \sec x = \frac{1}{\cos x} \qquad \cot x = \frac{1}{\tan x}$$

*2.4
ADDITIONAL
APPLICATIONS

Refraction

If n_1 and n_2 represent the **index of refraction** for mediums M_1 and M_2, respectively, α is the angle of incidence, and β is the angle of refraction (see Fig. 3), then according to **Snell's law:**

FIGURE 3

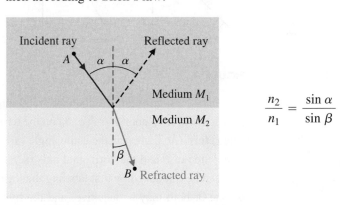

$$\frac{n_2}{n_1} = \frac{\sin \alpha}{\sin \beta}$$

The angle α for which $\beta = 90°$ is the **critical angle.**

Bow Waves

If S_b is the (uniform) speed of a boat, S_w is the speed of the **bow waves** generated by the boat, and θ is the angle between the bow waves, then

$$\sin\frac{\theta}{2} = \frac{S_w}{S_b} \qquad S_b > S_w$$

This relationship also applies to **sound waves** when planes travel faster than the speed of sound and to **nuclear particles** that travel faster than the speed of light in certain materials.

Perception

When individuals try to estimate the position of a horizontal line in a field of parallel lines tilted at an angle θ (in degrees), the difference between their estimate and the actual horizontal in degrees d can be approximated by

$$d = a + b \sin 4\theta$$

where a and b are constants associated with a particular individual.

2.5 EXACT VALUES AND PROPERTIES OF TRIGONOMETRIC FUNCTIONS

A **quadrantal angle** is an integer multiple of $\pi/2$ or $90°$. Because we know the coordinates of the four points on the unit circle that lie on the coordinate axes, we can easily give the exact values of the trigonometric functions for any quadrantal angle (some values may be undefined). Exact values can also be given for the trigonometric functions of angles that are multiples of $\pi/4$ ($45°$) or $\pi/6$ ($30°$), either by calculating the coordinates of points on the unit circle or by using reference triangles. It is helpful to remember the side lengths of two special right triangles (Fig. 4).

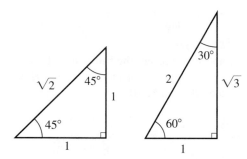

FIGURE 4

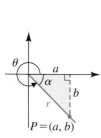

$(a, b) \neq (0, 0)$

FIGURE 5

A **reference triangle** for a nonquadrantal angle θ is formed by dropping a perpendicular from a point $P = (a, b)$ on the terminal side of θ to the horizontal axis. The **reference angle** α is the acute angle (always taken positive) between the terminal side of θ and the horizontal axis (see Fig. 5). The trigonometric functions of θ can be calculated by first labeling the legs of the reference triangle with a and b (one or both may be negative) and the hypotenuse with $r = \sqrt{a^2 + b^2} > 0$, and then using those labels to find "trigonometric ratios" (some may be negative) of the reference angle α.

A function f is **periodic** if there is a positive real number p such that $f(x + p) = f(x)$ for all x in the domain of f. The smallest such positive p, if it exists, is called **the period of** f. Both the sine function and the cosine function have a period of 2π.

The properties of trigonometric functions can be used to establish the following **fundamental identities:**

1. $\csc x = \dfrac{1}{\sin x}$ **2.** $\sec x = \dfrac{1}{\cos x}$

3. $\cot x = \dfrac{1}{\tan x}$

4. $\tan x = \dfrac{\sin x}{\cos x}$

5. $\cot x = \dfrac{\cos x}{\sin x}$

6. $\sin(-x) = -\sin x$

7. $\cos(-x) = \cos x$

8. $\tan(-x) = -\tan x$

9. $\sin^2 x + \cos^2 x = 1$

CHAPTER 2 REVIEW EXERCISE

A *Work through all the problems in this chapter review and check the answers. Answers to all review problems appear in the back of the book; following each answer is an italic number that indicates the section in which that type of problem is discussed. Where weaknesses show up, review the appropriate sections in the text. Review problems flagged with a star (☆) are from optional sections.*

1. Convert to radian measure in terms of π.

(A) 60°   (B) 45°   (C) 90°

2. Convert to degree measure.

(A) $\pi/6$ (B) $\pi/2$ (C) $\pi/4$

3. Explain the meaning of a central angle of radian measure 2.

4. Which is larger: an angle of radian measure 1.5 or an angle of degree measure 1.5? Explain.

5. (A) Find the degree measure of 15.26 rad.

(B) Find the radian measure of $-389.2°$.

☆ **6.** Find the velocity V of a point on the rim of a wheel if $r = 25$ ft and $\omega = 7.4$ rad/min.

☆ **7.** Find the angular velocity ω of a point on the rim of a wheel if $r = 5.2$ m and $V = 415$ m/hr.

8. Find the value of $\sin\theta$ and $\tan\theta$ if the terminal side of θ contains $P = (-4, 3)$.

9. Is it possible to find a real number x such that $\sin x$ is negative and $\csc x$ is positive? Explain.

Evaluate Problems 10–12 to four significant digits using a calculator.

10. (A) $\cot 53°40'$ (B) $\csc 67°10'$

11. (A) $\cos 23.5°$ (B) $\tan 42.3°$

12. (A) $\cos 0.35$ (B) $\tan 1.38$

13. Sketch the reference triangle and find the reference angle α for:

(A) $\theta = 120°$ (B) $\theta = -\dfrac{7\pi}{4}$

14. Evaluate exactly without a calculator.

(A) $\sin 60°$ (B) $\cos(\pi/4)$ (C) $\tan 0°$

In Problems 15 and 16, refer to the following figure.

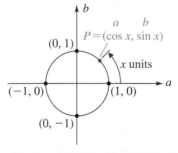

15. Refer to the figure and state the coordinates of P for the indicated values of x.

(A) $x = -2\pi$ (B) $x = \pi$

(C) $x = -3\pi/2$ (D) $x = \pi/2$

(E) $x = -5\pi$ (F) $x = 7\pi/2$

16. Refer to the figure: Given $y = \sin x$, how does y vary for the indicated variations in x?

(A) x varies from 0 to $\pi/2$

(B) x varies from $\pi/2$ to π

(C) x varies from π to $3\pi/2$

(D) x varies from $3\pi/2$ to 2π

(E) x varies from 2π to $5\pi/2$

(F) x varies from $5\pi/2$ to 3π

B 17. List all angles that are coterminal with $\theta = \pi/6$ rad, $-3\pi \le \theta \le 3\pi$. Explain how you arrived at your answer.

18. What is the degree measure of a central angle subtended by an arc exactly $\frac{7}{60}$ of the circumference of a circle?

19. If the radius of a circle is 4 cm, find the length of an arc intercepted by an angle of 1.5 rad.

20. Convert $212°$ to radian measure in terms of π.

21. Convert $\pi/12$ rad to degree measure.

22. Use a calculator with an automatic radian–degree conversion routine to find:

(A) The radian measure (to two decimal places) of $-213.23°$

(B) The degree measure (to two decimal places) of 4.62 rad

23. If the radian measure of an angle is tripled, is the degree measure of the angle tripled? Explain.

24. If $\sin \alpha = \sin \beta$, $\alpha \ne \beta$, are angles α and β necessarily coterminal? Explain.

25. From the following display on a graphing calculator, explain how you would find $\csc x$ without finding x. Then find $\csc x$ to four decimal places.

```
sin(X)
            .8594
```

26. Find the tangent of 0, $\pi/2$, π, and $3\pi/2$.

27. In which quadrant does the terminal side of each angle lie?

(A) $732°$ (B) -7 rad

Evaluate Problems 28–33 to three significant digits using a calculator.

28. $\cos 187.4°$ **29.** $\sec 103°20'$

30. $\cot(-37°40')$ **31.** $\sin 2.39$

32. $\cos 5$ **33.** $\cot(-4)$

In Problems 34–42, find the exact value of each without using a calculator.

34. $\cos \dfrac{5\pi}{6}$ **35.** $\cot \dfrac{7\pi}{4}$

36. $\sin \dfrac{3\pi}{2}$ **37.** $\cos \dfrac{3\pi}{2}$

38. $\sin \dfrac{-4\pi}{3}$ **39.** $\sec \dfrac{-4\pi}{3}$

40. $\cos 3\pi$ **41.** $\cot 3\pi$

42. $\sin \dfrac{-11\pi}{6}$

43. In the following graphing calculator display, indicate which value(s) are not exact and find the exact value.

```
cos(π/3)
            .5000
sin(180°)
            0.0000
tan(-60°)
           -1.7321
```

In Problems 44–47, use a calculator to evaluate each to five decimal places.

44. $\sin 384.0314°$ **45.** $\tan(-198°43'6'')$

46. $\cos 26$ **47.** $\cot(-68.005)$

48. If $\sin \theta = -\frac{4}{5}$ and the terminal side of θ does not lie in the third quadrant, find the exact values of $\cos \theta$ and $\tan \theta$ without finding θ.

49. Find the least positive exact value of θ in radian measure such that $\sin \theta = -\frac{1}{2}$.

50. Find the exact value of each of the other five trigonometric functions if

$$\sin \theta = -\tfrac{2}{5} \quad \text{and} \quad \tan \theta < 0$$

51. Find all the angles exactly between $0°$ and $360°$ for which $\tan \theta = -1$.

52. Find all the angles exactly between 0 and 2π for which $\cos \theta = -\sqrt{3}/2$.

53. In a circle of radius 12.0 cm, find the length of an arc subtended by a central angle of:

(A) 1.69 rad   (B) $22.5°$

54. In a circle with diameter 80 ft, find the area (to three significant digits) of the circular sector with central angle:

(A) 0.773 rad   (B) $135°$

55. Find the distance between Charleston, West Virginia ($38°21'$N latitude), and Cleveland, Ohio ($41°28'$N latitude). Both cities have the same longitude, and the radius of the earth is 3,964 mi.

☆ **56.** Find the angular velocity of a wheel turning through 6.43 rad in 15.24 sec.

☆**57.** What is meant by a rotating object having an angular velocity of 12π rad/sec?

58. Evaluate cos x to three significant digits for:

(A) $x = 7$ (B) $x = 7 + 2\pi$

(C) $x = 7 - 30\pi$

59. Evaluate $\tan(-x)$ and $-\tan x$ to three significant digits for:

(A) $x = 7$ (B) $x = -17.9$

(C) $x = -2{,}135$

60. One of the following is not an identity. Indicate which one.

(A) $\csc x = \dfrac{1}{\sin x}$ (B) $\cot x = \dfrac{1}{\tan x}$

(C) $\tan x = \dfrac{\sin x}{\cos x}$ (D) $\sec x = \dfrac{1}{\sin x}$

(E) $\sin^2 x + \cos^2 x = 1$ (F) $\cot x = \dfrac{\cos x}{\sin x}$

61. Simplify: $(\csc x)(\cot x)(1 - \cos^2 x)$

62. Simplify: $\cot(-x)\sin(-x)$

63. A point $P = (a, b)$ moves clockwise around a unit circle starting at $(1, 0)$ for a distance of 29.37 units. Explain how you would find the coordinates of the point P at its final position, and how you would determine which quadrant P is in. Find the coordinates to four decimal places and the quadrant.

C **64.** An angle in standard position intercepts an arc of length 1.3 units on a unit circle with center at the origin. Explain why the radian measure of the angle is also 1.3.

65. Which trigonometric functions are not defined for $x = k\pi$, k any integer? Explain.

66. In the following graphing calculator display, find the least positive exact value of x (in radian measure) that produces the indicated result.

```
-√(3)/2
           -.86602540
sin(X)
           -.86602540
```

67. The following graphing calculator display shows the coordinates of a point on a unit circle. Let s be the length of the least positive arc from $(1, 0)$ to the point. Find s to four decimal places.

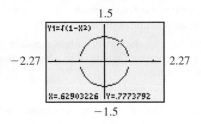

68. A circular sector has an area of 342.5 m^2 and a radius of 12 m. Calculate the arc length of the sector to the nearest meter.

69. A circle with its center at the origin in a rectangular coordinate system passes through the point $(4, 5)$. What is the length of the arc on the circle in the first quadrant between the positive horizontal axis and the point $(4, 5)$? Compute the answer to two decimal places.

Applications

70. **Engineering**   Through how many radians does a pulley with 10 cm diameter turn when 10 m of rope has been pulled through it without slippage? How many revolutions result? (Give answers to one decimal place.)

☆**71.** **Engineering**   If the large gear in the figure completes 5 revolutions, how many revolutions will the middle gear complete? How many revolutions will the small gear complete?

Figure for 71

☆**72.** **Engineering**   An automobile is traveling at 70 ft/sec. If the wheel diameter is 27 in., what is the angular velocity in radians per second?

73. **Space Science**   A satellite is placed in a circular orbit 1,000 mi above the earth's surface. If the satellite completes one orbit every 114 min and the radius of the earth is 3,964 mi, what is the linear velocity (to three significant digits) of the satellite in miles per hour?

74. Electric Current   An alternating current generator produces an electrical current (measured in amperes) that is described by the equation

$$I = 30 \sin(120\pi t - 60\pi)$$

where t is time in seconds. What is the current I when $t = 0.015$ sec? (Give the answer to one decimal place.)

75. Precalculus   A ladder of length L leaning against a building just touches a fence that is 10 ft high located 2 ft from the building (see the figure).

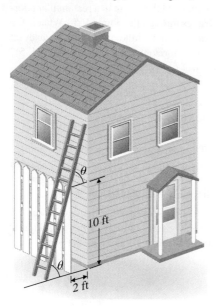

Figure for 75

(A) Express the length of the ladder in terms of θ.

(B) Describe what you think happens to the length of the ladder L as θ varies between 0 and $\pi/2$ radians.

(C) Complete the table (to two decimal places) using a calculator. (If handy, use a table-generating calculator.)

θ rad	0.70	0.80	0.90	1.00	1.10	1.20	1.30
L ft	18.14						

(D) From the table, select the angle θ that produces the shortest ladder that satisfies the conditions in the problem. [Calculus techniques can be used on the equation from part (A) to find the angle θ that produces the shortest ladder.]

☆**76. Light Waves**   A light wave passing through air strikes the surface of a pool of water so that the angle of incidence is $\alpha = 31.7°$. Find the angle of refraction. (Water has a refractive index of 1.33 and air has a refractive index of 1.00.)

☆**77. Light Waves**   A triangular crown glass prism is surrounded by air. If light inside the prism strikes one of its facets, what is the critical angle of incidence α for total reflection? (The refractive index for crown glass is 1.52, and the refractive index for air is 1.00.)

☆**78. Bow Waves**   If a boat traveling at 25 mph produces bow waves that separate at an angle of 51°, how fast are the bow waves traveling?

Graphing Trigonometric Functions

3

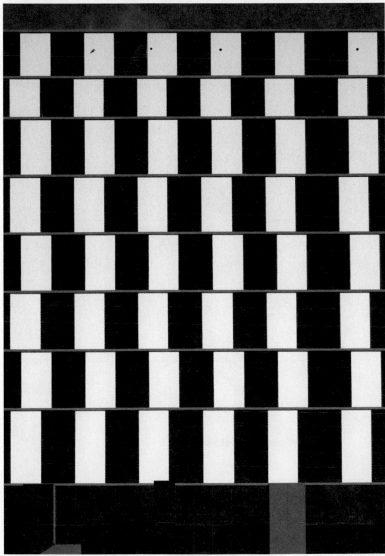

3.1 Basic Graphs

3.2 Graphing
$y = k + A \sin Bx$ and
$y = k + A \cos Bx$

3.3 Graphing
$y = k + A \sin(Bx + C)$
and
$y = k + A \cos(Bx + C)$

☆ **3.4** Additional Applications

3.5 Graphing Combined
Forms

3.6 Tangent, Cotangent,
Secant, and Cosecant
Functions Revisited

*Chapter 3 Group Activity:
Predator–Prey Analysis Involving
Coyotes and Rabbits*

Chapter 3 Review

*Cumulative Review Exercise,
Chapters 1–3*

☆ Sections marked with a star may be omitted
without loss of continuity.

Before you begin this chapter, briefly review Appendix B.2 on graphs and transformations. Many of the concepts discussed there will be used in our study of the graphs of trigonometric functions.

W ith the trigonometric functions defined, we are now in a position to consider a substantially expanded list of applications and properties. As a brief preview, look at Figure 1.

What feature seems to be shared by the different illustrations? All appear to be repetitive—that is, periodic. The trigonometric functions, as we will see shortly, can be used to describe such phenomena with remarkable precision. This is one of the main reasons that trigonometric functions have been widely studied for over

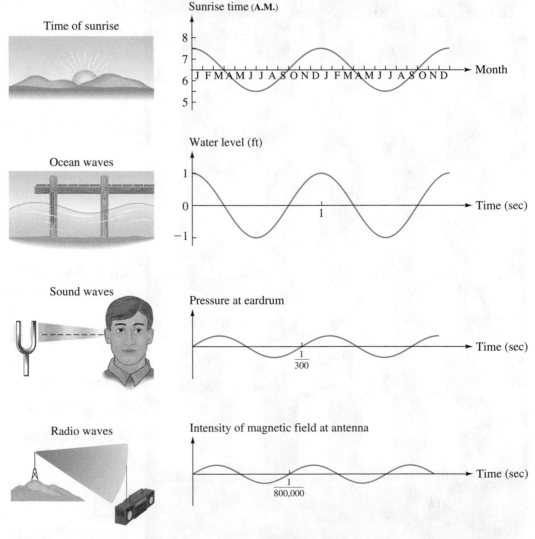

FIGURE 1

2,000 years! You may find it interesting to take a quick look through Section 3.4, "Additional Applications," to preview an even greater variety of applications.

In this chapter you will learn how to quickly and easily sketch graphs of the trigonometric functions. You will also learn how to recognize certain fundamental and useful properties of these functions.

3.1 Basic Graphs

- Graphs of $y = \sin x$ and $y = \cos x$
- Graphs of $y = \tan x$ and $y = \cot x$
- Graphs of $y = \csc x$ and $y = \sec x$
- Graphing with a Graphing Calculator

In this section we will discuss the graphs of the six trigonometric functions introduced in Chapter 2. We will also discuss the domains, ranges, and periodic properties of these functions.

Although it seems like there is a lot to remember in this section, you mainly need to be familiar with the graphs and properties of the sine, cosine, and tangent functions. The reciprocal relationships we discussed in Section 2.3 will enable you to determine the graphs and properties of the other three trigonometric functions based on the sine, cosine, and tangent functions.

■ Graphs of $y = \sin x$ and $y = \cos x$

First, we consider

$$y = \sin x \qquad x \text{ a real number} \tag{1}$$

The graph of the sine function is the graph of the set of all ordered pairs of real numbers (x, y) that satisfy equation (1). How do we find these pairs of numbers? To help make the process clear, we refer to a function machine with the unit circle definition inside (Fig. 1).

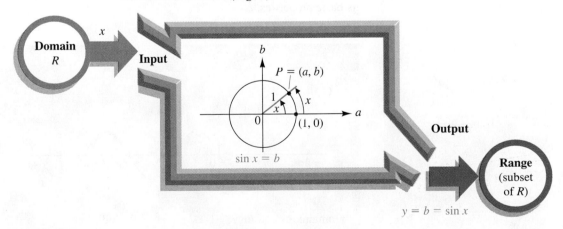

FIGURE 1
Sine function machine ($R = $ All real numbers)

We are interested in graphing, in an *xy* coordinate system, all ordered pairs of real numbers (x, y) produced by the function machine. We could resort to point-by-point plotting using a calculator, which becomes tedious and tends to obscure some important properties. Instead, we choose to speed up the process by using some of the properties discussed in Section 2.5 and by observing how $y = \sin x = b$ varies as $P = (a, b)$ moves around the unit circle. We know that the domain of the sine function is the set of all real numbers *R*, its range is the set of all real numbers *y* such that $-1 \le y \le 1$, and its period is 2π.

Because the sine function is periodic with period 2π, we will begin by finding the graph over one period, from 0 to 2π. Once we have the graph for one period, we can complete as much of the rest of the graph as we wish by drawing identical copies of the graph to the left and to the right.

Figure 2 illustrates how $y = \sin x = b$ varies as *x* increases from 0 to 2π and $P = (a, b)$ moves around the unit circle.

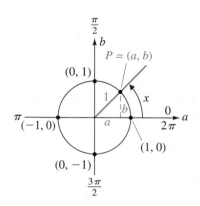

FIGURE 2
$y = \sin x = b$

As x increases	$y = \sin x = b$
from 0 to $\pi/2$	increases from 0 to 1
from $\pi/2$ to π	decreases from 1 to 0
from π to $3\pi/2$	decreases from 0 to -1
from $3\pi/2$ to 2π	increases from -1 to 0

The information in Figure 2 can be translated into a graph of $y = \sin x$ for *x* between 0 and 2π as shown in Figure 3 on the next page. (Where the graph is uncertain, if you like, you can fill in with calculator values.)

To complete the graph of $y = \sin x$ over any interval we like, we just need to repeat the final graph in Figure 3 to the left and to the right as far as we wish. The next box summarizes what we now know about the graph of $y = \sin x$ and its basic properties.

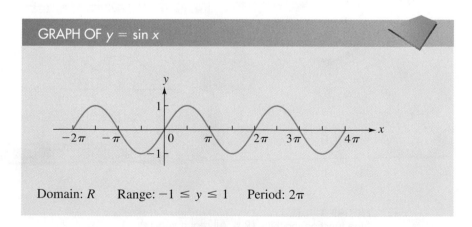

GRAPH OF $y = \sin x$

Domain: *R*   Range: $-1 \le y \le 1$ Period: 2π

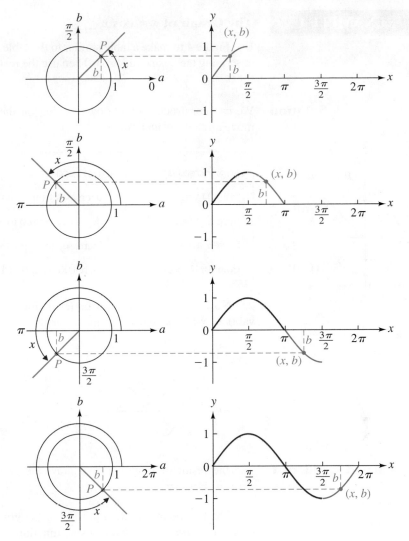

FIGURE 3
$y = \sin x,\ 0 \le x \le 2\pi$

Both the x and y axes are real number lines (see Appendix A.1). Because the domain of the sine function is all real numbers, the graph of $y = \sin x$ extends without limit in both horizontal directions. Also, because the range is $-1 \le y \le 1$, no point on the graph can have a y coordinate greater than 1 or less than -1.

In Example 1, we'll use a similar procedure to draw the graph of the cosine function.

EXAMPLE 1

The Graph of $y = \cos x$

Use Figure 4 to make a table similar to the table accompanying Figure 2, this time describing the cosine function. Then use the result to draw the graph of $y = \cos x$ for $0 \leq x \leq 2\pi$.

Solution   We can construct the table by focusing on the first coordinate of points as we move around the unit circle.

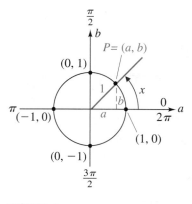

FIGURE 4
$y = \cos x = a$

As x increases	$y = \cos x = a$
from 0 to $\pi/2$	decreases from 1 to 0
from $\pi/2$ to π	decreases from 0 to -1
from π to $3\pi/2$	increases from -1 to 0
from $3\pi/2$ to 2π	increases from 0 to 1

Now we can use this information to draw the graph of $y = \cos x$ on the interval $0 \leq x \leq 2\pi$, starting at the point (1, 0).

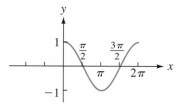

Matched Problem 1   Use the result of Example 1 to draw the graph of $y = \cos x$ on the interval $-3\pi \leq x \leq 3\pi$.

The box below shows a portion of the graph of $y = \cos x$ and summarizes some of the properties of the cosine function.

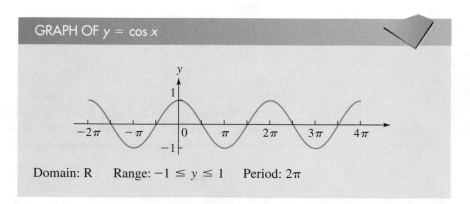

GRAPH OF $y = \cos x$

Domain: R   Range: $-1 \leq y \leq 1$ Period: 2π

If you take a few minutes to learn the basic characteristics of the sine and cosine graphs, you'll be able to sketch them quickly and easily. In particular, you should be able to answer the following questions:

(A) How often does the graph repeat (what is the period)?

(B) Where are the x intercepts?

(C) Where are the y intercepts?

(D) Where do the high and low points occur?

(E) What are the symmetry properties relative to the origin, y axis, and x axis?

EXPLORE/DISCUSS 1

(A) Discuss how the graphs of the sine and cosine functions are related.

(B) Can one be obtained from the other by a horizontal shift? Explain how.

■ Graphs of $y = \tan x$ and $y = \cot x$

First, we discuss the graph of $y = \tan x$. Later, because $\cot x = 1/(\tan x)$, we will be able to get the graph of $y = \cot x$ from the graph of $y = \tan x$ using reciprocals of second coordinates.

Take a look at Figure 5. Notice that for $x = 0, \pi, -\pi, 2\pi, -2\pi, \ldots$, the corresponding point on the unit circle has coordinates either $(1, 0)$ or $(-1, 0)$. In general, whenever $x = k\pi$ for some integer k, then $(a, b) = (\pm 1, 0)$ and $\tan x = b/a = 0/\pm 1 = 0$. We have just found the x intercepts of the graph of $y = \tan x$!

x **intercepts:** $x = k\pi$ k **an integer**

So as a first step in graphing $y = \tan x$, we locate the x intercepts with solid dots along the x axis, as indicated in Figure 6. (We'll explain the dashed vertical lines in a bit.)

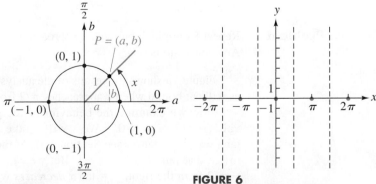

FIGURE 5

$y = \tan x = \dfrac{b}{a}$

FIGURE 6
x intercepts and vertical asymptotes for $y = \tan x$

Also, from Figure 5 we can see that whenever $P = (a, b)$ is on the vertical axis of the unit circle (that is, whenever $x = \pi/2 + k\pi$, k an integer), then $(a, b) = (0, \pm 1)$ and $\tan x = b/a = \pm 1/0$, which is not defined. This tells us that there can be no points plotted for these values of x. So, as a second step in graphing $y = \tan x$, we draw dashed vertical lines through each of these points on the x axis where $\tan x$ is not defined; the graph cannot touch these lines (see Fig. 6). These dashed lines will become guidelines, called *vertical asymptotes,* which are very helpful for sketching the graph of $y = \tan x$.

Vertical asymptotes: $\quad x = \dfrac{\pi}{2} + k\pi \qquad k$ **an integer**

We next investigate the behavior of the graph of $y = \tan x$ over the interval $0 \leq x < \pi/2$. Two points are easy to plot: $\tan 0 = 0$ and $\tan(\pi/4) = 1$. What happens to $\tan x$ as x approaches $\pi/2$ from the left? [Remember that $\tan(\pi/2)$ is not defined.] When x approaches $\pi/2$ from the left, $P = (a, b)$ approaches $(0, 1)$ and stays in the first quadrant. That is, a approaches 0 through positive values and b approaches 1. What happens to $y = \tan x$ in this process? In Example 2, we perform a calculator experiment that suggests an answer.

 **EXAMPLE 2**

Calculator Experiment

Form a table of values for $y = \tan x$ with x approaching $\pi/2 \approx 1.570\ 796$ from the left (through values less than $\pi/2$). Any conclusions?

Solution   We create a table as follows:

x	0	0.5	1	1.5	1.57	1.5707	1.570 796
tan x	0	0.5	1.6	14.1	1,256	10,381	3,060,023

Conclusion: As x approaches $\pi/2$ from the left, $y = \tan x$ appears to increase without bound. ∎

Matched Problem 2   Repeat Example 2, but with x approaching $-\pi/2 \approx -1.570\ 796$ from the right. Any conclusions? ∎

Figure 7a shows the results of the analysis in Example 2: $y = \tan x$ increases without bound when x approaches $\pi/2$ from the left.

Now we examine the behavior of the graph of $y = \tan x$ over the interval $-\pi/2 < x \leq 0$. We will take advantage of the identity $\tan(-x) = -\tan x$ (see Section 2.5). At the very least, we can use the identity to see that $\tan(-\pi/4) = -1$. But we can also conclude that as x approaches $-\pi/2$ from the right, $y = \tan x$ *decreases* without bound.

The result is the full graph over the interval $-\pi/2 < x < \pi/2$ shown in Figure 7b on the next page.

FIGURE 7
$y = \tan x$

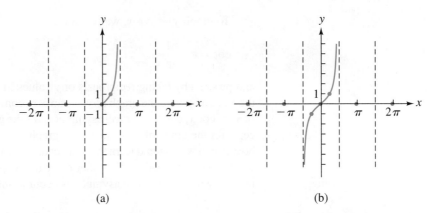

(a)   (b)

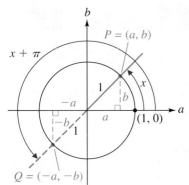

FIGURE 8
$\tan (x + \pi) = \tan x$

Proceeding in the same way for the other intervals between the asymptotes (the dashed vertical lines), it appears that the tangent function is periodic with period π. We confirm this as follows: If (a, b) are the coordinates of the point P associated with x (see Fig. 8), then $(-a, -b)$ are the coordinates of the point Q associated with $x + \pi$ (this is due to the symmetry of the unit circle and to congruent reference triangles). Consequently,

$$\tan(x + \pi) = \frac{-b}{-a} = \frac{b}{a} = \tan x$$

We conclude that the tangent function is periodic with period π. In general,

$$\tan(x + k\pi) = \tan x \qquad k \text{ an integer}$$

for all values of x for which both sides of the equation are defined.

Now, to complete as much of the general graph of $y = \tan x$ as we wish, all we need to do is to repeat the graph in Figure 7b to the left and to the right over intervals of π units. The main characteristics of the graph of $y = \tan x$ should be learned so that the graph can be sketched quickly. Most importantly, make note of the x intercepts and vertical asymptotes. The figure in the next box summarizes the preceding discussion.

GRAPH OF $y = \tan x$

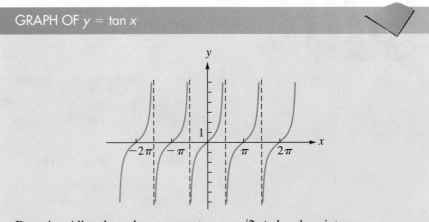

Domain: All real numbers x except $x = \pi/2 + k\pi$, k an integer
Range: R Period: π

To graph $y = \cot x$, we recall that

$$\cot x = \frac{1}{\tan x}$$

and proceed by taking reciprocals of y values in the graph of $y = \tan x$. Note that the x intercepts for the graph of $y = \tan x$ become vertical asymptotes for the graph of $y = \cot x$, and the vertical asymptotes for the graph of $y = \tan x$ become x intercepts for the graph of $y = \cot x$. The graph of $y = \cot x$ is shown in the following box. Again, you should learn the main characteristics of this function so that its graph can be sketched readily. Note that the x intercepts occur at $x = \pi/2 + k\pi$, k an integer, while the vertical asymptotes occur at integer multiples of π.

GRAPH OF $y = \cot x$

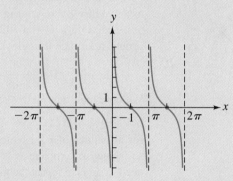

Domain: All real numbers x except $x = k\pi$, k an integer
Range: R Period: π

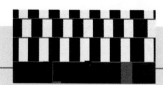

EXPLORE/DISCUSS 2

(A) Discuss how the graphs of the tangent and cotangent functions are related.

(B) Explain how the graph of one can be obtained from the graph of the other by horizontal shifts and/or reflections across an axis or through the origin.

■ Graphs of $y = \csc x$ and $y = \sec x$

Just as we obtained the graph of $y = \cot x$ by taking reciprocals of the y coordinates in the graph of $y = \tan x$, since

$$\csc x = \frac{1}{\sin x} \qquad \text{and} \qquad \sec x = \frac{1}{\cos x}$$

we can obtain the graphs of $y = \csc x$ and $y = \sec x$ by taking reciprocals of y coordinates in the respective graphs of $y = \sin x$ and $y = \cos x$.

The graphs of $y = \csc x$ and $y = \sec x$ are shown in the next two boxes. Note that, because of the reciprocal relationship, vertical asymptotes occur at the x intercepts of $\sin x$ and $\cos x$, respectively. It is helpful to first draw the graphs of $y = \sin x$ and $y = \cos x$ and then draw vertical asymptotes through the x intercepts.

GRAPH OF $y = \csc x$

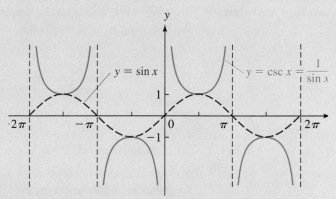

Domain: All real numbers x, except $x = k\pi$, k an integer
Range: All real numbers y such that $y \leq -1$ or $y \geq 1$
Period: 2π

GRAPH OF $y = \sec x$

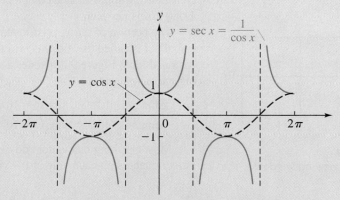

Domain: All real numbers x, except $x = \pi/2 + k\pi$, k an integer
Range: All real numbers y such that $y \leq -1$ or $y \geq 1$
Period: 2π

EXAMPLE 3   **Calculator Experiment**

To verify some points on the graph of $y = \csc x$, $0 < x < \pi$, make a table of values for $y = \csc x$. Use $x = 0.001, 0.01, 0.1, 1.57, 3.00, 3.13,$ and 3.14.

Solution

x	0.001	0.01	0.1	1.57	3.00	3.13	3.14
$\csc x$	1,000	100	10	1	7.1	86.3	628

Matched Problem 3   To verify some points on the graph of $y = \sec x$, $-\pi/2 < x < \pi/2$, complete the following table using a calculator:

x	-1.57	-1.56	-1.4	0	1.4	1.56	1.57
$\sec x$							

■ Graphing with a Graphing Calculator*

We determined the graphs of the six basic trigonometric functions by analyzing the behavior of the functions, exploiting relationships between the functions, and plotting only a few points. We refer to this process as curve sketching; one of the major objectives of this course is that you master this technique.

Graphing calculators also can be used to sketch graphs of functions; their accuracy depends on the screen resolution of the calculator. The smallest darkened rectangular area on the screen that the calculator can display is called a *pixel.* Most graphing calculators have a resolution of about 50 pixels per inch, which results in rough but useful sketches. Note that the graphs shown earlier in this section were created using sophisticated computer software and printed at a resolution of about 1,000 pixels per inch.

The portion of the xy coordinate plane displayed on the screen of a graphing calculator is called the **viewing window** and is determined by the **range** and **scale** for x and for y. Figure 9 illustrates a **standard viewing window** using the following range and scale:

$$\text{xmin} = -10 \qquad \text{xmax} = 10 \qquad \text{xscl} = 1$$
$$\text{ymin} = -10 \qquad \text{ymax} = 10 \qquad \text{yscl} = 1$$

Most graphing calculators do not display labels on the axes. We have added numeric labels on each side of the viewing window to make the graphs easier to

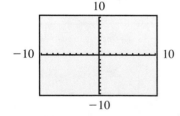

FIGURE 9
Standard viewing window

---

* Material that requires a graphing calculator is included in the text and exercise sets in this and subsequent chapters. This material is clearly identified with the icon shown in the margin. Any or all of the graphing calculator material may be omitted without loss of continuity. Treatments are generic in nature. If you need help with your specific calculator, refer to your user's manual.

read. Now we want to see what the graphs of the trigonometric functions will look like on a graphing calculator.

EXAMPLE 4

Trigonometric Graphs on a Graphing Calculator

Use a graphing calculator to graph the functions

$$y = \sin x \qquad y = \tan x \qquad y = \sec x$$

for $-2\pi \le x \le 2\pi, -5 \le y \le 5$. Display each graph in a separate viewing window.

Solution First, set the calculator to the radian mode. Most calculators remember this setting, so you should have to do this only once. Next, enter the following values:

$$\text{xmax} = 2\pi \qquad \text{xmax} = 2\pi \qquad \text{xscl} = 1$$
$$\text{ymin} = -5 \qquad \text{ymax} = 5 \qquad \text{yscl} = 1$$

This defines a viewing window ranging from -2π to 2π on the horizontal axis and from -5 to 5 on the vertical axis with tick marks one unit apart on each axis. Now enter the function $y = \sin x$ and draw the graph (see Fig. 10a). Repeat for $y = \tan x$ and $y = \sec x = 1/(\cos x)$ to obtain the graphs in Figures 10b and 10c. ∎

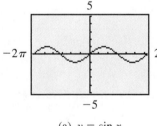

(a) $y = \sin x$

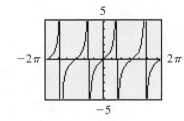

(b) $y = \tan x$

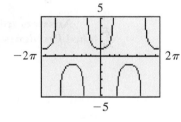

(c) $y = \sec x$

FIGURE 10
Graphing calculator graphs of trigonometric functions

Matched Problem 4 Repeat Example 4 for:

(A) $y = \cos x$ (B) $y = \cot x$ (C) $y = \csc x$ ∎

In Figure 10b, it appears that the calculator has drawn the vertical asymptotes for $y = \tan x$, but in Figure 10c they do not appear for $y = \sec x$. The difference is due to the fact that Figure 10b was drawn with a TI-83 and Figure 10c was drawn with a TI-84. Some calculator models will appear to draw vertical asymptotes, but this is not actually what they are doing. Most graphing calculators graph functions by calculating many points on a graph and connecting these points with line segments. The last point plotted to the left of the asymptote and the first plotted to the right of the asymptote will usually have very large *y* coordinates. If these *y* coordinates have opposite signs, then some calculators will connect the two points with a nearly vertical line segment, which gives the appearance of an asymptote. There is no harm in this as long as you understand that the calculator is not performing any analysis to identify asymptotes; it is simply connecting points with line segments. If the model you are using connects segments of the graph, as in Figure 10b, you can set the calculator in dot mode to plot the points without the connecting line segments, as illustrated in Figure 11.

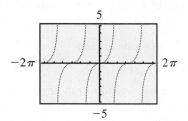

FIGURE 11
Graph of *y* = tan *x* in dot mode

Answers to Matched Problems

1.

2.

x	0	−0.5	−1	−1.5	−1.57	−1.5707	−1.570 796
tan *x*	0	−0.5	−1.6	−14.1	−1,256	−10,381	−3,060,023

Conclusion: As *x* approaches −π/2 from the right, *y* = tan *x* appears to decrease without bound.

3.

x	−1.57	−1.56	−1.4	0	1.4	1.56	1.57
sec *x*	1,256	92.6	5.9	1	5.9	92.6	1,256

4. (A) $y = \cos x$ (B) $y = \cot x$

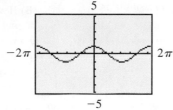
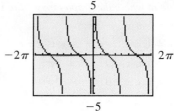

(C) $y = \csc x$

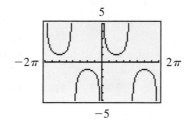

EXERCISE 3.1

A

1. What are the periods of the cosine, secant, and tangent functions?

2. What are the periods of the sine, cosecant, and cotangent functions?

3. Explain how to remember the x intercepts for the graph of $y = \sin x$ and $y = \cos x$ by thinking of the unit circle.

4. Explain how to remember the vertical asymptotes for the graph of $y = \tan x$ and $y = \cot x$ by thinking of the unit circle.

5. Explain how to obtain the graph of $y = \csc x$ from the graph of $y = \sin x$.

6. Explain how to obtain the graph of $y = \sec x$ from the graph of $y = \cos x$.

7. How far does the graph of each of the following functions deviate from the x axis?

(A) $\sin x$ (B) $\cot x$ (C) $\sec x$

8. How far does the graph of each of the following functions deviate from the x axis?

(A) $\cos x$ (B) $\tan x$ (C) $\csc x$

In Problems 9–14, what are the x intercepts for the graph of each function over the interval $-2\pi \le x \le 2\pi$?

9. $\cos x$ **10.** $\sin x$ **11.** $\tan x$

12. $\cot x$ **13.** $\sec x$ **14.** $\csc x$

B 15. For what values of x, $-2\pi \le x \le 2\pi$, are the following not defined?

(A) $\sin x$ (B) $\cot x$ (C) $\sec x$

16. For what values of x, $-2\pi \le x \le 2\pi$, are the following not defined?

(A) $\cos x$ (B) $\tan x$ (C) $\csc x$

17. Use a calculator and point-by-point plotting to produce an accurate graph of $y = \cos x$, $0 \le x \le 1.6$, using domain values 0, 0.1, 0.2, . . . , 1.5, 1.6.

18. Use a calculator and point-by-point plotting to produce an accurate graph of $y = \sin x$, $0 \le x \le 1.6$, using domain values 0, 0.1, 0.2, . . . , 1.5, 1.6.

In Problems 19–24, make a sketch of each trigonometric function without looking at the text or using a calculator. Label each point where the graph crosses the x axis in terms of π.

19. $y = \sin x$, $-2\pi \le x \le 2\pi$

20. $y = \cos x$, $-2\pi \le x \le 2\pi$

21. $y = \tan x$, $0 \le x \le 2\pi$

22. $y = \cot x$, $0 < x < 2\pi$

23. $y = \csc x$, $-\pi < x < \pi$

24. $y = \sec x$, $-\pi \le x \le \pi$

25. (A) With your graphing calculator set in radian mode, graph $y = \sin x$ on $-2\pi \le x \le 2\pi$, $-2 \le y \le 2$.

(B) Switch your calculator to degree mode and have it redraw the graph.

(C) What can you conclude about the importance of the mode setting in graphing trigonometric functions?

26. (A) With your graphing calculator set in radian mode, graph $y = \tan x$ on $-\pi/2 \le x \le \pi/2$, $-5 \le y \le 5$.

(B) Switch your calculator to degree mode and have it redraw the graph.

(C) What can you conclude about the importance of the mode setting in graphing trigonometric functions?

Problems 27–32 require the use of a graphing calculator. These problems provide a preliminary exploration into the relationships of the graphs of $y = A \sin x$, $y = A \cos x$, $y = \sin Bx$, $y = \cos Bx$, $y = \sin(x + C)$, and $y = \cos(x + C)$ relative to the graphs of $y = \sin x$ and $y = \cos x$. The topic is discussed in detail in the next section.

27. (A) Graph $y = A \sin x$ ($-2\pi \le x \le 2\pi$, $-3 \le y \le 3$) for $A = -2, 1, 3$, all in the same viewing window.

(B) Do the x intercepts change? If so, where?

(C) How far does each graph deviate from the x axis? Experiment with other values of A.

(D) Describe how the graph of $y = \sin x$ is changed by changing the values of A in $y = A \sin x$.

28. (A) Graph $y = A \cos x$ ($-2\pi \le x \le 2\pi$, $-3 \le y \le 3$) for $A = -3, 1, 2$, all in the same viewing window.

(B) Do the x intercepts change? If so, where?

(C) How far does each graph deviate from the x axis? Experiment with other values of A.

(D) Describe how the graph of $y = \cos x$ is changed by changing the values of A in $y = A \cos x$.

29. (A) Graph $y = \cos Bx$ ($-\pi \le x \le \pi, -2 \le y \le 2$) for $B = 1, 2, 3$, all in the same viewing window.

(B) How many periods of each graph appear in this viewing window? Experiment with additional positive values of B.

(C) Based on your experiments in part (B), how many periods of the graph of $y = \cos nx$, n a positive integer, would appear in this viewing window?

30. (A) Graph $y = \sin Bx$ ($-\pi \le x \le \pi, -2 \le y \le 2$) for $B = 1, 2, 3$, all in the same viewing window.

(B) How many periods of each graph appear in this viewing window? Experiment with additional positive values of B.

(C) Based on your experiments in part (B), how many periods of the graph of $y = \sin nx$, n a positive integer, would appear in this viewing window?

31. (A) Graph $y = \sin(x + C)$

($-2\pi \le x \le 2\pi$, $-2 \le y \le 2$) for $C = -\pi/2$, $0, \pi/2$, all in the same viewing window. Experiment with additional values of C.

(B) Describe how the graph of $y = \sin x$ is changed by changing the values of C in $y = \sin(x + C)$.

32. (A) Graph $y = \cos(x + C)$

($-2\pi \le x \le 2\pi$, $-2 \le y \le 2$) for $C = -\pi/2$, $0, \pi/2$, all in the same viewing window. Experiment with additional values of C.

(B) Describe how the graph of $y = \cos x$ is changed by changing the values of C in $y = \cos(x + C)$.

C **33.** Try to calculate each of the following on your calculator. Explain the problem.

(A) $\cot 0$ (B) $\tan(\pi/2)$ (C) $\csc \pi$

34. Try to calculate each of the following on your calculator. Explain the problem.

(A) $\tan(-\pi/2)$ (B) $\cot(-\pi)$ (C) $\sec(\pi/2)$

Problems 35–38 require the use of a graphing calculator.

35. In applied mathematics certain formulas, derivations, and calculations are simplified by replacing $\tan x$ with x for small x. What justifies this procedure? To find out, graph $y_1 = \tan x$ and $y_2 = x$ in the same viewing window for $-1 \le x \le 1$ and $-1 \le y \le 1$.

(A) What do you observe about the two graphs when x is close to 0, say $-0.5 \le x \le 0.5$?

(B) Complete the table to three decimal places (use the table feature on your graphing calculator if it has one):

x	-0.3	-0.2	-0.1	0.0	0.1	0.2	0.3
$\tan x$							

(C) Is it valid to replace $\tan x$ with x for small x if x is in degrees? Graph $y_1 = \tan x$ and $y_2 = x$ in the same viewing window with the calculator set in degree mode for $-45° \le x \le 45°$ and $-5 \le y \le 5$, and explain the results after exploring the graphs using $\boxed{\text{TRACE}}$.

36. In applied mathematics certain formulas, derivations, and calculations are simplified by replacing $\sin x$ with x for small x. What justifies this procedure? To find out, graph $y_1 = \sin x$ and $y_2 = x$ in the same viewing window for $-1 \le x \le 1$ and $-1 \le y \le 1$.

(A) What do you observe about the two graphs when x is close to 0, say $-0.5 \le x \le 0.5$?

(B) Complete the table to three decimal places (use the table feature on your graphing calculator if it has one):

x	-0.3	-0.2	-0.1	0.0	0.1	0.2	0.3
$\sin x$							

(C) Is it valid to replace $\sin x$ with x for small x if x is in degrees? Graph $y_1 = \sin x$ and $y_2 = x$ in the same viewing window with the calculator set in degree mode for $-10° \le x \le 10°$ and $-1 \le y \le 1$, and explain the results after exploring the graphs using $\boxed{\text{TRACE}}$.

37. Set your graphing calculator in radian and parametric (Par) modes. Make the entries as indicated in the figure to obtain the indicated graph (set Tmax and Xmax to 2π and set Xscl to $\pi/2$). The parameter T represents a central angle in the unit circle of T radians or an arc length of T on the unit circle starting at $(1, 0)$. As T moves from 0 to 2π, the point $P = (\cos T, \sin T)$ moves counterclockwise around the unit circle starting at $(1, 0)$ and ending at $(1, 0)$, and the point $P = (T, \sin T)$ moves along the sine curve from $(0, 0)$ to $(2\pi, 0)$. Use $\boxed{\text{TRACE}}$ to observe this behavior on each curve.

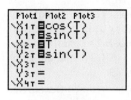

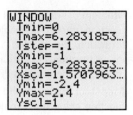

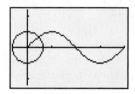

FIGURE FOR 37

Now use $\boxed{\text{TRACE}}$ and move back and forth between the unit circle and the graph of the sine function for various values of T as T increases from 0 to 2π. Discuss what happens in each case.

38. Repeat Problem 37 with $Y_{2T} = \cos T$.

In Problems 39–42, find the smallest positive number C that makes the statement true.

39. If the graph of the sine function is shifted C units to the right, it coincides with the graph of the cosine function.

40. If the graph of the cosine function is shifted C units to the right, it coincides with the graph of the sine function.

41. If the graph of the secant function is shifted C units to the left, it coincides with the graph of the cosecant function.

42. If the graph of the cosecant function is shifted C units to the left, it coincides with the graph of the secant function.

43. **Modeling Sunrise and Sunset Times** Table 1 on page 140 contains the sunrise and sunset times (using a 24-hour clock) from January 1, 2003, to December 1, 2004, in Milwaukee, Wisconsin. Plot the sunrise times by plotting the points $(1, 7.4)$, $(2, 7.1)$, . . ., $(24, 7.1)$, where 1 represents January 1, 2003, and 24 represents December 1, 2004. Connect the points with straight-line segments.

44. **Modeling Sunrise and Sunset Times** Refer to Problem 43. Plot the sunset times and connect the points with straight-line segments.

45. Modeling Sunrise and Sunset Times Denote the function graphed in Problem 43 by $a(x)$. Use the transformations discussed in Appendix B.2 to transform $a(x)$ into a function $f(x)$ with range $[-1, 1]$. Add the graph of $f(x)$ to the graph in Problem 43.

46. Modeling Sunrise and Sunset Times Denote the function graphed in Problem 44 by $b(x)$. Use the transformations discussed in Appendix B.2 to transform $b(x)$ into a function $g(x)$ with range $[-1, 1]$. Add the graph of $g(x)$ to the graph in Problem 44.

47. Modeling Sunrise and Sunset Times Refer to Problems 45 and 46. Interpret the function $d(x) = b(x) - a(x)$. Explain why the functions $a(x), b(x)$, and $d(x)$ are expected to be (approximately) periodic.

48. Modeling Sunrise and Sunset Times Show that if $f(x)$ and $g(x)$ are periodic functions with the same period p, then $h(x) = f(x) - g(x)$ is also periodic with period p.

TABLE 1		
Month	Sunrise	Sunset
1	7.4	16.4
2	7.1	17.1
3	6.5	17.7
4	5.6	18.3
5	4.8	18.9
6	4.2	19.4
7	4.3	19.6
8	4.7	19.2
9	5.3	18.4
10	5.8	17.6
11	6.4	16.7
12	7.0	16.3
13	7.4	16.4
14	7.1	17.1
15	6.4	17.7
16	5.6	18.3
17	4.8	18.9
18	4.2	19.4
19	4.3	19.6
20	4.7	19.2
21	5.3	18.4
22	5.8	17.5
23	6.4	16.7
24	7.1	16.3

3.2 Graphing $y = k + A \sin Bx$ and $y = k + A \cos Bx$

- Graphing $y = A \sin x$ and $y = A \cos x$
- Graphing $y = \sin Bx$ and $y = \cos Bx$
- Graphing $y = A \sin Bx$ and $y = A \cos Bx$
- Graphing $y = k + A \sin Bx$ and $y = k + A \cos Bx$
- Application: Sound Frequency
- Application: Floating Objects

Having completed Section 3.1, you should be familiar with the graphs of the basic functions $y = \sin x$ and $y = \cos x$. In this section we will look at graphs of more complicated trigonometric functions. In so doing, we will be able to accurately model real-world situations involving quantities that tend to repeat in cycles.

Fortunately, graphing equations of the form $y = k + A \sin Bx$ and $y = k + A \cos Bx$, where k, A, and B are real numbers, is not difficult if you have a clear understanding of the graphs of the basic equations $y = \sin x$ and $y = \cos x$ studied in Section 3.1.

EXPLORE/DISCUSS 1

Describe how the graph of y_2 is related to the graph of y_1 in each of the following graphing calculator displays.

(A)

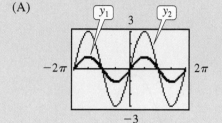

(B)

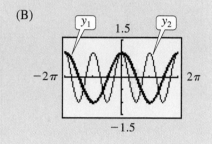

■ Graphing $y = A \sin x$ and $y = A \cos x$

We'll begin by studying the effect of multiplying the output of $y = \sin x$ by a real number A. The result depends on the particular value of A, as summarized in the table. (Transformations of this type are reviewed in Appendix B.2.)

A	How A in $y = A \sin x$ changes the graph of $y = \sin x$
$A > 1$	The graph of $y = \sin x$ is stretched vertically by a factor of A
$0 < A < 1$	The graph of $y = \sin x$ is shrunk vertically by a factor of A
$A < -1$	The graph of $y = \sin x$ is reflected in the x axis and then stretched vertically by a factor of $|A|$
$-1 < A < 0$	The graph of $y = \sin x$ is reflected in the x axis and then shrunk vertically by a factor of $|A|$

Note that any x intercept of $y = \sin x$ is also an x intercept of $y = A \sin x$, since multiplying an output of zero by the real number A will have no effect.

We know that the maximum deviation of the graph of $y = \sin x$ from the x axis is 1. Then the maximum deviation of the graph of $y = A \sin x$ from the x axis is $|A| \cdot 1 = |A|$. The constant $|A|$ is called the **amplitude** of the graph of $y = A \sin x$ and represents the maximum deviation of the graph from the x axis. Finally, the period of $y = A \sin x$ is also 2π, since $A \sin(x + 2\pi) = A \sin x$.

 **EXAMPLE 1**

Comparing Amplitudes

Compare the graphs of $y = \frac{1}{3}\sin x$ and $y = -3\sin x$ with the graph of $y = \sin x$ by graphing all three on the same coordinate system for $0 \le x \le 2\pi$.

Solution

We see that the graph of $y = \frac{1}{3}\sin x$ has an amplitude of $\left|\frac{1}{3}\right| = \frac{1}{3}$, the graph of $y = -3\sin x$ has an amplitude of $|-3| = 3$, and the graph of $y = \sin x$ has an amplitude of $|1| = 1$. The negative sign in $y = -3\sin x$ turns the graph of $y = 3\sin x$ upside down. That is, the graph of $y = -3\sin x$ is the same as the graph of $y = 3\sin x$ reflected in the x axis. The graphs of all three equations, $y = \frac{1}{3}\sin x$, $y = -3\sin x$, and $y = \sin x$, are shown in Figure 1.

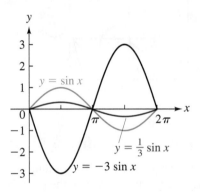

FIGURE 1
Comparing amplitudes

In summary, the results of Example 1 show that the effect of A in $y = A\sin x$ is to increase or decrease the y values of $y = \sin x$ without affecting the x values. A similar analysis applies to $y = A\cos x$; this function also has an amplitude of $|A|$ and a period of 2π.

Matched Problem 1

Compare the graphs of $y = \frac{1}{2}\cos x$ and $y = -2\cos x$ with the graph of $y = \cos x$ by graphing all three on the same coordinate system for $0 \le x \le 2\pi$.   ■

■ **Graphing $y = \sin Bx$ and $y = \cos Bx$**

We now examine the effect of B by comparing

$$y = \sin x \quad \text{and} \quad y = \sin Bx \quad B > 0$$

We know that $y = \sin x$ has a period of 2π, which means that the graph completes one full cycle as the input varies from 0 to 2π. For $y = \sin Bx$, the input is now Bx instead of x. So $y = \sin Bx$ completes one full cycle as Bx varies from

$$Bx = 0 \quad \text{to} \quad Bx = 2\pi$$

or as x varies from

$$x = \frac{0}{B} = 0 \quad \text{to} \quad x = \frac{2\pi}{B}$$

We can conclude that the period of sin Bx is $2\pi/B$. We can check this result as follows: If $f(x) = \sin Bx$, then

$$f\left(x + \frac{2\pi}{B}\right) = \sin\left[B\left(x + \frac{2\pi}{B}\right)\right] = \sin(Bx + 2\pi) = \sin Bx = f(x)$$

EXAMPLE 2

Comparing Periods

Compare the graphs of $y = \sin 2x$ and $y = \sin(x/2)$ with the graph of $y = \sin x$ by graphing all three on the same coordinate system for one period starting at the origin.

Solution   *Period for* sin $2x$:   sin $2x$ completes one cycle as $2x$ varies from

$$2x = 0 \quad \text{to} \quad 2x = 2\pi$$

or as x varies from

$$x = 0 \quad \text{to} \quad x = \pi$$

So the period for sin $2x$ is π.

Period for $\sin(x/2)$: $\sin(x/2)$ completes one cycle as $x/2$ varies from

$$\frac{x}{2} = 0 \quad \text{to} \quad \frac{x}{2} = 2\pi$$

or as x varies from

$$x = 0 \quad \text{to} \quad x = 4\pi$$

So, the period for $\sin(x/2)$ is 4π. Note that the output of all three functions varies from -1 to 1, which means that all three have amplitude 1. The graphs of all three equations, $y = \sin 2x$, $y = \sin(x/2)$, and $y = \sin x$, are shown in Figure 2.

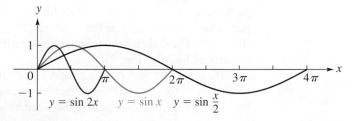

FIGURE 2
Comparing periods

Matched Problem 2   Compare the graphs of $y = \cos 2x$ and $y = \cos(x/2)$ with the graph of $y = \cos x$ by graphing all three on the same coordinate system for one period starting at the origin.   ■

We can see from Example 2 that for $y = \sin Bx$, if $B > 1$, the effect of B is to shrink the graph of $y = \sin x$ horizontally by a factor of $1/B$. If $0 < B < 1$, the effect of B is to stretch the graph of $y = \sin x$ horizontally by a factor of $1/B$. In either case, the period of $y = \sin x$ is $2\pi/B$. A similar analysis applies to the graph of $y = \cos Bx$; its period is also $2\pi/B$.

■ Graphing $y = A \sin Bx$ and $y = A \cos Bx$

We summarize the results of the discussions of amplitude and period in the following box.

AMPLITUDE AND PERIOD

For $y = A \sin Bx$ or $y = A \cos Bx$, $B > 0$:

$$\text{Amplitude} = |A| \qquad \text{Period} = \frac{2\pi}{B}$$

If $B > 1$, the basic sine or cosine curve is horizontally compressed.
If $0 < B < 1$, the basic sine or cosine curve is horizontally stretched.

We will now consider several examples where we show how graphs of $y = A \sin Bx$ and $y = A \cos Bx$ can be sketched rather quickly.

EXAMPLE 3

Graphing the Form $y = A \cos Bx$

State the amplitude and period for $y = 3 \cos 2x$, and graph the equation for $-\pi \le x \le 2\pi$.

Solution $\text{Amplitude} = |A| = |3| = 3 \qquad \text{Period} = \dfrac{2\pi}{B} = \dfrac{2\pi}{2} = \pi$

To sketch the graph, divide the interval of one period, from 0 to π, into four equal parts, locate x intercepts, and locate high and low points (see Fig. 3a on the next page). Since this is a cosine curve with $A > 0$, the graph is at its maximum value for $x = 0$. Then, sketch the graph for one period, and extend this graph to fill out the desired interval (Fig. 3b).   ■

The graph of $y = 3 \cos 2x$ is the result of stretching the graph of $y = \cos x$ vertically by a factor of 3 and then shrinking it horizontally by a factor of $1/2$. Notice that we scaled the x axis using the period divided by 4; that is, the basic unit on the x axis is $\pi/4$. Also, we adjusted the scale on the y axis to accommodate the amplitude 3.

The scales on both axes do not have to be the same.

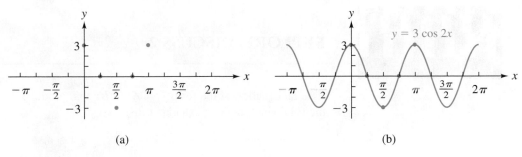

(a)   (b)

FIGURE 3

Matched Problem 3   State the amplitude and period for $y = \frac{1}{3} \sin(x/2)$, and graph the equation for $-2\pi \leq x \leq 6\pi$.   ■

EXAMPLE 4   **Graphing the Form $y = A \sin Bx$**

State the amplitude and period for $y = -\frac{1}{2} \sin(\pi x/2)$ and graph the equation for $-5 \leq x \leq 5$.

Solution   Amplitude $= |A| = \left|-\frac{1}{2}\right| = \frac{1}{2}$ Period $= \dfrac{2\pi}{B} = \dfrac{2\pi}{\pi/2} = 4$

Because of the factor $-\frac{1}{2}$, the graph of $y = -\frac{1}{2} \sin(\pi x/2)$ is the graph of $y = \frac{1}{2} \sin(\pi x/2)$ reflected through the x axis (turned upside down). As before, we divide one period, from 0 to 4, into four equal parts, locate x intercepts, and locate high and low points (see Fig. 4a). Since this is a sine curve, the graph is at height zero for $x = 0$. Then we graph the equation for one period and extend the graph over the desired interval (see Fig. 4b).

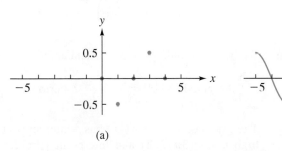

(a)   (b)

FIGURE 4   ■

Matched Problem 4   State the amplitude and period for $y = -2 \cos 2\pi x$, and graph the equation for $-2 \leq x \leq 2$.   ■

EXPLORE/DISCUSS 2

Find an equation of the form $y = A \sin Bx$ that produces the graph shown in the following graphing calculator display:

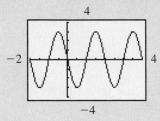

Is it possible for an equation of the form $y = A \cos Bx$ to produce the same graph? Explain.

■ **Graphing $y = k + A \sin Bx$ and $y = k + A \cos Bx$**

By adding a constant k to either $A \sin Bx$ or $A \cos Bx$, we are simply adding k to the y coordinates of the points on their graphs. That is, we are **vertically translating** the graphs of $y = A \sin Bx$ and $y = A \cos Bx$ up k units if $k > 0$ or down $|k|$ units if $k < 0$.

Graphing the Form $y = k + A \sin Bx$

Graph: $y = 4 + 2 \sin\left(\dfrac{1}{3}x\right), \quad -12\pi \le x \le 6\pi$

Solution   We begin by finding the period and amplitude, which are unaffected by $k = 4$.

$$\text{Amplitude} = |A| = 2 \qquad \text{Period} = \frac{2\pi}{B} = \frac{2\pi}{1/3} = 6\pi$$

The graph of $y = 2 \sin\left(\frac{1}{3}x\right)$ has x intercepts at $x = 0$, 3π, and 6π. It also has high point $(3\pi/2, 2)$ and low point $(9\pi/2, -2)$. The effect of $k = 4$ is to translate every point up by 4 units. This moves the x intercepts to height 4, the high points to height 6, and the low points to height 2 (see Fig. 5a). We can then connect these points to draw one period of the graph and extend to the rest of the interval $-12\pi \le x \le 6\pi$ (Fig. 5b). The horizontal dashed line at $y = 4$ was drawn to help you visualize the result of translating $y = 2 \sin\left(\frac{1}{3}x\right)$ up by 4 units.

FIGURE 5

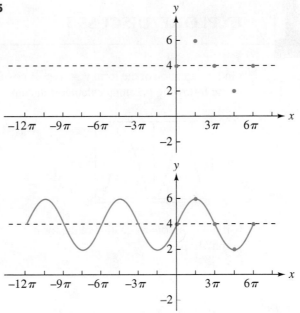

Matched Problem 5   Graph: $y = -1 - \sin \pi x,\ -4 \le x \le 4$

 **EXAMPLE 6**

Graphing the Form $y = k + A \cos Bx$

Graph: $y = -2 + 3\cos 2x,\quad -\pi \le x \le 2\pi$

Solution   The -2 indicates that the graph of $y = 3\cos 2x$ is translated 2 units down. So we first graph $y = 3\cos 2x$ (as we did in Example 3) and then move the graph down 2 units, as shown in Figure 6.

FIGURE 6

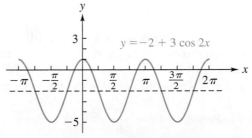

$$y = -2 + 3\cos 2x$$

Again, you may find it helpful to first draw the horizontal dashed line shown in Figure 6. In this case, the line is 2 units below the x axis, which represents a vertical translation of -2. Then the graph of $y = 3\cos 2x$ is drawn relative to the dashed line and the original y axis. Note that the high and low points are 3 units above and below the dashed line, respectively, because $A = 3$.

Matched Problem 6   Graph: $y = 3 - 2\cos 2\pi x,\quad -2 \le x \le 2$

EXPLORE/DISCUSS 3

Find an equation of the form $y = k + A \cos Bx$ that produces the graph shown in the following graphing calculator display:

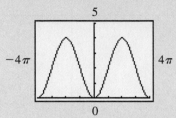

Is it possible for an equation of the form $y = k + A \sin Bx$ to produce the same graph? Explain.

EXAMPLE 7   **Finding the Equation for a Graph**

Find an equation of the form $y = A \sin Bx$ that produces the graph in Figure 7.

FIGURE 7

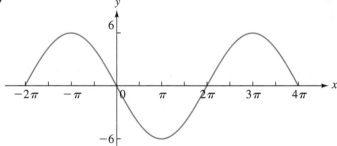

Solution   The amplitude is 6, so we know that $|A| = 6$. Also, the graph decreases on the first quarter-period to the right of $x = 0$: this looks like an upside-down sine curve, so A is negative, and $A = -6$. One full cycle is completed between $x = 0$ and $x = 4\pi$, so the period is 4π. We can use this fact to find B:

$$\text{Period} = 4\pi = \frac{2\pi}{B} \qquad \text{\textit{Multiply both sides by B.}}$$

$$4\pi B = 2\pi \qquad \text{\textit{Divide both sides by 4π.}}$$

$$B = \frac{2\pi}{4\pi} = \frac{1}{2}$$

The equation is $y = -6 \sin(\frac{1}{2}x)$. You may want to check this answer by graphing it using a graphing calculator.

Matched Problem 7   Find an equation of the form $y = A \cos Bx$ that produces the graph in Figure 8.

FIGURE 8

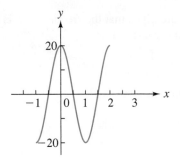

■ Application: Sound Frequency

In many applications involving periodic phenomena and time (sound waves, pendulum motion, water waves, electromagnetic waves, and so on), we speak of the **period** as the length of time taken for one complete cycle of motion. For example, since each complete vibration of an A440 tuning fork (see Fig. 9) lasts $\frac{1}{440}$ sec, we say that its period is $\frac{1}{440}$ sec.

Closely related to the period is the concept of **frequency,** which is the number of periods or cycles per second. The frequency of the A440 tuning fork is 440 cycles/sec, which is the reciprocal of the period. Instead of "cycles per second," we usually write "Hz," where Hz (read "hertz") is the standard unit for frequency (cycles per unit of time) and is named after the German physicist Heinrich Rudolph Hertz (1857–1894), who discovered and produced radio waves.

FIGURE 9
Tuning fork

PERIOD AND FREQUENCY

For any periodic phenomenon, if P is the period and f is the frequency, then:

$$P = \frac{1}{f} \qquad \text{Period is the time for one complete cycle.}$$

$$f = \frac{1}{P} \qquad \text{Frequency is the number of cycles per unit of time.}$$

EXAMPLE 8

An Equation of a Sound

Referring to the tuning fork in Figure 9, suppose that we want to model the motion of the tip of one prong using a sine or cosine function. We will choose deviation from the rest position to the right as positive and to the left as negative. If the prong motion has a frequency of 440 Hz (cycles/sec), an amplitude of 0.04 cm, and is 0.04 cm to the right when $t = 0$, find A and B so that $y = A \cos Bt$ is an approximate model for this motion.

Solution   *Find A.*   The amplitude $|A|$ is given to be 0.04. Since $y = 0.04$ when $t = 0$, $A = 0.04$ (and not -0.04).

Find B.   We are given that the frequency, f, is 440 Hz, so we can find the period using the reciprocal formula:

$$P = \frac{1}{f} = \frac{1}{440} \text{ sec}$$

From the earlier discussion, $P = 2\pi/B$, so, solving for B, we get

$$B = \frac{2\pi}{P} = \frac{2\pi}{\frac{1}{440}} = 880\pi$$

Write the equation:

$$y = A \cos Bt = 0.04 \cos 880\pi t$$

✔ Check   Amplitude $= |A| = |0.04| = 0.04$

$$\text{Period} = \frac{2\pi}{B} = \frac{2\pi}{880\pi} = \frac{1}{440} \text{ sec}$$

$$\text{Frequency} = \frac{1}{\text{Period}} = \frac{1}{\frac{1}{440}} = 440 \text{ Hz}$$

And when $t = 0$,

$$y = 0.04 \cos(880\pi \cdot 0)$$
$$= 0.04(1) = 0.04$$ ■

Matched Problem 8   Repeat Example 8 if the prong motion has a frequency of 262 Hz (cycles/sec), has an amplitude of 0.065 cm, and is 0.065 cm to the left of the rest position when $t = 0$.   ■

■ **Application: Floating Objects**

Did you know that it is possible to determine the mass of a floating object (Fig. 10) simply by making it bob up and down in the water and timing its period of oscillation?

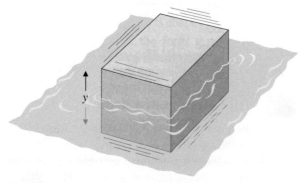

FIGURE 10
Floating object

If the rest position of the floating object is taken to be 0 and we start counting time as the object passes up through 0 when it is made to oscillate, then its equation of motion can be shown to be (neglecting water and air resistance)

$$y = D \sin \sqrt{\frac{1{,}000 g A}{M}}\, t$$

where y and D are in meters, t is time in seconds, $g = 9.75$ m/sec^2 (gravitational constant), A is the horizontal cross-sectional area in square meters, and M is mass in kilograms. The amplitude and period for the motion are given by

$$\text{Amplitude} = |D| \qquad \text{Period} = \frac{2\pi}{\sqrt{1{,}000 g A / M}}$$

EXAMPLE 9

Mass of a Buoy

A cylindrical buoy with cross-sectional area 1.25 m^2 is observed (after being pushed) to bob up and down with a period of 0.50 sec. Approximately what is the mass of the buoy (in kilograms)?

Solution We use the formula

$$\text{Period} = \frac{2\pi}{\sqrt{1{,}000 g A / M}}$$

with $g = 9.75$ m/sec^2, $A = 1.25$ m^2, and period $= 0.50$ sec, and solve for M. Thus,

$$0.50 = \frac{2\pi}{\sqrt{(1{,}000)(9.75)(1.25)/M}} \qquad \text{Simplify radical}$$

$$0.50 = \frac{2\pi}{\sqrt{12{,}187.5 / M}} \qquad \text{Multiply both sides by radical}$$

$$0.50\sqrt{12{,}187.5 / M} = 2\pi \qquad \text{Multiply both sides by 2}$$

$$\sqrt{12{,}187.5 / M} = 4\pi \qquad \text{Square both sides}$$

$$\frac{12{,}187.5}{M} = 16\pi^2 \qquad \text{Solve for M.}$$

$$M \approx 77 \text{ kg} \qquad\blacksquare$$

Matched Problem 9 A larger buoy with twice the cross-sectional area as the one in Example 9 bobs up and down with a period of 0.7 sec. Find its mass to the nearest kilogram. $\blacksquare$

Answers to
Matched Problems

1.
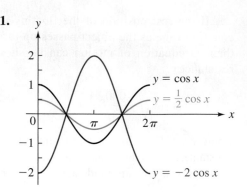

2.
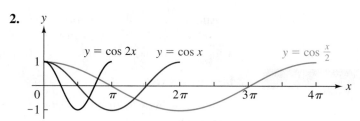

3. Amplitude = 1/3; Period = 4π

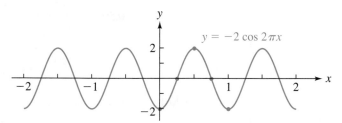

4. Amplitude = 2; Period = 1

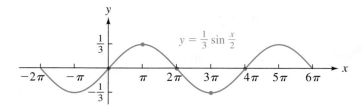

5.
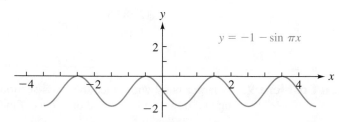

6.

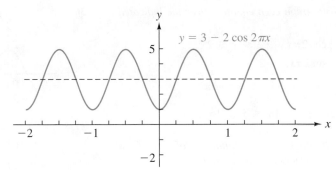

$$y = 3 - 2 \cos 2\pi x$$

7. $y = 20 \cos \pi x$ **8.** $y = -0.065 \cos 524\pi t$ **9.** 303 kg

EXERCISE 3.2

 1. Explain why the amplitude of $y = A \cos x$ is $|A|$.

2. How can you find the period of a function of the form $y = A \sin Bx$? Why does that procedure work?

3. For functions of the form $y = k + A \sin Bx$, does k affect the amplitude? The period? Explain.

4. Does every function of the form $y = A \cos Bx$ have a period that is a multiple of π? Explain.

In Problems 5 and 6, make a sketch of each trigonometric function. Try not to look at the text or use a calculator. Label each point where the graph crosses the x axis.

5. $y = \sin x$, $-2\pi \leq x \leq 2\pi$

6. $y = \cos x$, $-2\pi \leq x \leq 2\pi$

In Problems 7–14, state the amplitude and period for each equation and graph it over the indicated interval.

7. $y = -2 \sin x$, $0 \leq x \leq 4\pi$

8. $y = -3 \cos x$, $0 \leq x \leq 4\pi$

9. $y = \frac{1}{2} \sin x$, $0 \leq x \leq 2\pi$

10. $y = \frac{1}{3} \cos x$, $0 \leq x \leq 2\pi$

11. $y = \sin 2\pi x$, $-2 \leq x \leq 2$

12. $y = \cos 4\pi x$, $-1 \leq x \leq 1$

13. $y = \cos \frac{x}{4}$, $0 \leq x \leq 8\pi$

14. $y = \sin \frac{x}{2}$, $0 \leq x \leq 4\pi$

B *In Problems 15–20, state the amplitude and period for each equation and graph it over the indicated interval.*

15. $y = 2 \sin 4x$, $-\pi \leq x \leq \pi$

16. $y = 3 \cos 2x$, $-\pi \leq x \leq \pi$

17. $y = \frac{1}{3} \cos 2\pi x$, $-2 \leq x \leq 2$

18. $y = \frac{1}{2} \sin 2\pi x$, $-2 \leq x \leq 2$

19. $y = -\frac{1}{4} \sin \frac{x}{2}$, $-4\pi \leq x \leq 4\pi$

20. $y = -3 \cos \frac{x}{2}$, $-4\pi \leq x \leq 4\pi$

In Problems 21–24, y is the displacement of an oscillating object from a central position at time t. For each problem, find an equation of the form $y = A \sin Bt$ or $y = A \cos Bt$ that satisfies the given conditions.

21. Displacement from the t axis is 0 ft when t is 0, amplitude is 3 ft, and period is 4 sec.

22. Displacement from the t axis is 7 cm when t is 0, amplitude is 7 cm, and period is 0.4 sec.

23. Displacement from the t axis is 9 m when t is 0, amplitude is 9 m, and period is 0.2 sec.

24. Displacement from the t axis is 0 ft when t is 0, amplitude is 5 ft, and period is 2 sec.

25. Describe what happens to the size of the period of $y = A \sin Bx$ as B increases without bound.

26. Describe what happens to the size of the period of $y = A \cos Bx$ as B decreases through positive values toward 0.

In Problems 27–30, graph each equation over the indicated interval.

27. $y = -1 + \frac{1}{3} \cos 2\pi x$, $-2 \le x \le 2$

28. $y = -\frac{1}{2} + \frac{1}{2} \sin 2\pi x$, $-2 \le x \le 2$

29. $y = 2 - \frac{1}{4} \sin \frac{x}{2}$, $-4\pi \le x \le 4\pi$

30. $y = 3 - 3 \cos \frac{x}{2}$, $-4\pi \le x \le 4\pi$

In Problems 31–34, find the equation of the form $y = A \sin Bx$ that produces the given graph.

31.

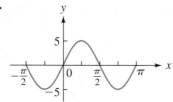

32.

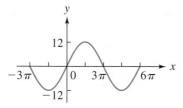

33.

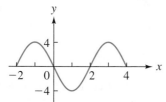

34.

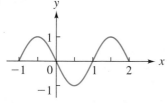

In Problems 35–38, find the equation of the form $y = A \cos Bx$ that produces the given graph.

35.

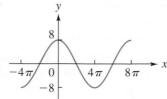

36.

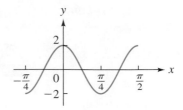

37.

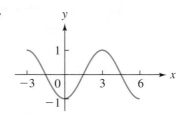

38.

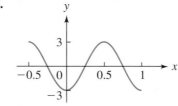

In Problems 39–44, graph the given equation on a graphing calculator. (Adjust the ranges in the viewing windows so that you see at least two periods of a particular function.) Find an equation of the form $y = k + A \sin Bx$ or $y = k + A \cos Bx$ that has the same graph. These problems suggest the existence of further identities in addition to the basic identities discussed in Section 2.5. Further identities are discussed in detail in Chapter 4.

39. $y = \sin x \cos x$ **40.** $y = \cos^2 x - \sin^2 x$

41. $y = 2 \cos^2 x$ **42.** $y = 2 \sin^2 x$

43. $y = 2 - 4 \sin^2 2x$ **44.** $y = 6 \cos^2 \frac{x}{2} - 3$

Problems 45 and 46 graphically explore the relationships of the graphs of $y = \sin(x + C)$ and $y = \cos(x + C)$ relative to the corresponding graphs of $y = \sin x$ and $y = \cos x$. This topic is discussed in detail in the next section.

45. (A) Graph $y = \sin(x + C)$, $-2\pi \le x \le 2\pi$, for $C = 0$ and $C = -\pi/2$ in one viewing window, and for $C = 0$ and $C = \pi/2$ in another viewing window. (Experiment with other positive and negative values of C.)

(B) Based on the graphs in part (A), describe how the graph of $y = \sin(x + C)$ is related to the graph of $y = \sin x$ for various values of C.

46. (A) Graph $y = \cos(x + C)$, $-2\pi \le x \le 2\pi$, for $C = 0$ and $C = -\pi/2$ in one viewing window, and for $C = 0$ and $C = \pi/2$ in another viewing window. (Experiment with other positive and negative values of C.)

(B) Based on the graphs in part (A), describe how the graph of $y = \cos(x + C)$ is related to the graph of $y = \cos x$ for various values of C.

C **47.** From the graph of $f(x) = \cos^2 x$, $-2\pi \le x \le 2\pi$, on a graphing calculator, determine the period of f.

48. From the graph of $f(x) = \sin^2 x$, $-2\pi \le x \le 2\pi$, on a graphing calculator, determine the period of f.

49. The table in the figure was produced using a table feature for a particular graphing calculator. Find a function of the form $y = A \sin Bx$ or $y = A \cos Bx$ that will produce the table.

X	Y₁	
0.0	2.0	
.5	1.0	
1.0	-1.0	
1.5	-2.0	
2.0	-1.0	
2.5	1.0	
3.0	2.0	
X=0		

50. The table in the figure was produced using a table feature for a particular graphing calculator. Find a function of the form $y = A \sin Bx$ or $y = A \cos Bx$ that will produce the table.

X	Y₁	
0.0	0.0	
1.0	-3.0	
2.0	0.0	
3.0	3.0	
4.0	0.0	
5.0	-3.0	
6.0	0.0	
X=0		

Applications

In these applications, assume all given values are exact unless indicated otherwise.

The following table gives the lowest and highest frequencies of five common sound sources. In Problems 51–54, use these frequencies to find the time it takes each sound to complete one cycle.

51. Sound   The lowest frequency of human hearing.

52. Sound   The highest frequency of human hearing.

Source	Lowest frequency (Hz)	Highest frequency (Hz)
Piano	28	4,186
Female speech	140	500
Male speech	80	240
Compact disc	0	22,050
Human hearing	20	20,000

53. Sound   The highest frequency of female speech.

54. Sound   The lowest frequency of male speech.

55. Electrical Circuits   The alternating voltage E in an electrical circuit is given by $E = 110 \sin 120\pi t$, where t is time in seconds. What are the amplitude and period of the function? What is the frequency of the function? Graph the function for $0 \le t \le \frac{3}{60}$.

56. Spring–Mass System   The equation $y = -4 \cos 8t$, where t is time in seconds, represents the motion of a weight hanging on a spring after it has been pulled 4 cm below its equilibrium point and released (see the figure). What are the amplitude, period, and frequency of the function? [Air resistance and friction (damping forces) are neglected.] Graph the function for $0 \le t \le 3\pi/4$.

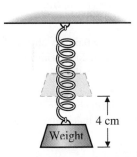

4 cm

Weight

Figure for 56

57. Electrical Circuits   If the alternating voltage E in an electrical circuit has an amplitude of 12 V and a frequency of 40 Hz, and if $E = 12$ V when $t = 0$ sec, find an equation of the form $E = A \cos Bt$ that gives the voltage at any time t.

58. Spring–Mass System   If the motion of the weight in Problem 56 has an amplitude of 6 in. and a frequency of 2 Hz, and if its position when $t = 0$ sec is 6 in. above its position at rest (above the rest position is positive and below is negative), find an equation of the form $y = A \cos Bt$ that describes the motion at any time t. (Neglect any damping forces—that is, air resistance and friction.)

59. Floating Objects A shipping company requires chemicals to be shipped in watertight cylinders with a mass of not more than 20 kg. A chemical company submits containers with a cross-sectional area of 0.4 sq m, and these containers are found to bob up and down with a period of 0.5 sec when placed into a testing tank. Are the containers within the 20 kg limit?

60. Floating Objects A professional stuntman plans a trip over Niagara Falls in a fiberglass barrel. An engineer calculates that a combined mass (stuntman plus barrel) over 95 kg could result in a disastrous failure of the barrel, which has a cross-sectional area of 1.4 sq m. In a test, the barrel with the stuntman in it bobs up and down with a period of 0.46 sec. Will our friend the stuntman be safe?

61. Floating Objects

(A) A 3 m $\times$ 3 m $\times$ 1 m float in the shape of a rectangular solid is observed to bob up and down with a period of 1 sec. What is the mass of the float (in kilograms, to three significant digits)?

(B) Write an equation of motion for the float in part (A) in the form $y = D \sin Bt$, assuming the amplitude of the motion is 0.2 m.

(C) Graph the equation found in part (B) for $0 \le t \le 2$.

62. Floating Objects

(A) A cylindrical buoy with diameter 0.6 m is observed (after being pushed) to bob up and down with a period of 0.4 sec. What is the mass of the buoy (to the nearest kilogram)?

(B) Write an equation of motion for the buoy in the form $y = D \sin Bt$, assuming the amplitude of the motion is 0.1 m.

(C) Graph the equation for $0 \le t \le 1.2$.

63. Physiology A normal seated adult breathes in and exhales about 0.80 L of air every 4.00 sec. The volume of air $V(t)$ in the lungs (in liters) t seconds after exhaling is modeled approximately by

$$V(t) = 0.45 - 0.40 \cos \frac{\pi t}{2} \qquad 0 \le t \le 8$$

(A) What is the maximum amount of air in the lungs, and what is the minimum amount of air in the lungs? Explain how you arrived at these numbers.

(B) What is the period of breathing?

(C) How many breaths are taken per minute? Show your computation.

(D) Graph the equation on a graphing calculator for $0 \le t \le 8$ and $0 \le V \le 1$. Find the maximum and minimum volumes of air in the lungs from the graph. [Compare with part (A).]

64. Physiology When you have your blood pressure measured during a physical examination, the nurse or doctor will record two numbers as a ratio—for example, 120/80. The first number (systolic pressure) represents the amount of pressure in the blood vessels when the heart contracts (beats) and pushes blood through the circulatory system. The second number (diastolic pressure) represents the pressure in the blood vessels at the lowest point between the heartbeats, when the heart is at rest. According to the National Institutes of Health, normal blood pressure is below 120/80, moderately high is from 140/90 to 159/99, and severe is higher than 160/100. The blood pressure of a particular person is modeled approximately by

$$P = 135 + 30 \cos 2.5\pi t \qquad t \ge 0$$

where P is pressure in millimeters of mercury t seconds after the heart contracts.

(A) Express the person's blood pressure as an appropriate ratio of two numbers. Explain how you arrived at these numbers.

(B) What is the period of the heartbeat?

(C) What is the pulse rate in beats per minute? Show your computation.

(D) Graph the equation using a graphing calculator for $0 \le t \le 4$, $100 \le P \le 170$. Explain how you would find the person's blood pressure as a ratio using the graph. Find the person's blood pressure using this method. [Compare with part (A).]

65. Rotary and Linear Motion A Ferris wheel with a diameter of 40 m rotates counterclockwise at 4 rpm (revolutions per minute), as indicated in the figure on the next page. It starts at $\theta = 0$ when time t is 0. At the end of t minutes, $\theta = 8\pi t$. Why? Explain why the position of the person's shadow (from the sun directly overhead) on the x axis is given by

$$x = 20 \sin 8\pi t$$

Graph this equation for $0 \le t \le 1$.

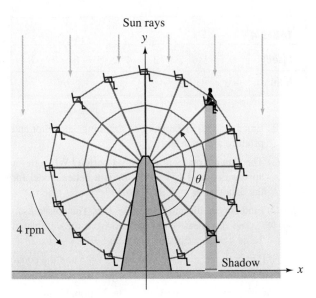

Sun rays

4 rpm

Shadow

Figure for 65 and 66

66. Rotary and Linear Motion Refer to Problem 65. Find the equation for the position of the person's shadow on the x axis for a Ferris wheel with a diameter of 30 m rotating at 6 rpm.

67. Bioengineering For a walking person (see the figure), a leg rotates back and forth relative to the hip socket through an angle θ. The angle θ is measured relative to the vertical, and is negative if the leg is behind the vertical and positive if the leg is in front of the vertical. Measurements of θ are calculated for a particular person taking one step per second, where t is time in seconds. The results for the first 4 sec are shown in Table 1.

Figure for 67 and 68

(A) Verbally describe how you can find the period and amplitude of the data in Table 1, and find each.

(B) Which of the equations, $\theta = A \cos Bt$ or $\theta = A \sin Bt$, is the more suitable model for the data in Table 1? Why? Find the more suitable equation.

(C) Plot the points in Table 1 and the graph of the equation found in part (B), in the same coordinate system.

68. Bioengineering Repeat Problem 67 for upper-arm oscillation, using the data in Table 2.

TABLE 1

Leg Oscillation

t (sec)	0.0	0.5	1.0	1.5	2.0	2.5	3.0	3.5	4.0
θ (deg)	$-25°$	$0°$	$25°$	$0°$	$-25°$	$0°$	$25°$	$0°$	$-25°$

TABLE 2

Upper-Arm Oscillation

t (sec)	0.0	0.5	1.0	1.5	2.0	2.5	3.0	3.5	4.0
θ (deg)	$0°$	$18°$	$0°$	$-18°$	$0°$	$18°$	$0°$	$-18°$	$0°$

Problems 69 and 70 require a graphing calculator that can produce scatter plots.

69. Engineering—Data Analysis A mill has a paddle wheel 30 ft in diameter that rotates counterclockwise at 5 rpm (see the figure). The bottom of the wheel is 2 ft below the surface of the water in the mill chase. We are interested in finding a function that will give the height h (in feet) of the point P above (or below) the water t seconds after P is at the top of the wheel.

Figure for 69

(A) Complete Table 3 using the information provided above:

TABLE 3

t (sec)	0	3	6	9	12	15	18	21	24
h (ft)			-2	13					

(B) Enter the data in Table 3 in a graphing calculator and produce a scatter plot.

(C) The data are periodic with what period? Why does it appear that $h = k + A \cos Bt$ is a better model for the data than $h = k + A \sin Bt$?

(D) Find the equation to model the data. The constants k, A, and B are easily determined from the completed Table 3 as follows:

$$|A| = (\text{Max } h - \text{Min } h)/2$$
$$B = 2\pi/\text{Period}$$
$$k = |A| + \text{Min } h$$

Refer to the scatter plot to determine the sign of A.

(E) Plot both the scatter plot from part (B) and the equation from part (D) in the same viewing window.

70. Engineering—Data Analysis The top of a turning propeller of a large, partially loaded ship protrudes 3 ft above the water as shown in the figure. The propeller is 12 ft in diameter and rotates clockwise at 15 rpm. We are interested in finding a function that will give the height h (in feet) of the point P above (or below) the water t seconds after P is at the bottom of the circle of rotation.

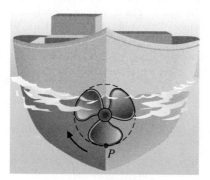

Figure for 70

(A) Complete Table 4 using the information provided above.

TABLE 4

t (sec)	0	1	2	3	4	5	6	7	8
h (ft)	−9		3						

(B) Enter the data in Table 4 in a graphing calculator and produce a scatter plot.

(C) The data are periodic with what period? Why does it appear that $h = k + A \cos Bt$ is a better model for the data than $h = k + A \sin Bt$?

(D) Find the equation to model the data. The constants k, A, and B are easily determined from the completed Table 4 as described in Problem 69D.

(E) Plot both the scatter plot from part (B) and the equation from part (D) in the same viewing window.

71. Modeling Sea Temperatures The National Data Buoy Center (NDBC) collects climatic observations from about 75 buoys. Table 5 contains the monthly mean sea temperatures from 2 of these buoys. Find a model of the form $t = k + A \sin Bx$ for the data from Buoy 41001. Graph the model for $1 \leq x \leq 12$.

72. Modeling Sea Temperatures Repeat Problem 71 for the data from Buoy 44004.

73. Modeling Gas Storage The Energy Information Administration (EIA) maintains data for the natural gas stored in underground facilities. Table 6 lists the monthly working natural gas storage for the eastern and the western regions of the United States from July 2005 to June 2007 in billions of cubic feet (Bcf). Find a model of the form $g = k + A \sin Bx$ for the data from the eastern region. Graph the model for $1 \leq x \leq 24$.

74. Modeling Gas Storage Find a model of the form $g = k + A \sin Bx$ for the data for the western region. Graph the model for $1 \leq x \leq 24$.

TABLE 5

Mean Sea Temperature (°C)

Month	1	2	3	4	5	6	7	8	9	10	11	12
Buoy 41001	20.4	20.1	19.5	20.0	21.8	24.1	26.3	27.0	26.3	24.6	22.7	21.5
Buoy 44004	15.9	20.3	23.6	24.0	21.9	18.0	14.4	10.8	8.5	8.0	9.2	11.8

TABLE 6
Monthly Storage of Natural Gas in Bcf (billions of cubic feet)

Month	Eastern region	Western region	Month	Eastern region	Western region
1	1,351	368	13	1,552	379
2	1,541	381	14	1,709	406
3	1,742	411	15	1,903	450
4	1,897	440	16	1,963	470
5	1,870	431	17	1,936	457
6	1,511	376	18	1,726	391
7	1,319	332	19	1,326	285
8	995	270	20	823	230
9	831	236	21	715	239
10	971	263	22	757	267
11	1,199	321	23	1,041	326
12	1,404	363	24	1,308	372

3.3 Graphing $y = k + A \sin (Bx + C)$ and $y = k + A \cos (Bx + C)$

- Graphing $y = A \sin(Bx + C)$ and $y = A \cos(Bx + C)$
- Graphing $y = k + A \sin(Bx + C)$ and $y = k + A \cos(Bx + C)$
- Finding the Equation for the Graph of a Simple Harmonic

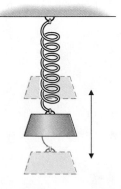

In the last section we learned a lot about graphing sine and cosine functions with varying periods and amplitudes. But for many situations, in order to obtain accurate models, we'll need to shift these graphs to the left or right as well.

A function of the form $y = A \sin(Bx + C)$ or $y = A \cos(Bx + C)$ is said to be a **simple harmonic.** Imagine an object suspended from the ceiling by a spring. If it is pulled down and released, then, assuming no air resistance or friction, it would oscillate up and down forever with the same amplitude and frequency. Such idealized motion can be described by a simple harmonic and is called **simple harmonic motion.** These functions are used extensively to model real-world phenomena; for example, see Problems 43–54 in Exercise 3.3 and the applications discussed in Section 3.4.

■ Graphing $y = A \sin(Bx + C)$ and $y = A \cos(Bx + C)$

We are interested in graphing equations of the form

$$y = A \sin(Bx + C) \quad \text{and} \quad y = A \cos(Bx + C)$$

We will find that the graphs of these equations are simply the graphs of

$$y = A \sin Bx \quad \text{or} \quad y = A \cos Bx$$

translated horizontally to the left or to the right. There is a simple process for determining how to shift the graph. Since $A \sin x$ has a period of 2π, it follows that $A \sin(Bx + C)$ completes one cycle as its input, $Bx + C$, varies from

$$Bx + C = 0 \quad \text{to} \quad Bx + C = 2\pi$$

This gives us two equations that we can solve for x; the results will be the x values that begin and end one full period of the graph.

$$x = -\frac{C}{B} \quad \text{to} \quad x = -\frac{C}{B} + \frac{2\pi}{B}$$

We can now see that $y = A \sin(Bx + C)$ has a period of $2\pi/B$, and its graph is the graph of $y = A \sin Bx$ translated horizontally $|-C/B|$ units to the right if $-C/B > 0$ and $|-C/B|$ units to the left if $-C/B < 0$. The horizontal translation, determined by the number $-C/B$, is often referred to as the **phase shift.**

In Example 1, we will illustrate this process with a specific function.

EXAMPLE 1

Finding Period and Phase Shift

Find the period and phase shift for $y = \sin(2x + \pi/2)$.

Solution   **Method I**   The graph will cover one full period as the input of sine, $2x + \pi/2$, varies from 0 to 2π. Find the corresponding x values:

$$2x + \frac{\pi}{2} = 0 \qquad\qquad 2x + \frac{\pi}{2} = 2\pi$$

$$2x = -\frac{\pi}{2} \qquad\qquad 2x = -\frac{\pi}{2} + 2\pi$$

$$x = -\frac{\pi}{4} \qquad\qquad x = -\frac{\pi}{4} + \pi \quad \text{or} \quad \frac{3\pi}{4}$$

The phase shift is $-\pi/4$, and the period is π.

Method II Using the formulas developed above, the period is $2\pi/B$ and the phase shift is $-C/B$. In this case, $B = 2$ and $C = \pi/2$:

$$P = \frac{2\pi}{2} = \pi$$

$$\text{Phase shift} = -\frac{\pi/2}{2} = -\frac{\pi}{4}$$ ■

Matched Problem 1   Find the period and phase shift for $y = \cos(\pi x - \pi/2)$.   ■

 Caution

At first glance, it might seem like using formulas to find the period and phase shift is simpler than Method 1 in Example 1. But Method 1 has a big advantage in graphing: It identifies the x values where one full period begins and ends. See Example 2.   □

A similar analysis applies to $y = A \cos(Bx + C)$. While Method 1 in Example 1 is probably the preferred method for graphing $y = A \sin(Bx + C)$ and $y = A \cos(Bx + C)$, it is still a good idea to summarize the key features of graphs in terms of the constants A, B, and C.   ■

PROPERTIES OF $y = A \sin(Bx + C)$ AND $y = A \cos(Bx + C)$

For $B > 0$,

$$\text{Amplitude} = |A| \qquad \text{Period} = \frac{2\pi}{B} \qquad \text{Phase shift} = -\frac{C}{B}$$

As we have already indicated, it is not necessary to memorize the formulas for period and phase shift unless you want to. The period and phase shift are easily found in the following steps for graphing:

STEPS FOR GRAPHING $y = A \sin(Bx + C)$ AND $y = A \cos(Bx + C)$

Step 1   Find the amplitude: $|A|$

Step 2   Solve $Bx + C = 0$ and $Bx + C = 2\pi$:

$$Bx + C = 0 \qquad \text{and} \qquad Bx + C = 2\pi$$

$$x = -\frac{C}{B} \qquad\qquad\qquad x = -\frac{C}{B} + \frac{2\pi}{B}$$

$$\underbrace{\qquad\qquad\qquad\qquad}_{\text{Phase shift}} \qquad \overset{\uparrow}{\text{Period}}$$

This identifies x values that begin and end one full cycle of the graph.

Step 3   Graph one cycle over the interval covering the x values found in step 2.

Step 4   Extend the graph in step 3 to the left or to the right as desired.

EXAMPLE 2

Graphing the Form $y = A \cos (Bx + C)$

Graph

$$y = 20 \cos\left(\pi x - \frac{\pi}{2} \right) \qquad -1 \le x \le 3$$

Solution   *Step 1*   Find the amplitude:

Amplitude $= |A| = |20| = 20$

Step 2   Solve $Bx + C = 0$ and $Bx + C = 2\pi$:

$$\pi x - \frac{\pi}{2} = 0 \qquad \pi x - \frac{\pi}{2} = 2\pi$$

$$x = \frac{1}{2} \qquad\qquad x = \frac{1}{2} + 2 = \frac{5}{2}$$

One full cycle of the graph is completed between $x = \frac{1}{2}$ and $x = \frac{5}{2}$, and the period is 2.

Step 3   Graph one cycle over the interval from $x = \frac{1}{2}$ to $x = \frac{5}{2}$ (Fig. 1). Note that this is a cosine graph with a positive value of A, so we begin and end one cycle at the maximum height, 20.

FIGURE 1

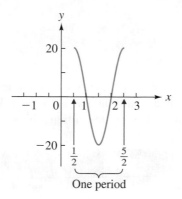

Step 4   Extend the graph from -1 to 3 (Fig. 2).

FIGURE 2

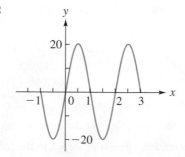

Matched Problem 2 State the amplitude, period, and phase shift for $y = -5 \sin(x/2 + \pi/2)$. Graph the equation for $-3\pi \le x \le 5\pi$. ■

Graphing $y = k + A \sin(Bx + C)$ and $y = k + A \cos(Bx + C)$

In order to graph an equation of the form $y = k + A \sin(Bx + C)$ or $y = k + A \cos(Bx + C)$, we first graph $y = A \sin(Bx + C)$ or $y = A \cos(Bx + C)$, as outlined previously, and then vertically translate the graph up k units if $k > 0$ or down $|k|$ units if $k < 0$. Graphing equations of the form $y = k + A \sin(Bx + C)$ or $y = k + A \cos(Bx + C)$, where $k \ne 0$ and $C \ne 0$, involves both a horizontal translation (phase shift) and a vertical translation of the basic equation $y = A \sin Bx$ or $y = A \cos Bx$.

EXAMPLE 3

Graphing the Form $y = k + A \cos(Bx + C)$

Graph three full periods of $y = 10 - 20 \sin(3x - \pi)$

Solution First, we'll graph $y = -20 \sin(3x - \pi)$ and then translate 10 units up. We begin by finding the amplitude and the beginning and ending x values of one cycle, as in Example 2:

$$\text{Amplitude} = |A| = |-20| = 20$$

$$3x - \pi = 0 \qquad\qquad\qquad 3x - \pi = 2\pi$$
$$3x = \pi \qquad\qquad\qquad 3x = \pi + 2\pi$$
$$x = \frac{\pi}{3} \qquad\qquad\qquad x = \frac{\pi}{3} + \frac{2\pi}{3} = \pi$$

This is a sine graph with a negative value of A, so the full cycle between $x = \pi/3$ and $x = \pi$ starts at height zero and proceeds downward to a low point with height -20. After drawing the first cycle (Fig. 3a), we extend to two more periods, one in each direction (Fig. 3b).

FIGURE 3a

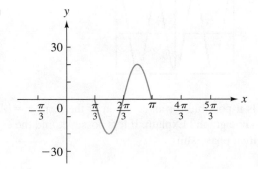

FIGURE 3b

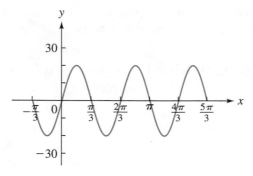

Finally, we translate the graph up by 10 units (Fig. 4). ■

FIGURE 4

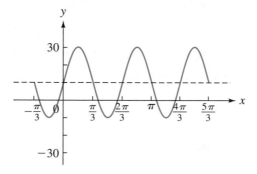

Matched Problem 3   Graph three full periods of $y = -8 - 4\cos(x/2 + \pi/4)$ ■

EXPLORE/DISCUSS 1

Find an equation of the form $y = A\cos(Bx + C)$ that produces the graph in the following graphing calculator display (choose the smallest positive phase shift):

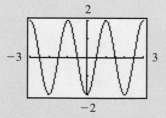

Is it possible for an equation of the form $y = A\sin(Bx + C)$ to produce the same graph? Explain. If it is possible, find the equation using the smallest positive phase shift.

■ Finding the Equation of the Graph of a Simple Harmonic

In order to model real-world situations with simple harmonics, it is very useful to be able to find an equation of the form $y = A \sin(Bx + C)$ or $y = A \cos(Bx + C)$ that produces a given graph. An example illustrates the process with the aid of a graphing calculator.

Finding the Equation for the Graph of a Simple Harmonic with the Aid of a Graphing Calculator

 Graph $y_1 = 4 \sin x - 3 \cos x$ on a graphing calculator and find an equation of the form $y_2 = A \sin(Bx + C)$ that has the same graph. Find A and B exactly and C to three decimal places.

Solution The graph of y_1 is shown in Figure 5. This graph appears to be a sine curve, with amplitude 5 and period 2π, that has been shifted to the right. We conclude that $A = 5$ and $B = 2\pi/P = 2\pi/2\pi = 1$.

To determine C, first find the phase shift from the graph. We choose the smallest positive phase shift, which is the first point to the right of the origin where the graph crosses the x axis. This is the first positive zero of y_1. Most graphing calculators have a built-in command for finding the zeros of a function. Figure 6 shows the result from a TI-84. The zero to three decimal places is $x = 0.644$. To find C, we substitute $B = 1$ and $x = 0.644$ in the phase shift equation $x = -C/B$ and solve for C:

$$x = -\frac{C}{B}$$

$$0.644 = -\frac{C}{1}$$

$$C = -0.644$$

The equation is

$$y_2 = 5 \sin(x - 0.644)$$

If greater accuracy is desired, choose the phase shift x from Figure 6 to more decimal places.

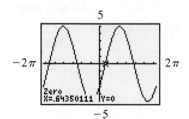

FIGURE 5

FIGURE 6

✔ Check Graph $y_1 = 4 \sin x - 3 \cos x$ and $y_2 = 5 \sin(x - 0.644)$ in the same viewing window. If the graphs are the same, it appears that only one graph is drawn—the second graph is drawn over the first. To check further that the graphs are the same, use $\boxed{\text{TRACE}}$ and switch back and forth between y_1 and y_2 at different values of x. Figure 7 on the next page shows a comparison at $x = 0$.

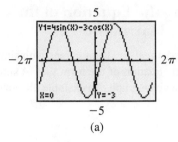

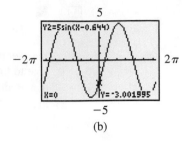

(a)   (b)

FIGURE 7   ■

 Matched Problem 4   Graph $y_1 = 3 \sin x + 4 \cos x$, find the x intercept closest to the origin (correct to three decimal places), and find an equation of the form $y_2 = A \sin(Bx + C)$ that has the same graph.   ■

EXPLORE/DISCUSS 2

Explain why any function of the form $y = A \sin(Bx + C)$ can also be written in the form $y = A \cos(Bx + D)$ for an appropriate choice of D.

 **EXAMPLE 5**   **Modeling Lunar Phases**

The phases of the moon are periodic, so we may be able to model them with a sine or cosine function. The moon was full, meaning 100% visible, at 3:45 P.M. Eastern Daylight Time (EDT) on September 26, 2007. Table 1 provides the moon's phases (percent visible) in terms of number of days after 10/1/07, 12 A.M. EDT.

TABLE 1	
Lunar Phases	
Phase	*Days after 10/1/07*
100% (full moon)	−4.34
50%	3.04
0% (new moon)	10.43
50%	17.84
100% (full moon)	25.19

(A) Use the information in the table to find a function of the form $y = k + A \cos(Bx + C)$ that models the lunar phases in terms of days after 10/1/07, 12 A.M. EDT. Round all constants to three significant digits.

(B) According to the model, what was the lunar phase at the exact beginning of 2008?

Solution   (A) The percentage ranges from 0% to 100%, so the amplitude will be 50, and the graph will be a cosine curve translated up by 50 units. This tells us that $|A| = 50$ and $k = 50$. To find the period, subtract to find the time between full moons:

$$\text{Period} = 25.19 - (-4.34) = 29.53$$

$$29.53 = \frac{2\pi}{B}$$

$$B = \frac{2\pi}{29.53} = 0.213$$

The high point occurs for $x = -4.34$, so with a phase shift of -4.34, this will be a cosine curve with positive A; that is, $A = 50$. Finally, we can use the phase shift formula to find C:

$$-4.34 = -\frac{C}{B} = -\frac{C}{0.213}$$

$$C = (0.213)(4.34) = 0.924$$

The model is $y = 50 + 50 \cos(0.213x + 0.924)$.

✔ Check   A graph of this equation on a graphing calculator shows that the model agrees well with the data in the table (Fig. 8).

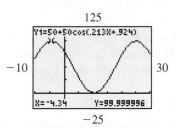

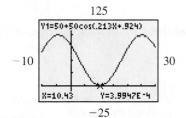

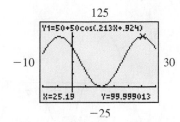

FIGURE 8

(B) The beginning of 2008 is exactly 92 days after 10/1/07, 12 A.M., so we substitute 92 for x in our model:

$$y(92) = 50 + 50\cos(0.213(92) + 0.924)$$

$$= 45\%$$ ∎

EXPLORE/DISCUSS 3

We can obtain an alternative solution to Example 5(A) as follows: Take the same values of $|A|$, k, and B, but note that the low point occurs at 10.43. With a phase shift of 10.43, the model will be a cosine curve with negative A, so $A = -50$.

(A) Use the phase shift formula to show that $C = -2.22$.

(B) Use a graphing calculator to show that the graph of the resulting model, $y = 50 - 50\cos(0.213x - 2.22)$, is apparently identical to the graph of the model of Figure 8.

(C) We often prefer to choose the phase shift that has the smallest absolute value [as in the solution to Example 5(A)]. Explain why it is possible to choose the phase shift so that

$$|\text{Phase shift}| \le \frac{\text{Period}}{4}$$

(D) Explain why it is possible to choose the constant C so that

$$|C| \le \frac{\pi}{2} = 1.57$$

Matched Problem 5   Find a function of the form $y = k + A\sin(Bx + C)$ that models the data in Table 1. Use an almanac or the Internet to find the lunar phase for today, then see if the phase predicted by this model is accurate. ∎

Answers to   1. Period $= 2$; phase shift $= 1/2$
Matched Problems   2. Amplitude $= 5$; period $= 4\pi$; phase shift $= -\pi$

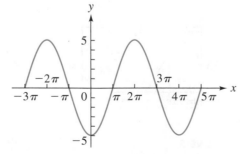

3.

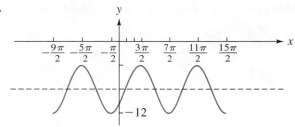

4. x intercept: -0.927; $y_2 = 5\sin(x + 0.927)$

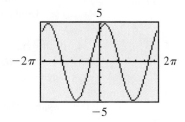

5. $y = 50 + 50\sin(0.213x - 3.80)$

EXERCISE 3.3

A

1. Explain what is meant by the graph of a simple harmonic having a phase shift of -2.

2. Explain what is meant by the graph of a simple harmonic having a phase shift of 3.

3. Identify a real-world quantity that you think might be modeled well by a function of the form $y = k + A\sin(Bx + C)$, and discuss why you chose that quantity.

4. Explain how you can find the period and phase shift of $y = A\cos(Bx + C)$ based on the fact that one full cycle of $y - \cos x$ begins at $x = 0$ and ends at $x = 2\pi$.

In Problems 5–8, find the phase shift for each equation, and graph it over the indicated interval.

5. $y = \cos\left(x + \dfrac{\pi}{2}\right)$, $\dfrac{-\pi}{2} \le x \le \dfrac{3\pi}{2}$

6. $y = \cos\left(x - \dfrac{\pi}{2}\right)$, $\dfrac{\pi}{2} \le x \le \dfrac{5\pi}{2}$

7. $y = \sin\left(x - \dfrac{\pi}{4}\right)$, $-\pi \le x \le 2\pi$

8. $y = \cos\left(x + \dfrac{\pi}{4}\right)$, $-\pi \le x \le 2\pi$

B *In Problems 9–12, state the amplitude, period, and phase shift for each equation, and graph it over the indicated interval.*

9. $y = 4\cos\left(\pi x + \dfrac{\pi}{4}\right)$, $-1 \le x \le 3$

10. $y = 2\sin\left(\pi x - \dfrac{\pi}{2}\right)$, $-2 \le x \le 2$

11. $y = -2\cos(2x + \pi)$, $-\pi \le x \le 3\pi$

12. $y = -3\sin(4x - \pi)$, $-\pi \le x \le \pi$

13. Graph

$$y = \cos\left(x - \dfrac{\pi}{2}\right) \quad \text{and} \quad y = \sin x$$

in the same coordinate system. Conclusion?

14. Graph

$$y = \sin\left(x + \dfrac{\pi}{2}\right) \quad \text{and} \quad y = \cos x$$

in the same coordinate system. Conclusion?

In Problems 15–18, graph three full periods of each equation.

15. $y = -2 + 4 \cos\left(\pi x + \dfrac{\pi}{4}\right)$

16. $y = -3 + 2 \sin\left(\pi x - \dfrac{\pi}{2}\right)$

17. $y = 3 - 2 \cos(2x + \pi)$

18. $y = 4 - 3 \sin(4x - \pi)$

In Problems 19–22, match each equation with one of the following graphing calculator displays. Explain how you made the choice relative to period and phase shift.

(A)

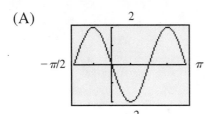

(B)

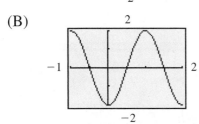

(C)

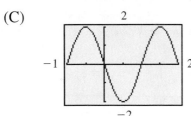

(D)

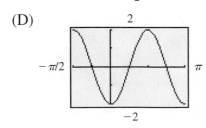

19. $y = 2 \sin\left(\pi x - \dfrac{\pi}{2}\right)$ **21.** $y = 2 \cos\left(2x + \dfrac{\pi}{2}\right)$

20. $y = 2 \cos\left(\pi x + \dfrac{\pi}{2}\right)$ **22.** $y = 2 \sin\left(2x - \dfrac{\pi}{2}\right)$

For Problems 23 and 24, refer to the graph:

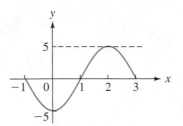

23. If the graph is a graph of an equation of the form $y = A \sin(Bx + C), 0 < -C/B < 2$, find the equation.

24. If the graph is a graph of an equation of the form $y = A \sin(Bx + C), -2 < -C/B < 0$, find the equation.

For Problems 25 and 26, refer to the graph:

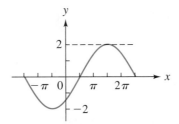

25. If the graph is a graph of an equation of the form $y = A \cos(Bx + C), -2\pi < -C/B < 0$, find the equation.

26. If the graph is a graph of an equation of the form $y = A \cos(Bx + C), 0 < -C/B < 2\pi$, find the equation.

C *In Problems 27 and 28, state the amplitude, period, and phase shift for each equation, and graph it over the indicated interval.*

27. $y = 2 \sin\left(3x - \dfrac{\pi}{2}\right), \quad -\dfrac{2\pi}{3} \le x \le \dfrac{5\pi}{3}$

28. $y = -4 \cos\left(4x + \dfrac{\pi}{2}\right), \quad -\dfrac{\pi}{2} \le x \le \pi$

In Problems 29 and 30, graph each equation over the indicated interval.

29. $y = 4 + 2 \sin\left(3x - \dfrac{\pi}{2}\right), \quad -\dfrac{2\pi}{3} \le x \le \dfrac{5\pi}{3}$

30. $y = 6 - 4 \cos\left(4x + \dfrac{\pi}{2}\right), \quad -\dfrac{\pi}{2} \le x \le \pi$

Problems 31–34 require the use of a graphing calculator. First, state the amplitude, period, and phase shift of each function; then graph the function with a graphing calculator.

31. $y = 2.3 \sin\left[\dfrac{\pi}{1.5}(x - 2)\right], \quad 0 \le x \le 6$

32. $y = -4.7 \sin\left[\dfrac{\pi}{2.2}(x + 3)\right], \quad 0 \le x \le 10$

33. $y = 18 \cos[4\pi(x + 0.137)], \quad 0 \le x \le 2$

34. $y = -48 \cos[2\pi(x - 0.205)], \quad 0 \le x \le 3$

Problems 35–42 require the use of a graphing calculaor. Graph the given equation and find the x intercept closest to the origin, correct to three decimal places. Use this intercept to find an equation of the form $y = A \sin(Bx + C)$ that has the same graph as the given equation.

35. $y = \sin x + \sqrt{3} \cos x$

36. $y = \sqrt{3} \sin x - \cos x$

37. $y = \sqrt{2} \sin x - \sqrt{2} \cos x$

38. $y = \sqrt{2} \sin x + \sqrt{2} \cos x$

39. $y = 1.4 \sin 2x + 4.8 \cos 2x$

40. $y = 4.8 \sin 2x - 1.4 \cos 2x$

41. $y = 2 \sin \dfrac{x}{2} - \sqrt{5} \cos \dfrac{x}{2}$

42. $y = \sqrt{5} \sin \dfrac{x}{2} + 2\cos \dfrac{x}{2}$

Applications

In these applications, assume all given values are exact unless indicated otherwise.

43. Water Waves (See Section 3.4) At a particular point in the ocean, the vertical change in the water due to wave action is given by

$$y = 5 \sin \dfrac{\pi}{6}(t + 3)$$

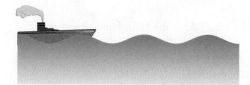

Figure for 43 and 44

where y is in meters and t is time in seconds. What are the amplitude, period, and phase shift? Graph the equation for $0 \le t \le 39$.

44. Water Waves Repeat Problem 43 if the wave equation is

$$y = 8 \cos \dfrac{\pi}{12}(t - 6) \quad 0 \le t \le 72$$

45. Electrical Circuit (See Section 3.4) The current I (in amperes) in an electrical circuit is given by $I = 30 \sin(120\pi t - \pi)$, where t is time in seconds. State the amplitude, period, frequency (cycles per second), and phase shift. Graph the equation for the interval $0 \le t \le \frac{3}{60}$.

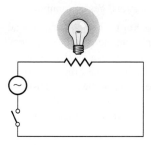

Figure for 45 and 46

46. Electrical Circuit Repeat Problem 45 for

$$I = 110 \cos\left(120\pi t + \dfrac{\pi}{2}\right) \quad 0 \le t \le \dfrac{2}{60}$$

47. Energy Consumption Energy consumption often depends on temperature and number of daylight hours, so it tends to repeat in annual cycles. Table 2 specifies total monthly consumption of natural gas in the United States for selected months between June 2006 and June 2007.

TABLE 2
Natural Gas Consumption (Trillion BTU)

Month	Residential consumption	Commerical consumption
June 2006	145	142
August 2006	111	130
October 2006	246	144
December 2006	637	263
February 2007	923	443
April 2007	420	363
June 2007	141	173

Source: Energy Information Administration.

(A) Use the information in Table 2 to find an equation of the form $y = k + A\cos(Bx + C)$ that describes residential gas consumption, y, in terms of number of months after June 2006, x. [*Hint:* Use half the difference between the highest and lowest consumptions for $|A|$, and choose the phase shift that has the smallest absolute value.]

(B) Visit the website http://www.eia.doe.gov/emeu/mer/consump.html to find residential consumption for the most recent available month. Then use your equation from part (A) to estimate the consumption in that month. Discuss the accuracy of your estimate.

48. Repeat Problem 47 for commercial gas consumption.

49. **An Experiment with Phase Shift**   The tip of the second hand on a wall clock is 6 in. from the center of the clock (see the figure). Let d be the (perpendicular) distance from the tip of the second hand at time t (in seconds) to the vertical line through the center of the clock. Distance to the right of the vertical line is positive and distance to the left is negative. Consider the following relations:

Figure for 49 and 50

    (1) Distance d, t seconds after the second hand points to 12

    (2) Distance d, t seconds after the second hand points to 9

(A) Complete Tables 3 and 4, giving values of d to one decimal place.

(B) Referring to Tables 3 and 4, and to the clock in the figure, discuss how much relation (2) is out of phase relative to relation (1).

(C) Relations (1) and (2) are simple harmonics. What are the amplitudes and periods of these relations?

(D) Both relations have equations of the form $y = A\sin(Bx + C)$. Find the equations and check that the equations give the values in the tables. (Choose the phase shift so that $|C|$ is minimum.)

TABLE 3

Distance d, t Seconds After the Second Hand Points to 12

t (sec)	0	5	10	15	20	25	30	35	40	45	50	55	60
d (in.)	0.0		6.0				−3.0				−5.2		

TABLE 4

Distance d, t Seconds After the Second Hand Points to 9

t (sec)	0	5	10	15	20	25	30	35	40	45	50	55	60
d (in.)	−6.0		0.0				5.2				−3.0		

(E) Write an equation of the form $y = A\sin(Bx + C)$ that represents the distance, d, t seconds after the second hand points to 3. What is the phase shift? (Choose the phase shift so that $|C|$ is minimum.)

(F) Graph the three equations found in parts (D) and (E) in the same viewing window on a graphing calculator for $0 \le t \le 120$.

50. **An Experiment with Phase Shift**   The tip of the hour hand on a wall clock is 4 in. from the center of the clock (see the figure). Let d be the distance from the tip of the hour hand to the vertical line through the center of the clock at time t (in hours). Distance to the right of the vertical line is positive and distance to the left is negative. Consider the following relations:

    (1) Distance d, t hours after the hour hand points to 12

    (2) Distance d, t hours after the hour hand points to 3

(A) Complete Tables 5 and 6, giving values of d to one decimal place.

(B) Referring to Tables 5 and 6, and to the clock in the figure, discuss how much relation (2) is out of phase relative to relation (1).

TABLE 5

Distance d, t Hours After the Hour Hand Points to 12

t (hr)	0	1	2	3	4	5	6	7	8	9	10	11	12
d (in.)	0.0		4.0				−2.0				−3.5		

TABLE 6

Distance d, t Hours After the Hour Hand Points to 3

t (hr)	0	1	2	3	4	5	6	7	8	9	10	11	12
d (in.)	4.0			0.0				−3.5				2.0	

(C) Relations (1) and (2) are simple harmonics. What are the amplitudes and periods of these relations?

(D) Both relations have equations of the form $y = A \sin(Bx + C)$. Find the equations, and check that the equations give the values in the tables. (Choose the phase shift so that $|C|$ is minimum.)

(E) Write an equation of the form $y = A \sin(Bx + C)$ that represents the distance, d, t hours after the hour hand points to 9. What is the phase shift? (Choose the phase shift so that $|C|$ is minimum.)

 (F) Graph the three equations found in parts (D) and (E) in the same viewing window on a graphing calculator for $0 \leq t \leq 24$.

Problems 51–54 require a graphing calculator with sinusoidal regression capabilities.

 51. Modeling Temperature Variation The 30 yr average monthly temperatures (in °F) for each month of the year for San Antonio, TX, are given in Table 7.

(A) Enter the data for a 2 yr period ($1 \leq x \leq 24$) in your graphing calculator and produce a scatter plot in the viewing window.

(B) From the scatter plot in part (A), it appears that a sine curve of the form

$$y = k + A \sin(Bx + C)$$

will closely model the data. Discuss how the constants k, $|A|$, and B can be determined from Table 7 and find them. To estimate C, visually estimate (to one decimal place) the smallest positive phase shift from the scatter plot in part (A). Write the equation, and plot it in the same viewing window as the scatter plot. Adjust C as necessary to produce a better visual fit.

(C) Fit the data you have entered into your graphing calculator from Table 7 with a **sinusoidal regression equation and curve.** This is a process many graphing calculators perform automatically (consult your user's manual for the process). The result is an equation of the form $y = k + A \sin(Bx + C)$ that gives the "best fit" to the data when graphed. Write the sinusoidal regression equation to one decimal place. Show the graphs of the scatter plot of the data in Table 7 and the regression equation that is already in your calculator in the same viewing window. (Do not reenter the regression equation.)

(D) What are the differences between the regression equation in part (C) and the equation obtained in part (B)? Which equation appears to give the better fit?

52. Modeling Sunrise Times Sunrise times for the fifth day of each month over a 1 yr period were taken from a tide booklet for San Francisco Bay to form Table 8 (daylight savings time was ignored). Repeat parts (A)–(D) in Problem 51 with Table 7 replaced by Table 8 on the next page. (Before entering the data, convert sunrise times from hours and minutes to decimal hours rounded to two decimal places.)

53. Modeling Temperature Variation—Collecting Your Own Data Replace Table 7 in Problem 51 with data for your own city or a city near you, using an appropriate reference. Then repeat parts (A)–(D).

TABLE 7												
x (month)	1	2	3	4	5	6	7	8	9	10	11	12
y (temperature, °F)	50	54	62	70	76	82	85	84	79	70	60	53

Source: World Almanac

TABLE 8												
x (month)	1	2	3	4	5	6	7	8	9	10	11	12
y (sunrise time, A.M.)	7:26	7:11	6:36	5:50	5:10	4:48	4:53	5:16	5:43	6:09	6:39	7:10

54. Modeling Sunrise Times—Collecting Your Own Data
Replace Table 8 in Problem 52 with data for your own city or a city near you, using an appropriate reference. Then repeat parts (A)–(D).

Biorhythms involve three rhythmic cycles that many believe affect human behavior. The physical cycle is 23 days long, the emotional cycle is 28 days long, and the intellectual cycle is 33 days long. All three cycles start at birth and continue throughout a person's lifetime. For modeling purposes, we assume that each cycle can be represented by the graph of $y = \sin(Bx + C)$.

55. Biorhythms   Graph the physical cycle during the first 23 days after birth.

56. Biorhythms   Graph the emotional cycle during the first 28 days after birth.

57. Biorhythms   Graph the intellectual cycle during the first 33 days after birth.

58. Biorhythms   Graph all three cycles during the first 33 days after birth on the same axes.

59. Biorhythms   John was born on July 4, 1987. On January 20, 2005, John had been alive for 6,411 days. Graph John's biorhythm cycles for the next 30 days.

60. Biorhythms   How old will John be (to the nearest year) on the first day that all three cycles return to their starting points?

61. Biorhythms   Graph your biorhythm cycles for 30 days starting with today. (When determining how many days you have been alive, don't forget to take leap years into consideration.)

☆3.4   Additional Applications

- Modeling Electric Current
- Modeling Light and Other Electromagnetic Waves
- Modeling Water Waves
- Simple and Damped Harmonic Motion; Resonance

Many types of applications of trigonometry have already been considered in this and the preceding chapters. This section provides a sampler of additional applications from several different fields. You don't need prior knowledge of any particular topic to understand the discussion or to work any of the problems. Several of the applications considered use important properties of the sine and cosine functions, which are restated next for convenient reference.

☆ Sections marked with a star may be omitted without loss of continuity.

> ## PROPERTIES OF SINE AND COSINE
>
> For $y = A \sin(Bt + C)$ or $y = A \cos(Bt + C)$:
>
> $$\text{Amplitude} = |A| \quad \text{Period} = \frac{2\pi}{B} \quad \text{Frequency} = \frac{1}{\text{Period}}$$
>
> $$\text{Phase shift} = -\frac{C}{B} \begin{cases} \text{Right} & \text{if } -C/B > 0 \\ \text{Left} & \text{if } -C/B < 0 \end{cases}$$
>
> As we indicated in the previous section, you do not need to memorize the formulas for period and phase shift. You can obtain the same results by solving the two equations
>
> $$Bt + C = 0 \qquad \text{and} \qquad Bt + C = 2\pi$$
>
> (We use the variable t, rather than x, because most of the applications here involve functions of time.)

Recall that functions of the form $y = A \sin(Bt + C)$ or $y = A \cos(Bt + C)$ are said to be **simple harmonics.**

■ Modeling Electric Current

In physics, a **field** is an area surrounding an object in which a gravitational or electromagnetic force is exerted on other objects. Fields were introduced by the British physicist Sir Isaac Newton (1642–1727) to explain gravitational forces. Around 140 years later, the British physicist-chemist Michael Faraday (1791–1867) used the concept of fields to explain electromagnetic forces. In the 19th century, the British physicist James Maxwell (1831–1879) developed equations that describe electromagnetic fields. And in the 20th century, the German-American physicist Albert Einstein (1879–1955) developed a set of equations for gravitational fields.

A particularly significant discovery by Faraday was the observation that a flow of electricity could be created by moving a wire in a magnetic field. This discovery was the start of the electronic revolution, which is continuing today with no slowdown in sight.

Suppose we bend a wire in the form of a rectangle, and we locate this wire between the south and north poles of magnets, as shown in Figure 1. We now rotate the wire at a constant counterclockwise speed, starting with side BC in its lowest position. As BC turns toward the horizontal, an electrical current will flow from C to B (see Fig. 1a). The strength of the current (measured in amperes) will be 0 at the lowest position and will increase to a maximum value at the horizontal position. As BC continues to turn from the horizontal to the top position, the current flow decreases to 0. As BC starts down from the top position in a counterclockwise

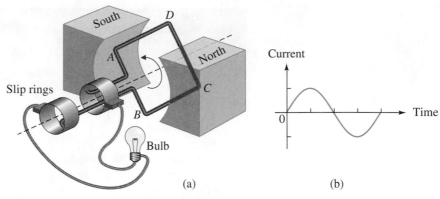

FIGURE 1
Alternating current

direction, the current flow reverses, going from *B* to *C,* and again reaches a maximum at the left horizontal position. When *BC* moves from this horizontal position back to the original bottom position, the current flow again decreases to 0. This pattern repeats itself for each revolution, so it is periodic.

Is it too much to expect that a trigonometric function can describe the relationship between current and time (as in Fig. 1b)? If we measure and graph the strength of the current *I* (in amperes) in the wire relative to time *t* (in seconds), we can indeed find an equation of the form

$$I = A \sin(Bt + C)$$

that will give us the relationship between current and time. For example,

$$I = 30 \sin 120\,\pi t$$

represents an alternating current flow with a frequency of 60 Hz (cycles per second) and a maximum value of 30 amperes.

EXAMPLE 1

Alternating Current Generator

An alternating current generator produces an electrical current (measured in amperes) that is described by the equation

$$I = 35 \sin(40\pi t - 10\pi)$$

where *t* is time in seconds.
(A) What are the amplitude, period, frequency, and phase shift for the current?
(B) Graph the equation on a graphing calculator for $0 \le t \le 0.2$.

Solution (A) $y = 35 \sin(40\pi t - 10\pi)$

Amplitude $= |35| = 35$ amperes

To find the period and phase shift, solve $Bt + C = 0$ and $Bt + C = 2\pi$:

$$40\pi t - 10\pi = 0 \qquad\qquad 40\pi t - 10\pi = 2\pi$$

$$40\pi t = 10\pi \qquad\qquad\qquad 40\pi t = 10\pi + 2\pi$$

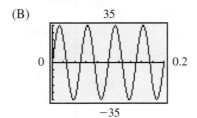

Phase shift $= \frac{1}{4}$ sec (right)   Period $= \frac{1}{20}$ sec

$$\text{Frequency} = \frac{1}{\text{Period}} = 20 \text{ Hz}$$

(B)

Matched Problem 1   If the alternating current generator in Example 1 produces an electrical current described by the equation $I = -25 \sin(30\pi t + 5\pi)$:

(A) Find the amplitude, period, frequency, and phase shift for the current.

(B) Graph the equation on a graphing calculator for $0 \le t \le 0.3$.

EXPLORE/DISCUSS 1

An electric garbage disposal unit in a house has a printed plate fastened to it that states: 120 V   60 Hz   7.6 A   (that is, 120 volts, 60 Hz, 7.6 amperes). Discuss how you can arrive at an equation for the current I in the form $I = A \sin(Bt + C)$, where t is time in seconds, if the current is 5 amperes when $t = 0$ sec. Find the equation.

■ Modeling Light and Other Electromagnetic Waves

Visible light is a transverse wave form with a frequency range between 4×10^{14} Hz (red) and 7×10^{14} Hz (violet). The retina of the eye responds to these vibrations, and through a complicated chemical process, the vibrations are eventually perceived by the brain as light in various colors. Light is actually a small part of a continuous spectrum of electromagnetic wave forms—most of which are not visible (see Fig. 2). Included in the spectrum are radio waves (AM and FM), microwaves, X rays, and gamma rays. All these waves travel at the speed of light, approx. 3×10^8 m/sec (186,000 mi/sec), and many of them are either partially or totally adsorbed in the atmosphere of the earth, as indicated in Figure 2. Their dis-

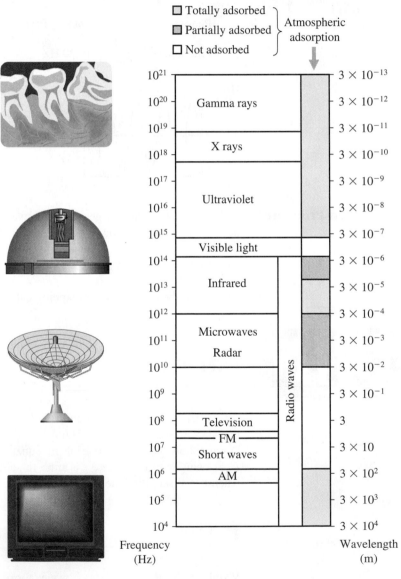

FIGURE 2
Electromagnetic wave spectrum

tinguishing characteristics are wavelength and frequency, which are related by the formula

$$\lambda \nu = c \tag{1}$$

where c is the speed of light, λ is the wavelength, and ν is the frequency.

Electromagnetic waves are traveling waves that can be described by an equation of the form

$$E = A \sin 2\pi \left(\nu t - \frac{r}{\lambda} \right) \qquad \text{Traveling wave equation} \tag{2}$$

where t is time and r is the distance from the source. This equation is a function of two variables, t and r. If we freeze time, then the graph of (2) looks something like Figure 3a. If we look at the electromagnetic field at a single point in space—that is, if we hold r fixed—then the graph of (2) looks something like Figure 3b.

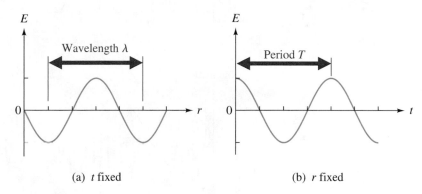

(a) t fixed   (b) r fixed

FIGURE 3
Electromagnetic wave

Electromagnetic waves are produced by electrons that have been excited into oscillatory motion. This motion creates a combination of electric and magnetic fields that move through space at the speed of light. The frequency of the oscillation of the electron determines the nature of the wave (see Fig. 2), and a receiver responds to the wave through induced oscillation of the same frequency (see Fig. 4).

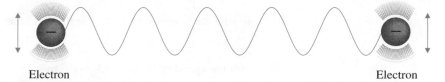

Electron   Electron

FIGURE 4
Electromagnetic field

EXPLORE/DISCUSS 2

Commonly received radio waves in the home are FM waves and AM waves—both are electromagnetic waves. FM means **frequency modulation,** and AM means **amplitude modulation.** Figure 5 illustrates AM and FM waves at fixed points in space. Identify which is which and explain your reasoning in your selection.

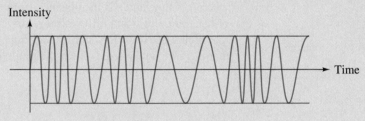

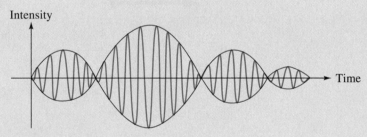

FIGURE 5
FM and AM electromagnetic waves

EXAMPLE 2

Electromagnetic Waves

If an electromagnetic wave has a frequency of $\nu = 10^{12}$ Hz, what is its period? What is its wavelength (in meters)?

Solution   Period $= \dfrac{1}{\nu} = \dfrac{1}{10^{12}} = 10^{-12}$ sec

To find the wavelength λ, we use the formula

$$\lambda\nu = c$$

with the speed of light $c \approx 3 \times 10^8$ m/sec:

$$\lambda = \frac{3 \times 10^8 \text{ m/sec}}{10^{12} \text{ Hz}} = 3 \times 10^{-4} \text{ m}$$ ∎

Matched Problem 2   Repeat Example 2 for $\nu = 10^6$ Hz.   ■

   **EXAMPLE 3**

Ultraviolet Waves

An ultraviolet wave has an equation of the form

$$y = A \sin Bt$$

Find B if the wavelength is $\lambda = 3 \times 10^{-9}$ m.

Solution   We will first use $\lambda\nu = c$ with $c = 3 \times 10^8$ m/sec to find the frequency:

$$\lambda\nu = c; \quad \nu = \frac{c}{\lambda}$$

$$\nu = \frac{3 \times 10^8 \text{ m/sec}}{3 \times 10^{-9} \text{ m}} = 10^{17} \text{ Hz}$$

Now we can find the period:

$$P = \frac{1}{\nu} = \frac{1}{10^{17}}$$

Finally, we can use $P = 2\pi/B$ to find B:

$$\frac{1}{10^{17}} = \frac{2\pi}{B}$$

$$B = 2\pi \times 10^{17}$$

Matched Problem 3   A gamma ray has an equation of the form $y = A \sin Bt$. Find B if the wavelength is $\lambda = 3 \times 10^{-12}$ m.   ■

■ **Modeling Water Waves**

Water waves are probably the most familiar wave form. You can actually see the wave in motion! Water waves are formed by particles of water rotating in circles (Fig. 6). A particle actually moves only a short distance as the wave passes through. These wave forms are moving waves, and it can be shown that in their simplest form they are sine curves.

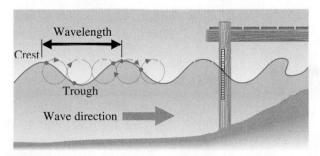

FIGURE 6
Water waves

The waves can be represented by an equation of the form

$$y = A \sin 2\pi \left(ft - \frac{r}{\lambda} \right) \qquad \text{Moving wave equation*} \tag{3}$$

where f is frequency, t is time, r is distance from the source, and λ is wavelength. So for a wave of a given frequency and wavelength, y is a function of the two variables t and r.

Equation (3) is typical for moving waves. If a wave passes a pier piling with a vertical scale attached (Fig. 6), the graph of the water level on the scale relative to time would look something like Figure 7a, since r (distance from the source) would be fixed. On the other hand, if we actually photograph a wave, freezing the motion in time, then the profile of the wave would look something like Figure 7b.

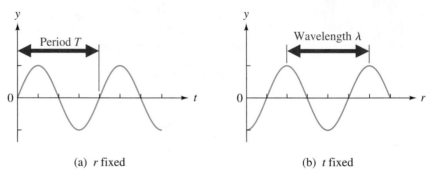

(a) r fixed   (b) t fixed

FIGURE 7
Water wave

Experiments have shown that the **wavelength** λ for water waves is given approximately by

$$\lambda = 5.12T^2 \quad \text{In feet}$$

where T is the period of the wave in seconds (see Fig. 7a), and the **speed** S of the wave is given approximately by

$$S = \sqrt{\frac{g\lambda}{2\pi}} \quad \text{In feet per second}$$

where $g = 32$ ft/sec^2 (gravitational constant).

EXAMPLE 4

Modeling Water Waves

A water wave at a fixed position has an equation of the form $y = 3 \sin[(\pi/4)t + 2]$, where t is time in seconds and y is in feet. How high is

---

* Actually, this general wave equation can be used to describe any longitudinal (compressional) or transverse wave motion of one frequency in a nondispersive medium. A sound wave is an example of a longitudinal wave; water waves and electromagnetic waves are examples of transverse waves.

the wave from trough to crest (see Fig. 6)? What are its period and wavelength? How fast is it traveling in feet per second? In miles per hour?

Solution   The amplitude is 3, so the distance from trough to crest is twice that, or 6 ft. We can find the period using $T = 2\pi/B$:

$$T = \frac{2\pi}{\pi/4} = 2\pi \cdot \frac{4}{\pi} = 8 \text{ sec}$$

Now we can calculate the wavelength using $\lambda = 5.12(T)^2$:

$$\lambda = 5.12(8)^2 \approx 328 \text{ ft}$$

Next, we can use the speed formula:

$$S = \sqrt{\frac{g\lambda}{2\pi}} = \sqrt{\frac{32(328)}{2\pi}} \approx 41 \text{ ft/sec}$$

Finally, we convert this to miles per hour:

$$41 \, \frac{\text{ft}}{\text{sec}} \cdot \frac{3{,}600 \text{ sec}}{1 \text{ hr}} \cdot \frac{1 \text{ mi}}{5{,}280 \text{ ft}} \approx 28 \text{ mi/hr}$$ ■

Matched Problem 4   Repeat Example 4 for a wave with equation $y = 11 \sin(2t)$.   ■

■ Simple and Damped Harmonic Motion; Resonance

An object of mass M hanging on a spring will produce simple harmonic motion when pulled down and released (if we neglect friction and air resistance; see Fig. 8).

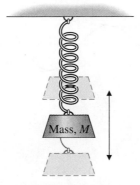

FIGURE 8

Figure 9 illustrates simple harmonic motion, damped harmonic motion, and resonance. **Damped harmonic motion** occurs when amplitude decreases to 0 as time increases. **Resonance** occurs when amplitude increases as time increases. Damped harmonic motion is essential in the design of suspension

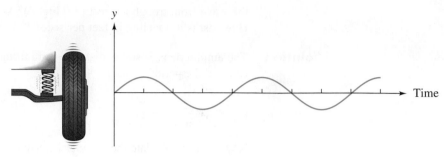

Spring only
Simple harmonic motion (neglect friction and air resistance)

(a)

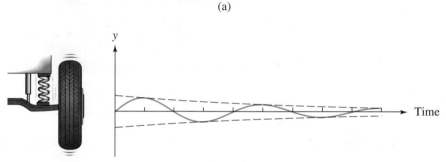

Spring and shock absorber
Damped harmonic motion

(b)

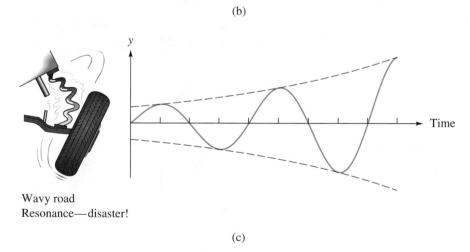

Wavy road
Resonance—disaster!

(c)

FIGURE 9

systems for cars, buses, trains, and motorcycles, as well as in the design of buildings, bridges, and aircraft. Resonance is useful in some electric circuits and in some mechanical systems, but it can be disastrous in bridges, buildings, and aircraft. Commercial jets have lost wings in flight, and large bridges have collapsed because of resonance. Marching soldiers must break step while cross-ing a bridge in order to avoid the creation of resonance—and the collapse of the bridge!

In one famous example of the destructive power of resonance, the Tacoma Narrows Bridge in Washington collapsed on November 7, 1940. If you're interested, an Internet search for "Tacoma Narrows Bridge" is likely to locate several videos of the collapse, which is an excellent illustration of resonance.

EXAMPLE 5

Damped Harmonic Motion

Graph

$$y = \frac{1}{t} \sin \frac{\pi}{2} t \quad 1 \le t \le 8$$

and indicate the type of motion.

Solution   The $1/t$ factor in front of $\sin(\pi t/2)$ affects the amplitude. Since the maximum and minimum values that $\sin(\pi t/2)$ can assume are 1 and -1, respectively, if we graph $y = 1/t$ first ($1 \le t \le 8$) and reflect the graph across the t axis, we will have upper and lower bounds for the graph of $y = (1/t) \sin(\pi t/2)$. We also note that $(1/t) \sin(\pi t/2)$ will still be 0 when $\sin(\pi t/2)$ is 0.

Step 1   Graph $y = 1/t$ and its reflection (Fig. 10)—called the **envelope** for the graph of $y = (1/t) \sin(\pi t/2)$.

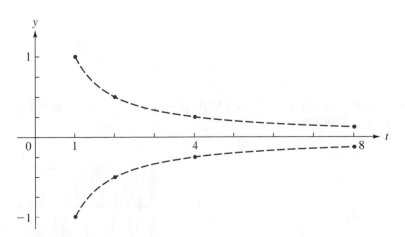

FIGURE 10

Step 2   Now sketch the graph of $y = \sin(\pi t/2)$, but keep high and low points within the envelope (Fig. 11).

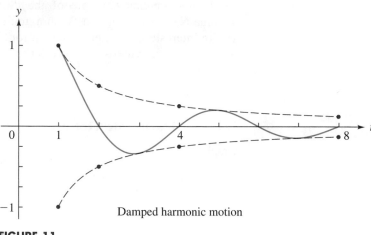

Damped harmonic motion

FIGURE 11

Matched Problem 5   Graph $y = t \sin(\pi t/2)$, $1 \le t \le 8$, and indicate the type of motion.

EXPLORE/DISCUSS 3

Graph the following equations in a graphing utility, along with their envelopes, and discuss whether each represents simple harmonic motion, damped harmonic motion, or resonance.

(A) $y = t^{0.7} \sin(\pi t/2)$, $1 \le t \le 10$

(B) $y = t^{-0.5} \sin(\pi t/2)$, $1 \le t \le 10$

Answers to Matched Problems

1. (A) Amplitude = 25 amperes; period = $\frac{1}{15}$ sec; frequency = 15 Hz; phase shift = $-\frac{1}{6}$ sec

(B)

25

0   0.3

−25

2. Period $= 10^{-6}$ sec; $\lambda = 300$ m

3. $B = 2\pi \times 10^{20}$

4. Height: 22 feet; period: π sec; wavelength: 51 ft; speed: 16 ft/sec or 11 mi/hr

5.

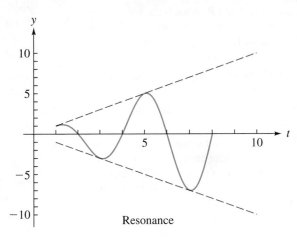

Resonance

EXERCISE 3.4

Applications

In these applications assume all given values are exact unless indicated otherwise.

1. Modeling Electric Current   An alternating current generator produces a current given by

$$I = 10 \sin(120\pi t - \pi/2)$$

where t is time in seconds and I is in amperes.

(A) What are the amplitude, frequency, and phase shift for this current?

(B) Explain how you would find the maximum current in this circuit and find it.

 (C) Graph the equation in a graphing calculator for $0 \le t \le 0.1$. How many periods are shown in the graph?

2. Modeling Electric Current   An alternating current generator produces a current given by

$$I = -50 \cos(80\pi t + 2\pi/3)$$

where t is time in seconds and I is in amperes.

(A) What are the amplitude, frequency, and phase shift for this current?

(B) Explain how you would find the maximum current in this circuit and find it.

(C) Graph the equation in a graphing calculator for $0 \le t \le 0.1$. How many periods are shown in the graph?

3. Modeling Electric Current   An alternating current generator produces a 30 Hz current flow with a maximum value of 20 amperes. Write an equation in the form $I = A \cos Bt$, $A > 0$, for this current.

4. Modeling Electric Current   An alternating current generator produces a 60 Hz current flow with a maximum value of 10 amperes. Write an equation of the form $I = A \sin Bt$, $A > 0$, for this current.

5. Modeling Water Waves   A water wave at a fixed position has an equation of the form

$$y = 15 \sin \frac{\pi}{8} t$$

where t is time in seconds and y is in feet. How high is the wave from trough to crest (see Fig. 6)? What is its wavelength in feet? How fast is it traveling in feet per second? Calculate your answers to the nearest foot.

6. Modeling Water Waves A water wave has an amplitude of 30 ft and a period of 14 sec. If its equation is given by

$$y = A \sin Bt$$

at a fixed position, find A and B. What is the wavelength in feet? How fast is it traveling in feet per second? How high would the wave be from trough to crest? Calculate your answers to the nearest foot.

7. Modeling Water Waves Graph the equation in Problem 5 for $0 \leq t \leq 32$.

8. Modeling Water Waves Graph the equation in Problem 6 for $0 \leq t \leq 28$.

9. Modeling Water Waves A tsunami is a sea wave caused by an earthquake. These are very long waves that can travel at nearly the rate of a jet airliner. (The speed of a tsunami, whose wavelength far exceeds the ocean's depth D, is approximately $\sqrt{gD}$, where g is the gravitational constant.) At sea, these waves have an amplitude of only 1 or 2 ft, so a ship would be unaware of the wave passing underneath. As a tsunami approaches a shore, however, the waves are slowed down and the water piles up to a virtual wall of water, sometimes over 100 ft high.

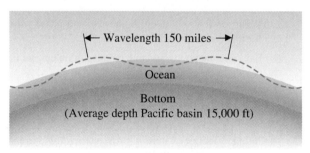

Tsunami (vertical exaggerated)

Figure for 9

When such a wave crashes into land, considerable destruction can result—more than 250,000 lives were lost in December 2004 when a tsunami, caused by an earthquake near Sumatra, hit the coastlines around the Indian Ocean.

(A) If at a particular moment in time, a tsunami has an equation of the form

$$y = A \sin Br$$

where y is in feet and r is the distance from the source in miles, find the equation if the amplitude is 2 ft and the wavelength is 150 mi.

(B) Find the period (in seconds) of the tsunami in part (A).

10. Modeling Water Waves Because of increased friction from the ocean bottom, a wave will begin to break when the ocean depth becomes less than one-half the wavelength. If an ocean wave at a fixed position has an equation of the form

$$y = 8 \sin \frac{\pi}{5} t$$

at what depth (to the nearest foot) will the wave start to break?

11. Modeling Water Waves A water wave has an equation of the form

$$y = 25 \sin \left[2\pi \left(\frac{t}{10} + \frac{r}{512} \right) \right]$$

where y and r are in feet and t is in seconds.

(A) Describe what the equation models if r (the distance from the source) is held constant at $r = 1,024$ ft and t is allowed to vary.

(B) Under the conditions in part (A), compute the period or wavelength, whichever is appropriate, and explain why you made your choice.

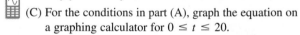

(C) For the conditions in part (A), graph the equation on a graphing calculator for $0 \leq t \leq 20$.

12. Modeling Water Waves Referring to the equation in Problem 11:

(A) Describe what the equation models if t (time in seconds) is frozen at $t = 0$ and r is allowed to vary.

(B) Under the conditions in part (A), compute the period or wavelength, whichever is appropriate, and explain why you made your choice. What is the height of the wave from trough to crest?

(C) For the conditions in part (A), graph the equation on a graphing calculator for $0 \leq r \leq 1,024$. What does the distance between two consecutive crests represent?

13. Modeling Electromagnetic Waves Suppose an electron oscillates at 10^8 Hz($\nu = 10^8$ Hz), creating an electromagnetic wave. What is its period? What is its wavelength (in meters)?

14. Modeling Electromagnetic Waves Repeat Problem 13 with $\nu = 10^{18}$ Hz.

15. Modeling Electromagnetic Waves An X ray has an equation of the form

$$y = A \sin Bt$$

Find B if the wavelength of the X ray, λ, is 3×10^{-10} m.
[*Note:* $c \approx 3 \times 10^8$ m/sec]

16. **Modeling Electromagnetic Waves** A microwave has an equation of the form

$$y = A \sin Bt$$

Find B if the wavelength is $\lambda = 0.003$ m.

17. **Modeling Electromagnetic Waves** An AM radio wave for a given station has an equation of the form

$$y = A[1 + 0.02 \sin(2\pi \cdot 1{,}200t)] \sin(2\pi \cdot 10^6 t)$$

for a given 1,200 Hz tone. The expression in brackets modulates the amplitude A of the carrier wave,

$$y = A \sin(2\pi \cdot 10^6 t)$$

(see Fig. 5, page 180). What are the period and frequency of the carrier wave for time t in seconds? Can a wave with this frequency pass through the atmosphere (see Fig. 2, page 178)?

18. **Modeling Electromagnetic Waves** An FM radio wave for a given station has an equation of the form

$$y = A \sin[2\pi \cdot 10^9 t + 0.02 \sin(2\pi \cdot 1{,}200t)]$$

for a given tone of 1,200 Hz. The second term within the brackets modulates the frequency of the carrier wave, $y = A \sin(2\pi \cdot 10^9 t)$ (see Fig. 5, page 180). What are the period and frequency of the carrier wave? Can a wave with this frequency pass through the atmosphere (see Fig. 2, page 178)?

19. **Spring–Mass Systems** Graph each of the following equations representing the vertical motion y, relative to time t, of a mass attached to a spring. Indicate whether the motion is simple harmonic, is damped harmonic, or illustrates resonance.

(A) $y = -5 \cos(4\pi t)$, $0 \le t \le 2$

(B) $y = -5e^{-1.2t} \cos(4\pi t)$, $0 \le t \le 2$

20. **Spring–Mass Systems** Graph each of the following equations representing the vertical motion y, relative to time t, of a mass attached to a spring. Indicate whether the motion is simple harmonic, is damped harmonic, or illustrates resonance.

(A) $y = \sin(4\pi t)$, $0 \le t \le 2$

(B) $y = e^{1.25t} \sin(4\pi t)$, $0 \le t \le 2$

21. **Bungee Jumping** The motion of a bungee jumper after reaching the lowest point in the jump and being pulled back up by the cord can be modeled by a damped harmonic. One particular jump can be described by

$$y = 42 - \frac{36}{t + 2} \cos(0.989t)$$

where y is height in meters above the ground t seconds after the low point of the jump is reached.

(A) Find the highest and lowest point described by the model.

(B) How many times does the jumper bounce in the first 30 sec, counting the initial bounce?

(C) We can consider the jump to be over when the jumper's maximum and minimum heights are within 1 ft of the resting height of 42 ft. How long does the jump last, to the nearest second?

22. **Bungee Jumping** Repeat Problem 21 for a different jumper whose jump, using the same equipment from the same height, is modeled by

$$y = 42 - \frac{50}{t + 2} \cos(0.786t)$$

Do you think this jumper is heavier or lighter than the one in Problem 21? Why?

23. **Modeling Body Temperatures** The body temperature of a healthy adult varies during the day, with the lowest temperature occurring about 6 A.M. and the highest about 6 P.M. Table 1 lists the daily maximum and minimum temperatures for both men and women using two different methods for measuring temperature. Find models of the form $t = k + A \sin Bx$ for the daily variation in body temperature of men and of women if an oral thermometer is used. Graph both models on the same axes for $0 \le x \le 24$.

24. **Modeling Body Temperatures** Find models of the form $t = k + A \sin Bx$ for the daily variation in body temperature of men and of women if an aural thermometer is used. Graph both models on the same axes for $0 \le x \le 24$.

TABLE 1

Daily Body Temperatures

	Oral (°C) Minimum Maximum		Aural (°C) Minimum Maximum	
	Minimum	Maximum	Minimum	Maximum
Men:	35.7	37.7	35.5	37.5
Women:	33.2	38.1	35.7	37.5

Source: Scand J Caring Sci 2002 16(2):122–8.

3.5   Graphing Combined Forms

- Graphing by Hand Using Addition of Ordinates
- Graphing Combined Forms on a Graphing Calculator
- Sound Waves
- Fourier Series: A Brief Look

Having considered the trigonometric functions individually, we will now consider them in combination with each other and with other functions. To help develop an understanding of basic ideas, we will start by hand graphing an easy combination using a technique called *addition of ordinates*. (Recall that the ordinate of a point is its second coordinate.) It soon becomes clear that for more complicated combinations, hand graphing is not practical, and we will focus our attention on the use of a graphing calculator.

■ Graphing by Hand Using Addition of Ordinates

The process of graphing by hand using addition of ordinates is best illustrated through an example. The process is used to introduce basic principles and is not practical for most problems of significance.

EXAMPLE 1

Addition of Ordinates

Graph

$$y = \frac{x}{2} + \sin x \qquad 0 \le x \le 2\pi$$

Solution   One method of hand graphing equations involving two or more terms is to graph each term separately on the same axes, and then add ordinates. In essence, we are adding the outputs (heights) of the individual terms to find the output (height) of the sum. In this case, we will work with the terms

$$y_1 = \frac{x}{2} \quad \text{and} \quad y_2 = \sin x$$

We can draw the graphs of both equations on the same axes, and then use a compass, ruler, or eye to add the ordinates $y_1 + y_2$ (see Fig. 1). The final graph of $y = (x/2) + \sin x$ is shown in Figure 2. If greater accuracy is needed, it would be a good idea to use a calculator to determine specific points on the graph.

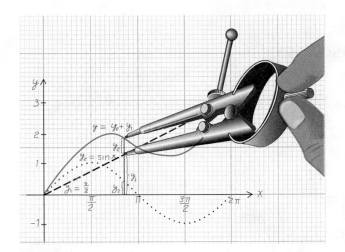

FIGURE 1
Addition of ordinates

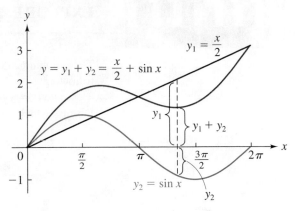

FIGURE 2
Addition of ordinates

Matched Problem 1   Graph $y = (x/2) + \cos x, 0 \leq x \leq 2\pi$, using addition of ordinates. ■

 ■ **Graphing Combined Forms on a Graphing Calculator**

We will now repeat Example 1 on a graphing calculator.

EXAMPLE 2   **Addition of Ordinates on a Graphing Calculator**

Graph the following equations from Example 1 on a graphing calculator. Graph all in the same viewing window for $0 \leq x \leq 2\pi$. Watch each graph as it is traced out. Plot y_3 with a darker line so that it stands out.

$$y_1 = \frac{x}{2} \qquad y_2 = \sin x \qquad y_3 = \frac{x}{2} + \sin x$$

Solution

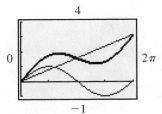

Matched Problem 2   Graph $y_1 = x/2, y_2 = \cos x$, and $y_3 = x/2 + \cos x$, $0 \leq x \leq 2\pi$, in the same viewing window. Plot y_3 with a darker line so that it stands out. ■

EXPLORE/DISCUSS 1

The graphing calculator display in Figure 3 illustrates the graph of one of the following:

$$(1)\ y_1 = y_2 + y_3 \qquad (2)\ y_2 = y_1 + y_3 \qquad (3)\ y_3 = y_1 + y_2$$

Discuss which one and why.

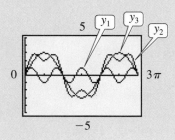

FIGURE 3

A graphing calculator can plot complicated combined forms almost as easily and quickly as it plots simple forms. It is, after all, a computer, which means it's really good at computing the locations of many points.

 **EXAMPLE 3**

Graphing Combined Forms on a Graphing Calculator

(A) Graph $y = 3 \sin x + \cos(3x), \quad 0 \le x \le 3\pi$.

(B) What does the period appear to be?

Solution   (A)

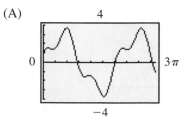

(B) We should be able to find the period by finding the distance between the two high points of the graph. We can locate them using the $\boxed{\text{MAXIMUM}}$ command:

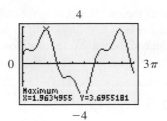

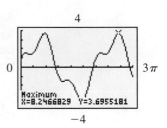

The distance between x values is $8.247 - 1.963 = 6.284$. This is a close approximation of 2π, so the period appears likely to be 2π. ■

Matched Problem 3   (A) Graph $y = 2 \cos x - 2 \sin(2x)$, $0 \le x \le 4\pi$.

(B) What does the period appear to be? ■

■ Sound Waves

Sound is produced by a vibrating object that, in turn, excites air molecules into motion. The vibrating air molecules cause a periodic change in air pressure that travels through air at about 1,100 ft/sec. (Sound is not transmitted in a vacuum.) When this periodic change in air pressure reaches your eardrum, the drum vibrates at the same frequency as the source, and the vibration is transmitted to the brain as sound. The range of audible frequencies covers about 10 octaves, extending from 20 Hz up to 20,000 Hz. Low frequencies are associated with low pitch and high frequencies with high pitch.

FIGURE 4
A simple sound wave

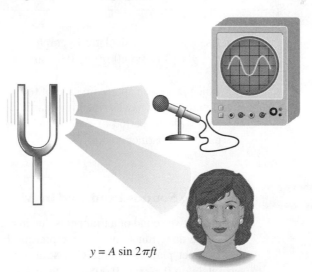

$y = A \sin 2\pi ft$

If in place of an eardrum we use a microphone, then the air disturbance can be changed into a pulsating electrical signal that can be visually displayed on an oscilloscope (Fig. 4). A pure tone from a tuning fork will look like a sine curve, also called a sine wave. Actually, the sound from a tuning fork can be accurately described by either the sine function or the cosine function. For example, a tuning fork vibrating at 264 Hz ($f = 264$) with an amplitude of 0.002 in. produces C on the musical scale, and the wave on an oscilloscope can be described by the simple harmonic

$$y = A \sin 2\pi f t$$
$$= 0.002 \sin 2\pi(264)t$$

Most sounds are more complex than that produced by a tuning fork. Figure 5 (page 195) illustrates a note produced by a guitar. Musical tones are typically composed of a number of pure tones, like those from a tuning fork. This is exactly why different musical instruments sound different when playing the same notes: the combinations of partial tones are different.

A *harmonic tone* is a sum of pure tones (called *harmonics*) with frequencies that are integer multiples of the tone with the lowest frequency, which is called the **fundamental tone**. The other harmonics of the fundamental tone are called **overtones** (see Fig. 5). Electronic synthesizers simulate the sounds of other instruments by electronically producing an instrument's harmonic tones.

One way of describing complex sound forms is with a combination of trigonometric functions known as a *Fourier series*, which we will touch on later in this section.

EXPLORE/DISCUSS 2

(A) Use a graphing calculator to graph
$$y = 0.12 \sin 400\pi t + 0.04 \sin 800\pi t + 0.02 \sin 1{,}200\pi t$$
$$0 \le t \le 0.01$$
which is a close approximation of the guitar note in Figure 5.

(B) Based on graphical evidence, what is the period of y? How does it compare with the period of each term? What do you suspect determines the period for harmonic tones?

EXAMPLE 4

Modeling Sounds Produced by Musical Instruments

The distinctive sound of a trumpet is due in part to the high amplitudes of its overtones. The three functions below represent the first three harmonics for a particular trumpet playing middle C. Use a graphing calculator to graph each tone separately on $0 \le t \le 0.008$, and then graph the sum of all three. What do you notice about the period of the combined form?

$$y_1 = 0.30 \sin(528\pi t)$$
$$y_2 = 0.28 \sin(1{,}056\pi t)$$
$$y_3 = 0.22 \sin(2{,}112\pi t)$$

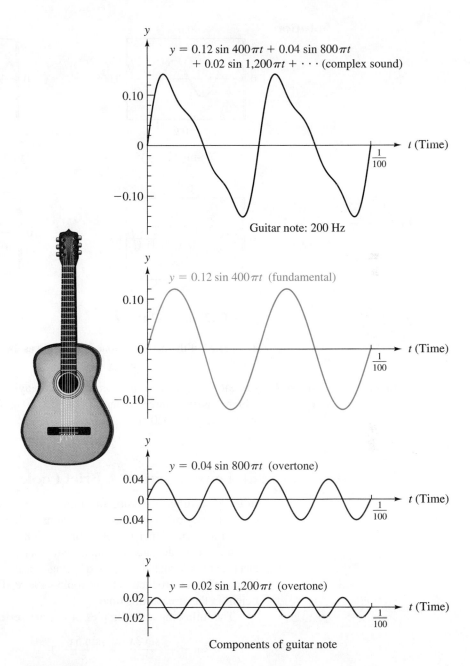

FIGURE 5
A complex sound wave—the guitar note shown at the top—can
be approximated very closely by adding the three simple
harmonics (pure tones) below

Solution

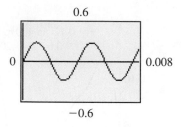

$y_1 = 0.30 \sin(528\pi t)$

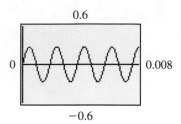

$y_2 = 0.28 \sin(1{,}056\pi t)$

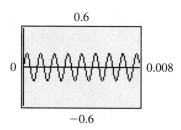

$y_3 = 0.22 \sin(2{,}112\pi t)$

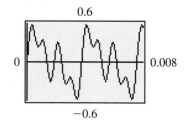

$y = y_1 + y_2 + y_3$

The period of the harmonic tone is the same as the period of the fundamental tone. ■

Matched Problem 4   Repeat Example 4 for the same trumpet playing a low A (frequency of 220 Hz), which has first three harmonics $y_1 = 0.30 \sin(440\pi t)$, $y_2 = 0.28 \sin(880\pi t)$, $y_3 = 0.22 \sin(1{,}320\pi t)$. ■

■ Fourier Series: A Brief Look

In the preceding discussion on sound waves, we mentioned a significant use of combinations of trigonometric functions called **Fourier series** (named after the French mathematician Joseph Fourier, 1768–1830). These combinations may be encountered in advanced applied mathematics in the study of sound, heat flow, electrical fields and circuits, and spring–mass systems. The following discussion is only a brief introduction to the topic—the reader is not expected to become proficient in this area at this time.

The following are examples of Fourier series:

$$y = \sin x + \frac{\sin 3x}{3} + \frac{\sin 5x}{5} + \cdots \tag{1}$$

$$y = \sin \pi x + \frac{\sin 2\pi x}{2} + \frac{\sin 3\pi x}{3} + \cdots \tag{2}$$

The three dots at the end of each series indicate that the pattern established in the first three terms continues indefinitely. (Note that the terms of each series are simple harmonics.) If we graph the first term of each series, then the sum of the first two terms, and so on, we will obtain a sequence of graphs that will get closer and closer to a **square wave** for (1) and a **sawtooth wave** for (2). The greater the number of terms we take in the series, the more the graph will look like the indicated wave form. Figures 6 and 7 illustrate these phenomena.

FIGURE 6
Square wave

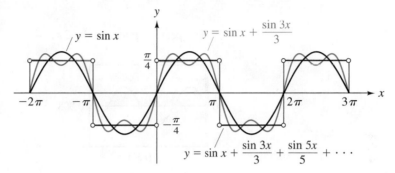
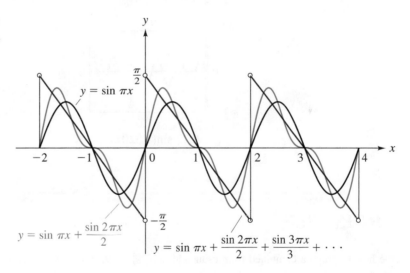

FIGURE 7
Sawtooth wave

Answers to Matched Problems

1.

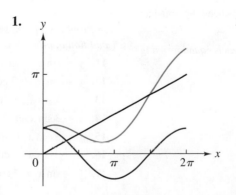

2.

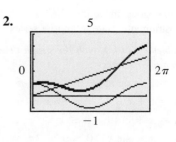

3. (A)

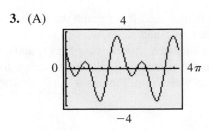

(B) The period appears to be 2π.

4.

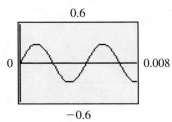

$y_1 = 0.30 \sin(440\pi t)$

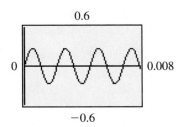

$y_2 = 0.28 \sin(880\pi t)$

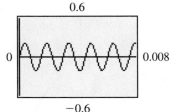

$y_3 = 0.22 \sin(1,320\pi t)$

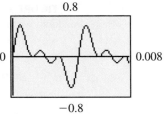

$y = y_1 + y_2 + y_3$

EXERCISE 3.5

A **1.** Describe how to graph a combined form using addition of ordinates.

  2. How does modeling the tones produced by musical instruments relate to Fourier series?

In Problems 3–10, sketch the graph of each equation using addition of ordinates.

  3. $y = 1 + \sin x$, $-\pi \le x \le \pi$

  4. $y = 1 + \cos x$, $-\pi \le x \le \pi$

  5. $y = x + \cos x$, $0 \le x \le 5\pi/2$

  6. $y = x + \sin x$, $0 \le x \le 2\pi$

  7. $y = x/2 + \cos \pi x$, $0 \le x \le 3$

  8. $y = x/2 - \sin 2\pi x$, $0 \le x \le 2$

  9. $y = x - \cos 2\pi x$, $-1 \le x \le 1$

  10. $y = x - \sin \pi x$, $-2 \le x \le 2$

B *In Problems 11–18, sketch the graph of each equation using addition of ordinates.*

  11. $y = \sin x + \cos x$, $0 \le x \le 2\pi$

  12. $y = \sin x + 2 \cos x$, $0 \le x \le 2\pi$

  13. $y = 3 \sin x + \cos x$, $0 \le x \le 2\pi$

  14. $y = 3 \sin x + 2 \cos x$, $0 \le x \le 2\pi$

  15. $y = 3 \cos x + \sin 2x$, $0 \le x \le 3\pi$

  16. $y = 3 \cos x + \cos 3x$, $0 \le x \le 2\pi$

  17. $y = \sin x + 2 \cos 2x$, $0 \le x \le 3\pi$

18. $y = \cos x + 3 \sin 2x, \quad 0 \le x \le 3\pi$

Problems 19–22 require the use of a graphing calculator.

19. Graph the following equation for $-2\pi \le x \le 2\pi$ and $-1 \le y \le 1$ (compare to the square wave in Fig. 6, page 197).

$$y = \sin x + \frac{\sin 3x}{3} + \frac{\sin 5x}{5} + \frac{\sin 7x}{7}$$
$$+ \frac{\sin 9x}{9}$$

20. Graph the following equation for $-4 \le x \le 4$ and $-2 \le y \le 2$ (compare to the sawtooth wave in Fig. 7, page 197).

$$y = \sin \pi x + \frac{\sin 2\pi x}{2} + \frac{\sin 3\pi x}{3} + \frac{\sin 4\pi x}{4}$$
$$+ \frac{\sin 5\pi x}{5}$$

21. A particular Fourier series is given as follows:

$$y = 1.15 + \cos x + \frac{\cos 3x}{3^2} + \frac{\cos 5x}{5^2} + \cdots$$

(A) Using the window dimensions $-4\pi \le x \le 4\pi$ and $0 \le y \le 4$, graph in three separate windows: the first two terms of the series, the first three terms of the series, and the first four terms of the series.

(B) Describe the shape of the wave form the graphs tend to approach as more terms are added to the series.

(C) As additional terms are added to the series, the graphs appear to approach the wave form you described in part (B), and that form is composed entirely of straight line segments. Hand-sketch a graph of this wave form.

22. A particular Fourier series is given as follows:

$$y = \cos \frac{x}{2} + \frac{1}{9} \cos \frac{3x}{2} + \frac{1}{25} \cos \frac{5x}{2} + \cdots$$

(A) Using the window dimensions $-4\pi \le x \le 4\pi$ and $-2 \le y \le 2$, graph in two separate windows: the first two terms of the series and the first three terms of the series.

(B) Describe the shape of the wave form the graphs tend to approach as more terms are added to the series.

(C) As additional terms are added to the series, the graphs appear to approach the wave form you described in part (B), and that form is composed entirely of straight line segments. Hand-sketch a graph of this wave form.

Applications

23. Physiology A normal seated adult breathes in and exhales about 0.80 L of air every 4.00 sec. See the figure. The volume V of air in the lungs t seconds after exhaling can be approximated by an equation of the form $V = k + A \cos Bt, 0 \le t \le 8$. Find the equation.

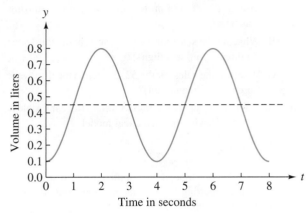

Figure for 23

24. Earth Science During the autumnal equinox, the surface temperature (in °C) of the water of a lake x hours after sunrise was recorded over a 24 hr period, and the results were recorded in the figure. The surface temperature can be approximated by an equation of the form $T = k + A \cos Bx$. Find the equation.

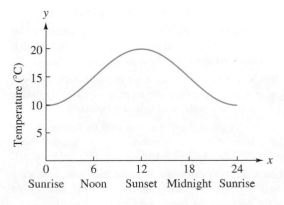

Figure for 24

25. *Modeling a Seasonal Business Cycle* A large soft drink company's sales vary seasonally, but overall the company is also experiencing a steady growth rate. The financial analysis department has developed the following mathematical model for sales:

$$S = 5 + \frac{t}{52} - 4 \cos \frac{\pi t}{26}$$

where S represents sales in millions of dollars for a week of sales t weeks after January 1.

(A) Graph this function for a 3 yr period starting January 1.

(B) What are the sales for the 26th week in the third year (to three significant digits)?

(C) What are the sales for the 52nd week in the third year (to one significant digit)?

26. *Modeling a Seasonal Business Cycle* Repeat Problem 25 for the following mathematical model:

$$S = 4 + \frac{t}{52} - 2 \cos \frac{\pi t}{26}$$

27. *Sound* A pure tone is transmitted through a speaker that is also emitting high-frequency, low-volume static. The resulting sound is given by

$$y = 0.1 \sin 500\pi t + 0.003 \sin 24{,}000\,\pi t$$

(A) Graph this equation, using a graphing calculator, for $0 \le t \le 0.004$ and $-0.3 \le y \le 0.3$. Notice that the result looks like a pure sine wave.

(B) Use the ZOOM feature (or its equivalent) to zoom in on the graph in the vicinity of $(0.001, 0.1)$. You should now *see* the static element of the sound.

28. *Sound* Another pure tone is transmitted through the same speaker as in Problem 27. The resulting sound is given by

$$y = 0.08 \cos 500\pi t + 0.004 \sin 26{,}000\,\pi t$$

(A) Graph this equation, using a graphing calculator, for $0 \le t \le 0.004$ and $-0.3 \le y \le 0.3$. Notice that the sound appears to be a pure tone.

(B) Use the ZOOM feature (or its equivalent) to zoom in on the graph in the vicinity of $(0.002, -0.08)$. Now you should be able to *see* the static.

29. *Music* A guitar playing a note with frequency 200 Hz has first three harmonics $y_1 = 0.08 \sin(400\pi t)$, $y_2 = 0.04 \sin(800\pi t)$, and $y_3 = 0.02 \sin(1{,}200\pi t)$. Use a graphing calculator to graph each tone separately on $0 \le t \le 0.011$, then graph the sum of all three. What is the period of the combined form?

30. *Music* A synthesizer playing a note with frequency 100 Hz has first three harmonics $y_1 = 0.06 \sin(200\pi t)$, $y_2 = 0.05 \sin(400\pi t)$, and $y_3 = 0.03 \sin(600\pi t)$. Use a graphing calculator to graph each tone separately on $0 \le t \le 0.022$, then graph the sum of all three. What is the period of the combined form?

31. *Experiment on Wave Interference* When wave forms are superposed, they may tend to reinforce or cancel each other. Such effects are called **interference.** If the resultant amplitude is increased, the interference is said to be **constructive;** if it is decreased, the interference is **destructive.** Interference applies to all wave forms, including sound, light, and water. In the following three equations, y_3 represents the superposition of the individual waves y_1 and y_2:

$$y_1 = 2 \sin t \quad y_2 = 2 \sin(t + C) \quad y_3 = y_1 + y_2$$

For each given value of C that follows, graph these three equations in the same viewing window $(0 \le t \le 4\pi)$ and indicate whether the interference shown by y_3 is constructive or destructive. (Graph y_3 with a darker line so it will stand out.)

(A) $C = 0$ (B) $C = -\pi/4$

(C) $C = -3\pi/4$ (D) $C = -\pi$

32. *Experiment on Wave Interference* Repeat Problem 31 with the following three equations:

$$y_1 = 2 \cos t \quad y_2 = 2 \cos(t + C) \quad y_3 = y_1 + y_2$$

33. *Noise Control* Jet engines, air-conditioning systems in commercial buildings, and automobile engines all have something in common—NOISE! Destructive interference (see Problem 31) provides an effective tool for noise control. Acoustical engineers have devised an

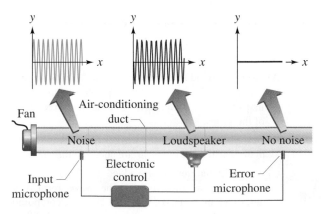

Figure for 33 and 34

Active Noise Control (ANC) system that is now widely used. The figure illustrates a system used to reduce noise in air-conditioning ducts. The input microphone senses the noise sound wave, processes it electronically, and produces the same sound wave in the loudspeaker but out of phase with the original. The resulting destructive interference effectively reduces the noise in the system.

(A) Suppose the input microphone indicates a sound wave given by $y_1 = 65 \sin 400\pi t$, where y_1 is sound intensity in decibels and t is time in seconds. Write an equation of the form $y_2 = A \sin(Bt + C)$, $C < 0$, that represents a sound wave that will have total destructive interference with y_1. [*Hint:* The results of Problem 31 should help.]

(B) Graph y_1, y_2, and $y_3 = y_1 + y_2$, $0 \leq t \leq 0.01$, in the same viewing window of a graphing calculator, and explain the results.

34. Noise Control Repeat Problem 33 with the noise given by $y_1 = 30 \cos 600\pi t$.

Problems 35 and 36 use the data provided in Tables 1 and 2, and require use of a graphing calculator with the ability to find sinusoidal regression curves. Round all numbers to three significant digits.

35. Modeling Carbon Dioxide in the Atmosphere The reporting station at Mauna Loa, Hawaii, has been recording CO_2 levels in the atmosphere for almost 50 years. Table 1 contains the levels for 2003–2006.

(A) Enter the data in Table 1 from January 2003 to December 2006 $(1 \leq x \leq 48)$ and produce a scatter plot of the data.

(B) Enter the annual averages for each year using $x = 6.5, 18.5, 30.5$, and 42.5 and find the linear regression line, $f(x)$, for these four points. Display the graph of $f(x)$ and the scatter plot from part (A) on the same graphing calculator screen.

(C) Subtract $f(x)$, $1 \leq x \leq 48$, from the corresponding data points in part (A) and produce a scatter plot for this modified data.

(D) Find a sinusoidal regression curve, $s(x)$, for the modified data in part (C). Display the graph of $s(x)$ and the scatter plot in part (C) on the same graphing calculator screen. Is $s(x)$ a good model for the modified data? Explain.

(E) Display the graph of $g(x) = f(x) + s(x)$ and the scatter plot in part (A) on the same graphing calculator screen. Does $g(x)$ appear to be a good model for the original data in Table 1?

36. Modeling Carbon Dioxide in the Atmosphere Table 2 on the next page contains the CO_2 levels from a reporting station located at Barrow, Alaska.

(A) Enter the data in Table 2 from January 2003 to December 2006 $(1 \leq x \leq 48)$ and produce a scatter plot of the data.

(B) Enter the annual averages for each year using $x = 6.5, 18.5, 30.5$, and 42.5 and find the linear regression line, $f(x)$, for these four points. Display the graph of $f(x)$ and the scatter plot from part (A) on the same graphing calculator screen.

(C) Subtract $f(x)$, $1 \leq x \leq 48$, from the corresponding data points in part (A) and produce a scatter plot for this modified data.

TABLE 1

CO$_2$ Levels at Mauna Loa, in Parts per Million (ppm)

	Jan	Feb	Mar	Apr	May	June	Jul	Aug	Sep	Oct	Nov	Dec	Average
2003	375	376	377	378	379	378	377	374	373	373	375	376	376
2004	377	378	379	380	381	380	377	376	374	374	376	378	378
2005	378	380	381	382	382	382	381	379	377	377	378	380	380
2006	381	382	383	385	385	384	382	380	379	379	380	382	382

TABLE 2

CO_2 Levels at Barrow, in Parts per Million (ppm)												
	Jan	Feb	Mar	Apr	May	June	Jul	Aug	Sep	Oct	Nov	Dec	Average
2003	381	383	381	382	384	379	372	366	368	373	378	380	377
2004	382	383	384	385	385	382	373	367	369	375	379	383	379
2005	384	385	386	386	385	383	378	371	372	378	382	383	381
2006	385	388	388	388	389	387	376	370	374	378	385	389	383

(D) Find a sinusoidal regression curve, $s(x)$, for the modified data in part (C). Display the graph of $s(x)$ and the scatter plot in part (C) on the same graphing calculator screen. Is $s(x)$ a good model for the modified data? Explain.

(E) Use the modified data to construct a model of the form $t(x) = A \sin(Bx) + C$. Display the graph of $t(x)$ and the scatter plot in part (C) on the same graphing calculator screen. Is $t(x)$ a good model for the modified data? Explain.

(F) Display the graph of $g(x) = f(x) + t(x)$ and the scatter plot in part (A) on the same graphing calculator screen. Does $g(x)$ appear to be a good model for the original data in Table 2?

3.6   Tangent, Cotangent, Secant, and Cosecant Functions Revisited

- Graphing $y = A \tan(Bx + C)$ and $y = A \cot(Bx + C)$
- Graphing $y = A \sec(Bx + C)$ and $y = A \csc(Bx + C)$

We will now look at graphing the more general forms of the tangent, cotangent, secant, and cosecant functions following essentially the same process we developed for graphing $y = A \sin(Bx + C)$ and $y = A \cos(Bx + C)$. The process is not difficult if you have a clear understanding of the basic graphs for these functions, including periodic properties. In each case, the key will be to locate the asymptotes, which are the most noticeable features of these graphs.

■ Graphing $y = A \tan(Bx + C)$ and $y = A \cot(Bx + C)$

The basic graphs for $y = \tan x$ and $y = \cot x$ that were developed in Section 3.1 are repeated here for convenient reference.

GRAPH OF $y = \tan x$

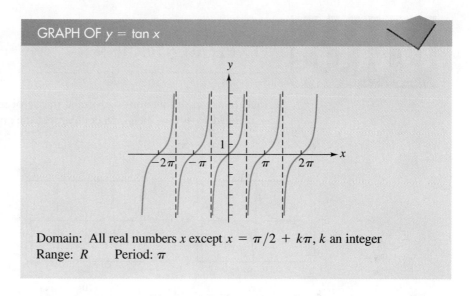

Domain: All real numbers x except $x = \pi/2 + k\pi$, k an integer
Range: R Period: π

Notice that two of the vertical asymptotes are $x = -\pi/2$ and $x = \pi/2$. This will be a big help in graphing.

GRAPH OF $y = \cot x$

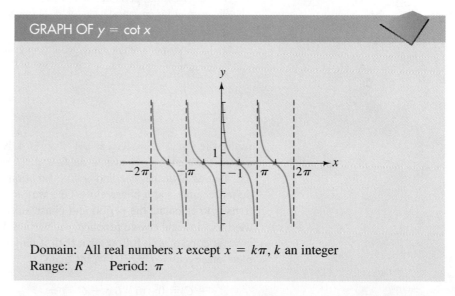

Domain: All real numbers x except $x = k\pi$, k an integer
Range: R Period: π

This time, note that two of the vertical asymptotes are $x = 0$ and $x = \pi$.

To quickly sketch the graphs of equations of the form $y = A \tan(Bx + C)$ and $y = A \cot(Bx + C)$, you need to know how the constants A, B, and C affect the basic graphs $y = \tan x$ and $y = \cot x$, respectively.

First note that amplitude is not defined for the tangent and cotangent functions, since the deviation from the x axis for both functions is indefinitely far in both directions. The effect of A is to make the graph steeper if $|A| > 1$ or to

EXPLORE/DISCUSS 1

(A) Match each function to its graph in one of the graphing calculator displays and discuss how the graph compares to the graph of $y = \tan x$ or $y = \cot x$.

    (1) $y = \cot 2x$ (2) $y = 3 \tan x$ (3) $y = \tan(x + \pi/2)$

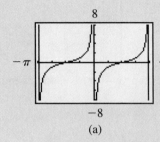
(a)

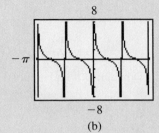
(b)

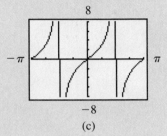
(c)

(B) Use a graphing calculator to explore the nature of the changes in the graphs of the following functions when the values of A, B, and C are changed. Discuss what happens in each case.

    (4) $y = A \tan x$ and $y = A \cot x$

    (5) $y = \tan Bx$ and $y = \cot Bx$

    (6) $y = \tan(x + C)$ and $y = \cot(x + C)$

make the curve less steep if $|A| < 1$. If A is negative, the graph is reflected across the x axis (turned upside down).

Just as with the sine and cosine functions, the constants B and C involve a period change and a phase shift. One way to attack the graphing of these functions is to calculate the period and phase shift using the same procedure that we used for sine and cosine functions in Section 3.3. Since both $A \tan x$ and $A \cot x$ have a period of π, it follows that both $A \tan(Bx + C)$ and $A \cot(Bx + C)$ complete one cycle as $Bx + C$ varies from

$$Bx + C = 0 \quad \text{to} \quad Bx + C = \pi$$

or, solving for x, as x varies from

$$x = -\frac{C}{B} \quad \text{to} \quad x = -\frac{C}{B} + \frac{\pi}{B}$$

                                    └── Phase shift ──┘   Period

This tells us that $y = A \tan(Bx + C)$ and $y = A \cot(Bx + C)$ each have a period of π/B, and their graphs are translated horizontally $-C/B$ units to the right if $-C/B > 0$ and $|-C/B|$ units to the left if $-C/B < 0$. As before, the horizontal translation determined by the number $-C/B$ is the phase shift.

There is a second approach to graphing general tangent and cotangent functions that is probably simpler than the first. We'll focus on locating the asymptotes, then we will fill in the graph between the asymptotes, using our knowledge of the standard graphs for tangent and cotangent.

Earlier we pointed out that $y = \tan x$ has asymptotes at $x = -\pi/2$ and $x = \pi/2$. We can then find two asymptotes for $y = A \tan(Bx + C)$ by solving $Bx + C = -\pi/2$ and $Bx + C = \pi/2$. Once we know two asymptotes, we can find as many as we like by repeating the distance between the first two that we found. The procedure is illustrated in Example 1.

EXAMPLE 1

Graphing $y = A \tan(Bx + C)$

Sketch the graph of

$$y = 3 \tan\left(\frac{\pi}{2}x + \frac{\pi}{4}\right) \qquad \text{for} \qquad -\frac{7}{2} < x < \frac{5}{2}$$

What are the period and phase shift?

Solution   To find two asymptotes, we set $(\pi/2)x + \pi/4$ equal to $-\pi/2$ and $\pi/2$ and solve for x.

$$\frac{\pi}{2}x + \frac{\pi}{4} = -\frac{\pi}{2} \qquad\qquad \frac{\pi}{2}x + \frac{\pi}{4} = \frac{\pi}{2}$$

$$\frac{\pi}{2}x = -\frac{\pi}{2} - \frac{\pi}{4} \qquad\qquad \frac{\pi}{2}x = \frac{\pi}{2} - \frac{\pi}{4}$$

$$x = -\frac{\pi}{2} \cdot \frac{2}{\pi} - \frac{\pi}{4} \cdot \frac{2}{\pi} \qquad\qquad x = \frac{\pi}{2} \cdot \frac{2}{\pi} - \frac{\pi}{4} \cdot \frac{2}{\pi}$$

$$x = -1 - \frac{1}{2} = -\frac{3}{2} \qquad\qquad x = 1 - \frac{1}{2} = \frac{1}{2}$$
$$\underset{\text{Phase shift}}{\Big\uparrow} \qquad\qquad\qquad \underset{\text{Phase shift}}{\Big\uparrow}$$

The two asymptotes we found, $x = -3/2$ and $x = 1/2$, are 2 units apart, so the period is 2. The phase shift is $-1/2$, which could also be computed using the formula $-C/B$. Now we sketch these asymptotes and use the period of 2 to draw in the next asymptote in each direction, $x = -7/2$ and $x = 5/2$ (see Fig. 1a on page 206). Once we have the asymptotes drawn, it's a relatively simple matter to fill in the graph (Fig. 1b).

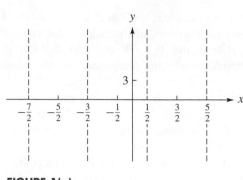

FIGURE 1(a)   **(b)**   ■

Matched Problem 1   Graph $y = 2\tan(2x - \pi/2)$ for $-7\pi < x < 3\pi$. Find the period and phase shift.   ■

For cotangent functions, the procedure is the same, but we use $x = 0$ and $x = \pi$ as the two beginning asymptotes, as illustrated in Example 2.

EXAMPLE 2

Graphing $y = A \cot(Bx + C)$

Sketch three full periods of $y = -\cot(4x + \pi)$. Find the period and phase shift.

Solution   We begin by setting $4x + \pi$ equal to 0 and π, which will allow us to find two of the asymptotes.

$$4x + \pi = 0 \qquad\qquad 4x + \pi = \pi$$
$$4x = -\pi \qquad\qquad 4x = \pi - \pi$$
$$x = -\frac{\pi}{4} \qquad\qquad x = \frac{\pi}{4} - \frac{\pi}{4} = 0$$
$$\uparrow \qquad\qquad\qquad \uparrow$$
$$\text{Phase shift} \qquad\qquad \text{Phase shift}$$

These two asymptotes are separated by $\pi/4$ units, so the period is $\pi/4$. The phase shift is $-\pi/4$. The next asymptotes in each direction are $x = -\pi/2$ and $\pi/4$. Note that $A = -1$ turns the standard cotangent graph upside down. The graph is shown in figure 2.

FIGURE 2

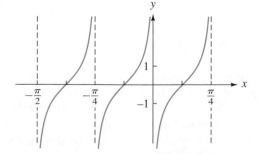

Matched Problem 2   Sketch three full periods of $y = -10\cot(20x)$. Find the period and phase shift.   ■

■ Graphing $y = A \sec(Bx + C)$ and $y = A \csc(Bx + C)$

For convenient reference, the basic graphs for $y = \sec x$ and $y = \csc x$ that were developed in Section 3.1 are repeated here.

GRAPH OF $y = \sec x$

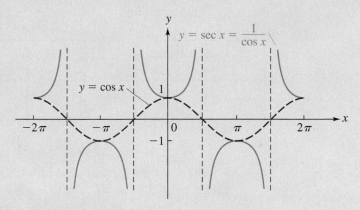

Domain: All real numbers x, except $x = \pi/2 + k\pi$, k an integer
Range: All real numbers y such that $y \leq -1$ or $y \geq 1$
Period: 2π Same asymptotes as tangent

GRAPH OF $y = \csc x$

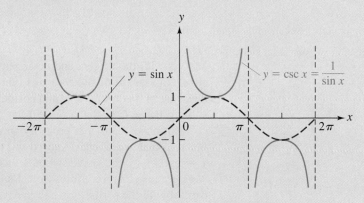

Domain: All real numbers x, except $x = k\pi$, k an integer
Range: All real numbers y such that $y \leq -1$ or $y \geq 1$
Period: 2π Same asymptotes as cotangent

EXPLORE/DISCUSS 2

(A) Match each function to its graph in one of the graphing calculator displays and discuss how the graph compares to the graph of $y = \csc x$ or $y = \sec x$.

(1) $y = \sec 2x$ (2) $y = \frac{1}{2}\csc x$ (3) $y = \csc(x + \pi/2)$

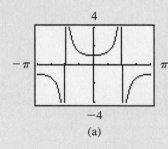
(a)

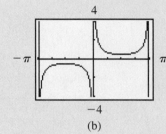
(b)

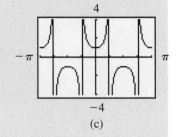
(c)

(B) Use a graphing calculator to explore the nature of the changes in the graphs of the following functions when the values of A, B, and C are changed. Discuss what happens in each case.

(4) $y = A\csc x$ and $y = A\sec x$

(5) $y = \csc Bx$ and $y = \sec Bx$

(6) $y = \csc(x + C)$ and $y = \sec(x + C)$

As with the tangent and cotangent functions, **amplitude is not defined for either the secant or the cosecant** functions. Since the period of each function is the same as that for the sine and cosine, 2π, we find the period and phase shift by solving $Bx + C = 0$ and $Bx + C = 2\pi$.

But once again, it is probably simpler to find the asymptotes, and the good news is that we already know how to do that! The functions $y = \sec x$ and $y = \csc x$ have the same asymptotes as $y = \tan x$ and $y = \cot x$, respectively, so we can reuse the procedures from Examples 1 and 2. We'll just need to be a bit careful with the period.

EXAMPLE 3

Graphing $y = A\,\sec(Bx + C)$

Graph $y = 5\sec(\frac{1}{2}x + \pi)$ for $-7\pi \le x \le 3\pi$. Find the period and phase shift.

Solution   We again begin by finding two asymptotes, noting that $y = \sec x$ has asymptotes at $x = -\pi/2$ and $\pi/2$.

$$\frac{1}{2}x + \pi = -\frac{\pi}{2} \qquad\qquad \frac{1}{2}x + \pi = \frac{\pi}{2}$$

$$\frac{1}{2}x = -\frac{\pi}{2} - \pi \qquad\qquad \frac{1}{2}x = \frac{\pi}{2} - \pi$$

$$x = -\pi - 2\pi = -3\pi \qquad\qquad x = \pi - 2\pi = -\pi$$

$$\underset{\text{Phase shift}}{\uparrow} \qquad\qquad\qquad \underset{\text{Phase shift}}{\uparrow}$$

So there are asymptotes at $x = -3\pi$ and $-\pi$, and the phase shift is -2π. Now we have to be careful. This might lead us to conclude that the period is 2π, but it isn't. Recall that there are two portions of a secant graph in each period: one that opens up and one that opens down. Only one portion will go between the asymptotes we found, so the actual period is 4π, not 2π. In any case, the separation between consecutive asymptotes is 2π, so the next two asymptotes in each direction are -5π, -7π, π, and 3π. We draw these in to begin our graph (Fig. 3a). Next, we make two important observations. Since $A = 5$, the high and low points on each portion of the graph will be at heights 5 and -5, not heights 1 and -1; also, the portion of the graph of $y = \sec x$ between $x = -\pi/2$ and $x = \pi/2$ opens up, so the portion on the graph in this example between $x = -3\pi$ and $x = -\pi$ opens up as well. This allows us to put in one part of the graph (Fig. 3a) and then repeat until we have the graph on the requested interval (Fig. 3b).

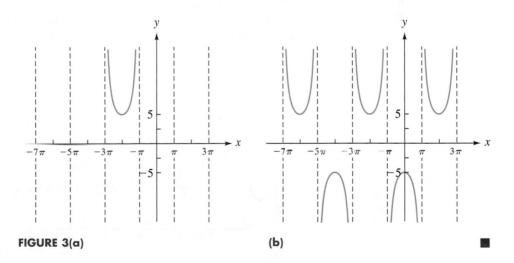

FIGURE 3(a)   **(b)**

Matched Problem 3   Graph $y = \frac{1}{2}\sec(2x + \pi)$ for $-3\pi/4 < x < 3\pi/4$. Find the period and phase shift.

EXAMPLE 4

Graphing $y = A \csc(Bx + C)$

Graph

$$y = 2 \csc\left(\frac{\pi}{2}x - \pi\right) \qquad \text{for } -2 < x < 10$$

Find the period and phase shift.

Solution   There are asymptotes for $y = \csc x$ at $x = 0$ and $x = \pi$, so we find the asymptotes for this function by setting $(\pi/2)x - \pi$ equal to 0 and π.

$$\frac{\pi}{2}x - \pi = 0 \qquad\qquad\qquad \frac{\pi}{2}x - \pi = \pi$$

$$\frac{\pi}{2}x = 0 + \pi \qquad\qquad\qquad \frac{\pi}{2}x = \pi + \pi$$

$$x = 0 \cdot \frac{2}{\pi} + \pi \cdot \frac{2}{\pi} \qquad\qquad x = \pi \cdot \frac{2}{\pi} + \pi \cdot \frac{2}{\pi}$$

$$x = 0 + 2 = 2 \qquad\qquad\qquad x = 2 + 2 = 4$$

$$\uparrow \qquad\qquad\qquad\qquad\qquad \uparrow$$
$$\text{Phase shift} \qquad\qquad\qquad \text{Phase shift}$$

There are asymptotes at $x = 2$ and $x = 4$, and a portion of the graph that opens up is between them. The low point of this portion is at height 2, since $A = 2$. The period is 4 (twice the distance between successive asymptotes) and the phase shift is 2. The remaining asymptotes on the requested interval are at $x = -2, 0, 6, 8,$ and 10. The graph is shown in Figure 4.

FIGURE 4

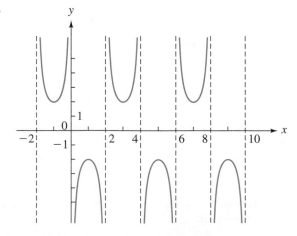

Matched Problem 4 Graph $y = \frac{1}{3}\csc(\frac{1}{4}x + \pi)$ for $-8\pi < x < 12\pi$. Find the period and phase shift. ■

Answers to Matched Problems

1. Period $= \pi/2$; phase shift $= \pi/4$

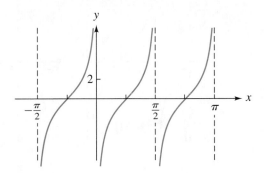

2. Period $= \pi/20$; phase shift $= 0$

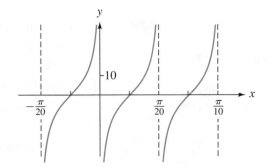

3. Period $= \pi$; phase shift $= -\pi/2$

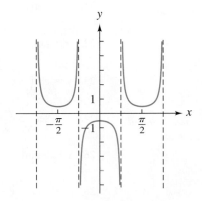

4. Period $= 8\pi$; phase shift $= -4\pi$

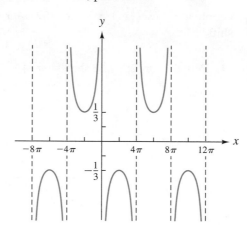

EXERCISE 3.6

A

1. Explain how to find the asymptotes for a function of the form $y = A \tan(Bx + C)$.

2. Explain how to find the asymptotes for a function of the form $y = A \csc(Bx + C)$.

3. How do we know that $y = A \tan(Bx + C)$ and $y = A \sec(Bx + C)$ have the same asymptotes?

4. Explain why $y = A \tan(Bx + C)$ and $y = A \sec(Bx + C)$ have different periods even though they have the same asymptotes.

5. What effect does the value of A have on the graphs of $y = A \tan(Bx + C)$ and $y = A \cot(Bx + C)$?

6. What effect does the value of A have on the graphs of $y = A \sec(Bx + C)$ and $y = A \csc(Bx + C)$?

In Problems 7–10, sketch a graph of each function over the indicated interval. Try not to look at the text or use a calculator.

7. $y = \tan x, \quad 0 \le x \le 2\pi$

8. $y = \cot x, \quad 0 < x < 2\pi$

9. $y = \csc x, \quad -\pi < x < \pi$

10. $y = \sec x, \quad -\pi \le x \le \pi$

B *In Problems 11–20, indicate the period of each function and sketch a graph of the function over the indicated interval.*

11. $y = 3 \tan 2x, \quad -\pi \le x \le \pi$

12. $y = 2 \cot 4x, \quad 0 < x < \pi/2$

13. $y = \frac{1}{2} \tan(x/2), \quad -\pi < x < 3\pi$

14. $y = \frac{1}{2} \cot(x/2), \quad 0 < x < 4\pi$

15. $y = 2 \csc(x/2), \quad 0 < x < 8\pi$

16. $y = 2 \sec \pi x, \quad -1 < x < 3$

17. $y = -10 \tan(\pi x), \quad -1 \le x \le 1$

18. $y = -\frac{1}{4} \cot(3x), \quad 0 < x < \pi$

19. $y = -\frac{1}{10} \sec(\frac{1}{4}x), \quad -6\pi < x < 6\pi$

20. $y = -5 \csc(\pi x), \quad -3 < x < 3$

In Problems 21–26, indicate the period and phase shift for each function, and sketch a graph of the function over the indicated interval.

21. $y = \cot(2x - \pi), \quad -\dfrac{\pi}{2} < x < \dfrac{\pi}{2}$

22. $y = \tan(2x + \pi), \quad -\dfrac{3\pi}{4} < x < \dfrac{3\pi}{4}$

23. $y = \csc\left(\pi x - \dfrac{\pi}{2}\right), \quad -\dfrac{1}{2} < x < \dfrac{5}{2}$

24. $y = \sec\left(\pi x + \dfrac{\pi}{2}\right), \quad -1 < x < 1$

25. $y = -\tan\left(x - \dfrac{\pi}{4}\right), \quad -\dfrac{7\pi}{4} < x < \dfrac{\pi}{4}$

26. $y = -\cot\left(x + \dfrac{\pi}{4}\right), \quad -\dfrac{\pi}{4} < x < \dfrac{3\pi}{4}$

Graph each equation in Problems 27–30 on a graphing calculator; then find an equation of the form $y = A \tan Bx$, $y = A \cot Bx$, $y = A \csc Bx$, or $y = A \sec Bx$ that has the same graph. (These problems suggest additional identities beyond the fundamental ones that were discussed in Section 2.5—additional important identities will be discussed in detail in Chapter 4.)

27. $y = \csc x - \cot x$ **28.** $y = \csc x + \cot x$

29. $y = \cot x + \tan x$ **30.** $y = \cot x - \tan x$

C   *In Problems 31–34, indicate the period and phase shift for each function and sketch a graph of the function over the indicated interval.*

31. $y = -3 \cot(\pi x - \pi)$, $-2 < x < 2$

32. $y = -2 \tan\left(\dfrac{\pi}{4} x - \dfrac{\pi}{4}\right)$, $-1 < x < 7$

33. $y = 2 \sec\left(\pi x - \dfrac{\pi}{2}\right)$, $-1 < x < 3$

34. $y = 3 \csc\left(\dfrac{\pi}{2} x + \dfrac{\pi}{2}\right)$, $-1 < x < 3$

Graph each equation in Problems 35–38 on a graphing calculator; then find an equation of the form $y = A \tan Bx$, $y = A \cot Bx$, $y = A \csc Bx$, or $y = A \sec Bx$ that has the same graph. (These problems suggest additional identities beyond the fundamental ones that were discussed in Section 2.5—additional important identities will be discussed in detail in Chapter 4.)

35. $y = \cos 2x + \sin 2x \tan 2x$

36. $y = \sin 3x + \cos 3x \cot 3x$

37. $y = \dfrac{\sin 6x}{1 - \cos 6x}$ **38.** $y = \dfrac{\sin 4x}{1 + \cos 4x}$

Applications

39. Precalculus A beacon light 15 ft from a wall rotates clockwise at the rate of exactly 1 revolution per second (rps), so that $\theta = 2\pi t$ (see the figure).

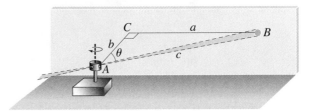

Figure for 39 and 40

(A) If we start counting time in seconds when the light spot is at C, write an equation for the distance a the light spot travels along the wall in terms of time t.

(B) Graph the equation found in part (A) for the time interval $0 \le t < 0.25$.

(C) Explain what happens to a as t approaches 0.25 sec.

40. Precalculus Refer to Problem 39.

(A) Write an equation for the length of the light beam c in terms of t.

(B) Graph the equation found in part (A) for the time interval $0 \le t < 0.25$.

(C) Explain what happens to c as t approaches 0.25 sec.

CHAPTER 3 GROUP ACTIVITY

Predator–Prey Analysis Involving Coyotes and Rabbits

In a national park in the western United States, natural scientists from a nearby university conducted a study on the interrelated populations of coyotes (sometimes called prairie wolves) and rabbits. They found that the population

continued

of each species goes up and down in cycles, but these cycles are out of phase with each other. Each year for 12 years the scientists estimated the number in each population with the results indicated in Table 1.

TABLE 1
Coyote/Rabbit Populations

Year	1	2	3	4	5	6	7	8	9	10	11	12
Coyotes (hundreds)	8.7	11.7	11.5	9.1	6.3	6.5	8.9	11.6	11.9	8.8	6.2	6.4
Rabbits (thousands)	54.5	53.7	38.0	25.3	26.8	41.2	55.4	53.1	38.6	26.2	28.0	40.9

(A) *Coyote population analysis*
  (1) Enter the data for the coyote population in a graphing calculator for the time interval $1 \leq t \leq 12$ and produce a scatter plot for the data.
  (2) A function of the form $y = K + A \sin(Bx + C)$ can be used to model the data. Use the data for the coyote population to determine K, A, and B. Use the graph from part (1) to visually estimate C to one decimal place. (If your graphing calculator has a sine regression feature, use it to check your work.)
  (3) Produce the scatter plot from part (1) and graph the equation from part (2) in the same viewing window. Adjust C, if necessary, for a better equation fit of the data.
  (4) Write a summary of the results, describing fluctuations and cycles of the coyote population.

(B) *Rabbit population analysis*
  (1) Enter the data for the rabbit population in a graphing calculator for the time interval $1 \leq t \leq 12$ and produce a scatter plot for the data.
  (2) A function of the form $y = K + A \sin(Bx + C)$ can be used to model the data. Use the data for the rabbit population to determine K, A, and B. Use the graph from part (1) to visually estimate C to one decimal place. (If your graphing calculator has a sine regression feature, use it to check your work.)
  (3) Produce the scatter plot from part (1) and graph the equation from part (2) in the same viewing window. Adjust C, if necessary, for a better equation fit of the data.
  (4) Write a summary of the results, describing fluctuations and cycles of the rabbit population.

(C) *Predator–prey interrelationship*
  (1) Discuss the relationship of the maximum predator populations to the maximum prey populations relative to time.
  (2) Discuss the relationship of the minimum predator populations to the minimum prey populations relative to time.
  (3) Discuss the dynamics of the fluctuations of the two interdependent populations. What causes the two populations to rise and fall, and why are they out of phase with one another?

CHAPTER 3 REVIEW

3.1 BASIC GRAPHS

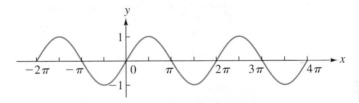

FIGURE 1
Graph of $y = \sin x$

Domain: R Range: $-1 \leq y \leq 1$ Period: 2π

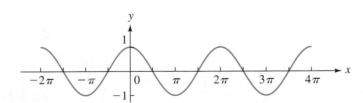

FIGURE 2
Graph of $y = \cos x$

Domain: R Range: $-1 \leq y \leq 1$ Period: 2π

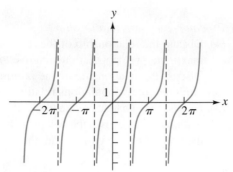

FIGURE 3
Graph of $y = \tan x$
Domain: All real numbers x except $x = \pi/2 + k\pi$, k an integer
Range: R Period: π

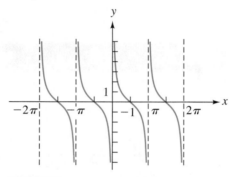

FIGURE 4
Graph of $y = \cot x$

Domain: All real numbers x except $x = k\pi$, k an integer
Range: R Period: π

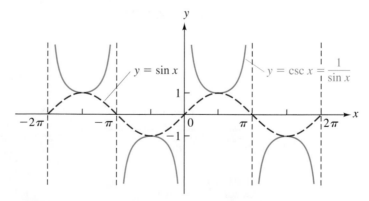

FIGURE 5
Graph of $y = \csc x$

Domain: All real numbers x, except $x = k\pi$, k an integer
Range: All real numbers y such that $y \leq -1$ or $y \geq 1$ Period: 2π

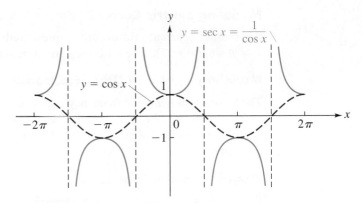

FIGURE 6
Graph of $y = \sec x$

Domain: All real numbers x, except $x = \pi/2 + k\pi$, k an integer
Range: All real numbers y such that $y \leq -1$ or $y \geq 1$
Period: 2π

**3.2
GRAPHING**
$y = k + A \sin Bx$
AND $y = k + A \cos Bx$

For functions of the form $y = k + A \sin Bx$ or $y = k + A \cos Bx$,

$$\text{Amplitude} = |A| \quad \text{Period} = \frac{2\pi}{B}$$

The basic sine or cosine curve is compressed if $B > 1$ and stretched if $0 < B < 1$. The graph is translated up k units if $k > 0$ and down $|k|$ units if $k < 0$.

If P is the period and f is the **frequency** of a periodic phenomenon, then

$$P = \frac{1}{f} \qquad \textit{Period is the time for one complete cycle.}$$

$$f = \frac{1}{P} \qquad \textit{Frequency is the number of cycles per unit of time.}$$

The period for an object bobbing up and down in water is given by $2\pi/\sqrt{1,000gA/M}$, where $g = 9.75$ m/sec^2, A is the horizontal cross-sectional area in square meters, and M is the mass in kilograms.

**3.3
GRAPHING**
$y = k + A \sin (Bx + C)$
AND
$y = k + A \cos (Bx + C)$

For functions of the form $y = k + A \sin(B + c)$ or $y = k + A \cos(Bx + C)$, to find the **period** and the **phase shift**, solve $Bx + C = 0$ and $Bx + C = 2\pi$:

$$x = -\frac{C}{B} \qquad x = -\frac{C}{B} + \frac{2\pi}{B}$$

$\underbrace{\qquad}_{\text{Phase shift}}$ $\underset{\text{Period}}{\uparrow}$

The graph completes one full cycle as x varies from $-C/B$ to $(-C/B) + (2\pi/B)$. A function of the form $y = A \sin(Bx + C)$ or $y = A \cos(Bx + C)$ is said to be a **simple harmonic**.

*3.4 ADDITIONAL APPLICATIONS

Modeling Electric Current

Alternating current flows are represented by equations of the form $I = A \sin(Bt + C)$, where I is the current in amperes and t is time in seconds.

Modeling Light and Other Electromagnetic Waves

The wavelength λ and the frequency ν of an electromagnetic wave are related by $\lambda\nu = c$, where c is the speed of light. These waves can be represented by an equation of the form $E = A \sin 2\pi(\nu t - r/\lambda)$, where t is time and r is distance from the source.

Modeling Water Waves

Water waves can be represented by equations of the form $y = A \sin 2\pi(ft - r/\lambda)$, where f is frequency, t is time, r is distance from the source, and λ is the wavelength.

Simple and Damped Harmonic Motion; Resonance

An object hanging on a spring that is set in motion is said to exhibit **simple harmonic motion** if its amplitude remains constant, **damped harmonic motion** if its amplitude decreases to 0 as time increases, and **resonance** if its amplitude increases as time increases.

Objects in simple harmonic motion are modeled well by functions of the form $y = A \sin(Bx + C)$ or $y = A \cos(Bx + C)$. Those exhibiting damped harmonic motion or resonance can often be modeled by multiplying a sine or cosine function by an appropriate decreasing or increasing function.

3.5 GRAPHING COMBINED FORMS

Functions combined using addition or subtraction, if simple enough, can be graphed by hand using **addition of ordinates;** but whether simple or not, such functions can always be graphed on a graphing calculator. In the analysis of sound waves, it is also useful to graph the separate components of the wave, called **partial tones.** The partial tone with the smallest frequency is called the **fundamental tone,** and the other partial tones are called **overtones. Fourier series** use sums of trigonometric functions to approximate periodic wave forms such as a **square wave** or a **sawtooth wave.**

3.6 TANGENT, COTANGENT, SECANT, AND COSECANT FUNCTIONS REVISITED

Amplitude is not defined for the tangent, cotangent, secant, and cosecant functions.

To find the period and the phase shift for the graph of $y = A \tan(Bx + C)$ or $y = A \cot(Bx + C)$, solve $Bx + C = 0$ and $Bx + C = \pi$:

$$x = -\frac{C}{B} \qquad x = -\frac{C}{B} + \frac{\pi}{B}$$

$$\underbrace{\qquad\qquad\qquad}_{\text{Phase shift}} \quad \overset{\uparrow}{\text{Period}}$$

The graph completes one full cycle as x varies from $-C/B$ to $(-C/B) + (\pi/B)$.

To graph $y = A \tan(Bx + C)$ easily, you can locate asymptotes by first solving $Bx + C = -\pi/2$ and $Bx + C = \pi/2$, then finding the distance between the resulting asymptotes to see where to locate others. For $y = A \cot(Bx + C)$, the asymptotes can be located by solving $Bx + C = 0$ and $Bx + C = \pi$. In each case, $|A|$ affects the steepness of the individual portions of the graph.

Find the period and the phase shift for the graph of $y = A \sec(Bx + C)$ or $y = A \csc(Bx + C)$ by solving $Bx + C = 0$ and $Bx + C = 2\pi$.

The asymptotes for $y = A \sec(Bx + C)$ are the same as those for $y = A \tan(Bx + C)$, and the asymptotes for $y = A \csc(Bx + C)$ are the same as those for $y = A \cot(Bx + C)$. In this case, A affects the height of the high and low points for each portion of the graph, as well as the orientation (opening up or down).

CHAPTER 3 REVIEW EXERCISE

Work through all the problems in this chapter review and check the answers. Answers to all review problems appear in the back of the book; following each answer is an italic number that indicates the section in which that type of problem is discussed. Where weaknesses show up, review the appropriate sections in the text. Review problems flagged with a star (☆) are from optional sections.

A *In Problems 1–6, sketch a graph of each function for* $-2\pi \le x \le 2\pi$.

1. $y = \sin x$

2. $y = \cos x$

3. $y = \tan x$

4. $y = \cot x$

5. $y = \sec x$

6. $y = \csc x$

In Problems 7–9, sketch a graph of each function for the indicated interval.

7. $y = 3 \cos \dfrac{x}{2}, \quad -4\pi \le x \le 4\pi$

8. $y = \frac{1}{2} \sin 2x, \quad -\pi \le x \le \pi$

9. $y = 4 + \cos x, \quad 0 \le x \le 2\pi$

10. Describe any properties that all six trigonometric functions share.

B **11.** Explain how increasing or decreasing the size of B, $B > 0$, in $y = A \cos Bx$ affects the period.

12. Explain how changing the value of C in $y = A \sin(Bx + C)$ affects the original graph, assuming A and $B > 0$ are not changed.

13. Match one of the basic trigonometric functions with each description:

(A) Not defined at $x = n\pi$, n an integer; period 2π

(B) Not defined at $x = n\pi$, n an integer; period π

(C) Amplitude 1; graph passes through $(0, 0)$

In Problems 14–19, state the period, amplitude (if applicable), and phase shift (if applicable) for each function.

14. $y = 3 \sin(3\pi x)$

15. $y = \frac{1}{4} \cos(\frac{1}{2} x - 2\pi)$

16. $y = -5 \tan\left(\dfrac{\pi}{3} x\right)$

17. $y = 4 \cot(x + 7)$

18. $y = -\frac{1}{2} \sec(2\pi x - 4\pi)$

19. $y = 2 \csc 5x$

In Problems 20–27, sketch a graph of each function for the indicated interval.

20. $y = -\frac{1}{3} \cos(2\pi x), \quad -2 \le x \le 2$

21. $y = -1 + \frac{1}{2} \sin 2x, \quad -\pi \le x \le \pi$

22. $y = 4 - 2 \sin(\pi x - \pi), \quad 0 \le x \le 2$

23. $y = \tan 2x, \quad -\pi \le x \le \pi$

24. $y = \cot \pi x, \quad -2 < x < 2$

25. $y = 3 \csc \pi x, \quad -1 < x < 2$

26. $y = 2 \sec \dfrac{x}{2}, \quad -\pi < x < 3\pi$

27. $y = \tan\left(x + \dfrac{\pi}{2}\right), \quad -\pi < x < \pi$

28. Explain how the graph of y_2 is related to the graph of y_1 in each of the following graphing calculator displays:

(A)

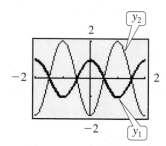

(B)

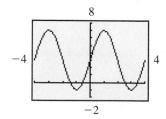

(C) Write the equations of the graphs of y_1 and y_2 in part (A) in the form $A \sin Bx$.

(D) Write the equations of the graphs of y_1 and y_2 in part (B) in the form $A \cos Bx$.

29. Find an equation of the form $y = k + A \sin Bx$ that produces the graph shown in the following graphing calculator display:

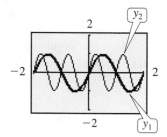

30. Find an equation of the form $y = A \sin Bt$ or $y = A \cos Bt$ for an oscillating system, if the displacement is 0 when $t = 0$, the amplitude is 65, and the period is 0.01.

31. The graphing calculator display in the figure illustrates the graph of one of the following:

(1) $y_1 = y_2 + y_3$ (2) $y_2 = y_1 + y_3$

(3) $y_3 = y_1 + y_2$

Discuss which one and why.

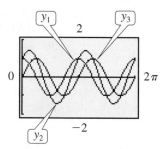

Figure for 31

Problems 32–36 require the use of a graphing calculator.

32. Graph $y = x + \sin \pi x, \quad 0 \le x \le 2$.

33. Graph $y = 2 \sin x + \cos 2x, \quad 0 \le x \le 4\pi$.

34. Graph $y = 1/(1 + \tan^2 x)$ in a viewing window that displays at least two full periods of the graph. Find an equation of the form $y = k + A \sin Bx$ or $y = k + A \cos Bx$ that has the same graph.

35. The sum of the first five terms of the Fourier series for a wave form is

$$y = -\frac{\pi}{4} + \frac{4}{\pi} \cos \frac{x}{2} + \frac{2}{\pi} \cos x$$
$$+ \frac{4}{9\pi} \cos \frac{3x}{2} + \frac{4}{25\pi} \cos \frac{5x}{2}$$

Use a graphing calculator to graph this equation for $-4\pi \le x \le 4\pi$ and $-3 \le y \le 3$. Then sketch by hand the corresponding wave form. (Assume that the graph of the wave form is composed entirely of straight line segments.)

36. Graph each of the following equations and find an equation of the form $y = A \tan Bx, \ y = A \cot Bx, \ y = A \sec Bx$, or $y = A \csc Bx$ that has the same graph as the given equation.

(A) $y = \dfrac{2 \sin x}{\sin 2x}$ (B) $y = \dfrac{2 \cos x}{\sin 2x}$

(C) $y = \dfrac{2 \cos^2 x}{\sin 2x}$ (D) $y = \dfrac{2 \sin^2 x}{\sin 2x}$

C **37.** Describe the smallest horizontal shift and/or reflection that transforms the graph of $y = \cot x$ into the graph of $y = \tan x$.

In Problems 38 and 39, sketch a graph of each function for the indicated interval.

38. $y = 2 \tan\left(\pi x + \dfrac{\pi}{2}\right), \quad -1 < x < 1$

39. $y = 2 \sec(2x - \pi), \quad 0 \le x < \dfrac{5\pi}{4}$

40. The table in the figure was produced using a table feature for a particular graphing calculator. Find a function of the form $y = A \sin Bx$ or $y = A \cos Bx$ that will produce the table.

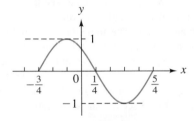

X	Y1	
0.00	0.00	
.25	2.00	
.50	0.00	
.75	-2.00	
1.00	0.00	
1.25	2.00	
1.50	0.00	

X=0

41. If the following graph is a graph of an equation of the form $y = A \sin(Bx + C), \; 0 < -C/B < 1$, find the equation.

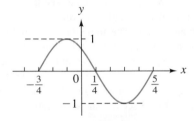

Problems 42–44 require the use of a graphing calculator.

42. Graph $y = 1.2 \sin 2x + 1.6 \cos 2x$ and approximate the x intercept closest to the origin (correct to three decimal places). Use this intercept to find an equation of the form $y = A \sin(Bx + C)$ that has the same graph as the given equation.

43. In calculus it is shown that

$$\sin x \approx x - \frac{x^3}{3!} + \frac{x^5}{5!} - \frac{x^7}{7!}$$

(A) Graph $\sin x$ and the first term of the series in the same viewing window.

(B) Graph $\sin x$ and the first two terms of the series in the same viewing window.

(C) Graph $\sin x$ and the first three terms of the series in the same viewing window.

(D) Explain what is happening as more terms of the series are used to approximate $\sin x$.

44. From the graph of $f(x) = |\sin x|$ on a graphing utility, determine the period of f; that is, find the smallest positive number p such that $f(x + p) = f(x)$.

 Applications

45. **Spring–Mass System** If the motion of a weight hung on a spring has an amplitude of 4 cm and a frequency of 8 Hz, and if its position when $t = 0$ sec is 4 cm below its position at rest (above the rest position is positive and below is negative), find an equation of the form $y = A \cos Bt$ that describes the motion at any time t (neglecting any damping forces such as air resistance and friction). Explain why an equation of the form $y = A \sin Bt$ cannot be used to model the motion.

46. **Pollution** In a large city the amount of sulfur dioxide pollutant released into the atmosphere due to the burning of coal and oil for heating purposes varies seasonally. If measurements over a 2 yr period produced the graph shown, find an equation of the form $P = k + A \cos Bn, 0 \le n \le 104$, where P is the number of tons of pollutants released into the atmosphere during the nth week after January 31. Can an equation of the form $P = k + A \sin Bn$ model the situation? If yes, find it. If no, explain why.

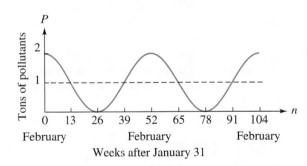

Weeks after January 31

47. **Floating Objects** A cylindrical buoy with diameter 1.2 m is observed (after being pushed down) to bob up and down with a period of 0.8 sec and an amplitude of 0.6 m.

(A) Find the mass of the buoy to the nearest kilogram.

(B) Write an equation of motion for the buoy in the form $y = D \sin Bt$.

(C) Graph the equation in part (B) for $0 \le t \le 1.6$.

48. **Sound Waves** The motion of one tip of a tuning fork is given by an equation of the form $y = 0.05 \cos Bt$.

(A) If the frequency of the fork is 280 Hz, what is the period? What is the value of B?

(B) If the period of the motion of the fork is 0.0025 sec, what is the frequency? What is the value of B?

(C) If $B = 700\pi$, what is the period? What is the frequency?

49. Electrical Circuits   If the voltage E in an electrical circuit has amplitude 18 and frequency 30 Hz, and $E = 18$ V when $t = 0$ sec, find an equation of the form $y = A \cos Bt$ that gives the voltage at any time t.

50. Water Waves   At a particular point in the ocean, the vertical change in the water due to wave action is given by

$$y = 6 \cos \frac{\pi}{10}(t - 5)$$

where y is in meters and t is time in seconds. Find the amplitude, period, and phase shift. Graph the equation for $0 \le t \le 80$.

☆**51. Water Waves**   A water wave at a fixed position has an equation of the form

$$y = 12 \sin \frac{\pi}{3} t$$

where t is time in seconds and y is in feet. How high is the wave from trough to crest? What is its wavelength (in feet)? How fast is it traveling (in feet per second)? Compute answers to the nearest foot.

☆**52. Electromagnetic Waves**   An ultraviolet wave has a frequency of $\nu = 10^{15}$ Hz. What is its period? What is its wavelength (in meters)? (The speed of light is $c \approx 3 \times 10^8$ m/sec.)

☆**53. Spring–Mass Systems**   Graph each of the following equations for $0 \le t \le 2$, and identify each as an example of simple harmonic motion, damped harmonic motion, or resonance.

(A) $y = \sin 2\pi t$ (B) $y = (1 + t) \sin 2\pi t$

(C) $y = \dfrac{1}{1 + t} \sin 2\pi t$

54. Modeling a Seasonal Business Cycle   The sales for a national chain of ice cream shops are growing steadily but are subject to seasonal variations. The following mathematical model has been developed to project the monthly sales for the next 24 months:

$$S = 5 + t + 5 \sin \frac{\pi t}{6}$$

where S represents the sales in millions of dollars and t is time in months.

(A) Graph the equation for $0 \le t \le 24$.

(B) Describe what the graph shows regarding monthly sales over the 2 yr period.

55. Rocket Flight   A camera recording the launch of a rocket is located 1,000 m from the launching pad (see the figure).

(A) Write an equation for the altitude h of the rocket in terms of the angle of elevation θ of the camera.

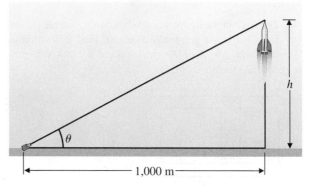

Figure for 55

(B) Graph this equation for $0 \le \theta < \pi/2$.

(C) Describe what happens to h as θ approaches $\pi/2$.

56. Bioengineering   For a running person (see the figure), the upper leg rotates back and forth relative to the hip socket through an angle θ. The angle θ is measured relative to the vertical and is negative if the leg is behind the vertical and positive if the leg is in front of the vertical. Measurements of θ were calculated for a person taking two strides per second, where t is the time in seconds. The results for the first 2 sec are shown in Table 1.

Figure for 56

(A) Verbally describe how you can find the period and amplitude of the data in Table 1 and find each.

TABLE 1

t (sec)	0.00	0.25	0.50	0.75	1.00	1.25	1.50	1.75	2.00
θ (deg)	0°	36°	0°	−36°	0°	36°	0°	−36°	0°

(B) Which of the equations, $\theta = A \cos Bt$ or $\theta = A \sin Bt$, is the more suitable model for the data in Table 1? Why? Use the model you chose to find the equation.

(C) Plot the points in Table 1 and the graph of the equation found in part (B) in the same coordinate system.

57. Modeling Sunset Times   Sunset times for the fifth day of each month over a 1 yr period were taken from a tide booklet for San Francisco Bay to form Table 2. Daylight savings time was ignored.

(A) Convert the data in Table 2 from hours and minutes to hours (to two decimal places). Enter the data for a

2 yr period in your graphing calculator and produce a scatter plot in the following viewing window: $1 \le x \le 24$, $16 \le y \le 20$.

(B) A function of the form $y = k + A \sin(Bx + C)$ can be used to model the data. Use the converted data from Table 2 to determine k and A (to two decimal places) and B (exact value). Use the graph from part (A) to visually estimate C (to one decimal place).

(C) Plot the data from part (A) and the equation from part (B) in the same viewing window. If necessary, adjust your value of C to produce a better fit.

TABLE 2

x (months)	1	2	3	4	5	6	7	8	9	10	11	12
y (sunset time, P.M.)	17:05	17:38	18:07	18:36	19:04	19:29	19:35	19:15	18:34	17:47	17:07	16:51

CUMULATIVE REVIEW EXERCISE CHAPTERS 1–3

Work through all the problems in this cumulative review and check the answers. Answers to all review problems are in the back of the book; following each answer is an italic number that indicates the section in which that type of problem is discussed. Where weaknesses show up, review the appropriate sections in the text. Review problems flagged with a star (☆) are from optional sections.

A 1. Change $43°32'$ to decimal degrees to two decimal places.

2. What is the radian measure of $88.73°$ to two decimal places?

3. What is the radian measure of $122°17'$ to two decimal places?

4. What is the decimal degree measure of 2.75 rad to two places?

5. What is the degree measure of -1.45 rad to the nearest minute?

6. The terminal side of θ contains $P = (7, 24)$. Find the value of each of the six trigonometric functions of θ.

7. The hypotenuse of a right triangle is 13 in. and one of the legs is 11 in. Find the other leg and the acute angles.

8. Evaluate to four significant digits using a calculator:

(A) $\sin 72.5°$ (B) $\cos 104°52'$ (C) $\tan 2.41$

9. Evaluate to four significant digits using a calculator:

(A) $\sec 246.8°$ (B) $\cot 23°15'$ (C) $\csc 1.83$

10. Sketch the reference triangle and find the reference angle α for:

(A) $\theta = \dfrac{11\pi}{6}$ (B) $\theta = -225°$

11. Sketch a graph of each function for $-2\pi \le x \le 2\pi$.

(A) $y = \sin x$ (B) $y = \tan x$ (C) $y = \sec x$

12. Explain what is meant by an angle of radian measure 1.5.

13. Is it possible to find a real number x such that $\cos x$ is negative and $\csc x$ is positive? Explain.

14. Is it possible to construct a triangle with more than one obtuse angle? Explain.

B 15. Find θ to the nearest $10'$ if $\tan \theta = 0.9465$ and $-90° \le \theta \le 90°$.

16. If the second hand of a clock is 5.00 cm long, how far does the tip of the hand travel in 40 sec?

☆ 17. How fast is the tip of the second hand in Problem 16 moving (in centimeters per minute)?

18. Find x in the following figure:

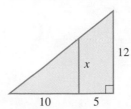

19. Convert $48°$ to exact radian measure in terms of π.

20. Which of the following graphing calculator displays is the result of the calculator being set in radian mode and which is the result of the calculator being set in degree mode?

(A)
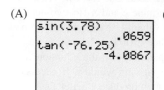

(B)
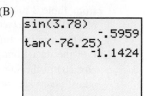

21. If the degree measure of an angle is doubled, is the radian measure of the angle doubled? Explain.

22. From the following graphing calculator display, explain how you would find cot x without finding x. Then find cot x to four decimal places.

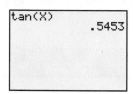

23. Find the exact value of each of the following without using a calculator.

(A) $\sin \dfrac{5\pi}{4}$ (B) $\cos \dfrac{7\pi}{6}$

(C) $\tan \left(-\dfrac{5\pi}{3} \right)$ (D) $\csc 3\pi$

24. Find the exact value of each of the other five trigonometric functions of θ if $\cos \theta = -\frac{2}{3}$ and $\tan \theta < 0$.

25. In the following graphing calculator display, indicate which value(s) are not exact and find the exact form:

```
cos(0°)
            1.0000
sin(-π/3)
            -.8660
tan(135°)
            -1.0000
```

In Problems 26–31, sketch a graph of each function for the indicated interval. State the period and, if applicable, the amplitude and phase shift for each function.

26. $y = 1 - \frac{1}{2} \cos 2x$, $-2\pi \le x \le 2\pi$

27. $y = 2 \sin \left(x - \dfrac{\pi}{4} \right)$, $-\pi \le x \le 3\pi$

28. $y = 5 \tan 4x$, $0 \le x \le \pi$

29. $y = \csc \dfrac{x}{2}$, $-4\pi < x < 4\pi$

30. $y = -2 \sec \pi x$, $-2 \le x \le 2$

31. $y = \cot \left(\pi x + \dfrac{\pi}{2} \right)$, $-1 \le x \le 3$

32. Find the equation of the form $y = A \sin Bx$ whose graph is shown in the figure that follows.

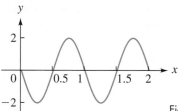

Figure for 32

33. Find an equation of the form $y = k + A \cos Bx$ that produces the graph shown in the graphing calculator display in the following figure.

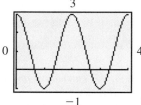

Figure for 33

34. Simplify: $(\tan x)(\sin x) + \cos x$

35. Find the exact value of all angles between $0°$ and $360°$ for which $\sin \theta = -\frac{1}{2}$.

36. If the sides of a right triangle are 23.5 in. and 37.3 in., find the hypotenuse and find the acute angles to the nearest $0.1°$.

Problems 37–39 require use of a graphing calculator.

37. Graph $y = \sin x + \sin 2x$ for $0 \le x \le 2\pi$.

38. Graph $y = (\tan^2 x)/(1 + \tan^2 x)$ in a viewing window that displays at least two full periods of the graph. Find an equation of the form $y = k + A \sin Bx$ or $y = k + A \cos Bx$ that has the same graph as the given equation.

39. The sum of the first four terms of the Fourier series for a wave form is

$$y = \frac{\pi}{2} - \frac{4}{\pi} \cos x - \frac{4}{9\pi} \cos 3x - \frac{4}{25\pi} \cos 5x$$

Use a graphing calculator to graph this equation for $-2\pi \le x \le 2\pi$ and $-2 \le y \le 4$ and sketch by hand the corresponding wave form. (Assume that the graph of the wave form is composed entirely of straight line segments.)

C 40. If θ is a first-quadrant angle and $\tan \theta = a$, express the other five trigonometric functions of θ in terms of a.

41. A circle with its center at the origin in a rectangular coordinate system passes through the point (8, 15). What is the length of the arc on the circle in the first quadrant between the positive horizontal axis and (8, 15)? Compute the answer to two decimal places.

42. A point moves clockwise around a unit circle, starting at (1, 0), for a distance of 53.077 units. Explain how you would find the coordinates of the point P at its final position and how you would determine which quadrant P is in. Find the coordinates to four decimal places.

43. The following graphing calculator display shows the coordinates of a point on a unit circle. Let s be the length of the least positive arc from (1, 0) to the point. Find s to four decimal places.

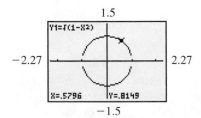

44. Find the equation of the form $y = A \sin(Bx + C)$, $0 < -C/B < 1$, whose graph is

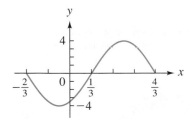

⊞ 45. Graph $y = 2.4 \sin(x/2) - 1.8 \cos(x/2)$ and approximate the x intercept closest to the origin, correct to three decimal places. Use this intercept to find an equation of the form $y = A \sin(Bx + C)$ that has the same graph as the given equation.

⊞ 46. Graph each of the following equations and find an equation of the form $y = A \tan Bx$, $y = A \cot Bx$, $y = A \sec Bx$, or $y = A \csc Bx$ that has the same graph as the given equation.

(A) $y = \dfrac{\sin 2x}{1 + \cos 2x}$ (B) $y = \dfrac{2 \cos x}{1 + \cos 2x}$

(C) $y = \dfrac{2 \sin x}{1 - \cos 2x}$ (D) $y = \dfrac{\sin 2x}{1 - \cos 2x}$

Applications

47. **Geography/Navigation** Find the distance (to the nearest mile) between Gary, IN, with latitude 41°36′N, and Pensacola, FL, with latitude 30°25′N. (Both cities have approximately the same longitude.) Use $r \approx 3{,}960$ mi for the radius of the earth.

48. **Volume of a Cone** A paper drinking cup has the shape of a right circular cone with altitude 9 cm and radius 4 cm (see the figure). Find the volume of the water in the cup (to the nearest cubic centimeter) when the water is 6 cm deep. [*Recall*: $V = \frac{1}{3} \pi r^2 h$.]

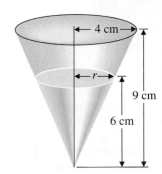

Figure for 48

49. **Construction** The base of a 40 ft arm on a crane is 10 ft above the ground (see the figure). The crane is picking up an object whose horizontal distance from the base of the arm is 30 ft. If the tip of the arm is directly above the object, find the angle of elevation of the arm and the altitude of the tip.

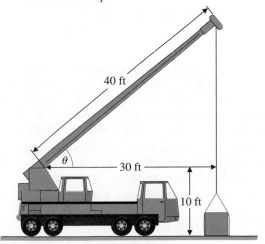

Figure for 49

50. Billiards   Under ideal conditions, when a ball strikes a cushion on a pocket billiards table at an angle, it bounces off the cushion at the same angle (see the figure). Where should the ball in the center of the table strike the upper cushion so that it bounces off the cushion into the lower right corner pocket? Give the answer in terms of the distance from the center of the side pocket to the point of impact.

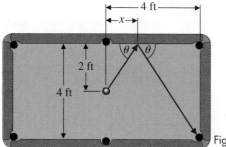

Figure for 50

51. Railroad Grades   The amount of inclination of a railroad track is referred to as its *grade* and is usually given as a percentage (see the figure). Most major railroads limit grades to a maximum of 3%, although some logging and mining railroads have grades as high as 7%, requiring the use of specially geared locomotives. Find the angle of inclination (in degrees, to the nearest 0.1°) for a 3% grade. Find the grade (to the nearest 1%) if the angle of inclination is 3°.

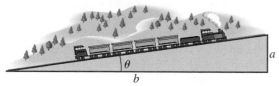

Angle of inclination: θ Grade: $\dfrac{a}{b}$

Figure for 51

52. Surveying   A house that is 10 m tall is located directly across the street from an office building (see the figure). The angle of elevation of the office building from the ground is 72° and from the top of the house is 68°. How tall is the office building? How wide is the street?

53. Forest Fires   Two fire towers are located 5 mi apart on a straight road. Observers at each tower spot a fire and report its location in terms of the angles in the figure.

(A) How far from tower B is the point on the road closest to the fire?

(B) How close is the fire to the road?

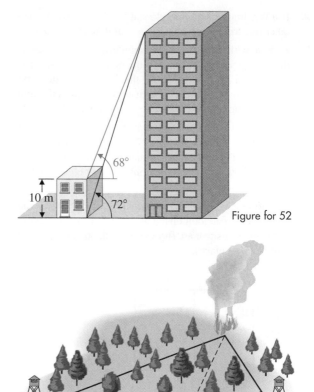

Figure for 52

Figure for 53

☆**54. Engineering**   A large saw in a sawmill is driven by a chain connected to a motor (see the figure). The drive wheel on the motor is rotating at 300 rpm.

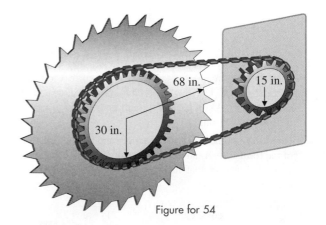

Figure for 54

(A) What is the angular velocity of the saw (in radians per minute)?

(B) What is the linear velocity of a point on the rim of the saw (in inches per minute)?

☆ **55. Light Waves**   A light ray passing through air strikes the surface of a pool of water so that the angle of incidence is $\alpha = 38.4°$. Find the angle of refraction β. (Water has a refractive index of 1.33 and air has a refractive index of 1.00.)

☆ **56. Sonic Boom**   Find the angle of the sound cone (to the nearest degree) for a plane traveling at 1.5 times the speed of sound.

57. Precalculus   In calculus, the ratio $(\sin x)/x$ comes up naturally, and the problem is to determine what the ratio approaches when x approaches 0. Can you guess? (Note that the ratio is not defined when x equals 0.) Here our approach to the problem is geometric; in Problem 58, we will use a graphic approach.

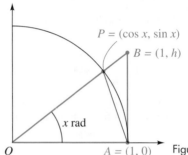

$P = (\cos x, \sin x)$

$B = (1, h)$

x rad

O $A = (1, 0)$ Figure for 57 and 58

(A) Referring to the figure, show the following:

$$A_1 = \text{Area of triangle } OAP = \frac{\sin x}{2}$$

$$A_2 = \text{Area of sector } OAP = \frac{x}{2}$$

$$A_3 = \text{Area of triangle } OAB = \frac{\tan x}{2}$$

(B) Using the fact that $A_1 < A_2 < A_3$, which we can see clearly in the figure, show that

$$\cos x < \frac{\sin x}{x} < 1 \qquad x > 0$$

(C) From the inequality in part (B), explain how you can conclude that for $x > 0$, $(\sin x)/x$ approaches 1 as x approaches 0.

58. Precalculus   Refer to Problem 57. We now approach the problem using a graphing calculator and we allow x to be either positive or negative.

(A) Graph $y_1 = \cos x$, $y_2 = (\sin x)/x$, and $y_3 = 1$ in the same viewing window for $-1 \le x \le 1$ and $0 \le y \le 1.2$.

(B) From the graph, write an inequality of the form $a < b < c$ using y_1, y_2, and y_3.

(C) Use TRACE to investigate the values of y_1, y_2, and y_3 for x close to 0. Explain how you can conclude that $(\sin x)/x$ approaches 1 as x approaches 0 from either side of 0. (Remember, the ratio is not defined when $x = 0$.)

59. Spring–Mass System   If the motion of a weight hung on a spring has an amplitude of 3.6 cm and a frequency of 6 Hz, and if its position when $t = 0$ sec is 3.6 cm above its rest position (above rest position is positive and below is negative), find an equation of the form $y = A \cos Bt$ that describes the motion at any time t (neglecting any damping forces such as air resistance and friction). Can an equation of the form $y = A \sin Bt$ be used to model the motion? Explain why or why not.

60. Electric Circuits   The current (in amperes) in an electrical circuit is given by $I = 12 \sin(60\pi t - \pi)$, where t is time in seconds.

(A) State the amplitude, period, frequency, and phase shift.

(B) Graph the equation for $0 \le t \le 0.1$.

☆ **61. Electromagnetic Waves**   An infrared wave has a wavelength of 6×10^{-5} m. What is its period? (The speed of light is $c \approx 3 \times 10^8$ m/sec.)

62. Precalculus: Surveying   A university has just constructed a new building at the intersection of two perpendicular streets (see the figure). Now it wants to connect the existing streets with a walkway that passes behind and just touches the building.

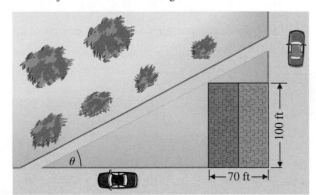

θ

100 ft

70 ft

Figure for 62

(A) Express the length L of the walkway in terms of θ.

(B) From the figure, describe what you think happens to L as θ varies between $0°$ and $90°$.

(C) Complete Table 1, giving values of L to one decimal place, using a calculator. (If you have a table-generating calculator, use it.) From the table, select the angle θ that produces the shortest walkway that satisfies the conditions of the problem. (In calculus, special techniques are used to find this angle.)

TABLE 1

θ (rad)	0.50	0.60	0.70	0.80	0.90	1.00	1.10
L (ft)	288.3						

(D) Graph the equation found in part (A) on a graphing calculator. Use the MINIMUM command to find the value of θ that produces the minimum value of L, then find the minimum value of L.

63. Emergency Vehicles An emergency vehicle is parked 50 ft from a building (see the figure). A warning light on top of the vehicle is rotating clockwise at exactly 20 rpm.

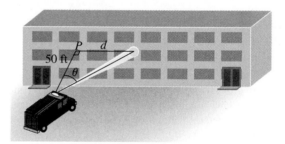

Figure for 63

(A) Write an equation for the distance d in terms of the angle θ.

(B) If $t = 0$ when the light is pointing at P, write an equation for θ in terms of t, where t is time in minutes.

(C) Write an equation for d in terms of t.

(D) Graph the equation from part (C) for $0 < t < \frac{1}{80}$, and describe what happens to d as t approaches $\frac{1}{80}$ min.

☆ **64. Spring–Mass System** Graph each of the following equations for $0 \leq t \leq 4$ and identify each as an example of simple harmonic motion, damped harmonic motion, or resonance.

(A) $y = t^{-0.4} \cos \pi t$ (B) $y = 2^{-0.5} \cos \pi t$

(C) $y = t^{0.8} \cos \pi t$

65. Modeling a Seasonal Business Cycle The sales for a national soft-drink company are growing steadily but are subject to seasonal variations. The following model has been developed to project the monthly sales for the next 36 months:

$$S = 30 + 0.5n + 8 \sin \frac{\pi n}{6}$$

where S represents the sales (in millions of dollars) for the nth month, $0 \leq n \leq 36$.

(A) Graph the equation for $0 \leq n \leq 36$.

(B) Describe what the graph shows regarding monthly sales over the 36 month period.

66. Modeling Temperature Variation The 30 year average monthly temperature (in °F) for each month of the year for Milwaukee, WI, is given in Table 2.

(A) Enter the data in Table 2 for a 2 yr period in your graphing calculator and produce a scatter plot in the viewing window $1 \leq x \leq 24$, $15 \leq y \leq 75$.

(B) A function of the form $y = k + A \sin(Bx + C)$ can be used to model the data. Use the data in Table 2 to determine k, A, and B. Use the graph in part (A) to visually estimate C (to one decimal place).

(C) Plot the data from part (A) and the equation from part (B) in the same viewing window. If necessary, adjust the value of C to produce a better fit.

TABLE 2

x (months)	1	2	3	4	5	6	7	8	9	10	11	12
y (temperature, °F)	19	23	32	45	55	65	71	69	62	51	37	25

Source: World Almanac

Identities

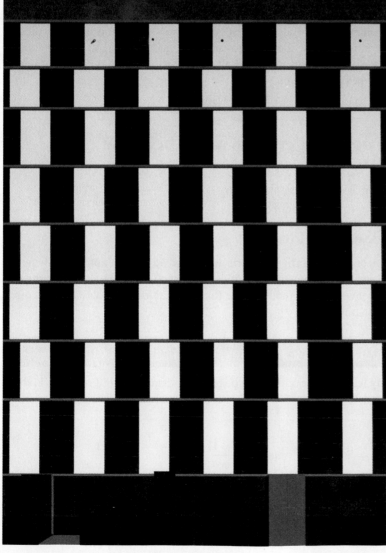

4.1 Fundamental Identities and Their Use

4.2 Verifying Trigonometric Identities

4.3 Sum, Difference, and Cofunction Identities

4.4 Double-Angle and Half-Angle Identities

☆**4.5** Product–Sum and Sum–Product Identities

Chapter 4 Group Activity: From M sin Bt + N cos Bt to A sin(Bt + C)

Chapter 4 Review

☆ Sections marked with a star may be omitted without loss of continuity.

T rigonometric functions have many uses. In addition to solving real-world problems, they are used in the development of mathematics—analytic geometry, calculus, and so on. Whatever their application, it is often useful to be able to change a trigonometric expression from one form to an equivalent form. This involves the use of *identities*. An equation in one or more variables is said to be an **identity** if the left side is equal to the right side for all replacements of the variables for which both sides are defined. The equation

$$x^2 - x - 6 = (x - 3)(x + 2)$$

is an identity, while

$$x^2 - x - 6 = 2x$$

is not. The latter is called a **conditional equation,** since it is true only for certain values of x and not for all values for which both sides are defined.

In this chapter we will develop a lot of trigonometric identities, and you will get practice in using these identities to convert a variety of trigonometric expressions into equivalent forms. You will also get practice in using the identities to solve other types of problems.

4.1   Fundamental Identities and Their Use

- Fundamental Identities
- Evaluating Trigonometric Functions
- Converting to Equivalent Forms

■ Fundamental Identities

Our first encounter with trigonometric identities was in Section 2.5, where we established several fundamental identities. We restate and name these identities in the box for convenient reference. These fundamental identities will be used often in the work that follows.

FUNDAMENTAL TRIGONOMETRIC IDENTITIES

For x any real number or angle in degree or radian measure for which both sides are defined:

Reciprocal identities

$$\csc x = \frac{1}{\sin x} \qquad \sec x = \frac{1}{\cos x} \qquad \cot x = \frac{1}{\tan x}$$

Quotient identities

$$\tan x = \frac{\sin x}{\cos x} \qquad \cot x = \frac{\cos x}{\sin x}$$

> **Identities for negatives**
>
> $$\sin(-x) = -\sin x \qquad \cos(-x) = \cos x \qquad \tan(-x) = -\tan x$$
>
> **Pythagorean identities**
>
> $$\sin^2 x + \cos^2 x = 1 \qquad \tan^2 x + 1 = \sec^2 x \qquad 1 + \cot^2 x = \csc^2 x$$

The second and third Pythagorean identities were established in Problems 91 and 92 in Exercise 2.5. An easy way to remember them is suggested in Explore/Discuss 1.

EXPLORE/DISCUSS 1

Discuss an easy way to remember the second and third Pythagorean identities based on the first. [*Hint:* Divide through the first Pythagorean identity by appropriate expressions.]

■ Evaluating Trigonometric Functions

Suppose we know that $\cos x = -\frac{4}{5}$ and $\tan x = \frac{3}{4}$. How can we find the exact values of the remaining trigonometric functions of x without finding x and without using reference triangles? We use fundamental identities.

EXAMPLE 1

Using Fundamental Identities

If $\cos x = -\frac{4}{5}$ and $\tan x = \frac{3}{4}$, use the fundamental identities to find the exact values of the remaining four trigonometric functions at x.

Solution   *Find* $\sec x$: $\sec x = \dfrac{1}{\cos x} = \dfrac{1}{\frac{4}{5}} = -\dfrac{5}{4}$

Find $\cot x$: $\cot x = \dfrac{1}{\tan x} = \dfrac{1}{\frac{3}{4}} = \dfrac{4}{3}$

Find $\sin x$:   We can start with either a Pythagorean identity or a quotient identity. We choose the quotient identity $\tan x = (\sin x)/(\cos x)$ changed to the form

$$\sin x = (\cos x)(\tan x) = \left(-\frac{4}{5}\right)\left(\frac{3}{4}\right) = -\frac{3}{5}$$

Find $\csc x$: $\csc x = \dfrac{1}{\sin x} = \dfrac{1}{-\frac{3}{5}} = -\dfrac{5}{3}$ ■

Matched Problem 1   If $\sin x = -\frac{4}{5}$ and $\cot x = -\frac{3}{4}$, use the fundamental identities to find the exact values of the remaining four trigonometric functions at x.   ■

   ## **Using Fundamental Identities**

Use the fundamental identities to find the exact values of the remaining trigonometric functions of x, given

$$\cos x = -\frac{4}{\sqrt{17}} \qquad \text{and} \qquad \tan x < 0$$

Solution   *Find* $\sin x$:   We start with the Pythagorean identity

$$\sin^2 x + \cos^2 x = 1$$

and solve for $\sin x$:

$$\sin x = \pm \sqrt{1 - \cos^2 x}$$

Since both $\cos x$ and $\tan x$ are negative, x is associated with the second quadrant, where $\sin x$ is positive:

$$\sin x = \sqrt{1 - \cos^2 x} \qquad \textit{Substitute } \cos x = -\frac{4}{\sqrt{17}}.$$

$$= \sqrt{1 - \left(-\frac{4}{\sqrt{17}}\right)^2} \qquad \textit{Simplify.}$$

$$= \sqrt{\frac{1}{17}} = \frac{1}{\sqrt{17}} \quad *$$

Find $\sec x$: $\sec x = \dfrac{1}{\cos x} = \dfrac{1}{-4/\sqrt{17}} = -\dfrac{\sqrt{17}}{4}$

Find $\csc x$: $\csc x = \dfrac{1}{\sin x} = \dfrac{1}{1/\sqrt{17}} = \sqrt{17}$

Find $\tan x$: $\tan x = \dfrac{\sin x}{\cos x} = \dfrac{1/\sqrt{17}}{-4/\sqrt{17}} = -\dfrac{1}{4}$

Find $\cot x$: $\cot x = \dfrac{1}{\tan x} = \dfrac{1}{-\frac{1}{4}} = -4$ ■

---

* An equivalent answer is $1/\sqrt{17} = \sqrt{17}/(\sqrt{17}\sqrt{17}) = \sqrt{17}/17$, a form in which we have rationalized (eliminated radicals in) the denominator. Whether we rationalize the denominator or not depends entirely on what we want to do with the answer—sometimes an unrationalized form is more useful than a rationalized form. For the remainder of this book, you should leave answers to matched problems and exercises unrationalized unless directed otherwise.

Matched Problem 2   Use the fundamental identities to find the exact values of the remaining trigonometric functions of x, given:

$$\tan x = -\frac{\sqrt{21}}{2} \qquad \text{and} \qquad \cos x > 0 \qquad\blacksquare$$

■ Converting to Equivalent Forms

One of the most important and frequent uses of the fundamental identities is the conversion of trigonometric forms into equivalent simpler or more useful forms. A couple of examples will illustrate the process.

EXAMPLE 3

Simplifying Trigonometric Expressions

Use fundamental identities and appropriate algebraic operations to simplify the following expression:

$$\frac{1}{\cos^2 \alpha} - 1$$

Solution   We start by using algebra to form a single fraction:

$$\frac{1}{\cos^2 \alpha} - 1 = \frac{1 - \cos^2 \alpha}{\cos^2 \alpha} \qquad \text{Use a Pythagorean identity.}$$

$$= \frac{\sin^2 \alpha}{\cos^2 \alpha} \qquad \text{Use algebra.}$$

$$= \left(\frac{\sin \alpha}{\cos \alpha}\right)^2 \qquad \text{Use a quotient identity.}$$

$$= \tan^2 \alpha$$

Key Algebraic Steps:

$$\frac{1}{b^2} \quad 1 = \frac{1}{b^2} - \frac{b^2}{b^2} = \frac{1 - b^2}{b^2} \quad \text{and} \quad \frac{a^2}{b^2} = \left(\frac{a}{b}\right)^2 \qquad\blacksquare$$

Matched Problem 3   Use fundamental identities and appropriate algebraic operations to simplify the following expression:

$$\frac{\sin^2 \theta}{\cos^2 \theta} + 1 \qquad\blacksquare$$

Because $\tan x$, $\cot x$, $\sec x$, and $\csc x$ can all be written in terms of $\sin x$ and/or $\cos x$, a trigonometric expression can be converted to a form, often simpler, that involves only sines and cosines. Example 4 illustrates this technique.

EXAMPLE 4

Converting a Trigonometric Expression to an Equivalent Form

Using fundamental identities, write the following expression in terms of sines and cosines and then simplify:

$$\frac{\tan x - \cot x}{\tan x + \cot x}$$

Write the final answer in terms of the cosine function.

Solution   We begin by using quotient identities to change to sines and cosines.

$$\frac{\tan x - \cot x}{\tan x + \cot x} = \frac{\dfrac{\sin x}{\cos x} - \dfrac{\cos x}{\sin x}}{\dfrac{\sin x}{\cos x} + \dfrac{\cos x}{\sin x}}$$ Multiply numerator and denominator by the least common denominator of all internal fractions.

$$= \frac{(\sin x \cos x)\left(\dfrac{\sin x}{\cos x} - \dfrac{\cos x}{\sin x}\right)}{(\sin x \cos x)\left(\dfrac{\sin x}{\cos x} + \dfrac{\cos x}{\sin x}\right)}$$ Use algebra.

$$= \frac{\sin^2 x - \cos^2 x}{\sin^2 x + \cos^2 x}$$ Use a Pythogrean identity twice.

$$= \frac{1 - \cos^2 x - \cos^2 x}{1}$$ Use algebra.

$$= 1 - 2\cos^2 x$$

Key Algebraic Steps:

$$\frac{\dfrac{a}{b} - \dfrac{b}{a}}{\dfrac{a}{b} + \dfrac{b}{a}} = \frac{ab\left(\dfrac{a}{b} - \dfrac{b}{a}\right)}{ab\left(\dfrac{a}{b} + \dfrac{b}{a}\right)} = \frac{a^2 - b^2}{a^2 + b^2}$$ ■

Matched Problem 4   Using fundamental identities, write the following expression in terms of sines and cosines and then simplify:

$$1 + \frac{\tan z}{\cot z}$$ ■

Answers to Matched Problems

1. $\csc x = -\frac{5}{4}$, $\tan x = -\frac{4}{3}$, $\cos x = \frac{3}{5}$, $\sec x = \frac{5}{3}$

2. $\cot x = -2/\sqrt{21}$, $\sec x = \frac{5}{2}$, $\cos x = \frac{2}{5}$, $\sin x = -\sqrt{21}/5$,
   $\csc x = -5/\sqrt{21}$

3. $\sec^2 \theta$ 4. $\sec^2 z$

EXERCISE 4.1

A

1. List the reciprocal identities and identities for negatives. Try not to look at the text.

2. List the quotient identities and Pythagorean identities. Try not to look at the text.

3. One of the following equations is an identity and the other is a conditional equation. Identify each and explain the difference between the two.

$$(1) \quad 3(2x - 3) = 3(3 - 2x)$$

$$(2) \quad 3(2x - 3) = 6x - 9$$

4. One of the following equations is an identity and the other is a conditional equation. Identify each and explain the difference between the two.

$$(1) \quad \sin x + \cos x = 1$$

$$(2) \quad \sin^2 x = 1 - \cos^2 x$$

In Problems 5–10, use the fundamental identities to find the exact values of the remaining trigonometric functions of x, given the following:

5. $\sin x = 2/\sqrt{5}$ and $\cos x = 1/\sqrt{5}$

6. $\sin x = \sqrt{5}/3$ and $\tan x = -\sqrt{5}/2$

7. $\cos x = 1/\sqrt{10}$ and $\csc x = -\sqrt{10}/3$

8. $\cos x = \sqrt{7}/4$ and $\cot x = -\sqrt{7}/3$

9. $\tan x = 1/\sqrt{15}$ and $\sec x = -4/\sqrt{15}$

10. $\cot x = 2/\sqrt{21}$ and $\csc x = 5/\sqrt{21}$

In Problems 11–22, simplify each expression using the fundamental identities.

11. $\tan u \cot u$

12. $\sec x \cos x$

13. $\tan x \csc x$

14. $\sec \theta \cot \theta$

15. $\dfrac{\sec^2 x - 1}{\tan x}$

16. $\dfrac{\csc^2 v - 1}{\cot v}$

17. $\dfrac{\sin^2 \theta}{\cos \theta} + \cos \theta$

18. $\dfrac{1}{\csc^2 x} + \dfrac{1}{\sec^2 x}$

19. $\dfrac{1}{\sin^2 \beta} - 1$

20. $\dfrac{1 - \sin^2 u}{\cos u}$

21. $\dfrac{(1 - \cos x)^2 + \sin^2 x}{1 - \cos x}$

22. $\dfrac{\cos^2 x + (\sin x + 1)^2}{\sin x + 1}$

B **23.** If an equation has an infinite number of solutions, is it an identity? Explain.

24. Does an identity have an infinite number of solutions? Explain.

In Problems 25–30, use the fundamental identities to find the exact values of the remaining trigonometric functions of x, given the following:

25. $\sin x = 2/5$ and $\cos x < 0$

26. $\cos x = 3/4$ and $\tan x < 0$

27. $\tan x = -1/2$ and $\sin x > 0$

28. $\cot x = -3/2$ and $\csc x > 0$

29. $\sec x = 4$ and $\cot x > 0$

30. $\csc x = -3$ and $\sec x < 0$

In Problems 31–34, is it possible to use the given information to find the exact values of the remaining trigonometric functions? Explain.

31. $\sin x = 1/3$ and $\csc x > 0$

32. $\tan x = 2$ and $\cot x > 0$

33. $\cot x = -\sqrt{3}$ and $\tan x < 0$

34. $\sec x = -\sqrt{2}$ and $\cos x < 0$

35. For the following graphing calculator displays, find the value of the final expression without finding x or using a calculator:

(A)   (B)

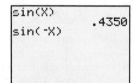   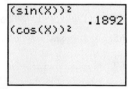

36. For the following graphing calculator displays, find the value of the final expression without finding x or using a calculator:

(A)   (B)

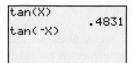   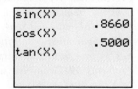

Using fundamental identities, write the expressions in Problems 37–46 in terms of sines and cosines and then simplify.

37. $\csc(-y)\cos(-y)$ **38.** $\sin(-\alpha)\sec(-\alpha)$

39. $\cot x \cos x + \sin x$ **40.** $\cos u + \sin u \tan u$

41. $\dfrac{\cot(-\theta)}{\csc \theta} + \cos \theta$ **42.** $\sin y - \dfrac{\tan(-y)}{\sec y}$

43. $\dfrac{\cot x}{\tan x} + 1$ **44.** $\dfrac{1 + \cot^2 y}{\cot^2 y}$

45. $\sec w \csc w - \sec w \sin w$

46. $\csc \theta \sec \theta - \cot \theta \cos \theta$

In Problems 47–58, is the equation an identity? Explain.

47. $(x + 3)^2 = x^2 + 9$ **48.** $\dfrac{x^2 - 25}{x + 5} = x - 5$

49. $\dfrac{1}{x + 1} = \dfrac{x - 1}{x^2 - 1}$ **50.** $(1 + 2x)^3 = 1 + 8x^3$

51. $\sin x \cot x = \cos x$ **52.** $\sec x \tan x = \csc x$

53. $\dfrac{\sec(-x)}{\sec x} = -1$ **54.** $\dfrac{\cot x}{\cot(-x)} = -1$

55. $\sin(-x) + \sin x = 0$ **56.** $\cos(-x) + \cos x = 0$

57. $\tan^2 x + \sec^2 x = 1$ **58.** $\cot^2 x - \csc^2 x = -1$

59. If $\sin x = \frac{2}{5}$, find:

    (A) $\sin^2(x/2) + \cos^2(x/2)$ (B) $\csc^2(2x) - \cot^2(2x)$

60. If $\cos x = \frac{3}{7}$, find:

    (A) $\sin^2(2x) + \cos^2(2x)$ (B) $\sec^2(x/2) - \tan^2(x/2)$

In Problems 61–68, the equation is an identity in certain quadrants. Indicate which quadrants.

61. $\sqrt{1 - \cos^2 x} = \sin x$ **62.** $\sqrt{1 - \sin^2 x} = \cos x$

63. $\sqrt{1 - \sin^2 x} = -\cos x$ **64.** $\sqrt{1 - \cos^2 x} = -\sin x$

65. $\sqrt{1 - \sin^2 x} = |\cos x|$ **66.** $\sqrt{1 - \cos^2 x} = |\sin x|$

67. $\dfrac{\sin x}{\sqrt{1 - \sin^2 x}} = \tan x$ **68.** $\dfrac{\sin x}{\sqrt{1 - \sin^2 x}} = -\tan x$

Applications

Precalculus: Trigonometric Substitution *In calculus, problems are frequently encountered that involve radicals of the forms $\sqrt{a^2 - u^2}$ and $\sqrt{a^2 + u^2}$. It is very useful to be able to make trigonometric substitutions and use fundamental identities to transform these expressions into non-radical forms. Problems 69–72 involve such transformations. (Recall that $\sqrt{N^2} = N$ if $N \geq 0$ and $\sqrt{N^2} = -N$ if $N < 0$.)*

69. In the expression $\sqrt{a^2 - u^2}$, $a > 0$, let $u = a \sin x$, $-\pi/2 < x < \pi/2$. After using an appropriate fundamental identity, write the given expression in a final form free of radicals.

70. In the expression $\sqrt{a^2 - u^2}$, $a > 0$, let $u = a \cos x$, $0 < x < \pi$. After using an appropriate fundamental identity, write the given expression in a final form free of radicals.

71. In the expression $\sqrt{a^2 + u^2}$, $a > 0$, let $u = a \tan x$, $0 < x < \pi/2$. After using an appropriate fundamental identity, write the given expression in a final form free of radicals.

72. In the expression $\sqrt{a^2 + u^2}$, $a > 0$, let $u = a \cot x$, $0 < x < \pi/2$. After using an appropriate fundamental identity, write the given expression in a final form free of radicals.

Precalculus: Parametric Equations *Suppose we are given the parametric equations of a curve,*

$$\begin{cases} x = \cos t \\ y = \sin t \end{cases} \qquad 0 \leq t \leq 2\pi$$

[The parameter t is assigned values and the corresponding points (cos t, sin t) are plotted in a rectangular coordinate system.] These parametric equations can be transformed into a standard rectangular form free of the parameter t by use of the fundamental identities as follows:

$$x^2 + y^2 = \cos^2 t + \sin^2 t = 1$$

Thus,

$$x^2 + y^2 = 1$$

is the nonparametric equation for the curve. The latter is the equation of a circle with radius 1 and center at the origin. Refer to this discussion for Problems 73–76.

73. **Spacecraft Orbits** The satellites of a planet usually travel in elliptical orbits. However, if a spacecraft has sufficient velocity when it comes under the gravitational influence of a planet, its orbit may be hyperbolic, and the spacecraft can escape the gravitational influence of the planet (see the figure on page 237).

    If the planet, in its own orbit around the sun, has greater velocity than the spacecraft, the spacecraft can actually gain velocity from the gravitational assist. *Voyagers I* and *II*, launched in 1977 and scheduled to return data until 2020, used this "slingshot effect" in their remarkable planetary tours. Orbits, whether elliptical or hyperbolic, can be described by parametric equations.

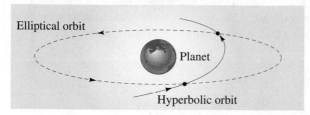

Figure for 73

(A) Transform the parametric equations (by suitable use of a fundamental identity) into nonparametric form.

$$\begin{cases} x = 5\cos t \\ y = 4\sin t \end{cases} \quad 0 \le t \le 360$$

 (B) Using a graphing calculator in parametric and degree modes, graph the parametric equations from part (A) and determine whether the orbit is elliptical or hyperbolic.

74. Spacecraft Orbits Refer to Problem 73.

(A) Transform the parametric equations (by suitable use of a fundamental identity) into nonparametric form.

$$\begin{cases} x = 2\sec t \\ y = \sqrt{5}\,\tan t \end{cases} \quad -90 \le t \le 90$$

 (B) Using a graphing calculator in parametric and degree modes, graph the parametric equations from part (A) and determine whether the orbit is elliptical or hyperbolic.

75. Spacecraft Orbits Refer to Problem 73.

(A) Transform the parametric equations (by suitable use of a fundamental identity) into nonparametric form.

$$\begin{cases} x = 5\cot t \\ y = -3\csc t \end{cases} \quad 0 \le t \le 180$$

 (B) Using a graphing calculator in parametric and degree modes, graph the parametric equations from part (A) and determine whether the orbit is elliptical or hyperbolic.

76. Spacecraft Orbits Refer to Problem 73.

(A) Transform the parametric equations (by suitable use of a fundamental identity) into nonparametric form.

$$\begin{cases} x = 2\sin t \\ y = 9\cos t \end{cases} \quad -180 \le t \le 180$$

 (B) Using a graphing calculator in parametric and degree modes, graph the parametric equations from part (A) and determine whether the orbit is elliptical or hyperbolic.

4.2 Verifying Trigonometric Identities

- Verifying Identities
- 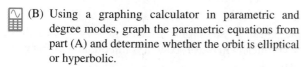 Testing Identities Using a Graphing Calculator

If we know that a certain equation is satisfied for several values of its variables, we might suspect that it is an identity. But to verify that it is an identity, we must prove that the equation is satisfied by *all* values of the variables for which both sides of the equation are defined. In this section we will discuss techniques for verifying trigonometric identities. We will also show that a graphing calculator can be helpful in demonstrating that certain equations are *not* identities.

■ Verifying Identities

We will now verify (prove) some given trigonometric identities; this process will be helpful to you if you want to convert a trigonometric expression into a form that may be more useful. Verifying a trigonometric identity is different from solving an equation. When solving an equation you use properties of equality such as adding the same quantity to each side or multiplying both sides by a nonzero quantity. These operations are not valid in the process of verifying identities because, at the start, we do not know that the left and right expressions are equal.

VERIFYING AN IDENTITY

To verify an identity, start with the expression on one side and, through a sequence of valid steps involving the use of known identities or algebraic manipulation, convert that expression into the expression on the other side.

⚠ **Caution**   When verifying an identity, *do not* add the same quantity to each side, multiply both sides by the same nonzero quantity, or square (or take the square root of) both sides.   □

The following examples illustrate some of the techniques used to establish certain identities. To become proficient in the process, it is important that you work many problems on your own.

EXAMPLE 1   **Identity Verification**

Verify the identity: $\csc(-x) = -\csc x$

Verification

$$\text{Left side} = \csc(-x) \qquad \text{Use a reciprocal identity.}$$

$$= \frac{1}{\sin(-x)} \qquad \text{Use an identity for negatives.}$$

$$= \frac{1}{-\sin x} \qquad \text{Use algebra.}$$

$$= -\frac{1}{\sin x} \qquad \text{Use a reciprocal identity.}$$

$$= -\csc x \qquad \text{Right side}$$

Matched Problem 1   Verify the identity: $\sec(-x) = \sec x$

EXAMPLE 2   **Identity Verification**

Verify the identity: $\tan x \sin x + \cos x = \sec x$

Verification

$$\text{Left side} = \tan x \sin x + \cos x \qquad \text{Use a quotient identity.}$$

$$= \frac{\sin x}{\cos x} \sin x + \cos x \qquad \text{Use algebra.}$$

$$= \frac{\sin^2 x + \cos^2 x}{\cos x} \qquad \text{Use a Pythagorean identity.}$$

$$= \frac{1}{\cos x} \qquad \text{Use a reciprocal identity.}$$

$$= \sec x \qquad \text{Right side}$$

Key Algebraic Steps:

$$\frac{a}{b} a + b = \frac{a^2}{b} + b = \frac{a^2 + b^2}{b}$$

Matched Problem 2   Verify the identity: $\cot x \cos x + \sin x = \csc x$

To verify an identity, proceed from one side to the other, making sure all steps are reversible. Even though there is no fixed method of verification that works for all identities, there are certain steps that help in many cases.

SOME SUGGESTIONS FOR VERIFYING IDENTITIES

Step 1   Start with the more complicated side of the identity and transform it into the simpler side.

Step 2   Try using basic or other known identities.

Step 3   Try algebraic operations such as multiplying, factoring, combining fractions, or splitting fractions.

Step 4   If other steps fail, try expressing each function in terms of sine and cosine functions; then perform appropriate algebraic operations.

Step 5   At each step, keep the other side of the identity in mind. This often reveals what you should do in order to get there.

EXAMPLE 3   **Identity Verification**

Verify the identity: $\dfrac{\cot^2 x - 1}{1 + \cot^2 x} = 1 - 2 \sin^2 x$

Verification   Left side $= \dfrac{\cot^2 x - 1}{1 + \cot^2 x}$ Convert to sines and cosines.

$$= \frac{\dfrac{\cos^2 x}{\sin^2 x} - 1}{1 + \dfrac{\cos^2 x}{\sin^2 x}} \qquad \begin{array}{l}\text{Multiply numerator and}\\ \text{denominator by } \sin^2 x, \text{ the LCD}\\ \text{of all secondary fractions.}\end{array}$$

$$= \frac{(\sin^2 x)\left(\dfrac{\cos^2 x}{\sin^2 x} - 1\right)}{(\sin^2 x)\left(1 + \dfrac{\cos^2 x}{\sin^2 x}\right)} \qquad \text{Use algebra.}$$

$$= \frac{\cos^2 x - \sin^2 x}{\sin^2 x + \cos^2 x}$$ Use a Pythagorean identity
twice.

$$= \frac{1 - \sin^2 x - \sin^2 x}{1}$$ Use algebra.

$$= 1 - 2 \sin^2 x$$ Right side

Key Algebraic Steps:

$$\frac{\dfrac{b^2}{a^2} - 1}{1 + \dfrac{b^2}{a^2}} = \frac{a^2\left(\dfrac{b^2}{a^2} - 1\right)}{a^2\left(1 + \dfrac{b^2}{a^2}\right)} = \frac{b^2 - a^2}{a^2 + b^2}$$ ∎

Matched Problem 3   Verify the identity: $\dfrac{\tan^2 x - 1}{1 + \tan^2 x} = 1 - 2 \cos^2 x$ ∎

EXPLORE/DISCUSS 1

Can you verify the identity in Example 3,

$$\frac{\cot^2 x - 1}{1 + \cot^2 x} = 1 - 2 \sin^2 x$$

using another sequence of steps? The following start, using Pythagorean identities, leads to a shorter verification:

$$\frac{\cot^2 x - 1}{1 + \cot^2 x} = \frac{\csc^2 x - 1 - 1}{\csc^2 x}$$

If an expression involves two or more fractions, it can be helpful to combine fractions by finding a common denominator, as illustrated in Example 4.

EXAMPLE 4   **Identity Verification**

Verify the identity: $\dfrac{1 + \cos x}{\sin x} + \dfrac{\sin x}{1 + \cos x} = 2 \csc x$

Verification

Left side $= \dfrac{1 + \cos x}{\sin x} + \dfrac{\sin x}{1 + \cos x}$

Rewrite with a common denomimnator.

$= \dfrac{(1 + \cos x)^2 + \sin^2 x}{(\sin x)(1 + \cos x)}$

Use algebra.

$= \dfrac{1 + 2\cos x + \cos^2 x + \sin^2 x}{(\sin x)(1 + \cos x)}$

Use a Pythagorean identity.

$= \dfrac{1 + 2\cos x + 1}{(\sin x)(1 + \cos x)}$

Use algebra.

$= \dfrac{2 + 2\cos x}{(\sin x)(1 + \cos x)}$

Use algebra.

$= \dfrac{2(1 + \cos x)}{(\sin x)(1 + \cos x)}$

Cancel common factor.

$= \dfrac{2}{\sin x}$

Use a reciprocal identity.

$= 2\csc x$

Right side

Key Algebraic Steps:

$$\frac{1 + b}{a} + \frac{a}{1 + b} = \frac{(1 + b)^2 + a^2}{a(1 + b)} = \frac{1 + 2b + b^2 + a^2}{a(1 + b)}$$

and

$$\frac{2 + 2b}{a(1 + b)} = \frac{2(1 + b)}{a(1 + b)} = \frac{2}{a}$$ ■

Matched Problem 4

Verify the identity: $\dfrac{1 + \sin x}{\cos x} + \dfrac{\cos x}{1 + \sin x} = 2\sec x$ ■

Any identity can be verified working either from left to right or from right to left. In practice, one of the two directions may seem easier or more natural than the other, so if one direction leads to difficulty, try the other direction.

EXAMPLE 5

Identity Verification

Verify the identity

$$\csc x + \cot x = \frac{\sin x}{1 - \cos x}$$

(A) Going from left to right   (B) Going from right to left

Verification   (A) Going from left to right:

$$\text{Left side} = \csc x + \cot x \qquad \textit{Convert to sines and cosines.}$$

$$= \frac{1}{\sin x} + \frac{\cos x}{\sin x} \qquad \textit{Use algebra.}$$

$$= \frac{1 + \cos x}{\sin x} \qquad \textit{The right side has sin x in the numerator, so we multiply numerator and denominator by sin x.}$$

$$= \frac{(\sin x)(1 + \cos x)}{\sin^2 x} \qquad \textit{Use a Pythagorean identity.}$$

$$= \frac{(\sin x)(1 + \cos x)}{1 - \cos^2 x} \qquad \textit{Factor denominator.}$$

$$= \frac{(\sin x)(1 + \cos x)}{(1 - \cos x)(1 + \cos x)} \qquad \textit{Cancel common factor.}$$

$$= \frac{\sin x}{1 - \cos x} \qquad \textit{Right side}$$

Key Algebraic Steps:

$$\frac{1}{a} + \frac{b}{a} = \frac{1 + b}{a} = \frac{a(1 + b)}{a^2} \quad \text{and} \quad \frac{a(1 + b)}{1 - b^2} = \frac{a(1 + b)}{(1 - b)(1 + b)} = \frac{a}{1 - b}$$

(B) Going from right to left:

$$\text{Right side} = \frac{\sin x}{1 - \cos x} \qquad \textit{Multiply numerator and denominator by 1 + cos x so that a Pythagorean identity can be used.}$$

$$= \frac{(\sin x)(1 + \cos x)}{(1 - \cos x)(1 + \cos x)} \qquad \textit{Use algebra.}$$

$$= \frac{(\sin x)(1 + \cos x)}{1 - \cos^2 x} \qquad \textit{Use a Pythagorean identity.}$$

$$= \frac{(\sin x)(1 + \cos x)}{\sin^2 x} \qquad \textit{Cancel common factor.}$$

$$= \frac{1 + \cos x}{\sin x} \qquad \textit{Use algebra.}$$

$$= \frac{1}{\sin x} + \frac{\cos x}{\sin x} \qquad \textit{Use fundamental identities.}$$

$$= \csc x + \cot x \qquad \textit{Left side}$$

Key Algebraic Steps:

$$\frac{a}{1 - b} = \frac{a(1 + b)}{(1 - b)(1 + b)} = \frac{a(1 + b)}{1 - b^2} \quad \text{and} \quad \frac{a(1 + b)}{a^2} = \frac{1 + b}{a} = \frac{1}{a} + \frac{b}{a} \quad ■$$

Matched Problem 5   Verify the identity

$$\sec m + \tan m = \frac{\cos m}{1 - \sin m}$$

(A) Going from left to right   (B) Going from right to left   ■

■ **Testing Identities Using a Graphing Calculator**

Given an equation, it is not always easy to tell whether it is an identity or a conditional equation. A graphing calculator puts us on the right track with little effort.

EXAMPLE 6

Testing Identities Using a Graphing Calculator

Use a graphing calculator to test whether each equation is an identity. If an equation appears to be an identity, verify it. If the equation does not appear to be an identity, find a value of x for which both sides are defined but are not equal.

(A) $\dfrac{\sin x}{1 - \cos^2 x} = \sec x$ (B) $\dfrac{\sin x}{1 - \cos^2 x} = \csc x$

Solution   (A) Graph both sides of the equation in the same viewing window (Fig. 1). The graphs do not match; therefore, the equation is not an identity. The left side is not equal to the right side for $x = 1$, for example.

(B) Graph both sides of the equation in the same viewing window (Fig. 2). Use $\boxed{\text{TRACE}}$ and check the values of each function for different values of x. The equation appears to be an identity, which we now verify:

$$\frac{\sin x}{1 - \cos^2 x} = \frac{\sin x}{\sin^2 x} = \frac{1}{\sin x} = \csc x$$

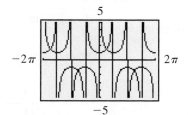

FIGURE 1

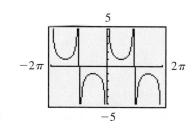

FIGURE 2   ■

 Caution   A graphing calculator, used as in Example 6(B), might *suggest* that a certain equation is an identity, but a graphing calculator *cannot prove* that an equation is an identity (it is possible that the graph of the left-hand side of a conditional equation is indistinguishable from the graph of the right-hand side in certain viewing windows—see Problems 81 and 82 in Exercise 4.2). Always verify an identity using known identities and algebraic manipulations.   □

Matched Problem 6   Repeat Example 6 for the following two equations:

(A) $\tan x + 1 = (\sec x)(\sin x - \cos x)$

(B) $\tan x - 1 = (\sec x)(\sin x - \cos x)$ ∎

Answers to
Matched Problems

1. $\sec(-x) = \dfrac{1}{\cos(-x)} = \dfrac{1}{\cos x} = \sec x$

2. $\cot x \cos x + \sin x = \dfrac{\cos^2 x}{\sin x} + \sin x = \dfrac{\cos^2 x + \sin^2 x}{\sin x}$

$$= \dfrac{1}{\sin x} = \csc x$$

3. $\dfrac{\tan^2 x - 1}{1 + \tan^2 x} = \dfrac{\dfrac{\sin^2 x}{\cos^2 x} - 1}{1 + \dfrac{\sin^2 x}{\cos^2 x}} = \dfrac{(\cos^2 x)\left(\dfrac{\sin^2 x}{\cos^2 x} - 1\right)}{(\cos^2 x)\left(1 + \dfrac{\sin^2 x}{\cos^2 x}\right)}$

$$= \dfrac{\sin^2 x - \cos^2 x}{\cos^2 x + \sin^2 x} = \dfrac{1 - \cos^2 x - \cos^2 x}{1}$$

$$= 1 - 2\cos^2 x$$

4. $\dfrac{1 + \sin x}{\cos x} + \dfrac{\cos x}{1 + \sin x} = \dfrac{(1 + \sin x)^2 + \cos^2 x}{(\cos x)(1 + \sin x)}$

$$= \dfrac{1 + 2\sin x + \sin^2 x + \cos^2 x}{(\cos x)(1 + \sin x)}$$

$$= \dfrac{2 + 2\sin x}{(\cos x)(1 + \sin x)} = \dfrac{2}{\cos x} = 2\sec x$$

5. (A) Going from left to right:

$$\sec m + \tan m = \dfrac{1}{\cos m} + \dfrac{\sin m}{\cos m} = \dfrac{1 + \sin m}{\cos m}$$

$$= \dfrac{(\cos m)(1 + \sin m)}{\cos^2 m} = \dfrac{(\cos m)(1 + \sin m)}{1 - \sin^2 m}$$

$$= \dfrac{(\cos m)(1 + \sin m)}{(1 - \sin m)(1 + \sin m)} = \dfrac{\cos m}{1 - \sin m}$$

(B) Going from right to left:

$$\dfrac{\cos m}{1 - \sin m} = \dfrac{(\cos m)(1 + \sin m)}{(1 - \sin m)(1 + \sin m)} = \dfrac{(\cos m)(1 + \sin m)}{1 - \sin^2 m}$$

$$= \dfrac{(\cos m)(1 + \sin m)}{\cos^2 m} = \dfrac{1 + \sin m}{\cos m}$$

$$= \dfrac{1}{\cos m} + \dfrac{\sin m}{\cos m} = \sec m + \tan m$$

6. (A) Not an identity; the left side is not equal to the right side for $x = 0$, for example:

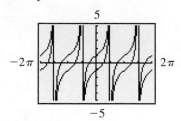

(B)

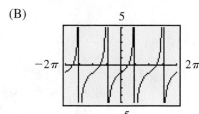

The equation appears to be an identity, which is verified as follows:

$$(\sec x)(\sin x - \cos x) = \frac{1}{\cos x}(\sin x - \cos x)$$

$$= \frac{\sin x}{\cos x} - \frac{\cos x}{\cos x}$$

$$= \tan x - 1$$

EXERCISE 4.2

A *In Problems 1–10, replace each ? with one of the six trigonometric functions so that the resulting equation is an identity. Verify the identity.*

1. $\dfrac{\cos x}{\cot x} = ?$

2. $\dfrac{\sec x}{\csc x} = ?$

3. $\dfrac{\sin x}{\tan x} = ?$

4. $\dfrac{\cot x}{\cos x} = ?$

5. $\dfrac{\csc x}{\cot x} = ?$

6. $\dfrac{\tan x}{\sec x} = ?$

7. $\dfrac{\sec x}{\tan x} = ?$

8. $\dfrac{\cot x}{\csc x} = ?$

9. $\dfrac{\csc x}{\sec x} = ?$

10. $\dfrac{\tan x}{\sin x} = ?$

In Problems 11–36, verify each identity.

11. $\cos x \sec x = 1$

12. $\sin x \csc x = 1$

13. $\tan x \cos x = \sin x$

14. $\cot x \sin x = \cos x$

15. $\tan x = \sin x \sec x$

16. $\cot x = \cos x \csc x$

17. $\csc(-x) = -\csc x$

18. $\sec(-x) = \sec x$

19. $\dfrac{\sin \alpha}{\cos \alpha \tan \alpha} = 1$

20. $\dfrac{\cos \alpha}{\sin \alpha \cot \alpha} = 1$

21. $\dfrac{\cos \beta \sec \beta}{\tan \beta} = \cot \beta$

22. $\dfrac{\tan \beta \cot \beta}{\sin \beta} = \csc \beta$

23. $(\sec \theta)(\sin \theta + \cos \theta) = \tan \theta + 1$

24. $(\csc \theta)(\cos \theta + \sin \theta) = \cot \theta + 1$

25. $\dfrac{\cos^2 t - \sin^2 t}{\sin t \cos t} = \cot t - \tan t$

26. $\dfrac{\cos \alpha - \sin \alpha}{\sin \alpha \cos \alpha} = \csc \alpha - \sec \alpha$

27. $\dfrac{\cos \beta}{\cot \beta} + \dfrac{\sin \beta}{\tan \beta} = \sin \beta + \cos \beta$

28. $\dfrac{\tan u}{\sin u} - \dfrac{\cot u}{\cos u} = \sec u - \csc u$

29. $\sec^2 \theta - \tan^2 \theta = 1$

30. $\csc^2 \theta - \cot^2 \theta = 1$

31. $(\sin^2 x)(1 + \cot^2 x) = 1$

32. $(\cos^2 x)(\tan^2 x + 1) = 1$

33. $(\csc \alpha + 1)(\csc \alpha - 1) = \cot^2 \alpha$

34. $(\sec \beta - 1)(\sec \beta + 1) = \tan^2 \beta$

35. $\dfrac{\sin t}{\csc t} + \dfrac{\cos t}{\sec t} = 1$

36. $\dfrac{1}{\sec^2 m} + \dfrac{1}{\csc^2 m} = 1$

B *In Problems 37–48, is the equation an identity? Explain.*

37. $\dfrac{x^2 + 5x + 6}{x + 2} = x + 3$

38. $\dfrac{1}{x^2 + 4x} = \dfrac{1}{x^2} + \dfrac{1}{4x}$

39. $\dfrac{x^2}{x + 1} = x + x^2$

40. $\dfrac{x^2 - 7x + 12}{x - 3} = x - 4$

41. $\sin^2 x + \csc^2 x = 1$

42. $1 - \cos^2(-x) = \sin^2(-x)$

43. $(\sin x + \cos x)^2 = 1$

44. $(1 - \sec x)^2 = \tan^2 x$

45. $\sin^4 x + \cos^4 x = 1$

46. $\sin^4 x + 2 \sin^2 x \cos^2 x + \cos^4 x = 1$

47. $\cos(-x) \sec x = 1$

48. $\sin(-x) \csc x = 1$

In Problems 49–80, verify each identity.

49. $\dfrac{1 - (\cos \theta - \sin \theta)^2}{\cos \theta} = 2 \sin \theta$

50. $\dfrac{1 - (\sin \theta - \cos \theta)^2}{\sin \theta} = 2 \cos \theta$

51. $\dfrac{\tan w + 1}{\sec w} = \sin w + \cos w$

52. $\dfrac{\cot y + 1}{\csc y} = \cos y + \sin y$

53. $\dfrac{1}{1 - \cos^2 \theta} = 1 + \cot^2 \theta$

54. $\dfrac{1}{1 - \sin^2 \theta} = 1 + \tan^2 \theta$

55. $\dfrac{\sin^2 \beta}{1 - \cos \beta} = 1 + \cos \beta$

56. $\dfrac{\cos^2 \beta}{1 + \sin \beta} = 1 - \sin \beta$

57. $\dfrac{2 - \cos^2 \theta}{\sin \theta} = \csc \theta + \sin \theta$

58. $\dfrac{2 - \sin^2 \theta}{\cos \theta} = \sec \theta + \cos \theta$

59. $\tan x + \cot x = \sec x \csc x$

60. $\dfrac{\csc x}{\cot x + \tan x} = \cos x$

61. $\dfrac{1 - \csc x}{1 + \csc x} = \dfrac{\sin x - 1}{\sin x + 1}$

62. $\dfrac{1 - \cos x}{1 + \cos x} = \dfrac{\sec x - 1}{\sec x + 1}$

63. $\csc^2 \alpha - \cos^2 \alpha - \sin^2 \alpha = \cot^2 \alpha$

64. $\sec^2 \alpha - \sin^2 \alpha - \cos^2 \alpha = \tan^2 \alpha$

65. $(\sin x + \cos x)^2 - 1 = 2 \sin x \cos x$

66. $\sec x - 2 \sin x = \dfrac{(\sin x - \cos x)^2}{\cos x}$

67. $(\sin u - \cos u)^2 + (\sin u + \cos u)^2 = 2$

68. $(\tan x - 1)^2 + (\tan x + 1)^2 = 2 \sec^2 x$

69. $\sin^4 x - \cos^4 x = 1 - 2 \cos^2 x$

70. $\sin^4 x + 2 \sin^2 x \cos^2 x + \cos^4 x = 1$

71. $\dfrac{\sin \alpha}{1 - \cos \alpha} - \dfrac{1 + \cos \alpha}{\sin \alpha} = 0$

72. $\dfrac{1 + \cos \alpha}{\sin \alpha} + \dfrac{\sin \alpha}{1 + \cos \alpha} = 2 \csc \alpha$

73. $\dfrac{\cos^2 n - 3 \cos n + 2}{\sin^2 n} = \dfrac{2 - \cos n}{1 + \cos n}$

74. $\dfrac{\sin^2 n + 4 \sin n + 3}{\cos^2 n} = \dfrac{3 + \sin n}{1 - \sin n}$

75. $\dfrac{1 - \cot^2 x}{\tan^2 x - 1} = \cot^2 x$

76. $\dfrac{\tan^2 x - 1}{1 - \cot^2 x} = \tan^2 x$

77. $\sec^2 x + \csc^2 x = \sec^2 x \csc^2 x$

78. $\tan^2 x - \sin^2 x = \tan^2 x \sin^2 x$

79. $\dfrac{1 + \sin t}{\cos t} = \dfrac{\cos t}{1 - \sin t}$

80. $\dfrac{\sin t}{1 - \cos t} = \dfrac{1 + \cos t}{\sin t}$

81. (A) Graph both sides of the following equation in the same viewing window for $-\pi \le x \le \pi$. Is the equation an identity over the interval $-\pi \le x \le \pi$? Explain. (Recall that 3!, read "3 factorial", is equal to $3 \cdot 2 \cdot 1 = 6$ and, similarly, $5! = 120$ and $7! = 5{,}040$.)

$$\sin x = x - \frac{x^3}{3!} + \frac{x^5}{5!} - \frac{x^7}{7!}$$

(B) Extend the interval in part (A) to $-2\pi \le x \le 2\pi$. Now, does the equation appear to be an identity? What do you observe?

 82. (A) Graph both sides of the following equation in the same viewing window for $-\pi \le x \le \pi$. Is the equation an identity over the interval $[-\pi, \pi]$? Explain.

$$\cos x = 1 - \frac{x^2}{2!} + \frac{x^4}{4!} - \frac{x^6}{6!} + \frac{x^8}{8!}$$

(B) Extend the interval in part (A) to $-2\pi \le x \le 2\pi$. Now does the equation appear to be an identity? What do you observe?

 In Problems 83–90, use a graphing calculator to test whether each equation is an identity. If an equation appears to be an identity, verify it. If an equation does not appear to be an identity, find a value of x for which both sides are defined but are not equal.

83. $\dfrac{\cos x}{\sin(-x)\cot(-x)} = 1$ **84.** $\dfrac{\sin x}{\cos x \tan(-x)} = -1$

85. $\dfrac{\cos(-x)}{\sin x \cot(-x)} = 1$ **86.** $\dfrac{\sin(-x)}{\cos(-x)\tan(-x)} = -1$

87. $\dfrac{\cos x}{\sin x + 1} - \dfrac{\cos x}{\sin x - 1} = 2\csc x$

88. $\dfrac{\tan x}{\sin x + 2\tan x} = \dfrac{1}{\cos x - 2}$

89. $\dfrac{\cos x}{1 - \sin x} + \dfrac{\cos x}{1 + \sin x} = 2\sec x$

90. $\dfrac{\tan x}{\sin x - 2\tan x} = \dfrac{1}{\cos x - 2}$

C *In Problems 91–96, verify each identity.*

91. $\dfrac{\sin x}{1 - \cos x} - \cot x = \csc x$

92. $\dfrac{\cos x}{1 - \sin x} - \tan x = \sec x$

93. $\dfrac{\cot \beta}{\csc \beta + 1} = \dfrac{\csc \beta - 1}{\cot \beta}$

94. $\dfrac{\tan \beta}{\sec \beta - 1} = \dfrac{\sec \beta + 1}{\tan \beta}$

95. $\dfrac{3\cos^2 m + 5\sin m - 5}{\cos^2 m} = \dfrac{3\sin m - 2}{1 + \sin m}$

96. $\dfrac{2\sin^2 z + 3\cos z - 3}{\sin^2 z} = \dfrac{2\cos z - 1}{1 + \cos z}$

In Problems 97 and 98, verify each identity. (The problems involve trigonometric functions with two variables. Be careful with the terms you combine and simplify.)

97. $\dfrac{\sin x \cos y + \cos x \sin y}{\cos x \cos y - \sin x \sin y} = \dfrac{\tan x + \tan y}{1 - \tan x \tan y}$

98. $\dfrac{\tan \alpha + \tan \beta}{1 - \tan \alpha \tan \beta} = \dfrac{\cot \alpha + \cot \beta}{\cot \alpha \cot \beta - 1}$

4.3 Sum, Difference, and Cofunction Identities

- Sum and Difference Identities for Cosine
- Cofunction Identities
- Sum and Difference Identities for Sine and Tangent
- Summary and Use

▪ Sum and Difference Identities for Cosine

The fundamental identities we discussed in Section 4.1 involve only one variable. We now consider an important identity, called a **difference identity for cosine**, that involves two variables:

$$\cos(x - y) = \cos x \cos y + \sin x \sin y \tag{1}$$

Many other useful identities can be readily established from this particular one. We will sketch a proof of identity (1) in which we assume that x and y are restricted as follows: $0 < y < x < 2\pi$. Identity (1) holds, however, for all real numbers and angles in radian or degree measure. In the proof, we will make use of the **formula for the distance between two points,**

$$d(P_1, P_2) = \sqrt{(x_2 - x_1)^2 + (y_2 - y_1)^2}$$

for points $P_1 = (x_1, y_1)$ and $P_2 = (x_2, y_2)$ in a rectangular coordinate system.

We associate x and y with arcs and angles on a unit circle, as indicated in Figure 1a. Using the definitions of the trigonometric functions in Section 2.3, the terminal points of x and y are labeled as indicated in Figure 1a.

Now, if we rotate the triangle AOB clockwise about the origin until the terminal point A coincides with $D = (1, 0)$, then terminal point B will be at C (see Fig. 1b). Since rotation preserves lengths, we have

FIGURE 1

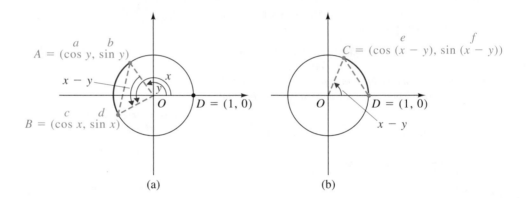

(a)   (b)

$$d(A, B) = d(C, D)$$
$$\sqrt{(c - a)^2 + (d - b)^2} = \sqrt{(1 - e)^2 + (0 - f)^2} \qquad \text{Square both sides.}$$
$$(c - a)^2 + (d - b)^2 = (1 - e)^2 + f^2 \qquad \text{Expand.}$$
$$c^2 - 2ac + a^2 + d^2 - 2db + b^2 = 1 - 2e + e^2 + f^2 \qquad \text{Rearrange terms.}$$
$$(c^2 + d^2) + (a^2 + b^2) - 2ac - 2db = 1 - 2e + (e^2 + f^2) \qquad \text{(2)}$$

Since $c^2 + d^2 = 1$, $a^2 + b^2 = 1$, and $e^2 + f^2 = 1$ (why?), equation (2) becomes

$$e = ac + bd \qquad (3)$$

Replacing e, a, c, b, and d with $\cos(x - y)$, $\cos y$, $\cos x$, $\sin y$, and $\sin x$, respectively (see Fig. 1), we obtain

$$\cos(x - y) = \cos y \cos x + \sin y \sin x$$
$$= \cos x \cos y + \sin x \sin y \qquad (4)$$

If we replace y with $-y$ in (4) and use the identities for negatives, we obtain the **sum identity for cosine:**

$$\cos(x + y) = \cos x \cos y - \sin x \sin y \tag{5}$$

EXPLORE/DISCUSS 1

Find values of x and y such that:
(A) $\cos(x - y) \neq \cos x - \cos y$
(B) $\cos(x + y) \neq \cos x + \cos y$

■ Cofunction Identities

To obtain sum and difference identities for the sine and tangent functions, we first derive **cofunction identities** directly from identity (1), the difference identity for cosine:

$$\cos(x - y) = \cos x \cos y + \sin x \sin y \qquad \text{Let } x = \pi/2.$$

$$\cos\left(\frac{\pi}{2} - y\right) = \cos\frac{\pi}{2}\cos y + \sin\frac{\pi}{2}\sin y$$

$$= 0\cos y + 1\sin y$$

$$= \sin y$$

Thus,

$$\cos\left(\frac{\pi}{2} - y\right) = \sin y \tag{6}$$

for y any real number or angle in radian measure. If y is in degree measure, replace $\pi/2$ with $90°$. Now, if in (6) we let $y = \pi/2 - x$, then we have

$$\cos\left[\frac{\pi}{2} - \left(\frac{\pi}{2} - x\right)\right] = \sin\left(\frac{\pi}{2} - x\right)$$

$$\cos x = \sin\left(\frac{\pi}{2} - x\right)$$

or

$$\sin\left(\frac{\pi}{2} - x\right) = \cos x \tag{7}$$

where x is any real number or angle in radian measure. If x is in degree measure, replace $\pi/2$ with $90°$.

Finally, we state the cofunction identities for tangent and secant (and leave the derivations to Problems 9 and 11 in Exercise 4.3):

$$\tan\left(\frac{\pi}{2} - x\right) = \cot x \qquad \sec\left(\frac{\pi}{2} - x\right) = \csc x \tag{8}$$

for x any real number or angle in radian measure. If x is in degree measure, replace $\pi/2$ with 90°.

Remark   If $0 < x < 90°$, then x and $90° - x$ are complementary angles. Originally, *cosine, cotangent,* and *cosecant* meant, respectively, *complements sine, complements tangent,* and *complements secant.* Now we simply refer to cosine, cotangent, and cosecant as **cofunctions of** sine, tangent, and secant, respectively.   □

■ **Sum and Difference Identities for Sine and Tangent**

To derive a **difference identity for sine,** we use (7), (1), and (6) as follows. Start by restating (6) in terms of z, z any real number:

$$\sin z = \cos\left(\frac{\pi}{2} - z\right) \qquad \text{Substitute } x - y \text{ for } z.$$

$$\sin(x - y) = \cos\left[\frac{\pi}{2} - (x - y)\right] \qquad \text{Use algebra.}$$

$$= \cos\left[\left(\frac{\pi}{2} - x\right) - (-y)\right] \qquad \text{Use (1).}$$

$$= \cos\left(\frac{\pi}{2} - x\right)\cos(-y) + \sin\left(\frac{\pi}{2} - x\right)\sin(-y) \quad \begin{array}{l}\text{Use (6), (7), and} \\ \text{identities for} \\ \text{negatives.}\end{array}$$

$$= \sin x \cos y - \cos x \sin y$$

The same result is obtained by replacing $\pi/2$ with 90°. Thus,

$$\sin(x - y) = \sin x \cos y - \cos x \sin y \tag{9}$$

Now, if we replace y with $-y$ (a good exercise to do), we obtain

$$\sin(x + y) = \sin x \cos y + \cos x \sin y \tag{10}$$

It is not difficult to derive sum and difference identities for the tangent function. See if you can supply the reason for each step:

$$\tan(x - y) = \frac{\sin(x - y)}{\cos(x - y)}$$

$$= \frac{\sin x \cos y - \cos x \sin y}{\cos x \cos y + \sin x \sin y}$$

$$= \cfrac{\cfrac{\sin x \cos y}{\cos x \cos y} - \cfrac{\cos x \sin y}{\cos x \cos y}}{\cfrac{\cos x \cos y}{\cos x \cos y} + \cfrac{\sin x \sin y}{\cos x \cos y}}$$

$$= \cfrac{\tan x - \tan y}{1 + \tan x \tan y}$$

Thus, for all angles or real numbers x and y,

$$\tan(x - y) = \frac{\tan x - \tan y}{1 + \tan x \tan y} \qquad (11)$$

And if we replace y in (11) with $-y$ (another good exercise to do), we obtain

$$\tan(x + y) = \frac{\tan x + \tan y}{1 - \tan x \tan y} \qquad (12)$$

EXPLORE/DISCUSS 2

Find values of x and y such that:

(A) $\tan(x - y) \neq \tan x - \tan y$

(B) $\tan(x + y) \neq \tan x + \tan y$

■ Summary and Use

Before proceeding with examples that illustrate the use of these new identities, we list them and the other cofunction identities for convenient reference.

SUMMARY OF IDENTITIES

For x and y any real numbers or angles in degree or radian measure for which both sides are defined:

Sum identities

$$\sin(x + y) = \sin x \cos y + \cos x \sin y$$

$$\cos(x + y) = \cos x \cos y - \sin x \sin y$$

$$\tan(x + y) = \frac{\tan x + \tan y}{1 - \tan x \tan y}$$

continued

Difference identities

$$\sin(x - y) = \sin x \cos y - \cos x \sin y$$

$$\cos(x - y) = \cos x \cos y + \sin x \sin y$$

$$\tan(x - y) = \frac{\tan x - \tan y}{1 + \tan x \tan y}$$

Cofunction identities

(Replace $\pi/2$ with 90° if x is in degree measure.)

$$\sin\left(\frac{\pi}{2} - x\right) = \cos x \qquad \cos\left(\frac{\pi}{2} - x\right) = \sin x$$

$$\tan\left(\frac{\pi}{2} - x\right) = \cot x \qquad \cot\left(\frac{\pi}{2} - x\right) = \tan x$$

$$\sec\left(\frac{\pi}{2} - x\right) = \csc x \qquad \csc\left(\frac{\pi}{2} - x\right) = \sec x$$

EXAMPLE 1

Using a Difference Identity

Simplify $\sin(x - \pi)$ using a difference identity.

Solution Use the difference identity for sine, replacing y with π:

$$\sin(x - y) = \sin x \cos y - \cos x \sin y$$
$$\sin(x - \pi) = \sin x \cos \pi - \cos x \sin \pi$$
$$= (\sin x)(-1) - (\cos x)(0)$$
$$= -\sin x \qquad \blacksquare$$

Matched Problem 1 Simplify $\cos(x + 3\pi/2)$ using a sum identity. ∎

EXAMPLE 2

Checking the Use of an Identity on a Graphing Calculator

Simplify $\cos(x + \pi)$ using an appropriate identity. Then check the result using a graphing calculator.

Solution
$$\cos(x + \pi) = \cos x \cos \pi - \sin x \sin \pi$$
$$= (\cos x)(-1) - (\sin x)(0)$$
$$= -\cos x$$

To check with a graphing calculator, graph $y_1 = \cos(x + \pi)$ and $y_2 = -\cos x$ in the same viewing window to see if they produce the same graph (Fig. 2). Use TRACE and move back and forth between y_1 and y_2 for different values of x to see that the corresponding y values are the same, or nearly the same. ■

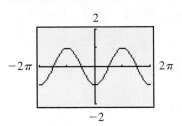

FIGURE 2
$y_1 = \cos(x + \pi)$, $y_2 = -\cos x$

Matched Problem 2   Simplify $\sin(x - \pi)$ using an appropriate identity. Then check the result using a graphing calculator. ■

In Section 2.5 we found exact values of the trigonometric functions for angles that are multiples of 30° or 45°. Exact values of the trigonometric functions for certain other angles can be found by using the sum and difference identities.

Finding Exact Values

Find the exact value of cos 15° in radical form.

Solution   Note that 15° = 45° − 30°, the difference of two special angles.

$$\cos 15° = \cos(45° - 30°)$$ *Use the identity for* $\cos(x - y)$.

$$= \cos 45° \cos 30° + \sin 45° \sin 30°$$ *Evaluate each trigonometric function of a special angle.*

$$= \frac{1}{\sqrt{2}} \cdot \frac{\sqrt{3}}{2} + \frac{1}{\sqrt{2}} \cdot \frac{1}{2}$$ *Use algebra.*

$$= \frac{\sqrt{3} + 1}{2\sqrt{2}}$$ ■

Matched Problem 3   Find the exact value of tan 75° in radical form. ■

Finding Exact Values

Find the exact value of $\cos(x + y)$, given $\sin x = \frac{3}{5}$, $\cos y = \frac{4}{5}$, x in quadrant II, and y in quadrant I. Do not use a calculator.

Solution   We start with the sum identity for cosine:

$$\cos(x + y) = \cos x \cos y - \sin x \sin y$$

We know $\sin x$ and $\cos y$, but not $\sin y$ and $\cos x$. We can find the latter two values using Pythagorean identities or, equivalently, by using reference triangles and the Pythagorean theorem. From Figure 3,

$$a = -\sqrt{5^2 - 3^2} = -4$$

$$\cos x = -\frac{4}{5}$$

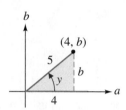

FIGURE 3

From Figure 4,

$$b = \sqrt{5^2 - 4^2} = 3$$

$$\sin y = \frac{3}{5}$$

FIGURE 4

Using the sum identity,

$$\cos(x + y) = \cos x \cos y - \sin x \sin y$$

$$= \left(-\frac{4}{5}\right)\left(\frac{4}{5}\right) - \left(\frac{3}{5}\right)\left(\frac{3}{5}\right) = \frac{-25}{25} = -1 \qquad ■$$

Matched Problem 4   Find the exact value of $\sin(x - y)$, given $\sin x = -\frac{2}{3}$, $\cos y = \sqrt{5}/3$, x in quadrant III, and y in quadrant IV. ■

EXAMPLE 5

Verifying an Identity

Verify the identity: $\cot y - \cot x = \dfrac{\sin(x - y)}{\sin x \sin y}$

Verification   We start with the right side because it involves $x - y$, while the left side involves only x and y:

$$\text{Right side} = \frac{\sin(x - y)}{\sin x \sin y} \qquad \textit{Use a difference identity.}$$

$$= \frac{\sin x \cos y - \cos x \sin y}{\sin x \sin y} \qquad \textit{Use algebra.}$$

$$= \frac{\sin x \cos y}{\sin x \sin y} - \frac{\cos x \sin y}{\sin x \sin y} \qquad \textit{Cancel common factors.}$$

$$= \frac{\cos y}{\sin y} - \frac{\cos x}{\sin x} \qquad \textit{Use a quotient identity.}$$

$$= \cot y - \cot x \qquad \textit{Left side} \qquad ■$$

Matched Problem 5   Verify the identity: $\tan x + \cot y = \dfrac{\cos(x-y)}{\cos x \sin y}$ ■

Answers to   **1.** $\sin x$
Matched Problems   **2.** $y_1 = \sin(x-\pi)$; $y_2 = -\sin x$

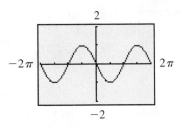

3. $2 + \sqrt{3}$

4. $\dfrac{-4\sqrt{5}}{9}$

5. $\dfrac{\cos(x-y)}{\cos x \sin y} = \dfrac{\cos x \cos y + \sin x \sin y}{\cos x \sin y}$

$= \dfrac{\cos x \cos y}{\cos x \sin y} + \dfrac{\sin x \sin y}{\cos x \sin y}$

$= \cot y + \tan x = \tan x + \cot y$

EXERCISE 4.3

A *We can use sum identities to verify periodic properties for the trigonometric functions. In Problems 1–8, verify the identities using sum identities.*

1. $\cos(x + 2\pi) = \cos x$ **2.** $\sin(x + 2\pi) = \sin x$

3. $\cot(x + \pi) = \cot x$ **4.** $\tan(x + \pi) = \tan x$

5. $\sin(x + 2k\pi) = \sin x$, k an integer

6. $\cos(x + 2k\pi) = \cos x$, k an integer

7. $\tan(x + k\pi) = \tan x$, k an integer

8. $\cot(x + k\pi) = \cot x$, k an integer

In Problems 9–12, verify each identity using cofunction identities for sine *and* cosine *and the fundamental identities discussed in Section 4.1.*

9. $\tan\left(\dfrac{\pi}{2} - x\right) = \cot x$ **10.** $\cot\left(\dfrac{\pi}{2} - x\right) = \tan x$

11. $\sec\left(\dfrac{\pi}{2} - x\right) = \csc x$ **12.** $\csc\left(\dfrac{\pi}{2} - x\right) = \sec x$

In Problems 13–20, use the formulas developed in this section to convert the indicated expression to a form involving $\sin x$, $\cos x$, *and/or* $\tan x$.

13. $\sin(x + 30°)$ **14.** $\cos(x - 45°)$

15. $\cos(x + 60°)$ **16.** $\sin(x - 60°)$

17. $\tan\left(x + \dfrac{\pi}{4}\right)$ **18.** $\tan\left(\dfrac{\pi}{3} - x\right)$

19. $\cot\left(\dfrac{\pi}{6} - x\right)$ **20.** $\cot\left(x - \dfrac{\pi}{4}\right)$

B *In Problems 21–28, use appropriate identities to find the exact value of the indicated expression. Check your results with a calculator.*

21. $\sin 15°$ **22.** $\cos 15°$

23. $\sin 20° \cos 25° + \cos 20° \sin 25°$

24. $\sin 55° \cos 10° - \cos 55° \sin 10°$

25. $\cos 81° \cos 21° + \sin 81° \sin 21°$

26. $\cos 12° \cos 18° - \sin 12° \sin 18°$

27. $\dfrac{\tan 17° + \tan 28°}{1 - \tan 17° \tan 28°}$ **28.** $\dfrac{\tan 52° - \tan 22°}{1 + \tan 52° \tan 22°}$

In Problems 29–34, use the given information and appropriate identities to find the exact value of the indicated expression.

29. Find $\sin(x + y)$ if $\sin x = 1/3$, $\cos y = -3/4$, x is in quadrant II, and y is in quadrant III.

30. Find $\sin(x - y)$ if $\sin x = -2/5$, $\sin y = 2/3$, x is in quadrant IV, and y is in quadrant I.

31. Find $\cos(x - y)$ if $\tan x = -1/4$, $\tan y = -1/5$, x is in quadrant II, and y is in quadrant IV.

32. Find $\cos(x + y)$ if $\tan x = 2/3$, $\tan y = 1/3$, x is in quadrant I, and y is in quadrant III.

33. Find $\tan(x + y)$ if $\sin x = -1/4$, $\cos y = -1/3$, $\cos x < 0$, and $\sin y < 0$.

34. Find $\tan(x - y)$ if $\sin x = 1/5$, $\cos y = 2/5$, $\tan x < 0$, and $\tan y > 0$.

In Problems 35–42, is the equation an identity? Explain.

35. $\tan(x + \pi) = -\tan x$

36. $\sin(x + \pi) = -\sin x$

37. $\cos(x - \pi/2) = \sin x$

38. $\sin(x - \pi/2) = \cos x$

39. $\sin(2\pi - x) = \sin x$

40. $\cos(2\pi - x) = \cos x$

41. $\csc(\pi - x) = \csc x$

42. $\cot(\pi - x) = \cot x$

In Problems 43–56, verify each identity.

43. $\sin 2x = 2 \sin x \cos x$

44. $\cos 2x = \cos^2 x - \sin^2 x$

45. $\cot(x - y) = \dfrac{\cot x \cot y + 1}{\cot y - \cot x}$

46. $\cot(x + y) = \dfrac{\cot x \cot y - 1}{\cot x + \cot y}$

47. $\cot 2x = \dfrac{\cot^2 x - 1}{2 \cot x}$

48. $\tan 2x = \dfrac{2 \tan x}{1 - \tan^2 x}$

49. $\dfrac{\tan \alpha + \tan \beta}{\tan \alpha - \tan \beta} = \dfrac{\sin(\alpha + \beta)}{\sin(\alpha - \beta)}$

50. $\dfrac{\cot \alpha + \cot \beta}{\cot \alpha - \cot \beta} = \dfrac{\sin(\beta + \alpha)}{\sin(\beta - \alpha)}$

51. $\tan x - \tan y = \dfrac{\sin(x - y)}{\cos x \cos y}$

52. $\cot x - \tan y = \dfrac{\cos(x + y)}{\sin x \cos y}$

53. $\tan(x + y) = \dfrac{\cot x + \cot y}{\cot x \cot y - 1}$

54. $\tan(x - y) = \dfrac{\cot y - \cot x}{\cot x \cot y + 1}$

55. $\dfrac{\sin(x + h) - \sin x}{h}$
$= (\sin x)\left(\dfrac{\cos h - 1}{h}\right) + (\cos x)\left(\dfrac{\sin h}{h}\right)$

56. $\dfrac{\cos(x + h) - \cos x}{h}$
$= (\cos x)\left(\dfrac{\cos h - 1}{h}\right) - (\sin x)\left(\dfrac{\sin h}{h}\right)$

57. How would you show that $\csc(x - y) = \csc x - \csc y$ is not an identity?

58. How would you show that $\sec(x + y) = \sec x + \sec y$ is not an identity?

59. Show that $\sin(x - 2) = \sin x - \sin 2$ is not an identity by using a graphing calculator. Then explain what you did.

60. Show that $\cos(x + 1) = \cos x + \cos 1$ is not an identity by using a graphing calculator. Then explain what you did.

In Problems 61–64, use sum or difference identities to convert each equation to a form involving $\sin x$, $\cos x$, and/or $\tan x$. To check your result, enter the original equation in a graphing calculator as y_1 and the converted form as y_2. Then graph y_1 and y_2 in the same viewing window. Use TRACE *to compare the two graphs.*

61. $y = \cos(x + 5\pi/6)$ **62.** $y = \sin(x - \pi/3)$

63. $y = \tan(x - \pi/4)$ **64.** $y = \tan(x + 2\pi/3)$

In Problems 65–68, write each equation in terms of a single trigonometric function. Check the result by entering the original equation in a graphing calculator as y_1 and the converted form as y_2. Then graph y_1 and y_2 in the same viewing window. Use TRACE *to compare the two graphs.*

65. $y = \sin 3x \cos x - \cos 3x \sin x$

66. $y = \cos 3x \cos x - \sin 3x \sin x$

67. $y = \sin\dfrac{\pi x}{4}\cos\dfrac{3\pi x}{4} + \cos\dfrac{\pi x}{4}\sin\dfrac{3\pi x}{4}$

68. $y = \cos(\pi x)\cos\dfrac{\pi x}{2} + \sin(\pi x)\sin\dfrac{\pi x}{2}$

C *Verify the identities in Problems 69 and 70.*
[*Hint:* $\sin(x + y + z) = \sin[(x + y) + z]$.]

69. $\sin(x + y + z) = \sin x \cos y \cos z$
$\qquad\qquad\qquad + \cos x \sin y \cos z + \cos x \cos y \sin z$
$\qquad\qquad\qquad - \sin x \sin y \sin z$

70. $\cos(x + y + z) = \cos x \cos y \cos z$
$\qquad\qquad\qquad - \sin x \sin y \cos z - \sin x \cos y \sin z$
$\qquad\qquad\qquad - \cos x \sin y \sin z$

Applications

71. Precalculus: Angle of Intersection of Two Lines   Use the information in the figure to show that

$$\tan(\theta_2 - \theta_1) = \frac{m_2 - m_1}{1 + m_1 m_2}$$

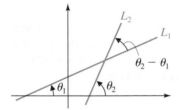

Figure for 71 and 72
$\tan\theta_1 = \text{Slope of } L_1 = m_1$
$\tan\theta_2 = \text{Slope of } L_2 = m_2$

72. Precalculus: Angle of Intersection of Two Lines   Find the acute angle of intersection between the two lines $y = 3x + 1$ and $y = \frac{1}{2}x - 1$. (Use the results of Problem 71.)

73. Analytic Geometry   Find the radian measure of the angle θ in the figure (to three decimal places), if A has coordinates $(2, 4)$ and B has coordinates $(3, 3)$. [*Hint:* Label the angle between OB and the x axis as α; then use an appropriate sum identity.]

74. Analytic Geometry   Find the radian measure of the angle θ in the figure (to three decimal places), if A has coordinates $(3, 9)$ and B has coordinates $(6, 3)$.

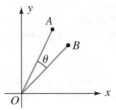

Figure for 73 and 74

75. Surveying   A prominent geological feature of Yosemite National Park is the large monolithic granite peak called Half Dome. The dome rises straight up from the valley floor, where Mirror Lake provides an early morning mirror image of the dome. How can the height H of Half Dome be determined by using only a sextant h feet high to measure the angle of elevation, β, to the top of the dome, and the angle of depression, α, to the reflected dome top in the lake? (See the figure, which is not to scale.) [*Note: AB* and *BC* are not measured as in earlier problems of this type.]*

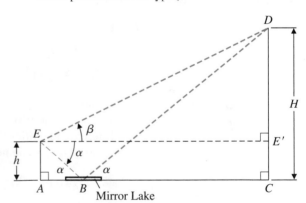

Figure for 75

(A) Using right triangle relationships, show that

$$H = h\left(\frac{1 + \tan\beta \cot\alpha}{1 - \tan\beta \cot\alpha}\right)$$

(B) Using sum identities, show that the result in part (A) can be written in the form

$$H = h\left(\frac{\sin(\alpha + \beta)}{\sin(\alpha - \beta)}\right)$$

(C) If a sextant of height 5.50 ft measures α to be 45.00° and β to be 44.92°, compute the height H of Half Dome above Mirror Lake to three significant digits.

---

* The solution outlined in parts (A) and (B) is credited to Richard J. Palmaccio of Fort Lauderdale, Florida.

76. Light Refraction   Light rays passing through a plate glass window are refracted when they enter the glass and again when they leave to continue on a path parallel to the entering rays (see the figure).

(A) If the plate glass is A inches thick, the parallel displacement of the light rays is B inches, the angle of incidence is α, and the angle of refraction is β, show that

$$\tan \beta = \tan \alpha - \frac{B}{A} \sec \alpha$$

[*Hint:* First use geometric relationships to obtain

$$\frac{A}{\sin(90° - \beta)} = \frac{B}{\sin(\alpha - \beta)}$$

Then use sum identities and fundamental identities to complete the task.]

(B) Using the results in part (A), find β to the nearest degree if $\alpha = 45°$, $A = 0.5$ in., and $B = 0.2$ in.

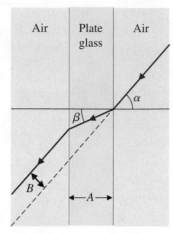

Figure for 76

4.4   Double-Angle and Half-Angle Identities

- Double-Angle Identities
- Half-Angle Identities

We now develop another important set of identities called **double-angle** and **half-angle identities.** We can obtain these identities directly from the sum and difference identities that were found in Section 4.3. In spite of names involving the word *angle,* the new identities hold for real numbers as well.

■ Double-Angle Identities

If we start with the sum identity for sine,

$$\sin(x + y) = \sin x \cos y + \cos x \sin y$$

and let $y = x$, we obtain

$$\sin(x + x) = \sin x \cos x + \cos x \sin x$$

or

$$\sin 2x = 2 \sin x \cos x \tag{1}$$

Similarly, if we start with the sum identity for cosine,

$$\cos(x + y) = \cos x \cos y - \sin x \sin y$$

and let $y = x$, we obtain

$$\cos(x + x) = \cos x \cos x - \sin x \sin x$$

or

$$\cos 2x = \cos^2 x - \sin^2 x \tag{2}$$

Now, using the Pythagorean identities in the two forms

$$\cos^2 x = 1 - \sin^2 x \tag{3}$$
$$\sin^2 x = 1 - \cos^2 x \tag{4}$$

and substituting (3) into (2), we obtain

$$\cos 2x = 1 - \sin^2 x - \sin^2 x$$
$$\cos 2x = 1 - 2 \sin^2 x \tag{5}$$

Substituting (4) into (2), we obtain

$$\cos 2x = \cos^2 x - (1 - \cos^2 x)$$
$$\cos 2x = 2 \cos^2 x - 1 \tag{6}$$

A double-angle identity can be developed for the tangent function in the same way by starting with the sum identity for tangent. This is left as an exercise for you to do. We list these double-angle identities for convenient reference.

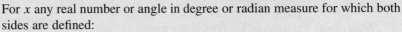

DOUBLE-ANGLE IDENTITIES

For x any real number or angle in degree or radian measure for which both sides are defined:

$$\sin 2x = 2 \sin x \cos x \qquad \cos 2x = \cos^2 x - \sin^2 x$$
$$\tan 2x = \frac{2 \tan x}{1 - \tan^2 x} \qquad\qquad = 1 - 2 \sin^2 x$$
$$= 2 \cos^2 x - 1$$

The double-angle formulas for cosine, written in the following forms, are used in calculus to transform power forms to nonpower forms:

$$\sin^2 x = \frac{1 - \cos 2x}{2} \qquad \cos^2 x = \frac{1 + \cos 2x}{2}$$

EXPLORE/DISCUSS 1

(A) Show that the following equations are *not* identities:

$$\sin 2x = 2 \sin x \qquad \cos 2x = 2 \cos x \qquad \tan 2x = 2 \tan x$$

(B) Graph $y_1 = \sin 2x$ and $y_2 = 2 \sin x$ in the same viewing window. What can you conclude? Repeat the process for the other two equations in part (A).

EXAMPLE 1

Verifying an Identity

Verify the identity: $\sin 2x = \dfrac{2 \tan x}{1 + \tan^2 x}$

Verification   We start with the right side:

$$\text{Right side} = \frac{2 \tan x}{1 + \tan^2 x} \qquad \text{Use a quotient identity.}$$

$$= \frac{2\left(\dfrac{\sin x}{\cos x}\right)}{1 + \dfrac{\sin^2 x}{\cos^2 x}} \qquad \text{Multiply numerator and denominator by } \cos^2 x.$$

$$= \frac{2 \sin x \cos x}{\cos^2 x + \sin^2 x} \qquad \text{Use double-angle and Pythagorean identities.}$$

$$= \frac{\sin 2x}{1}$$

$$= \sin 2x \qquad \text{Left side} \qquad \blacksquare$$

Matched Problem 1   Verify the identity: $\cos 2x = \dfrac{1 - \tan^2 x}{1 + \tan^2 x}$ ■

EXAMPLE 2

Using Double-Angle Identities

Find the exact value of $\cos 2x$ and $\tan 2x$ if $\sin x = \frac{4}{5}$, $\pi/2 < x < \pi$.

Solution   First draw a reference triangle in the second quadrant, and find $\tan x$ (see Fig. 1).

$$a = -\sqrt{5^2 - 4^2} = -3$$

$$\sin x = \tfrac{4}{5}$$

$$\tan x = -\tfrac{4}{3}$$

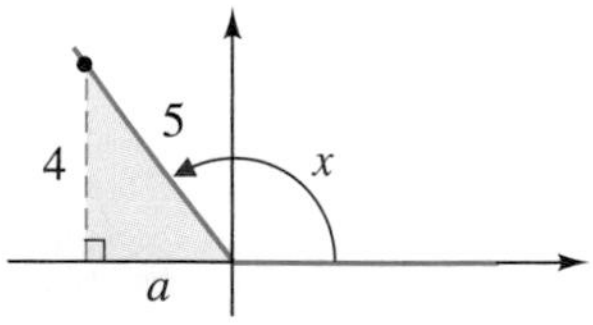

$$\cos 2x = 1 - 2 \sin^2 x \qquad \text{Use double-angle identity and the preceding results.}$$

$$= 1 - 2(\tfrac{4}{5})^2$$

$$= -\tfrac{7}{25}$$

$$\tan 2x = \frac{2 \tan x}{1 - \tan^2 x} \qquad \text{Use double-angle identity and the preceding results.}$$

$$= \frac{2(-\tfrac{4}{3})}{1 - (-\tfrac{4}{3})^2}$$

$$= \frac{24}{7} \qquad\qquad\qquad\qquad\qquad \blacksquare$$

FIGURE 1

(Figure 1: reference triangle in the second quadrant with hypotenuse 5, vertical side 4, horizontal side a, and angle x.)

Matched Problem 2   Find the exact value of $\sin 2x$ and $\cos 2x$ if $\tan x = -\frac{3}{4}$, $-\pi/2 < x < 0$.   ■

■ Half-Angle Identities

Half-angle identities are simply double-angle identities in an alternative form. We start with the double-angle identity for cosine in the form

$$\cos 2u = 1 - 2 \sin^2 u$$

and let $u = x/2$. Then

$$\cos x = 1 - 2 \sin^2 \frac{x}{2}$$

Now solve for $\sin(x/2)$ to obtain a half-angle formula for the sine function:

$$2 \sin^2 \frac{x}{2} = 1 - \cos x$$

$$\sin^2 \frac{x}{2} = \frac{1 - \cos x}{2}$$

$$\sin \frac{x}{2} = \pm \sqrt{\frac{1 - \cos x}{2}} \tag{7}$$

In identity (7), the choice of the sign is determined by the quadrant in which $x/2$ lies.

Now we start with the angle identity for cosine in the form

$$\cos 2u = 2 \cos^2 u - 1$$

and let $u = x/2$. We then obtain a half-angle formula for the cosine function:

$$\cos x = 2 \cos^2 \frac{x}{2} - 1$$

$$2 \cos^2 \frac{x}{2} = 1 + \cos x$$

$$\cos^2 \frac{x}{2} = \frac{1 + \cos x}{2}$$

$$\cos \frac{x}{2} = \pm \sqrt{\frac{1 + \cos x}{2}} \tag{8}$$

In identity (8), the choice of the sign is again determined by the quadrant in which $x/2$ lies.

To obtain a half-angle identity for the tangent function, we can use the quotient identity and the half-angle formulas for sine and cosine:

$$\tan \frac{x}{2} = \frac{\sin \dfrac{x}{2}}{\cos \dfrac{x}{2}} = \frac{\pm \sqrt{\dfrac{1 - \cos x}{2}}}{\pm \sqrt{\dfrac{1 + \cos x}{2}}} = \pm \sqrt{\frac{1 - \cos x}{1 + \cos x}} \tag{9}$$

where the sign is determined by the quadrant in which $x/2$ lies.

On the next page we list all the half-angle identities for convenient reference. Two additional half-angle identities for tangent, which can be obtained by rewriting the right-hand side of (9), are left as Problems 29 and 30 in Exercise 4.4.

HALF–ANGLE IDENTITIES

For x any real number or angle in degree or radian measure for which both sides are defined:

$$\sin \frac{x}{2} = \pm \sqrt{\frac{1 - \cos x}{2}} \qquad \cos \frac{x}{2} = \pm \sqrt{\frac{1 + \cos x}{2}}$$

$$\tan \frac{x}{2} = \pm \sqrt{\frac{1 - \cos x}{1 + \cos x}} = \frac{\sin x}{1 + \cos x} = \frac{1 - \cos x}{\sin x}$$

where the sign is determined by the quadrant in which $x/2$ lies.

EXPLORE/DISCUSS 2

(A) Show that the following equations are *not* identities:

$$\sin \frac{x}{2} = \frac{1}{2} \sin x \qquad \cos \frac{x}{2} = \frac{1}{2} \cos x \qquad \tan \frac{x}{2} = \frac{1}{2} \tan x$$

 (B) Graph $y_1 = \sin(x/2)$ and $y_2 = \frac{1}{2} \sin x$ in the same viewing window. What can you conclude? Repeat the process for the other two equations in part (A).

EXAMPLE 3

Using a Half-Angle Identity

Find $\cos 165°$ exactly by means of a half-angle identity.

Solution $\cos 165° = \cos \dfrac{330°}{2} = -\sqrt{\dfrac{1 + \cos 330°}{2}}$

The negative square root is used since $165°$ is in the second quadrant and cosine is negative there. We complete the evaluation by noting that the reference triangle for $330°$ is a $30°$-$60°$-$90°$ triangle in the fourth quadrant (see Fig. 2).

$$\cos 330° = \cos 30°$$

$$= \frac{\sqrt{3}}{2}$$

Therefore,

$$\cos 165° = -\sqrt{\frac{1 + \sqrt{3}/2}{2}} = -\frac{\sqrt{2 + \sqrt{3}}}{2}$$

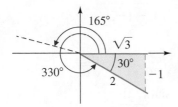

FIGURE 2

Matched Problem 3   Find the exact value of sin 165° using a half-angle identity.   ■

Using Half-Angle Identities

Find the exact value of $\sin(x/2)$, $\cos(x/2)$, and $\tan(x/2)$ if $\sin x = -\frac{3}{5}$, $\pi < x < 3\pi/2$.

Solution   Note that the half-angle identities for $\sin(x/2)$, $\cos(x/2)$, and $\tan(x/2)$ involve $\cos x$. We can find $\cos x$ using the Pythagorean identity $\sin^2 x + \cos^2 x = 1$ or, equivalently, by drawing a reference triangle (see Fig. 3). Because $\pi < x < 3\pi/2$, the reference triangle is in the third quadrant.

$$a = -\sqrt{5^2 - (-3)^2} = -4$$

$$\cos x = -\frac{4}{5}$$

If $\pi < x < 3\pi/2$, then

$$\pi/2 < x/2 < 3\pi/4 \qquad \text{Divide each member of } \pi < x < 3\pi/2 \text{ by 2.}$$

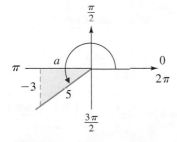

FIGURE 3

This tells us that $x/2$ is in the second quadrant, where sine is positive and cosine and tangent are negative. Using half-angle identities, we obtain

$$\sin \frac{x}{2} = \sqrt{\frac{1 - \cos x}{2}} \qquad\qquad \cos \frac{x}{2} = -\sqrt{\frac{1 + \cos x}{2}}$$

$$= \sqrt{\frac{1 - (-\frac{4}{5})}{2}} \qquad\qquad = -\sqrt{\frac{1 + (-\frac{4}{5})}{2}}$$

$$= \sqrt{\frac{9}{10}} \quad \text{or} \quad \frac{3}{\sqrt{10}} \qquad\qquad = -\sqrt{\frac{1}{10}} \quad \text{or} \quad -\frac{1}{\sqrt{10}}$$

$$\tan \frac{x}{2} = \frac{\sin(x/2)}{\cos(x/2)}$$

$$= \frac{3/\sqrt{10}}{-1/\sqrt{10}} = -3 \qquad\qquad\qquad ■$$

Matched Problem 4   Find the exact values for $\sin(x/2)$, $\cos(x/2)$, and $\tan(x/2)$ if $\cot x = -\frac{4}{3}$, $\pi/2 < x < \pi$. ■

EXAMPLE 5

Verifying an Identity

Verify the identity: $\cos^2 \dfrac{x}{2} = \dfrac{\tan x + \sin x}{2 \tan x}$

Verification   We start with the left side:

$$\text{Left side} = \cos^2 \frac{x}{2} \qquad\qquad \textit{Use the half-angle identity for cosine.}$$

$$= \left(\pm \sqrt{\frac{1 + \cos x}{2}} \right)^2 \qquad \textit{Use algebra.}$$

$$= \frac{1 + \cos x}{2} \qquad\qquad \textit{Multiply numerator and denominator by tan x.}$$

$$= \frac{\tan x}{\tan x} \cdot \frac{1 + \cos x}{2} \qquad\qquad \textit{Use algebra.}$$

$$= \frac{\tan x + \tan x \cos x}{2 \tan x} \qquad\qquad \textit{Use a quotient identity and algebra.}$$

$$= \frac{\tan x + \sin x}{2 \tan x} \qquad\qquad \textit{Right side} \qquad ■$$

Matched Problem 5   Verify the identity: $\sin^2 \dfrac{x}{2} = \dfrac{\tan x - \sin x}{2 \tan x}$ ■

Answers to Matched Problems

1. $\dfrac{1 - \tan^2 x}{1 + \tan^2 x} = \dfrac{1 - \dfrac{\sin^2 x}{\cos^2 x}}{1 + \dfrac{\sin^2 x}{\cos^2 x}} = \cos^2 x - \sin^2 x = \cos 2x$

2. $\sin 2x = -\frac{24}{25}$; $\cos 2x = \frac{7}{25}$

3. $\dfrac{\sqrt{2 - \sqrt{3}}}{2}$

4. $\sin(x/2) = 3/\sqrt{10},\ \cos(x/2) = 1/\sqrt{10},\ \tan(x/2) = 3$

5. $\sin^2 \dfrac{x}{2} = \dfrac{1 - \cos x}{2} = \dfrac{\tan x}{\tan x} \cdot \dfrac{1 - \cos x}{2} = \dfrac{\tan x - \sin x}{2 \tan x}$

EXERCISE 4.4

A *In Problems 1–6, replace each ? with one of the six trigonometric functions so that the resulting equation is an identity. Verify the identity.*

1. $\dfrac{\sin 2x}{2 \sin x} = ?$ **2.** $(1 + \cos x) \tan \dfrac{x}{2} = ?$

3. $1 - \sin x \tan \dfrac{x}{2} = ?$ **4.** $\dfrac{1 + \cos 2x}{2 \cos x} = ?$

5. $\dfrac{1 - \cos 2x}{2 \sin x} = ?$ **6.** $\dfrac{\sin 2x}{2 \cos x} = ?$

Use half-angle identities to find the exact values for Problems 7–10. Do not use a calculator.

7. $\sin 105°$ **8.** $\cos 105°$

9. $\tan 15°$ **10.** $\tan 75°$

In Problems 11–14, graph y_1 and y_2 in the same viewing window. Then use TRACE *to compare the two graphs.*

11. $y_1 = 2 \sin x \cos x,\quad y_2 = \sin 2x,\quad -2\pi < x < 2\pi$

12. $y_1 = \cos^2 x - \sin^2 x,\quad y_2 = \cos 2x,\quad -2\pi < x < 2\pi$

13. $y_1 = \dfrac{2 \tan x}{1 - \tan^2 x},\quad y_2 = \tan 2x,\quad -\pi < x < \pi$

14. $y_1 = \dfrac{\sin x}{1 + \cos x},\quad y_2 = \tan \dfrac{x}{2},\quad -2\pi < x < 2\pi$

B *In Problems 15–32, verify each identity.*

15. $\sin 2x = (\tan x)(1 + \cos 2x)$

16. $(\sin x + \cos x)^2 = 1 + \sin 2x$

17. $2 \sin^2 \dfrac{x}{2} = \dfrac{\sin^2 x}{1 + \cos x}$

18. $2 \cos^2 \dfrac{x}{2} = \dfrac{\sin^2 x}{1 - \cos x}$

19. $(\sin \theta - \cos \theta)^2 = 1 - \sin 2\theta$

20. $\sin 2\theta = (\sin \theta + \cos \theta)^2 - 1$

21. $\cos^2 \dfrac{w}{2} = \dfrac{1 + \cos w}{2}$ **22.** $\sin^2 \dfrac{w}{2} = \dfrac{1 - \cos w}{2}$

23. $\cot \dfrac{\alpha}{2} = \dfrac{1 + \cos \alpha}{\sin \alpha}$ **24.** $\cot \dfrac{\alpha}{2} = \dfrac{\sin \alpha}{1 - \cos \alpha}$

25. $\dfrac{\cos 2t}{1 - \sin 2t} = \dfrac{1 + \tan t}{1 - \tan t}$

26. $\cos 2t = \dfrac{1 - \tan^2 t}{1 + \tan^2 t}$

27. $\tan 2x = \dfrac{2 \tan x}{1 - \tan^2 x}$ **28.** $\sin 2x = \dfrac{2 \tan x}{1 + \tan^2 x}$

29. $\tan \dfrac{x}{2} = \dfrac{\sin x}{1 + \cos x}$ **30.** $\tan \dfrac{x}{2} = \dfrac{1 - \cos x}{\sin x}$

31. $\sec^2 x = (\sec 2x)(2 - \sec^2 x)$

32. $2 \csc 2x = \dfrac{1 + \tan^2 x}{\tan x}$

In Problems 33–44, is the equation an identity? Explain.

33. $\sqrt{x^2 + 6x + 9} = x + 3$

34. $\sqrt{x^3 + 3x^2 + 3x + 1} = x + 1$

35. $\sqrt{x - 1} = \dfrac{x - 1}{\sqrt{x - 1}}$

36. $\sqrt{x^2} = x$

37. $\sin 3x = 3 \sin x \cos x$

38. $\cos 4x = 2 \cos^2 2x - 1$

39. $\tan(-2x) = \dfrac{2}{\tan x - \cot x}$

40. $\sin(-2x) = 2 \sin x \cos x$

41. $\sin \dfrac{x}{2} = \sqrt{\dfrac{1 - \cos x}{2}}$

42. $\cos\left(-\dfrac{x}{2}\right) = -\sqrt{\dfrac{1 + \cos x}{2}}$

43. $\left|\cos\dfrac{x}{4}\right| = \sqrt{\dfrac{1 + \cos x}{4}}$

44. $\left|\tan\dfrac{x}{4}\right| = \dfrac{\sin x}{1 + \cos x}$

In Problems 45–48, use the given information to find the exact value of sin 2x, cos 2x, *and* tan 2x. *Check your answer with a calculator.*

45. $\sin x = \frac{7}{25}, \quad \pi/2 < x < \pi$

46. $\cos x = -\frac{8}{17}, \quad \pi/2 < x < \pi$

47. $\cot x = -\frac{12}{35}, \quad -\pi/2 < x < 0$

48. $\tan x = -\frac{20}{21}, \quad -\pi/2 < x < 0$

In Problems 49–54, use the given information to find the exact value of sin (x/2) *and* cos (x/2). *Check your answer with a calculator.*

49. $\cos x = \frac{1}{4}, \quad 0° < x < 90°$

50. $\sin x = \dfrac{\sqrt{21}}{5}, \quad 0° < x < 90°$

51. $\tan x = -\sqrt{8}, \quad 90° < x < 180°$

52. $\cot x = -\dfrac{3}{\sqrt{7}}, \quad 90° < x < 180°$

53. $\csc x = -\dfrac{5}{\sqrt{24}}, \quad -90° < x < 0°$

54. $\sec x = \frac{3}{2}, \quad -90° < x < 0°$

Your friend is having trouble finding exact values of sin θ *and* cos θ *from the information given in Problems 55 and 56, and comes to you for help. Instead of just working the problems, you guide your friend through the solution process using the following questions (A)–(E). What is the correct response to each question for each problem?*

(A) *The angle* 2θ *is in which quadrant? How do you know?*

(B) *How can you find* sin 2θ *and* cos 2θ? *Find each.*

(C) *Which identities relate* sin θ *and* cos θ *with either* sin 2θ *or* cos 2θ?

(D) *How would you use the identities in part (C) to find* sin θ *and* cos θ *exactly, including the correct sign?*

(E) *What are the exact values for* sin θ *and* cos θ?

55. Find the exact values of sin θ and cos θ, given $\sec 2\theta = -\frac{5}{4}, 0° < \theta < 90°$.

56. Find the exact values of sin θ and cos θ, given $\tan 2\theta = -\frac{4}{3}, 0° < \theta < 90°$.

57. In applied mathematics, approximate forms are often substituted for exact forms to simplify formulas or

computations. Graph each side of each statement below in the same viewing window, $-\pi/2 < x < \pi/2$, to show that the approximation is valid for x close to 0. Use TRACE and describe what happens to the approximation as x gets closer to 0.

(A) $\sin 2x \approx 2\sin x$ (B) $\sin\dfrac{x}{2} \approx \dfrac{1}{2}\sin x$

58. Repeat Problem 57 for:

(A) $\tan 2x \approx 2\tan x$ (B) $\tan\dfrac{x}{2} \approx \dfrac{1}{2}\tan x$

In Problems 59–62, graph y_1 *and* y_2 *in the same viewing window for* $-2\pi \le x \le 2\pi$, *and state the intervals for which* y_1 *and* y_2 *are identical.*

59. $y_1 = \sin\dfrac{x}{2}, \quad y_2 = \sqrt{\dfrac{1 - \cos x}{2}}$

60. $y_1 = \sin\dfrac{x}{2}, \quad y_2 = -\sqrt{\dfrac{1 - \cos x}{2}}$

61. $y_1 = \cos\dfrac{x}{2}, \quad y_2 = -\sqrt{\dfrac{1 + \cos x}{2}}$

62. $y_1 = \cos\dfrac{x}{2}, \quad y_2 = \sqrt{\dfrac{1 + \cos x}{2}}$

C *In Problems 63–68, use the given information to find the exact value of* sin x, cos x, *and* tan x. *Check your answer with a calculator.*

63. $\sin 2x = \frac{55}{73}, \quad 0 < x < \pi/4$

64. $\cos 2x = -\frac{28}{53}, \quad \pi/4 < x < \pi/2$

65. $\tan 2x = -\frac{28}{45}, \quad \pi/4 < x < \pi/2$

66. $\cot 2x = -\frac{55}{48}, \quad -\pi/4 < x < 0$

67. $\sec 2x = \frac{65}{33}, \quad -\pi/4 < x < 0$

68. $\csc 2x = \frac{65}{33}, \quad 0 < x < \pi/4$

In Problems 69–74, verify each identity.

69. $\sin 3x = 3\sin x - 4\sin^3 x$

70. $\cos 3x = 4\cos^3 x - 3\cos x$

71. $\sin 4x = (\cos x)(4\sin x - 8\sin^3 x)$

72. $\cos 4x = 8\cos^4 x - 8\cos^2 x + 1$

73. $\tan 3x = \dfrac{3\tan x - \tan^3 x}{1 - 3\tan^2 x}$

74. $4\sin^4 x = 1 - 2\cos 2x + \cos^2 2x$

In Problems 75–80, graph f(x), find a simpler function g(x) that has the same graph as f(x), and verify the identity

$f(x) = g(x)$. *[Assume $g(x) = k + A \cdot t(Bx)$, where $t(x)$ is one of the six basic trigonometric functions.]*

75. $f(x) = \csc x + \cot x$

76. $f(x) = \csc x - \cot x$

77. $f(x) = \dfrac{\cot x}{1 + \cos 2x}$

78. $f(x) = \dfrac{1}{\cot x \sin 2x - 1}$

79. $f(x) = \dfrac{1 + 2 \cos 2x}{1 + 2 \cos x}$

80. $f(x) = \dfrac{1 - 2 \cos 2x}{2 \sin x - 1}$

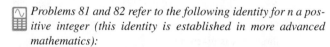 *Problems 81 and 82 refer to the following identity for n a positive integer (this identity is established in more advanced mathematics):*

$$\frac{1}{2} + \cos x + \cdots + \cos nx = \frac{\sin\left(\dfrac{2n+1}{2}x\right)}{2 \sin\left(\dfrac{1}{2}x\right)}$$

81. Graph each side of the equation for $n = 2$, $-2\pi \le x \le 2\pi$.

82. Graph each side of the equation for $n = 3$, $-2\pi \le x \le 2\pi$.

 Applications

83. **Sports: Javelin Throw**   In physics it can be shown that the theoretical horizontal distance d a javelin will travel (see the figure) is given approximately by

$$d = \frac{v_0^2 \sin \theta \cos \theta}{16}$$

where v_0 is the initial velocity of the javelin (in feet per second). (Air resistance and athlete height are ignored.)

(A) Write the formula in terms of the sine function only by using a suitable identity.

(B) Use the resulting equation from part (A) to determine the angle θ that will produce the maximum horizontal distance d for a given initial speed v_0. Explain your reasoning. This result is an important consideration for shot-putters, archers, and javelin and discus throwers.

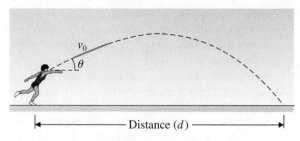

Figure for 83

(C) A world-class javelin thrower can throw the javelin with an initial velocity of 100 ft/sec. Graph the equation from part (A), and use the [MAXIMUM] command to find the maximum distance d and the angle θ (in degrees) that produces the maximum distance. Describe what happens to d as θ goes from 0° to 90°.

84. **Precalculus: Geometry**   An n-sided regular polygon is inscribed in a circle of radius r (see the figure).

Figure for 84

(A) Show that the area of the n-sided polygon is given by

$$A_n = \frac{1}{2} nr^2 \sin \frac{2\pi}{n}$$

[Hint: Area of triangle = (Base)(Height)/2. A double-angle identity is useful.]

(B) For a circle of radius 1, use the formula from part (A) to complete Table 1 (to six decimal places).

TABLE 1				
n	10	100	1,000	10,000
A_n				

(C) What does A_n seem to approach as n increases without bound?

(D) How close can A_n come to the actual area of the circle? Will A_n ever equal the exact area of the circle for any chosen n, however large? Explain. (In calculus, the area of the circumscribed circle, π, is called the *limit* of A_n as n increases without bound. Symbolically, we write $\lim_{n\to\infty} A_n = \pi$. The limit concept is fundamental to the development of calculus.)

85. Construction   An animal shelter is to be constructed using two 4 ft by 8 ft sheets of exterior plywood for the roof (see the figure). We are interested in approximating the value of θ that will give the maximum interior volume. [*Note:* The interior volume is the area of the triangular end, $bh/2$, times the length of the shelter, 8 ft.]

(A) Show that the interior volume is given by

$$V = 64 \sin 2\theta$$

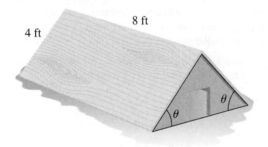

4 ft

8 ft

Figure for 85

(B) Explain how you can determine from the equation in part (A) the value of θ that produces the maximum volume, and find θ and the maximum volume.

(C) Complete Table 2 (to one decimal place) and find the maximum volume in the table and the value of θ that produces it. (Use the table feature in your calculator if it has one.)

TABLE 2

θ (deg)	30	35	40	45	50	55	60
V (ft³)	55.4						

(D) Graph the equation in part (A) on a graphing calculator, $0° \le \theta \le 90°$, and use the MAXIMUM command to find the maximum volume and the value of θ that produces it.

86. Construction   A new road is to be constructed from resort P to resort Q, turning at a point R on the horizontal line through resort P, as indicated in the figure.

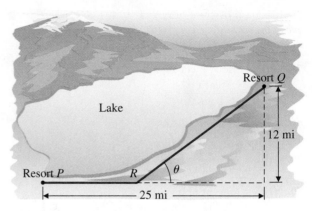

Figure for 86

(A) Show that the length of the road from P to Q through R is given by

$$d = 25 + 12 \tan \frac{\theta}{2}$$

(B) Because of the lake, θ is restricted to $40° \le \theta \le 90°$. What happens to the length of the road as θ varies between $40°$ and $90°$?

(C) Complete Table 3 (to one decimal place) and find the maximum and minimum length of the road. (Use a table-generating feature on your calculator if it has one.)

TABLE 3

θ (deg)	40	50	60	70	80	90
d (mi)	29.4					

(D) Graph the equation in part (A) for the restrictions in part (B); then use the MAXIMUM and MINIMUM commands to determine the maximum and minimum length of the road.

87. Engineering   Find the exact value of x in the figure; then find x and θ to three decimal places. [*Hint:* Use $\tan 2\theta = (2 \tan \theta)/(1 - \tan^2 \theta)$.]

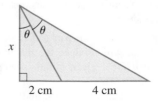

x

2 cm   4 cm

Figure for 87

88. Engineering Find the exact value of x in the figure; then find x and θ to three decimal places. [*Hint:* Use $\cos 2\theta = 2 \cos^2 \theta - 1$.]

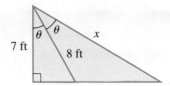

7 ft   8 ft x

Figure for 88

89. Geometry In part (a) of the figure, M and N are the midpoints of the sides of a square. Find the exact value of $\cos \theta$. [*Hint:* The solution uses the Pythagorean theorem, the definitions of sine and cosine, a half-angle identity, and some auxiliary lines as drawn in part (b) of the figure.]

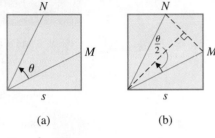

(a)   (b)

Figure for 89

☆4.5 Product–Sum and Sum–Product Identities

- Product–Sum Identities
- Sum–Product Identities
- Application: Music

In this section, we will develop identities for converting the product of two trigonometric functions into the sum of two trigonometric functions, and vice versa. These identities have many uses, both theoretical and practical. In calculus, knowing how to convert a product into a sum will allow easy solutions to some problems that would otherwise be difficult to solve. A discussion of beat frequencies in music, which appears at the end of this section, demonstrates how converting a sum into a product aids in the analysis of the beat phenomenon in sound. A similar type of analysis is used in modeling certain types of long-range underwater sound propagation for submarine detection.

■ Product–Sum Identities

The product–sum identities are easily derived from the sum and difference identities developed in Section 4.3. To obtain a product–sum identity, we add, left side to left side and right side to right side, the sum and difference identities for sine:

$$\sin(x + y) = \sin x \cos y + \cos x \sin y$$
$$\sin(x - y) = \sin x \cos y - \cos x \sin y$$
$$\sin(x + y) + \sin(x - y) = 2 \sin x \cos y$$

or

$$\sin x \cos y = \tfrac{1}{2}[\sin(x + y) + \sin(x - y)]$$

☆ Sections marked with a star may be omitted without loss of continuity.

Similarly, by adding or subtracting appropriate sum and difference identities, we can obtain three other product identities for sines and cosines. These identities are listed in the following box for convenient reference.

PRODUCT–SUM IDENTITIES

For x and y any real numbers or angles in degree or radian measure:

$$\sin x \cos y = \tfrac{1}{2}[\sin(x + y) + \sin(x - y)]$$

$$\cos x \sin y = \tfrac{1}{2}[\sin(x + y) - \sin(x - y)]$$

$$\sin x \sin y = \tfrac{1}{2}[\cos(x - y) - \cos(x + y)]$$

$$\cos x \cos y = \tfrac{1}{2}[\cos(x + y) + \cos(x - y)]$$

EXAMPLE 1

Using a Product–Sum Identity

Write the product $\cos 3t \sin t$ as a sum or difference.

Solution

$$\cos x \sin y = \tfrac{1}{2}[\sin(x + y) - \sin(x - y)] \qquad \text{Let } x = 3t \text{ and } y = t.$$

$$\cos 3t \sin t = \tfrac{1}{2}[\sin(3t + t) - \sin(3t - t)] \qquad \text{Simplify.}$$

$$= \tfrac{1}{2}\sin 4t - \tfrac{1}{2}\sin 2t$$

Matched Problem 1   Write the product $\cos 5\theta \cos 2\theta$ as a sum or difference.

EXAMPLE 2

Using a Product–Sum Identity

Evaluate $\sin 105° \sin 15°$ exactly using a product–sum identity.

Solution

$$\sin x \sin y = \tfrac{1}{2}[\cos(x - y) - \cos(x + y)] \qquad \text{Let } x = 105° \text{ and } y = 15°.$$

$$\sin 105° \sin 15° = \tfrac{1}{2}[\cos(105° - 15°) - \cos(105° + 15°)] \qquad \text{Simplify.}$$

$$= \tfrac{1}{2}[\cos 90° - \cos 120°] \qquad \text{Evaluate.}$$

$$= \tfrac{1}{2}[0 - (-\tfrac{1}{2})] = \tfrac{1}{4}$$

Matched Problem 2   Evaluate $\cos 165° \sin 75°$ exactly using a product–sum identity.

 **Sum–Product Identities**

The product–sum identities can be transformed into equivalent forms called sum–product identities. These identities are used to express sums and differences involving sines and cosines as products involving sines and cosines. We illustrate the transformation for one identity. The other three identities can be obtained by following the same procedure.

Let us start with the product–sum identity

$$\sin \alpha \cos \beta = \tfrac{1}{2}[\sin(\alpha + \beta) + \sin(\alpha - \beta)] \tag{1}$$

We would like

$$\alpha + \beta = x \qquad \alpha - \beta = y$$

Solving this system, we have

$$\alpha = \frac{x + y}{2} \qquad \beta = \frac{x - y}{2} \tag{2}$$

By substituting (2) into identity (1) and simplifying, we obtain

$$\sin x + \sin y = 2 \sin \frac{x + y}{2} \cos \frac{x - y}{2}$$

All four sum–product identities are listed in the following box for convenient reference.

SUM–PRODUCT IDENTITIES

For x and y any real numbers or angles in degree or radian measure:

$$\sin x + \sin y = 2 \sin \frac{x + y}{2} \cos \frac{x - y}{2}$$

$$\sin x - \sin y = 2 \cos \frac{x + y}{2} \sin \frac{x - y}{2}$$

$$\cos x + \cos y = 2 \cos \frac{x + y}{2} \cos \frac{x - y}{2}$$

$$\cos x - \cos y = -2 \sin \frac{x + y}{2} \sin \frac{x - y}{2}$$

EXAMPLE 3

Using a Sum–Product Identity

Write the difference $\sin 7\theta - \sin 3\theta$ as a product.

Solution

$$\sin x - \sin y = 2 \cos \frac{x + y}{2} \sin \frac{x - y}{2} \qquad \text{Let } x = 7\theta \text{ and } y = 3\theta.$$

$$\sin 7\theta - \sin 3\theta = 2 \cos \frac{7\theta + 3\theta}{2} \sin \frac{7\theta - 3\theta}{2} \qquad \text{Simplify.}$$

$$= 2 \cos 5\theta \sin 2\theta \qquad \blacksquare$$

Matched Problem 3   Write the sum $\cos 3t + \cos t$ as a product. $\blacksquare$

EXPLORE/DISCUSS 1

Proof without Words: Sum–Product Identities

Discuss how the relationships following Figure 1 can be verified from the figure.

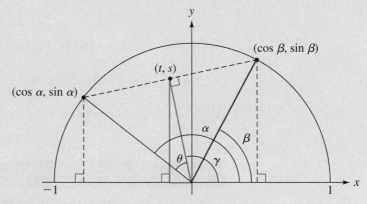

FIGURE 1

$$\theta = \frac{\alpha - \beta}{2}, \qquad \gamma = \frac{\alpha + \beta}{2}$$

$$\frac{\sin \alpha}{2} + \frac{\sin \beta}{2} = s = \cos \frac{\alpha - \beta}{2} \sin \frac{\alpha + \beta}{2}$$

$$\frac{\cos \alpha}{2} + \frac{\cos \beta}{2} = t = \cos \frac{\alpha - \beta}{2} \cos \frac{\alpha + \beta}{2}$$

(The above is based on a similar "proof" by Sidney H. Kung, Jacksonville University, printed in the October 1996 issue of *Mathematics Magazine*.)

 **EXAMPLE 4**

Using a Sum–Product Identity

Find the exact value of $\sin 105° - \sin 15°$ using an appropriate sum–product identity.

Solution

$$\sin x - \sin y = 2 \cos \frac{x + y}{2} \sin \frac{x - y}{2} \qquad \textit{Let x = 105° and y = 15°.}$$

$$\sin 105° - \sin 15° = 2 \cos \frac{105° + 15°}{2} \sin \frac{105° - 15°}{2} \qquad \textit{Simplify.}$$

$$= 2 \cos 60° \sin 45° \qquad \textit{Evaluate.}$$

$$= 2 \left(\frac{1}{2} \right) \left(\frac{\sqrt{2}}{2} \right) \qquad \text{Simplify.}$$

$$= \frac{\sqrt{2}}{2} \qquad\qquad\qquad \blacksquare$$

Matched Problem 4   Find the exact value of cos 165° − cos 75° by using an appropriate sum–product identity.   ■

■ **Application: Music**

If two tones that have the same loudness and that are close in pitch (frequency) are sounded, one following the other, most people have difficulty recognizing that the tones are different. However, if the tones are sounded simultaneously, they will react with each other, producing a low warbling sound called a **beat.** The beat or warble will be slow or rapid, depending on how far apart the initial frequencies are. Musicians, when tuning an instrument with other instruments or a tuning fork, listen for these lower beat frequencies and try to eliminate them by adjusting their instruments. The more rapid the beat frequency (warbling) when two instruments play together, the greater the difference in their frequencies and the more out of tune they are. If, after adjustments, no beats are heard, the two instruments are in tune.

What is behind this beat phenomenon? Figure 2a shows a tone of 64 Hz (cycles per second), and Figure 2b shows a tone of the same loudness but with a frequency of 72 Hz. If both tones are sounded simultaneously and time is started

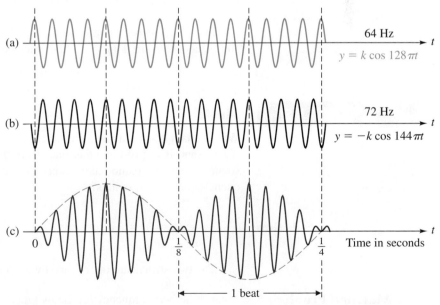

FIGURE 2
Beats

when the two tones are completely out of phase, waves (a) and (b) will interact to form wave (c).

Sum–product identities are useful in the mathematical analysis of the beat phenomenon. In particular, we will use the sum–product identity

$$\cos x - \cos y = -2 \sin \frac{x + y}{2} \sin \frac{x - y}{2}$$

in a brief mathematical discussion of the three waves in Figure 2.

We start with the fact that wave (c) is the sum of waves (a) and (b):

$$
\begin{aligned}
y &= k \cos 128\pi t - k \cos 144\pi t \\
&= k(\cos 128\pi t - \cos 144\pi t) \\
&= k \left(-2 \sin \frac{128\pi t + 144\pi t}{2} \sin \frac{128\pi t - 144\pi t}{2} \right) \\
&= -2k \sin 136\pi t \sin(-8\pi t) \\
&= 2k \sin 136\pi t \sin 8\pi t \\
&= 2k \sin 8\pi t \sin 136\pi t
\end{aligned}
$$

The factor $\sin 136\pi t$ represents a sound of frequency 68 Hz, the average of the frequencies of the individual tones. The first factor, $2k \sin 8\pi t$, can be thought of as a time-varying amplitude of the second factor, $\sin 136\pi t$. It is the first factor that describes the slow warble of volume, or the beat, of the combined sounds. This pulsation of sound is described in terms of frequency—that is, it tells how many times per second the maximum loudness or number of beats occurs. Referring to Figure 2c, we see that there are 2 beats in $\frac{1}{4}$ sec, or 8 beats/sec. So, the beat frequency is 8 beats/sec (which is the difference in frequencies of the two original waves).

In general, if two equally loud tones of frequencies f_1 and f_2 are produced simultaneously, and if $f_1 > f_2$, then the beat frequency, f_b, is given by

$$f_b = f_1 - f_2 \qquad \textit{Beat frequency}$$

EXAMPLE 5

Music

When certain keys on a piano are struck, a felt-covered hammer strikes two strings. If the piano is out of tune, the tones from the two strings create a beat, and the sound is sour. If a piano tuner counts 15 beats in 5 sec, how far apart are the frequencies of the two strings?

Solution

$$f_b = \frac{15}{5} = 3 \text{ beats/sec}$$
$$f_1 - f_2 = f_b = 3 \text{ Hz}$$

The two strings are out of tune by 3 cycles/sec.   ■

Matched Problem 5   What is the beat frequency for the two tones in Figure 2 on page 273?   ■

Answers to Matched Problems

1. $\cos 5\theta \cos 2\theta = \frac{1}{2}\cos 7\theta + \frac{1}{2}\cos 3\theta$

2. $(-\sqrt{3} - 2)/4$

3. $\cos 3t + \cos t = 2\cos 2t \cos t$

4. $-\sqrt{6}/2$

5. $f_b = 8$ Hz

EXERCISE 4.5

A *In Problems 1–8, write each product as a sum or difference involving sines and cosines.*

1. $\cos 4w \cos w$
2. $\sin 2t \sin t$
3. $\cos 2u \sin u$
4. $\sin 4B \cos B$
5. $\sin 2B \cos 5B$
6. $\cos 3\theta \cos 5\theta$
7. $\sin 3m \sin 4m$
8. $\cos 2A \sin 3A$

In Problems 9–16, write each sum or difference as a product involving sines and cosines.

9. $\cos 5\theta + \cos 3\theta$
10. $\sin 6A + \sin 4A$
11. $\sin 6u - \sin 2u$
12. $\cos 3t - \cos t$
13. $\sin 3B + \sin 5B$
14. $\cos 2m + \cos 4m$
15. $\cos w - \cos 5w$
16. $\sin 3C - \sin 7C$

In Problems 17–24, evaluate exactly using an appropriate identity.

17. $\cos 75° \sin 15°$
18. $\sin 195° \cos 75°$
19. $\sin 105° \sin 165°$
20. $\cos 15° \cos 75°$
21. $\sin 195° + \sin 105°$
22. $\cos 285° + \cos 195°$
23. $\sin 75° - \sin 165°$
24. $\cos 15° - \cos 105°$

B *In Problems 25 and 26, use sum and difference identities from Section 4.3 to establish each of the following:*

25. $\sin x \sin y = \frac{1}{2}[\cos(x - y) - \cos(x + y)]$

26. $\cos x \cos y = \frac{1}{2}[\cos(x + y) + \cos(x - y)]$

27. Explain how you can transform the product–sum identity

$$\cos u \cos v = \frac{1}{2}[\cos(u + v) + \cos(u - v)]$$

into the sum–product identity

$$\cos x + \cos y = 2\cos\frac{x + y}{2}\cos\frac{x - y}{2}$$

by a suitable substitution.

28. Explain how you can transform the product–sum identity

$$\sin u \sin v = \frac{1}{2}[\cos(u - v) - \cos(u + v)]$$

into the sum–product identity

$$\cos x - \cos y = -2\sin\frac{x + y}{2}\sin\frac{x - y}{2}$$

by a suitable substitution.

In Problems 29–36, verify each identity.

29. $\dfrac{\cos t - \cos 3t}{\sin t + \sin 3t} = \tan t$

30. $\dfrac{\sin 2t + \sin 4t}{\cos 2t - \cos 4t} = \cot t$

31. $\dfrac{\sin x + \sin y}{\cos x + \cos y} = \tan\dfrac{x + y}{2}$

32. $\dfrac{\sin x - \sin y}{\cos x - \cos y} = -\cot\dfrac{x + y}{2}$

33. $\dfrac{\cos x - \cos y}{\sin x + \sin y} = -\tan\dfrac{x - y}{2}$

34. $\dfrac{\cos x + \cos y}{\sin x - \sin y} = \cot\dfrac{x - y}{2}$

35. $\dfrac{\sin x + \sin y}{\sin x - \sin y} = \dfrac{\tan\frac{1}{2}(x + y)}{\tan\frac{1}{2}(x - y)}$

36. $\dfrac{\cos x + \cos y}{\cos x - \cos y} = -\cot\dfrac{x + y}{2}\cot\dfrac{x - y}{2}$

In Problems 37–44, is the equation an identity? Explain.

37. $\cos 2x \sin x = \frac{1}{2}(\sin 3x + \sin x)$
38. $\sin x \sin 3x = \frac{1}{2}(\cos 2x - \cos 4x)$
39. $\sin 2x \cos 2x = \frac{1}{2}\sin 4x$
40. $\cos 4x \cos 2x = \frac{1}{2}(\cos 8x + \cos 2x)$

41. $\sin 5x + \sin x = 2 \sin 3x \cos 2x$

42. $\cos 4x + \cos 2x = 2 \cos 6x \cos 2x$

43. $\cos x - \cos 3x = -2 \sin 2x \sin x$

44. $\sin 3x - \sin x = 2 \cos 2x \sin x$

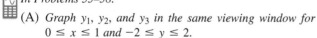

In Problems 45–52, write each as a sum or difference if y is a product, or as a product if y is a sum or difference. Enter the original equation in a graphing calculator as y_1, the converted form as y_2, and graph y_1 and y_2 in the same viewing window. Use $\boxed{\text{TRACE}}$ *to compare the two graphs.*

45. $y = \cos 5x \cos 3x$

46. $y = \sin 3x \cos x$

47. $y = \cos 1.9x \sin 0.5x$

48. $y = \sin 2.3x \sin 0.7x$

49. $y = \cos 3x + \cos x$

50. $y = \sin 2x + \sin x$

51. $y = \sin 2.1x - \sin 0.5x$

52. $y = \cos 1.7x - \cos 0.3x$

C *In Problems 53 and 54, verify each identity.*

53. $\sin x \sin y \sin z = \frac{1}{4}[\sin(x + y - z)$
$+ \sin(y + z - x)$
$+ \sin(z + x - y)$
$- \sin(x + y + z)]$

54. $\cos x \cos y \cos z = \frac{1}{4}[\cos(x + y - z)$
$+\cos(y + z - x)$
$+\cos(z + x - y)$
$+\cos(x + y + z)]$

In Problems 55–58:

(A) *Graph y_1, y_2, and y_3 in the same viewing window for $0 \le x \le 1$ and $-2 \le y \le 2$.*

(B) *Convert y_1 to a sum or difference and repeat part (A).*

55. $y_1 = 2 \cos(16\pi x) \sin(2\pi x)$
$y_2 = 2 \sin(2\pi x)$
$y_3 = -2 \sin(2\pi x)$

56. $y_1 = 2 \sin(20\pi x) \cos(2\pi x)$
$y_2 = 2 \cos(2\pi x)$
$y_3 = -2 \cos(2\pi x)$

57. $y_1 = 2 \sin(24\pi x) \sin(2\pi x)$
$y_2 = 2 \sin(2\pi x)$
$y_3 = -2 \sin(2\pi x)$

58. $y_1 = 2 \cos(28\pi x) \cos(2\pi x)$
$y_2 = 2 \cos(2\pi x)$
$y_3 = -2 \cos(2\pi x)$

Applications

59. Music   If one tone is described by $y = k \sin 522\pi t$ and another by $y = k \sin 512\pi t$, write their sum as a product. What is the beat frequency if both notes are sounded together?

60. Music   If one tone is described by $y = k \cos 524\pi t$ and another by $y = k \cos 508\pi t$, write their sum as a product. What is the beat frequency if both notes are sounded together?

61. Music

$$y = 0.3 \cos 72\pi t \quad \text{and} \quad y = -0.3 \cos 88\pi t$$

are equations of sound waves with frequencies 36 and 44 Hz, respectively. If both sounds are emitted simultaneously, a beat frequency results. Use the viewing window $0 \le t \le 0.25$, $-0.8 \le y \le 0.8$ for parts (A)–(D).

(A) Graph $y = 0.3 \cos 72 \pi t$.

(B) Graph $y = -0.3 \cos 88\pi t$.

(C) Graph $y_1 = 0.3 \cos 72\pi t - 0.3 \cos 88\pi t$ and $y_2 = 0.6 \sin 8\pi t$ in the same viewing window.

(D) Convert y_1 in part (C) to a product, and graph the new y_1 along with y_2 from part (C) in the same viewing window.

62. Music

$$y = 0.4 \cos 132\pi t \quad \text{and} \quad y = -0.4 \cos 152\pi t$$

are equations of sound waves with frequencies 66 and 76 Hz, respectively. If both sounds are emitted simultaneously, a beat frequency results. Use the viewing window $0 \le t \le 0.2$, $-0.8 \le y \le 0.8$ for parts (A)–(D).

(A) Graph $y = 0.4 \cos 132\pi t$.

(B) Graph $y = -0.4 \cos 152\pi t$.

(C) Graph $y_1 = 0.4 \cos 132\pi t - 0.4 \cos 152\pi t$ and $y_2 = 0.8 \sin 10\pi t$ in the same viewing window.

(D) Convert y_1 in part (C) to a product, and graph the new y_1 along with y_2 from part (C) in the same viewing window.

CHAPTER 4 GROUP ACTIVITY

From $M \sin Bt + N \cos Bt$ to $A \sin(Bt + C)$

Solving certain kinds of problems dealing with electrical circuits, spring–mass systems, heat flow, fluid flow, and so on, requires more advanced mathematics (differential equations), but often the solution process leads naturally to functions of the form

$$y = M \sin Bt + N \cos Bt \tag{1}$$

In the following investigation, you will show that function (1) is a simple harmonic; that is, it can be represented by an equation of the form

$$y = A \sin(Bt + C) \tag{2}$$

One of the objectives of this project is to transform equation (1) into the form of equation (2). Then it will be easy to determine the amplitude, period, frequency, and phase shift of the phenomenon that produced equation (1).

I. EXPLORATION

Use a graphing calculator to explore the nature of the graph of equation (1) for various values of M, N, and B. Do the graphs appear to be graphs of simple harmonics?

II. GEOMETRIC SOLUTION

The graph of $y = \sin(\pi t) - \sqrt{3} \cos(\pi t)$ is shown in Figure 1. The graph looks like the graph of a simple harmonic.

(A) From the graph, find A, B, and C so that $y = A \sin(Bt + C)$ produces the same graph. To find C, estimate the phase shift by finding the first zero (x intercept) to the right of the origin using the $\boxed{\text{ZERO}}$ command on a graphing calculator.

(B) Graph the equation obtained in part (A) and the original equation in the same viewing window and compare y values for various values of x using $\boxed{\text{TRACE}}$. Conclusions? Write the results as an identity.

A geometric solution to the problem is straightforward as long as M, N, and B are simple enough; often they are not. We now turn to an analytical solution that can always be used no matter how complicated M, N, and B are.

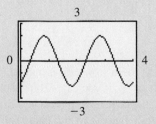

FIGURE 1

$y = \sin(\pi t) - \sqrt{3} \cos(\pi t)$

continued

III. GENERAL ANALYTICAL SOLUTION

Equations (1) and (2) are stated again for convenient reference:

$$y = M \sin Bt + N \cos Bt \tag{1}$$

$$y = A \sin(Bt + C) \tag{2}$$

The problem is: Given M, N, and B in equation (1), find A, B, and C in equation (2) so that equation (2) produces the same graph as equation (1), that is, so that $M \sin Bt + N \cos Bt = A \sin(Bt + C)$ is an identity. The transformation identity, which you are to establish, is stated in the following box:

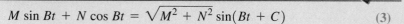

TRANSFORMATION IDENTITY

$$M \sin Bt + N \cos Bt = \sqrt{M^2 + N^2} \sin(Bt + C) \tag{3}$$

where C is any angle (in radians if t is real) having $P = (M, N)$ on its terminal side.

(A) *Establish the transformation identity.*   The process of finding A, B, and C, given M, N, and B, requires a little ingenuity and the use of the sum identity

$$\sin(x + y) = \sin x \cos y + \cos x \sin y \tag{4}$$

Start by trying to get $M \sin Bt + N \cos Bt$ to look like the right side of the sum identity (4). Then we use (4), from right to left, to obtain (2).

Hint: A first step is the following:

$$M \sin Bt + N \cos Bt = \frac{\sqrt{M^2 + N^2}}{\sqrt{M^2 + N^2}} (M \sin Bt + N \cos Bt)$$

(B) *Example 1.* Use equation (3) to transform

$$y_1 = \sin(\pi t) - \sqrt{3} \cos(\pi t)$$

into the form $y_2 = A \sin(Bt + C)$, where C is chosen so that $|C|$ is minimum. Compute C to three decimal places, or find C exactly, if you can. (You should get the same results that you got in part II using the geometric approach.) From the new equation determine the amplitude, period, frequency, and phase shift.

Check by graphing y_1 and y_2 in the same viewing window and comparing the graphs using $\boxed{\text{TRACE}}$.

(C) *Example 2.*   Use equation (3) to transform

$$y_1 = -3 \sin(2\pi t) - 4 \cos(2\pi t)$$

into the form $y_2 = A \sin(Bt + C)$, where C is chosen so that $-\pi < C \leq \pi$. Compute C to three decimal places. From the new equation determine the amplitude, period, frequency, and phase shift.

Check by graphing y_1 and y_2 in the same viewing window and comparing the graphs using TRACE.

(D) *Application 1: Spring–mass system.* A weight suspended from a spring (spring constant 64) is pulled 4 cm below its equilibrium position and is then given a downward thrust to produce an initial downward velocity of 24 cm/sec. In more advanced mathematics (differential equations) the equation of motion (neglecting air resistance and friction) is found to be given approximately by

$$y_1 = -3 \sin 8t - 4 \cos 8t$$

where y_1 is the position of the bottom of the weight on the scale in Figure 2 at time t (y is in centimeters and t is in seconds). Transform the equation into the form

$$y_2 = A \sin(Bt + C)$$

and indicate the amplitude, period, frequency, and phase shift of the simple harmonic motion. Choose C so that $-\pi < C \leq \pi$.

To check, graph y_1 and y_2 in the same viewing window, $0 \leq t \leq 6$, and compare the graphs using TRACE.

(E) *Application 2: Music.* A musical tone is described by

$$y_1 = 0.04 \sin 200\pi t - 0.03 \cos 200\pi t$$

where t is time in seconds. Write the equation in the form $y_2 = A \sin(Bt + C)$. Compute C (to three decimal places) so that $-\pi < C \leq \pi$. Indicate the amplitude, period, frequency, and phase shift.

To check, graph y_1 and y_2 in the same viewing window, $0 \leq t \leq 0.06$, and compare the graphs using TRACE.

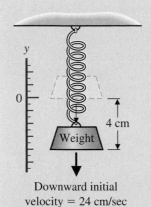

y

0

4 cm

Weight

Downward initial velocity = 24 cm/sec

FIGURE 2
Spring–mass system

CHAPTER 4 REVIEW

An equation in one or more variables is said to be an **identity** if the left side is equal to the right side for all replacements of the variables for which both sides are defined. If the left side is equal to the right side only for certain values of the variables and not for all values for which both sides are defined, then the equation is called a **conditional equation**.

4.1
FUNDAMENTAL
IDENTITIES AND
THEIR USE

Fundamental Trigonometric Identities

For x any real number or angle in degree or radian measure for which both sides are defined:

Reciprocal identities

$$\csc x = \frac{1}{\sin x} \qquad \sec x = \frac{1}{\cos x} \qquad \cot x = \frac{1}{\tan x}$$

Quotient identities

$$\tan x = \frac{\sin x}{\cos x} \qquad \cot x = \frac{\cos x}{\sin x}$$

Identities for negatives

$$\sin(-x) = -\sin x \qquad \cos(-x) = \cos x \qquad \tan(-x) = -\tan x$$

Pythagorean identities

$$\sin^2 x + \cos^2 x = 1 \qquad \tan^2 x + 1 = \sec^2 x \qquad 1 + \cot^2 x = \csc^2 x$$

4.2
VERIFYING
TRIGONOMETRIC
IDENTITIES

When **verifying an identity,** start with the expression on one side and, through a sequence of valid steps involving the use of known identities or algebraic manipulation, convert that expression into the expression on the other side. *Do not* add the same quantity to each side, multiply each side by the same nonzero quantity, or square or take the square root of both sides.

Some suggestions for verifying identities

Step 1   Start with the more complicated side of the identity and transform it into the simpler side.

Step 2   Try using basic or other known identities.

Step 3   Try algebraic operations such as multiplying, factoring, combining fractions, or splitting fractions.

Step 4   If other steps fail, try expressing each function in terms of sine and cosine functions; then perform appropriate algebraic operations.

Step 5   At each step, keep the other side of the identity in mind. This often reveals what you should do in order to get there.

4.3
SUM, DIFFERENCE,
AND COFUNCTION
IDENTITIES

For x and y any real numbers or angles in degree or radian measure for which both sides are defined:

Sum identities

$$\sin(x + y) = \sin x \cos y + \cos x \sin y$$

$$\cos(x + y) = \cos x \cos y - \sin x \sin y$$

$$\tan(x + y) = \frac{\tan x + \tan y}{1 - \tan x \tan y}$$

Difference identities

$$\sin(x - y) = \sin x \cos y - \cos x \sin y$$

$$\cos(x - y) = \cos x \cos y + \sin x \sin y$$

$$\tan(x - y) = \frac{\tan x - \tan y}{1 + \tan x \tan y}$$

Cofunction identities
(Replace $\pi/2$ with $90°$ if x is in degree measure.)

$$\sin\left(\frac{\pi}{2} - x\right) = \cos x \qquad \cos\left(\frac{\pi}{2} - x\right) = \sin x$$

$$\tan\left(\frac{\pi}{2} - x\right) = \cot x \qquad \cot\left(\frac{\pi}{2} - x\right) = \tan x$$

$$\sec\left(\frac{\pi}{2} - x\right) = \csc x \qquad \csc\left(\frac{\pi}{2} - x\right) = \sec x$$

**4.4
DOUBLE-ANGLE AND
HALF-ANGLE
IDENTITIES**

Double-angle identities
For x any real number or angle in degree or radian measure for which both sides are defined:

$$\sin 2x = 2 \sin x \cos x \qquad \cos 2x = \cos^2 x - \sin^2 x$$
$$\tan 2x = \frac{2 \tan x}{1 - \tan^2 x} \qquad \qquad = 1 - 2 \sin^2 x$$
$$= 2 \cos^2 x - 1$$

Half-angle identities
For x any real number or angle in degree or radian measure for which both sides are defined:

$$\sin \frac{x}{2} = \pm \sqrt{\frac{1 - \cos x}{2}} \qquad \cos \frac{x}{2} = \pm \sqrt{\frac{1 + \cos x}{2}}$$

$$\tan \frac{x}{2} = \pm \sqrt{\frac{1 - \cos x}{1 + \cos x}} = \frac{\sin x}{1 + \cos x} = \frac{1 - \cos x}{\sin x}$$

where the sign is determined by the quadrant in which $x/2$ lies.

**☆4.5
PRODUCT-SUM AND
SUM-PRODUCT
IDENTITIES**

Product–sum identities
For x and y any real numbers or angles in degree or radian measure:

$$\sin x \cos y = \tfrac{1}{2}[\sin(x + y) + \sin(x - y)]$$

$$\cos x \sin y = \tfrac{1}{2}[\sin(x + y) - \sin(x - y)]$$

$$\sin x \sin y = \tfrac{1}{2}[\cos(x - y) - \cos(x + y)]$$

$$\cos x \cos y = \tfrac{1}{2}[\cos(x + y) + \cos(x - y)]$$

Sum–product identities

For x and y any real numbers or angles in degree or radian measure:

$$\sin x + \sin y = 2 \sin \frac{x+y}{2} \cos \frac{x-y}{2}$$

$$\sin x - \sin y = 2 \cos \frac{x+y}{2} \sin \frac{x-y}{2}$$

$$\cos x + \cos y = 2 \cos \frac{x+y}{2} \cos \frac{x-y}{2}$$

$$\cos x - \cos y = -2 \sin \frac{x+y}{2} \sin \frac{x-y}{2}$$

CHAPTER 4 REVIEW EXERCISE

Work through all the problems in this chapter review and check the answers. Answers to all review problems appear in the back of the book; following each answer is an italic number that indicates the section in which that type of problem is discussed. Where weaknesses show up, review the appropriate sections in the text. Review problems flagged with a star ($\star$) are from optional sections.

A 1. One of the following equations is an identity and the other is a conditional equation. Identify which, and explain the difference between the two.

$$(1)\ (x-3)(x+2) = x^2 - x - 6$$

$$(2)\ (x-3)(x+2) = 0$$

Verify each identity in Problems 2–10. Try not to look at a table of identities.

2. $\csc x \sin x = \sec x \cos x$

3. $\cot x \sin x = \cos x$

4. $\tan x = -\tan(-x)$

5. $\dfrac{\sin^2 x}{\cos x} = \sec x - \cos x$

6. $\dfrac{\csc x}{\cos x} = \tan x + \cot x$

7. $(\cos^2 x)(\cot^2 x + 1) = \cot^2 x$

8. $\dfrac{\sin \alpha \csc \alpha}{\cot \alpha} = \tan \alpha$

9. $\dfrac{\sin^2 u - \cos^2 u}{\sin u \cos u} = \tan u - \cot u$

10. $\dfrac{\sec \theta - \csc \theta}{\sec \theta \csc \theta} = \sin \theta - \cos \theta$

11. Using $\cos(x+y) = \cos x \cos y - \sin x \sin y$, show that $\cos(x + 2\pi) = \cos x$.

12. Using $\sin(x+y) = \sin x \cos y + \cos x \sin y$, show that $\sin(x + \pi) = -\sin x$.

In Problems 13 and 14, verify each identity for the indicated value.

13. $\cos 2x = 1 - 2 \sin^2 x,\ x = 30°$

14. $\sin \dfrac{x}{2} = \pm \sqrt{\dfrac{1 - \cos x}{2}},\ x = \dfrac{\pi}{2}$

$\star$ 15. Write $\sin 8t \sin 5t$ as a sum or difference.

$\star$ 16. Write $\sin w + \sin 5w$ as a product.

Verify each identity in Problems 17–20.

17. $\dfrac{1 - \cos^2 t}{\sin^3 t} = \csc t$

18. $\dfrac{(\cos \alpha - 1)^2}{\sin^2 \alpha} = \dfrac{1 - \cos \alpha}{1 + \cos \alpha}$

19. $\dfrac{1 - \tan^2 x}{1 - \tan^4 x} = \cos^2 x$

20. $\cot^2 x \cos^2 x = \cot^2 x - \cos^2 x$

21. The equation $\sin x = 0$ is true for an infinite number of values ($x = k\pi$, k any integer). Is this equation an identity? Explain.

22. Explain how you would use a graphing calculator to show that $\sin x = 0$ is not an identity; then do it.

B *In Problems 23–30, is the equation an identity? Explain.*

23. $\sin^2 2x + \cos^2 2x = 2$

24. $\dfrac{\tan x}{\cot x} = 1$

25. $\tan(x - \pi/2) = -\cot x$

26. $\sec(\pi/2 - x) = \cos x$

27. $\cos 2x = 2\sin^2 x - 1$

28. $\dfrac{\tan 2x}{\tan x} = \dfrac{2}{(1 - \tan x)(1 + \tan x)}$

29. $\cos^2 x = \frac{1}{2}\cos 2x$

30. $2\sin x \cos 5x = \sin 6x - \sin 4x$

Verify the identities in Problems 31–45. Use the list of identities inside the front cover if necessary.

31. $\dfrac{\sin x}{1 - \cos x} = (\csc x)(1 + \cos x)$

32. $\dfrac{1 - \tan^2 x}{1 - \cot^2 x} = 1 - \sec^2 x$

33. $\tan(x + \pi) = \tan x$

34. $1 - (\cos \beta - \sin \beta)^2 = \sin 2\beta$

35. $\dfrac{\sin 2x}{\cot x} = 1 - \cos 2x$

36. $\dfrac{2\tan x}{1 + \tan^2 x} = \sin 2x$

37. $2\csc 2x = \tan x + \cot x$

38. $\csc x = \dfrac{\cot(x/2)}{1 + \cos x}$

39. $\dfrac{\sin(x - y)}{\sin(x + y)} = \dfrac{\tan x - \tan y}{\tan x + \tan y}$

40. $\csc 2x = \dfrac{\tan x + \cot x}{2}$

41. $\dfrac{2 - \sec^2 x}{\sec^2 x} = \cos 2x$

42. $\tan \dfrac{x}{2} = \dfrac{\sec x - 1}{\tan x}$

☆**43.** $\dfrac{\sin t + \sin 5t}{\cos t + \cos 5t} = \tan 3t$

☆**44.** $\dfrac{\sin x + \sin y}{\cos x - \cos y} = -\cot \dfrac{x - y}{2}$

☆**45.** $\dfrac{\cos x - \cos y}{\cos x + \cos y} = -\tan \dfrac{x + y}{2} \tan \dfrac{x - y}{2}$

Evaluate Problems 46 and 47 exactly using an appropriate identity.

☆**46.** $\sin 165° \sin 15°$

☆**47.** $\cos 165° - \cos 75°$

48. Use fundamental identities to find the exact values of the remaining trigonometric functions of x, given

$$\cos x = -\tfrac{2}{3} \quad \text{and} \quad \tan x < 0$$

49. Find the exact values of $\sin 2x$, $\cos 2x$, and $\tan 2x$, given $\tan x = \tfrac{4}{3}$ and $0 < x < \pi/2$. Do not use a calculator.

50. Find the exact values of $\sin(x/2)$, $\cos(x/2)$, and $\tan(x/2)$, given $\cos x = -\tfrac{5}{13}$ and $-\pi < x < -\pi/2$. Do not use a calculator.

51. Use a sum or difference identity to convert $y = \tan(x + \pi/4)$ into a form involving $\sin x$, $\cos x$, and/or $\tan x$. Check the results using a graphing calculator and $\boxed{\text{TRACE}}$.

52. Write $y = \cos 1.5x \cos 0.3x - \sin 1.5x \sin 0.3x$ in terms of a single trigonometric function. Check the result by entering the original equation in a graphing calculator as y_1 and the converted form as y_2. Then graph y_1 and y_2 in the same viewing window. Use $\boxed{\text{TRACE}}$ to compare the two graphs.

53. Graph $y_1 = \sin(x/2)$ and $y_2 = -\sqrt{(1 - \cos x)/2}$ in the same viewing window for $-2\pi \le x \le 2\pi$, and indicate the subinterval(s) for which y_1 and y_2 are identical.

54. Use a graphing calculator to test whether each equation that follows is an identity. If the equation appears to be an identity, verify it. If the equation does not appear to be an identity, find a value of x for which both sides are defined but are not equal.

(A) $\dfrac{\sin^2 x}{1 + \sin x} = 1 - \sin x$

(B) $\dfrac{\cos^2 x}{1 + \sin x} = 1 - \sin x$

C **55.** Find the exact values of $\sin x$, $\cos x$, and $\tan x$, given $\sec 2x = -\tfrac{13}{12}$ and $-\pi/2 < x < 0$. Do not use a calculator.

Verify the identities in Problems 56 and 57.

56. $\dfrac{\cot x}{\csc x + 1} = \dfrac{\csc x - 1}{\cot x}$

57. $\cot 3x = \dfrac{3\tan^2 x - 1}{\tan^3 x - 3\tan x}$

58. Use the definition of sine, cosine, and tangent on a unit circle to prove that

$$\tan x = \frac{\sin x}{\cos x}$$

59. Prove that the cosine function has a period of 2π.

60. Prove that the cotangent function has a period of π.

61. By letting

$$x + y = u \quad \text{and} \quad x - y = v$$

in $\sin x \sin y = \frac{1}{2}[\cos(x - y) - \cos(x + y)]$, show that

$$\cos v - \cos u = 2 \sin \frac{u + v}{2} \sin \frac{u - v}{2}$$

In Problems 62–66, graph f(x), find a simpler function g(x) that has the same graph as f(x), and verify the identity $f(x) = g(x)$. [Assume $g(x) = k + A \cdot t(Bx)$, where t(x) is one of the six trigonometric functions.]

62. $f(x) = \dfrac{3 \sin^2 x}{1 - \cos x} + \dfrac{\tan^2 x \cos^2 x}{1 + \cos x}$

63. $f(x) = \dfrac{\sin x}{\cos x - \sin x} + \dfrac{\sin x}{\cos x + \sin x}$

64. $f(x) = 3 \sin^2 x + \cos^2 x$

65. $f(x) = \dfrac{3 - 4 \cos^2 x}{1 - 2 \sin^2 x}$

66. $f(x) = \dfrac{2 + \sin x - 2 \cos x}{1 - \cos x}$

Problems 67–70 require the use of a graphing calculator. In Problems 67 and 68, graph y_1 and y_2 in the same viewing window for $-2\pi \le x \le 2\pi$, and state the interval(s) where the graphs of y_1 and y_2 coincide. Use TRACE.

67. $y_1 = \tan \dfrac{x}{2}, \quad y_2 = \sqrt{\dfrac{1 - \cos x}{1 + \cos x}}$

68. $y_1 = \tan \dfrac{x}{2}, \quad y_2 = -\sqrt{\dfrac{1 - \cos x}{1 + \cos x}}$

☆ **69.** Graph $y_1 = 2 \cos 30\pi x \sin 2\pi x$ and $y_2 = 2 \sin 2\pi x$ for $0 \le x \le 1$ and $-2 \le y \le 2$.

☆ **70.** Repeat Problem 69 after converting y_1 to a sum or difference.

Applications

71. **Precalculus: Trigonometric Substitution**   In the expression $\sqrt{u^2 - a^2}, a > 0$, let $u = a \sec x, 0 < x < \pi/2$, simplify, and write in a form that is free of radicals.

72. **Precalculus: Angle of Intersection of Two Lines**   Use the results of Problem 71 in Exercise 4.3 to find the acute angle of intersection (to the nearest $0.1°$) between the two lines $y = 4x + 5$ and $y = \frac{1}{3}x - 2$.

73. **Engineering**   Find the exact value of x in the figure; then find x and θ to three decimal places.

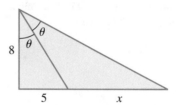

Figure for 73

74. **Analytic Geometry**   Find the radian measure of the angle θ in the figure (to three decimal places) if A has coordinates $(2, 6)$ and B has coordinates $(4, 4)$. [*Hint*: Label the angle between OB and the x axis as α; then use an appropriate sum identity.]

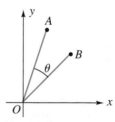

Figure for 74

75. **Architecture**   An art museum is being designed with a triangular skylight on the roof, as indicated in the figure. The facing of the museum is light granite and the top edge, excluding the skylight ridge, is to receive a black granite trim as shown. The total length of the trim depends on the choice of the angle θ for the skylight. All other dimensions are fixed.

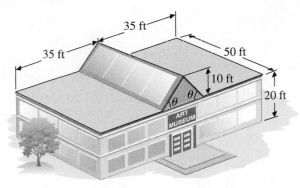

Figure for 75

(A) Show that the total length L of the black granite trim is given by

$$L = 240 + 40 \tan \frac{\theta}{2}$$

(B) Because of lighting considerations, θ is restricted to $30° \leq \theta \leq 60°$. Describe what you think happens to L as θ varies from $30°$ to $60°$.

(C) Complete Table 1 (to one decimal place) and select the maximum and minimum length of the trim. (Use the table-generating feature on your calculator if it has one.)

TABLE 1

θ	30	35	40	45	50	55	60
L (ft)	250.7						

 (D) Graph the equation in part (A) for the restrictions in part (B). Then determine the maximum and minimum lengths of the trim.

☆ **76. Music**   One tone is given by $y = 0.3 \cos 120\pi t$ and another by $y = -0.3 \cos 140\pi t$. Write their sum as a product. What is the beat frequency if both notes are sounded together?

☆ **77. Music**   Use a graphing calculator with the viewing window set to $0 \leq t \leq 0.2$, $-0.8 \leq y \leq 0.8$, to graph the indicated equations.

(A) $y_1 = 0.3 \cos 120\pi t$

(B) $y_2 = -0.3 \cos 140 \pi t$

(C) $y_3 = y_1 + y_2$ and $y_4 = 0.6 \sin 10\pi t$

(D) Repeat part (C) using the product form of y_3 from Problem 76.

☆ **78. Physics**   The equation of motion for a weight suspended from a spring is given by

$$y = -8 \sin 3t - 6 \cos 3t$$

where y is displacement of the weight from its equilibrium position in centimeters and t is time in seconds. Write this equation in the form $y = A \sin(Bt + C)$; keep A positive, choose C positive and as small as possible, and compute C to two decimal places. Indicate the amplitude, period, frequency, and phase shift.

☆ **79. Physics**   Use a graphing calculator to graph

$$y = -8 \sin 3t - 6 \cos 3t$$

for $-2\pi/3 \leq t \leq 2\pi/3$, approximate the t intercepts in this interval to two decimal places, and identify the intercept that corresponds to the phase shift determined in Problem 78.

Inverse Trigonometric Functions; Trigonometric Equations and Inequalities

5

5.1 Inverse Sine, Cosine, and Tangent Functions

☆**5.2** Inverse Cotangent, Secant, and Cosecant Functions

5.3 Trigonometric Equations: An Algebraic Approach

5.4 Trigonometric Equations and Inequalities: A Graphing Calculator Approach

Chapter 5 Group Activity:
$$\sin \frac{1}{x} = 0 \text{ and } \sin^{-1} \frac{1}{x} = 0$$

Chapter 5 Review

Cumulative Review Exercise, Chapters 1–5

☆ Sections marked with a star may be omitted without loss of continuity.

Before you begin this chapter, it would be very helpful to briefly review Appendix B.3, on the general concept of the inverse of a function. We will make use of much of the material discussed there in our investigation of inverse trigonometric functions.

I n Chapter 1, recall that we learned how to solve a right triangle (see Fig. 1) for an angle α as follows:

$$\sin \alpha = \tfrac{3}{6} = \tfrac{1}{2}$$
$$\alpha = \sin^{-1} \tfrac{1}{2} \quad \text{or} \quad \arcsin \tfrac{1}{2}$$
$$\alpha = \pi/6 \text{ rad} \quad \text{or} \quad 30°$$

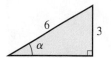

FIGURE 1

In this context, both $\sin^{-1}(1/2)$ and $\arcsin(1/2)$ represent the acute angle (in either radian or degree measure) whose sine is $1/2$. In Chapter 1, we said that the concepts behind the inverse function symbols $\sin^{-1}$ (or arcsin), $\cos^{-1}$ (or arccos), and $\tan^{-1}$ (or arctan) would be discussed in greater detail in this chapter. Now the time has come, and we will extend the meaning of these symbols so that they apply not only to triangle problems but also to a wide variety of problems that have nothing to do with triangles or angles. That is, we will create another set of tools—a new set of functions—that you can put in your mathematical toolbox for general use on a wide variety of new problems.

After we complete our discussion of inverse trigonometric functions, we will be in a position to solve many types of equations involving trigonometric functions. Trigonometric equations form the subject matter of the last two sections of this chapter.

5.1   Inverse Sine, Cosine, and Tangent Functions

- Inverse Sine Function
- Inverse Cosine Function
- Inverse Tangent Function
- Summary

In this section we will define the inverse sine, cosine, and tangent functions; look at their graphs; present some basic and useful identities; and consider some applications in Exercise 5.1.

◼ Inverse Sine Function

In solving for angle α in Figure 1 above, what we are really looking for is an angle whose sine is $1/2$. The key question to consider in this section is: How many such angles are there? Figure 2 shows a graph of $y = \sin x$ with height $1/2$ indicated by the dashed line. From this graph, we can see that there are infinitely many such angles!

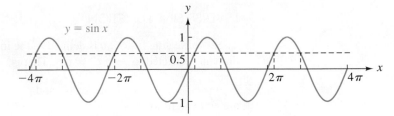

FIGURE 2
$y = \sin x$ and $y = 0.5$

This presents a major difficulty in defining the inverse of sine as a function. For any potential input, such as $1/2$, there is more than one potential output (infinitely many, in fact). This violates the definition of a function. When we were solving triangles this was not a problem because only one angle α with $\sin \alpha = 1/2$ is acute. This provides a good basis for how to define the inverse sine as a function: Of all possible outputs, we will choose exactly one for each input from -1 to 1.

In Appendix B.3 we note that a function that has exactly one domain value (input) for each range value (output) is called **one-to-one**. We also note that only one-to-one functions have inverses. This is exactly why $y = \sin x$ does not: There are many x values that correspond to a given y value.

But we can overcome this problem by restricting the domain of the sine function as shown in Figure 3.

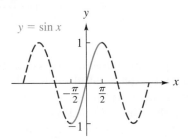

FIGURE 3
$y = \sin x$ is one-to-one for
$-\pi/2 \leq x \leq \pi/2$

On the domain $[-\pi/2, \pi/2]$, $y = \sin x$ *is* one-to-one, so we can define its inverse.* In essence, we are saying that there are many x values for which $\sin x = 1/2$, for example, but to define **the inverse sine function** at $1/2$ [denoted $\sin^{-1}(1/2)$ or arcsin $(1/2)$], we will choose the one that is between $-\pi/2$ and $\pi/2$—in this case, $\pi/6$.

---

* There are many different ways to restrict the domain of $y = \sin x$ to make it one-to-one. We have chosen the way that is generally accepted as standard.

INVERSE SINE FUNCTION

The **inverse sine function** is defined as the inverse of the sine function with domain restricted to $[-\pi/2, \pi/2]$. That is,

$$y = \sin^{-1}x \quad \text{and} \quad y = \arcsin x$$

are equivalent to

$$\sin y = x \quad \text{for } y \text{ in } [-\pi/2, \pi/2]$$

The inverse sine of x is the number or angle y, $-\pi/2 \le y \le \pi/2$, whose sine is x. The y values in this interval are called the **principal values** of the inverse sine function.

EXPLORE/DISCUSS 1

Suppose that $y = \sin x$ is restricted to each of the domains below. What would be the potential values for $\sin^{-1}(1/2)$ in each case? Which would lead to a legitimate definition of inverse sine as a function?

(A) $[0, \pi]$ (B) $[\pi, 2\pi]$ (C) $[\pi/2, 3\pi/2]$ (D) $[\pi, 3\pi/2]$

The symbol arcsin is used in some contexts, $\sin^{-1}$ in others. In Chapter 1, we used the $\sin^{-1}x$ key on a calculator to solve for an angle x, given $\sin x = y$. Although this equation has infinitely many solutions, the calculator displayed a single number, the principal value $\sin^{-1} y$.

How do we sketch the graph of $y = \sin^{-1} x$? We begin by constructing a table of ordered pairs for $y = \sin x$ (Table 1). Then we reverse the coordinates of each ordered pair to obtain a table for $y = \sin^{-1} x$ (Table 2).

TABLE 1
$y = \sin x$
$(-\pi/2, -1)$
$(-\pi/4, -\sqrt{2}/2)$
$(0, 0)$
$(\pi/4, \sqrt{2}/2)$
$(\pi/2, 1)$

TABLE 2
$y = \sin^{-1} x$
$(-1, -\pi/2)$
$(-\sqrt{2}/2, -\pi/4)$
$(0, 0)$
$(\sqrt{2}/2, \pi/4)$
$(1, \pi/2)$

Using the points in these tables, we quickly sketch graphs of $y = \sin x$ and $y = \sin^{-1} x$ (Fig. 4). A more accurate graph can be obtained using a calculator in radian mode and a set of domain values from -1 to 1 (see Problem 43 in Exercise 5.1). An instant graph can be obtained using a graphing calculator.

FIGURE 4

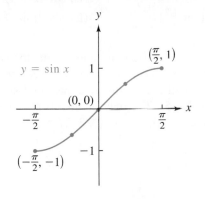

Domain: $-\dfrac{\pi}{2} \le x \le \dfrac{\pi}{2}$
Range: $-1 \le y \le 1$

(a) Sine function restricted to domain $\left[-\dfrac{\pi}{2}, \dfrac{\pi}{2}\right]$

Domain: $-1 \le x \le 1$
Range: $-\dfrac{\pi}{2} \le y \le \dfrac{\pi}{2}$

(b) Inverse sine function

Note that the graph of $y = \sin^{-1} x$ is the reflection of the graph of $y = \sin x$ through the line $y = x$.

We will now state the important sine–inverse sine identities, which follow from the general properties of inverse functions (see Appendix B.3).

SINE–INVERSE SINE IDENTITIES

$$\sin(\sin^{-1} x) = x \qquad -1 \le x \le 1$$
$$\sin^{-1}(\sin x) = x \qquad -\pi/2 \le x \le \pi/2$$

Each identity reinforces the basic idea that $y = \sin x$ and $y = \sin^{-1} x$ are inverses.

EXPLORE/DISCUSS 2

Use a calculator to evaluate each of the following. Which illustrate a sine–inverse sine identity and which do not? Explain.

(A) $\sin(\sin^{-1} 0.5)$

(B) $\sin(\sin^{-1} 1.5)$

(C) $\sin^{-1}[\sin(-1.3)]$

(D) $\sin^{-1}[\sin(-3)]$

EXAMPLE 1

Exact Values

Find exact values if possible without using a calculator:

(A) $\sin^{-1}(\sqrt{3}/2)$ (B) $\arcsin(-\frac{1}{2})$ (C) $\sin^{-1}(\sin 1.2)$

(D) $\cos(\sin^{-1}\frac{2}{3})$ (E) $\sin^{-1}(\sin \frac{3\pi}{4})$

Solution

(A) $y = \sin^{-1}(\sqrt{3}/2)$ is equivalent to $\sin y = \sqrt{3}/2$, $-\pi/2 \le y \le \pi/2$. What y between $-\pi/2$ and $\pi/2$ has sine $\sqrt{3}/2$?

We will use a unit circle diagram like the ones we studied in Section 2.5 (see Fig. 5). The only number between $-\pi/2$ and $\pi/2$ with sine equal to $\sqrt{3}/2$ is $\pi/3$, so $\sin^{-1}(\sqrt{3}/2) = \pi/3$.

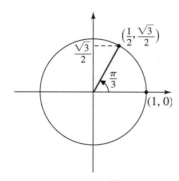

FIGURE 5

(B) $y = \arcsin(-\frac{1}{2})$ is equivalent to $\sin y = -\frac{1}{2}$, $-\pi/2 \le y \le \pi/2$. What y between $-\pi/2$ and $\pi/2$ has sine $-\frac{1}{2}$?

Again, we consult the unit circle (Fig. 6). The only number between $-\pi/2$ and $\pi/2$ with sine equal to $-1/2$ is $-\pi/6$. [*Note*: Even though $\sin 11\pi/6$ is $-1/2$ as well, $\sin^{-1}(-1/2)$ can't be $11\pi/6$. Why?]

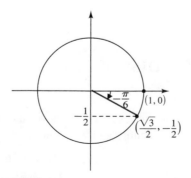

FIGURE 6

(C) $\sin^{-1}(\sin 1.2) = 1.2$ *Sine–inverse sine identity (1.2 is in $\left[-\frac{\pi}{2}, \frac{\pi}{2}\right]$)*

(D) Let $y = \sin^{-1}\frac{2}{3}$; then $\sin y = \frac{2}{3}$, $-\pi/2 \le y \le \pi/2$. Draw the reference triangle associated with y; then $\cos y = \cos(\sin^{-1}\frac{2}{3})$ can be determined directly from the triangle (after finding the third side) without the need to actually find y (see Fig. 7).

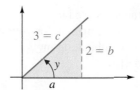

$$a^2 + b^2 = c^2$$
$$a = \sqrt{3^2 - 2^2}$$
$$= \sqrt{5}$$

FIGURE 7

So,

$$\cos\left(\sin^{-1}\frac{2}{3}\right) = \cos y = \frac{\sqrt{5}}{3}$$

(E) It is tempting to use the second sine–inverse sine identity, but it doesn't apply because $3\pi/4$ is not in $[-\pi/2, \pi/2]$. Instead, we note that $\sin(3\pi/4) = 1/\sqrt{2}$ (Fig. 8), so $\sin^{-1}(\sin(3\pi/4))$ is the number in $[-\pi/2, \pi/2]$ with sine equal to $1/\sqrt{2}$. That number is $\pi/4$, so $\sin^{-1}(\sin(3\pi/4)) = \pi/4$.

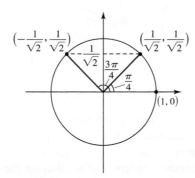

FIGURE 8

Matched Problem 1   Find exact values if possible without using a calculator:

(A) $\arcsin(\sqrt{2}/2)$ (B) $\sin^{-1}(-1)$

(C) $\sin[\sin^{-1}(-0.4)]$ (D) $\tan[\sin^{-1}(-1/\sqrt{5})]$

(E) $\sin^{-1}[\sin(-3\pi/4)]$

EXAMPLE 2

Calculator Values

Find to four significant digits if possible using a calculator:

(A) $\sin^{-1}(0.8432)$ (B) $\arcsin(-0.3042)$ (C) $\sin^{-1} 1.357$

(D) $\cot[\sin^{-1}(-0.1087)]$ (E) $\sin^{-1}[\sin(2)]$

Solution   [*Note:*   Recall that the keys used to obtain $\sin^{-1}$ vary among different brands of calculators. (Read the user's manual for your calculator.) Two common designations are $\boxed{\sin^{-1}}$ and the combination $\boxed{\text{inv}}\,\boxed{\sin}$. For all these problems set your calculator in radian mode.]

(A) $\sin^{-1}(0.8432) = 1.003$

(B) $\arcsin(-0.3042) = -0.3091$

(C) $\sin^{-1} 1.357 = $ Error*   *1.357 is not in the domain of $\sin^{-1}$, so $\sin^{-1}$ 1.357 is undefined.*

(D) $\cot[\sin^{-1}(-0.1087)] = -9.145$

(E) $\sin^{-1}[\sin(2)] = 1.142$ *Sine–inverse sine identity does not apply since $2 > \pi/2$.*   ■

Matched Problem 2   Find to four significant digits if possible using a calculator:

(A) $\arcsin 0.2903$ (B) $\sin^{-1}(-0.7633)$

(C) $\arcsin(-2.305)$ (D) $\sec[\sin^{-1}(-0.3446)]$

(E) $\sin^{-1}[\sin(-1.7)]$ ■

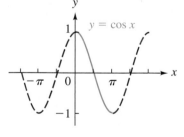

FIGURE 9
$y = \cos x$ is one-to-one for
$0 \le x \le \pi$

■ Inverse Cosine Function

Like the sine function, cosine must be restricted for us to be able to define inverse cosine as a function. But restricting it to the domain $[-\pi/2, \pi/2]$ won't work since cosine is not one-to-one on that interval.

The generally accepted restriction on the domain of the cosine function is $0 \le x \le \pi$. This choice ensures that the inverse of the restricted cosine function will exist (Fig. 9).

---

* Some calculators use a more advanced definition of the inverse sine function involving complex numbers and will display an ordered pair of real numbers as the value of $\sin^{-1}$ 1.357. You should interpret such a result as an indication that the number entered is not in the domain of the inverse sine function as we have defined it.

INVERSE COSINE FUNCTION

The **inverse cosine function** is defined as the inverse of the cosine function with domain restricted to $[0, \pi]$. That is,

$$y = \cos^{-1} x \quad \text{and} \quad y = \arccos x$$

are equivalent to

$$\cos y = x \quad \text{for } y \text{ in } [0, \pi].$$

The inverse cosine of x is the number or angle y, $0 \leq y \leq \pi$, whose cosine is x. The y values in this interval are called the **principal values** of the inverse cosine function.

Figure 10 compares the graphs of the restricted cosine function and its inverse. Notice that $(0, 1)$, $(\pi/4, \sqrt{2}/2)$, $(\pi/2, 0)$, $(3\pi/4, -\sqrt{2}/2)$, and $(\pi, -1)$ are on the restricted cosine graph. Reversing the coordinates gives us five points on the graph of the inverse cosine function.

FIGURE 10

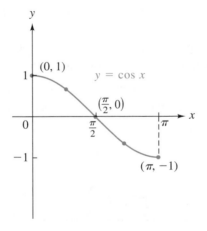

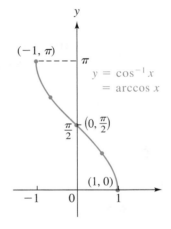

Domain: $0 \leq x \leq \pi$
Range: $-1 \leq y \leq 1$

Domain: $-1 \leq x \leq 1$
Range: $0 \leq y \leq \pi$

(a) Cosine function restricted to domain
    $[0, \pi]$

(b) Inverse cosine function

Note that the graph of $y = \cos^{-1} x$ is the reflection of the graph of $y = \cos x$ through the line $y = x$.

We will complete our discussion of inverse cosine by giving the cosine–inverse cosine identities.

COSINE–INVERSE COSINE IDENTITIES

$$\cos(\cos^{-1} x) = x \qquad -1 \leq x \leq 1$$
$$\cos^{-1}(\cos x) = x \qquad 0 \leq x \leq \pi$$

EXPLORE/DISCUSS 3

Use a calculator to evaluate each of the following. Which illustrate a cosine–inverse cosine identity and which do not? Explain.

(A) $\cos(\cos^{-1} 0.5)$ (B) $\cos(\cos^{-1} 1.1)$

(C) $\cos^{-1}(\cos 1.1)$ (D) $\cos^{-1}[\cos(-1)]$

 EXAMPLE 3

Exact Values

Find exact values if possible without using a calculator:

(A) $\cos^{-1} \frac{1}{2}$ (B) $\arccos(-\sqrt{3}/2)$ (C) $\cos(\cos^{-1} 0.7)$

(D) $\sin[\cos^{-1}(-\frac{1}{3})]$ (E) $\cos^{-1}[\cos(5\pi/4)]$

Solution   In Example 1, we used a unit circle approach to finding exact values. In this example, we will construct reference triangles. You can use either method.

(A) $y = \cos^{-1} \frac{1}{2}$ is equivalent to $\cos y = \frac{1}{2}$, $0 \le y \le \pi$. What y between 0 and π has cosine $\frac{1}{2}$? This y must be associated with a first-quadrant reference triangle (see Fig. 11):

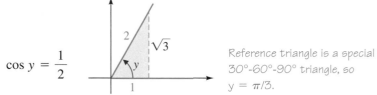

$$\cos y = \frac{1}{2}$$

Reference triangle is a special 30°-60°-90° triangle, so $y = \pi/3$.

FIGURE 11

$$\cos^{-1} \frac{1}{2} = \frac{\pi}{3}$$

(B) $y = \arccos(-\sqrt{3}/2)$ is equivalent to $\cos y = -\sqrt{3}/2$, $0 \le y \le \pi$. What y between 0 and π has cosine $-\sqrt{3}/2$? This y must be associated with a second quadrant reference triangle (see Fig. 12):

$$\cos y = -\frac{\sqrt{3}}{2}$$

Reference triangle is a special 30°-60°-90° triangle, so $y = 5\pi/6$.

FIGURE 12

$$\arccos\left(-\frac{\sqrt{3}}{2}\right) = \frac{5\pi}{6}$$

[*Note:* y cannot be $-5\pi/6$, even though $\cos(-5\pi/6) = -\sqrt{3}/2$. Why?]

(C) $\cos(\cos^{-1} 0.7) = 0.7$ *Cosine–inverse cosine identity (0.7 is in $[-1, 1]$)*

(D) Let $y = \cos^{-1}(-\frac{1}{3})$; then $\cos y = -\frac{1}{3}$, $0 \le y \le \pi$. Draw a reference triangle associated with y; then we can determine $\sin y = \sin[\cos^{-1}(-\frac{1}{3})]$ directly from the triangle (after finding the third side) without actually finding y (see Fig. 13):

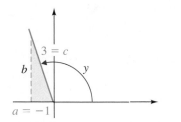

FIGURE 13

$$\sin\left[\cos^{-1}\left(-\frac{1}{3}\right)\right] = \sin y = \frac{2\sqrt{2}}{3}$$

(E) The second cosine–inverse cosine identity doesn't apply since $5\pi/4$ is not between 0 and π. Instead, we first note that $\cos(5\pi/4) = -1/\sqrt{2}$ (see the lower reference triangle in Fig. 14); then we use the upper reference triangle to see that the angle between 0 and π whose cosine is $-1/\sqrt{2}$ is $3\pi/4$. So $\cos^{-1}(\cos(5\pi/4)) = 3\pi/4$.

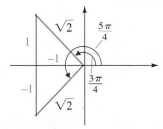

FIGURE 14   ■

Matched Problem 3   Find exact values if possible without using a calculator:

(A) $\arccos(\sqrt{2}/2)$ (B) $\cos^{-1}(-1)$ (C) $\cos^{-1}(\cos 3.05)$

(D) $\cot[\cos^{-1}(-1/\sqrt{5})]$ (E) $\cos^{-1}[\cos(-\pi/4)]$ ■

EXAMPLE 4

Calculator Values

Find to four significant digits if possible using a calculator:

(A) $\cos^{-1} 0.4325$ (B) $\arccos(-0.8976)$

(C) $\cos^{-1} 2.137$ (D) $\csc[\cos^{-1}(-0.0349)]$

(E) $\cos^{-1}[\cos(3.5)]$

Solution   Set your calculator in radian mode.

(A) $\cos^{-1} 0.4325 = 1.124$

(B) $\arccos(-0.8976) = 2.685$

(C) $\cos^{-1} 2.137 = $ Error   *2.137 is not in the domain of $\cos^{-1}$,*

(D) $\csc[\cos^{-1}(-0.0349)] = 1.001$ *so $\cos^{-1} 2.137$ is undefined.*

(E) $\cos^{-1}[\cos(3.5)] = 2.783$ *Cosine–inverse cosine identity does not apply since $3.5 > \pi$.* ■

Matched Problem 4   Find to four significant digits if possible using a calculator:

(A) $\arccos 0.6773$ (B) $\cos^{-1}(-0.8114)$ (C) $\arccos(-1.003)$

(D) $\cot[\cos^{-1}(-0.5036)]$ (E) $\cos^{-1}[\cos(-1)]$ ■

EXAMPLE 5

Exact Values

Find the exact value of $\cos(\sin^{-1}\frac{3}{5} - \cos^{-1}\frac{4}{5})$ without using a calculator.

Solution   We can use the difference identity for cosine and the procedure outlined in Examples 1D and 3D to obtain

$$\cos(x - y) = \cos x \cos y + \sin x \sin y$$

$$\cos(\sin^{-1}\tfrac{3}{5} - \cos^{-1}\tfrac{4}{5}) = \cos(\sin^{-1}\tfrac{3}{5})\cos(\cos^{-1}\tfrac{4}{5}) + \sin(\sin^{-1}\tfrac{3}{5})\sin(\cos^{-1}\tfrac{4}{5})$$

$$= \left(\tfrac{4}{5}\right) \cdot \left(\tfrac{4}{5}\right) + \left(\tfrac{3}{5}\right) \cdot \left(\tfrac{3}{5}\right)$$

$$= 1 \qquad\qquad ■$$

Matched Problem 5   Find the exact value of $\sin(2\cos^{-1}\frac{3}{5})$ without using a calculator. ■

■ Inverse Tangent Function

The customary choice for restricting the domain of the tangent function is $-\pi/2 < x < \pi/2$ (Fig. 15). Since the tangent function is one-to-one for this restriction, we can define inverse tangent as the inverse of this restricted function.

FIGURE 15
$y = \tan x$ is one-to-one for
$-\pi/2 < x < \pi/2$

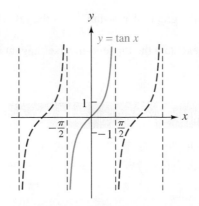

INVERSE TANGENT FUNCTION

The **inverse tangent function** is defined as the inverse of the tangent function with domain restricted to $(-\pi/2, \pi/2)$. That is,

$$y = \tan^{-1} x \quad \text{and} \quad y = \arctan x$$

are equivalent to

$$\tan y = x \quad \text{for } y \text{ in } (-\pi/2, \pi/2)$$

The inverse tangent of x is the number or angle y, $-\pi/2 < y < \pi/2$, whose tangent is x. The y values in this interval are called the **principal values** of the inverse tangent function.

Figure 16 compares the graphs of the restricted tangent function and its inverse. Notice that $(-\pi/4, -1)$, $(0, 0)$, and $(\pi/4, 1)$ are on the restricted tangent graph. Reversing the coordinates gives us three points on the graph of the inverse tangent function. Also note that the vertical asymptotes become horizontal asymptotes.

FIGURE 16

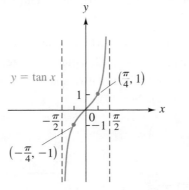

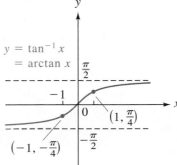

Domain: $-\frac{\pi}{2} < x < \frac{\pi}{2}$
Range: All real numbers

Domain: All real numbers
Range: $-\frac{\pi}{2} < y < \frac{\pi}{2}$

(a) Tangent function restricted to domain
$\left(-\frac{\pi}{2}, \frac{\pi}{2}\right)$

(b) Inverse tangent function

Note that the graph of $y = \tan^{-1} x$ is the reflection of $y = \tan x$ through the line $y = x$.

We next state the tangent–inverse tangent identities.

TANGENT–INVERSE TANGENT IDENTITIES

$$\tan(\tan^{-1} x) = x \quad \text{for all } x$$

$$\tan^{-1}(\tan x) = x \quad -\pi/2 < x < \pi/2$$

EXPLORE/DISCUSS 4

Use a calculator to evaluate each of the following. Which illustrate a tangent–inverse tangent identity and which do not? Explain.

(A) $\tan(\tan^{-1} 25)$ (B) $\tan[\tan^{-1}(-325)]$

(C) $\tan^{-1}(\tan 1.2)$ (D) $\tan^{-1}[\tan(-\pi)]$

EXAMPLE 6

Exact Values

Find exact values without using a calculator:

(A) $\tan^{-1}(-1/\sqrt{3})$ (B) $\tan^{-1}[\tan(-1.2)]$ (C) $\tan^{-1}(\tan \pi)$

Solution

(A) $y = \tan^{-1}(-1/\sqrt{3})$ is equivalent to $\tan y = -1/\sqrt{3}$, where y satisfies $-\pi/2 < y < \pi/2$. What y between $-\pi/2$ and $\pi/2$ has tangent $-1/\sqrt{3}$? This y must be negative and associated with a fourth-quadrant reference triangle (see Fig. 17):

$$\tan y = \frac{-1}{\sqrt{3}}$$

Special 30°-60°-90° triangle, so $y = -\pi/6$.

FIGURE 17

$$\tan^{-1}\left(-\frac{1}{\sqrt{3}}\right) = -\frac{\pi}{6}$$

[*Note:* y cannot be $11\pi/6$. Why?]

(B) $\tan^{-1}[\tan(-1.2)] = -1.2$ *Tangent–inverse tangent identity*

(C) The second tangent–inverse tangent identity doesn't apply since π is not in $(-\pi/2, \pi/2)$. Instead, we first note that $\tan\pi = 0$ (since $\sin\pi = 0$ and $\cos\pi = -1$). We also note that $\tan^{-1}(0)$ is the number between $-\pi/2$ and $\pi/2$ whose tangent is 0, which is 0. That is, $\tan^{-1}(\tan\pi) = 0$. ∎

Matched Problem 6 Find exact values without using a calculator:

(A) $\arctan\sqrt{3}$ (B) $\tan(\tan^{-1}35)$ (C) $\tan^{-1}[\tan(3\pi/4)]$ ∎

EXAMPLE 7

Calculator Values

Find to four significant digits using a calculator:

(A) $\tan^{-1}3$ (B) $\arctan(-25.45)$ (C) $\tan^{-1}1{,}435$

(D) $\sec[\tan^{-1}(-0.1308)]$ (E) $\tan^{-1}(\tan 2)$

Solution Set calculator in radian mode.

(A) $\tan^{-1}3 = 1.249$ (B) $\arctan(-25.45) = -1.532$

(C) $\tan^{-1}1{,}435 = 1.570$ (D) $\sec[\tan^{-1}(-0.1308)] = 1.009$

(E) $\tan^{-1}(\tan 2) = -1.142$ ∎

Matched Problem 7 Find to four significant digits using a calculator:

(A) $\tan^{-1}7$ (B) $\arctan(-13.08)$ (C) $\tan^{-1}735$

(D) $\csc[\tan^{-1}(-1.033)]$ (E) $\tan^{-1}[\tan(-3)]$ ∎

EXAMPLE 8

Finding an Equivalent Algebraic Expression

Express $\sin(\tan^{-1}x)$ as an algebraic expression in x.

Solution Let

$$y = \tan^{-1}x \qquad -\frac{\pi}{2} < y < \frac{\pi}{2}$$

or, equivalently,

$$\tan y = x = \frac{x}{1} \qquad -\frac{\pi}{2} < y < \frac{\pi}{2}$$

The two possible reference triangles for y are shown in Figure 18.

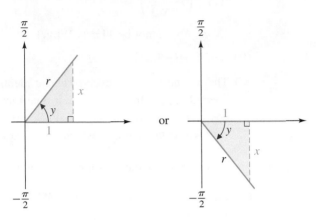

or

FIGURE 18
Reference triangles for $y = \tan^{-1}x$

In either case,

$$r = \sqrt{x^2 + 1}$$

Then

$$\sin(\tan^{-1} x) = \sin y = \frac{x}{r} = \frac{x}{\sqrt{x^2 + 1}}$$ ∎

Matched Problem 8   Express $\tan(\arccos x)$ as an algebraic expression in x.   ∎

The inverse trigonometric functions can also be used if angles are measured in degrees. In that case, we simply convert values in the range of the inverse trigonometric function from radians to degrees. For example, $\tan^{-1} 1$ is written as $45°$ rather than as $\pi/4$.

■ Summary

We will conclude this section by summarizing the definitions of the inverse trigonometric functions in the box below for convenient reference.

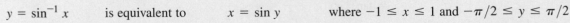

INVERSE SINE, COSINE, AND TANGENT FUNCTIONS

$y = \sin^{-1} x$	is equivalent to	$x = \sin y$	where $-1 \leq x \leq 1$ and $-\pi/2 \leq y \leq \pi/2$
$y = \cos^{-1} x$	is equivalent to	$x = \cos y$	where $-1 \leq x \leq 1$ and $0 \leq y \leq \pi$
$y = \tan^{-1} x$	is equivalent to	$x = \tan y$	where x is any real number and $-\pi/2 < y < \pi/2$

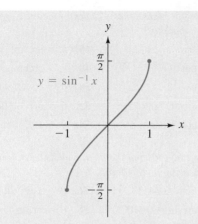

Domain: $-1 \le x \le 1$
Range: $-\frac{\pi}{2} \le y \le \frac{\pi}{2}$

(a)

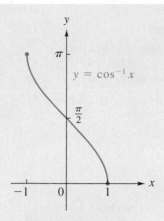

Domain: $-1 \le x \le 1$
Range: $0 \le y \le \pi$

(b)

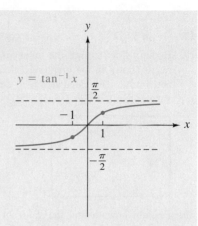

Domain: All real numbers
Range: $-\frac{\pi}{2} < y < \frac{\pi}{2}$

(c)

Answers to Matched Problems

1. (A) $\pi/4$ (B) $-\pi/2$ (C) -0.4 (D) $-\frac{1}{2}$ (E) Not defined
2. (A) 0.2945 (B) -0.8684 (C) Not defined   (D) 1.065 (E) -1.442
3. (A) $\pi/4$ (B) π (C) 3.05 (D) $-\frac{1}{2}$ (E) Not defined
4. (A) 0.8267 (B) 2.517 (C) Not defined   (D) -0.5829 (E) 1
5. $\frac{24}{25}$ 6. (A) $\pi/3$ (B) 35 (C) Not defined
7. (A) 1.429 (B) -1.494 (C) 1.569 (D) -1.392 (E) 0.1416
8. $\dfrac{\sqrt{1 - x^2}}{x}$

EXERCISE 5.1

A

1. Explain why there are many potential answers to the question "For what number x is $\sin x$ equal to $1/2$?"

2. Of the many possible numbers x for which $\sin x = 1/2$, why do we have to choose one of them in defining $\sin^{-1}(1/2)$? Which do we choose?

3. Why is it not sufficient to restrict the domain of $y = \cos x$ to $[-\pi/2, \pi/2]$ when defining inverse cosine?

4. Explain why the domain of inverse sine and inverse cosine is $[-1, 1]$, but the domain of inverse tangent is all real numbers.

In Problems 5–12, Find exact real number values without using a calculator.

5. $\sin^{-1} 0$
6. $\cos^{-1} 0$
7. $\arccos(\sqrt{3}/2)$
8. $\arcsin(\sqrt{3}/2)$
9. $\tan^{-1} 1$
10. $\arctan \sqrt{3}$
11. $\cos^{-1} \frac{1}{2}$
12. $\sin^{-1}(1/\sqrt{2})$

In Problems 13–18, evaluate to four significant digits using a calculator.

13. $\cos^{-1}(-0.9999)$
14. $\sin^{-1}(-0.0289)$

15. $\tan^{-1} 4.056$ **16.** $\tan^{-1}(-52.77)$

17. $\arcsin 3.142$ **18.** $\arccos 1.001$

19. Explain how to find the value of x that produces the result shown in the graphing calculator window below, and find it. The calculator is in degree mode. Give the answer to six decimal places.

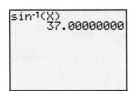

20. Explain how to find the value of x that produces the result shown in the graphing calculator window below, and find it. The calculator is in radian mode. Give the answer to six decimal places.

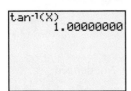

B *In Problems 21–36, find exact real number values, if possible without using a calculator.*

21. $\arccos\left(-\frac{1}{2}\right)$ **22.** $\arcsin(-1/\sqrt{2})$

23. $\tan^{-1}(-1)$ **24.** $\arctan(-\sqrt{3})$

25. $\sin^{-1}(-\sqrt{3}/2)$ **26.** $\cos^{-1}(-1)$

27. $\cos^{-1}(-\sqrt{3}/2)$ **28.** $\sin^{-1}(-1)$

29. $\sin[\sin^{-1}(-0.6)]$ **30.** $\tan(\tan^{-1} 25)$

31. $\cos[\sin^{-1}(-1/\sqrt{2})]$ **32.** $\sec[\sin^{-1}(-\sqrt{3}/2)]$

33. $\tan(\sin^{-1}\frac{2}{3})$ **34.** $\sin(\cos^{-1}\frac{1}{4})$

35. $\cos[\tan^{-1}(-2)]$ **36.** $\sin(\tan^{-1} 10)$

In Problems 37–42, evaluate to four significant digits using a calculator.

37. $\tan^{-1}(-4.038)$ **38.** $\arctan(-10.04)$

39. $\sec[\sin^{-1}(-0.0399)]$ **40.** $\cot[\cos^{-1}(-0.7003)]$

41. $\tan^{-1}(\tan 3)$ **42.** $\cos^{-1}[\cos(-2)]$

Graph the function in Problems 43 and 44 with the aid of a calculator. Plot points using x values $-1.0, -0.8, -0.6, -0.4, -0.2, 0.0, 0.2, 0.4, 0.6, 0.8,$ and 1.0; then join the points with a smooth curve.

43. $y = \sin^{-1} x$ **44.** $y = \cos^{-1} x$

In Problems 45–54, find the exact degree measure of θ if possible without a calculator.

45. $\theta = \arccos(-1/2)$ **46.** $\theta = \arcsin(-1/\sqrt{2})$

47. $\theta = \tan^{-1}(-1)$ **48.** $\theta = \arctan(-\sqrt{3})$

49. $\theta = \sin^{-1}(-\sqrt{3}/2)$ **50.** $\theta = \cos^{-1}(-1)$

51. $\theta = \arcsin(\tan 60°)$ **52.** $\theta = \arccos[\tan(-45°)]$

53. $\theta = \cos^{-1}[\cos(-60°)]$ **54.** $\theta = \sin^{-1}(\sin 135°)$

In Problems 55–60, find the degree measure of θ to two decimal places using a calculator.

55. $\theta = \tan^{-1} 3.0413$ **56.** $\theta = \cos^{-1} 0.7149$

57. $\theta = \arcsin(-0.8107)$ **58.** $\theta = \arccos(-0.7728)$

59. $\theta = \arctan(-17.305)$ **60.** $\theta = \tan^{-1}(-0.3031)$

61. Evaluate $\cos^{-1}[\cos(-0.3)]$ with a calculator set in radian mode. Explain why this does or does not illustrate a cosine–inverse cosine identity.

62. Evaluate $\sin^{-1}[\sin(-2)]$ with a calculator set in radian mode. Explain why this does or does not illustrate a sine–inverse sine identity.

63. The identity $\sin(\sin^{-1} x) = x$ is valid for $-1 \leq x \leq 1$.

  (A) Graph $y = \sin(\sin^{-1} x)$ for $-1 \leq x \leq 1$.

  (B) What happens if you graph $y = \sin(\sin^{-1} x)$ over a wider interval, say $-2 \leq x \leq 2$? Explain.

64. The identity $\cos(\cos^{-1} x) = x$ is valid for $-1 \leq x \leq 1$.

  (A) Graph $y = \cos(\cos^{-1} x)$ for $-1 \leq x \leq 1$.

  (B) What happens if you graph $y = \cos(\cos^{-1} x)$ over a wider interval, say $-2 \leq x \leq 2$? Explain.

C *In Problems 65–74, find exact real number values without using a calculator. [Hint: Use identities from Chapter 4.]*

65. $\sin[\arccos\frac{1}{2} + \arcsin(-1)]$

66. $\cos[\cos^{-1}(-\sqrt{3}/2) - \sin^{-1}(-\frac{1}{2})]$

67. $\sin[2\sin^{-1}(-\frac{4}{5})]$

68. $\cos\left(\dfrac{\cos^{-1}\frac{1}{3}}{2}\right)$

69. $\cos(\arctan 2 + \arcsin\frac{1}{3})$

70. $\sin(\arccos\frac{2}{5} - \arctan 5)$

71. $\tan(\sin^{-1}\frac{1}{4} - \cos^{-1}\frac{1}{4})$

72. $\tan(\cos^{-1}\frac{3}{4} + \tan^{-1} 3)$

73. $\sin\left[\dfrac{\arctan(-2)}{2}\right]$

74. $\tan(2\arcsin\frac{3}{5})$

In Problems 75–78, write each as an algebraic expression in x free of trigonometric or inverse trigonometric functions.

75. $\sin(\cos^{-1} x)$, $-1 \le x \le 1$

76. $\cos(\sin^{-1} x)$, $-1 \le x \le 1$

77. $\tan(\arcsin x)$, $-1 \le x \le 1$

78. $\cos(\arctan x)$

Verify each identity in Problems 79 and 80.

79. $\tan^{-1}(-x) = -\tan^{-1} x$

80. $\sin^{-1}(-x) = -\sin^{-1} x$

81. Let $f(x) = \cos^{-1}(2x - 3)$.

(A) Explain how you would find the domain of f and find it.

(B) Graph f over the interval $0 \le x \le 3$ and explain the result.

82. Let $g(x) = \sin^{-1}\left(\dfrac{x + 1}{2}\right)$.

(A) Explain how you would find the domain of g and find it.

(B) Graph g over the interval $-4 \le x \le 2$ and explain the result.

83. Let $h(x) = 3 + 5\sin(x - 1)$, $-\pi/2 \le x \le 1 + \pi/2$.

(A) Find $h^{-1}(x)$.

(B) Explain how x must be restricted in $h^{-1}(x)$.

84. Let $f(x) = 4 + 2\cos(x - 3)$, $-\pi/2 \le x \le 3 + \pi/2$.

(A) Find $f^{-1}(x)$.

(B) Explain how x must be restricted in $f^{-1}(x)$.

85. The identity $\sin^{-1}(\sin x) = x$ is valid for $-\pi/2 \le x \le \pi/2$.

(A) Graph $y = \sin^{-1}(\sin x)$ for $-\pi/2 \le x \le \pi/2$.

(B) What happens if you graph $y = \sin^{-1}(\sin x)$ over a larger interval, say $-2\pi \le x \le 2\pi$? Explain.

86. The identity $\cos^{-1}(\cos x) = x$ is valid for $0 \le x \le \pi$.

(A) Graph $y = \cos^{-1}(\cos x)$ for $0 \le x \le \pi$.

(B) What happens if you graph $y = \cos^{-1}(\cos x)$ over a larger interval, say $-2\pi \le x \le 2\pi$? Explain.

Verify each identity in Problems 87–90.

87. $\sin^{-1}x = \tan^{-1}\left(\dfrac{x}{\sqrt{1 - x^2}}\right)$

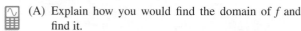

88. $\cos^{-1}x = \tan^{-1}\left(\dfrac{\sqrt{1 - x^2}}{x}\right)$

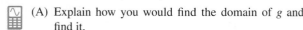

89. $\sin(2 \arcsin x) = 2x\sqrt{1 - x^2}$

90. $\tan(2 \arctan x) = \dfrac{2x}{1 - x^2}$

Applications

91. Viewing Angle A family is hoping to see a space shuttle launch from their hotel balcony, which is 5.2 mi from the launch site. The viewing angle (in degrees) between the horizontal from their balcony and the shuttle is given by the equation $\theta = \tan^{-1}(x/5.2)$, where x is the altitude of the shuttle in miles. Due to cloud cover, launch engineers announce that the shuttle's best visibility will be when it reaches a height of 3 mi. At what angle should the family look to see the shuttle at that height?

92. Viewing Angle A photographer is set up on a bridge over a river because she hears that a celebrity is sunbathing on a boat coming down the river. The boat will come around a bend in the river that is 0.2 mi from the bridge. The viewing angle (in degrees) below the horizontal to a boat on the river is given by the equation $\theta = 90 - \tan^{-1}(x/74)$, where x is distance (in feet) from the base of the bridge to the boat. At what angle should the photographer set up her camera to be ready for a shot when the boat first comes into view?

93. Sonic Boom An aircraft flying faster than the speed of sound produces sound waves that pile up behind the aircraft in the form of a cone. The cone of a level flying aircraft intersects the ground in the form of a hyperbola, and along this curve we experience a *sonic boom* (see the figure on the next page). In Section 2.4 it was shown that

$$\sin\frac{\theta}{2} = \frac{\text{Speed of sound}}{\text{Speed of aircraft}} \tag{1}$$

where θ is the cone angle. The ratio

$$M = \frac{\text{Speed of aircraft}}{\text{Speed of sound}} \tag{2}$$

is the Mach number. A Mach number of 2.3 would indicate an aircraft moving at 2.3 times the speed of sound. From equations (1) and (2) we obtain

$$\sin\frac{\theta}{2} = \frac{1}{M}$$

(A) Write θ in terms of M.

(B) Find θ to the nearest degree for $M = 1.7$ and for $M = 2.3$.

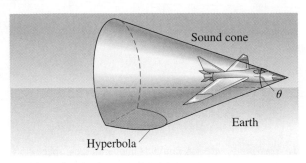

Figure for 93

94. **Space Science**   A spacecraft traveling in a circular orbit h miles above the earth observes horizons in each direction (see the figure).

(A) Express θ in terms of h and r, where r is the radius of the earth (3,959 mi).

(B) Find θ, in degrees (to one decimal place), for $h = 425.4$ mi.

(C) Find the length (to the nearest mile) of the arc subtended by angle θ found in part (B). What percentage (to one decimal place) of the great circle containing the arc does the arc represent? (A **great circle** is any circle on the surface of the earth having the center of the earth as its center.)

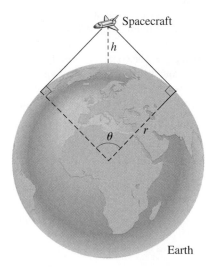

Figure for 94

95. **Precalculus: Sports**   A particular soccer field is 110 yd by 60 yd, and the goal is 8 yd wide at the end of the field (see the figure). A player is dribbling the ball along a line

parallel to and 5 yd inside the sideline. Assuming the player has a clear shot all along this line, is there an optimal distance x from the end of the field where the shot should be taken? That is, is there a distance x for which θ is maximum? Parts (A)–(D) will attempt to answer this question.

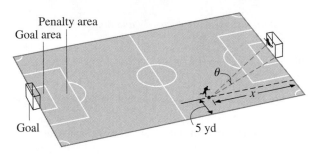

Figure for 95

(A) Discuss what you think happens to θ as x varies from 0 yd to 55 yd.

(B) Show that

$$\theta = \tan^{-1}\left(\frac{8x}{x^2 + 609}\right)$$

$$\left[\textit{Hint: } \tan(\alpha + \beta) = \frac{\tan \alpha + \tan \beta}{1 - \tan \alpha \tan \beta} \text{ is useful.}\right]$$

(C) Complete Table 1 (to two decimal places) and select the value of x that gives the maximum value of θ in the table. (If your calculator has a table-generating feature, use it.)

TABLE 1						
x (yd)	10	15	20	25	30	35
θ (deg)	6.44					

(D) Graph the equation in part (B) in a graphing calculator for $0 \le x \le 55$. Describe what the graph shows. Use the $\boxed{\text{MAXIMUM}}$ command to find the maximum θ and the distance x that produces it.

96. **Precalculus: Related Rates**   The figure on the next page represents a circular courtyard surrounded by a high stone wall. A floodlight, located at E, shines into the courtyard. A person walks from the center C along CD to D, at the rate of 6 ft/sec.

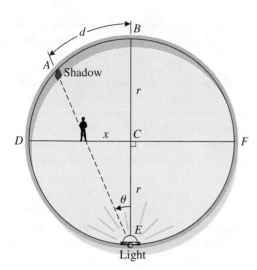

Figure for 96

(A) Do you think that the shadow moves along the circular wall at a constant rate, or does it speed up or slow down as the person walks from C to D?

(B) Show that if the person walks x feet from C along CD, then the shadow will move a distance d given by

$$d = 2r\theta = 2r \tan^{-1}\frac{x}{r} \qquad (1)$$

where θ is in radians. [*Hint:* Draw a line from A to C.] Express x in terms of time t, then for a courtyard of radius $r = 60$ ft, rewrite equation (1) in the form

$$d = 120 \tan^{-1}\frac{t}{10} \qquad (2)$$

(C) Using equation (2), complete Table 2 (to one decimal place). (If you have a calculator with a table-generating feature, use it.)

TABLE 2

t (sec)	0	1	2	3	4	5	6	7	8	9	10
d (ft)	0.0	12.0									

(D) From Table 2, determine how far the shadow moves during the first second, during the fifth second, and during the tenth second. Is the shadow speeding up, slowing down, or moving uniformly? Explain.

 (E) Graph equation (2) in a graphing calculator for $0 \le t \le 10$. Describe what the graph shows.

97. **Engineering** Horizontal cylindrical tanks are buried underground at service stations to store fuel. To determine the amount of fuel in the tank, a "dip stick" is often used to find the depth of the fuel (see the figure).

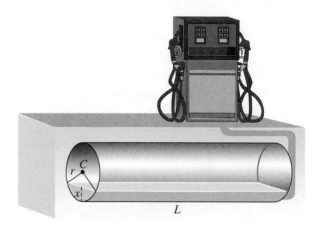

Figure for 97

(A) Show that the volume of fuel x feet deep in a horizontal circular tank L feet long with radius r, $x < r$, is given by (see the figure)

$$V = \left[r^2 \cos^{-1}\frac{r-x}{r} - (r-x)\sqrt{r^2-(r-x)^2} \right] L$$

(B) If the fuel in a tank 30 ft long with radius 3 ft is found to be 2 ft deep, how many cubic feet (to the nearest cubic foot) of fuel are in the tank?

(C) The function

$$y_1 = 30 \left[9 \cos^{-1}\frac{3-x}{3} - (3-x)\sqrt{9-(3-x)^2} \right]$$

represents the volume of fuel x feet deep in the tank in part (B). Graph y_1 and $y_2 = 350$ in the same viewing window and use the INTERSECT command to find the depth (to one decimal place) when the tank contains 350 ft³ of fuel.

98. **Engineering** In designing mechanical equipment, it is sometimes necessary to determine the length of the belt around two pulleys of different diameters (see the figure on the next page).

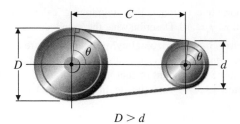

$D > d$

Figure for 98

(A) Show that the length of the belt around the two pulleys in the figure is given by

$$L = \pi D + (d - D)\theta + 2C \sin \theta$$

where θ (in radians) is given by

$$\theta = \cos^{-1} \frac{D - d}{2C}$$

(B) Find the length of the belt (to one decimal place) if $D = 6$ in., $d = 4$ in., and $C = 10$ in.

(C) The function

$$y_1 = 6\pi - 2 \cos^{-1} \frac{1}{x} + 2x \sin\left(\cos^{-1} \frac{1}{x}\right)$$

represents the length of the belt around the two pulleys in part (B) when the centers of the pulleys are x inches apart. Graph y_1 and $y_2 = 40$ in the same viewing window, and use the $\boxed{\text{INTERSECT}}$ command to find the distance between the centers of the pulleys (to one decimal place) when the belt is 40 in. long.

99. **Statistics** Statisticians often use functions, such as $y = \log x$, to transform data before applying various tests to the data. A popular choice is the **arcsin transformation** $y = \arcsin \sqrt{x}$, where x, $0 \leq x \leq 1$, is in the original data set and y is in the transformed data set. After the analysis is completed, the inverse of the arcsin transformation is used to return to the original data. Find the inverse of $y = \arcsin \sqrt{x}$.

100. **Statistics** A variation of the arcsin transformation is $y = \arcsin \sqrt{x/B}$ where B is a constant satisfying

$0 \leq x \leq B$ for x in the original data set and y is in the transformed data set. Find the inverse of this transformation.

The population of some species of songbirds has been declining while others are thriving. Since it is impossible to take a census of birds, scientists devise population indexes to estimate bird populations using data from a variety of sources.

101. **Bird Populations** Complete parts (A)–(E) to construct a model for the population index for Henslow's sparrows.

(A) Produce a scatter plot of the data in Table 1, where x is years since 1988, $2 \leq x \leq 18$, and y is the population index, $0 \leq y \leq 1$.

(B) Use the arcsin transformation in Problem 99 to transform the data. That is, evaluate the arcsin transformation at 0.657, 0.642, and so on.

(C) Produce a scatter plot for the transformed data using $2 \leq x \leq 18$ and $0 \leq y' \leq 1$ (where y' represents the transformed data).

(D) Fit the transformed data with a linear regression line and add the graph of this line to the scatter plot in part (C).

(E) If $y = ax + b$ is the regression line and $f(x)$ is the inverse of the arcsin transformation from Problem 99, add the graph of $y = f(ax + b)$ to the scatter plot in part (A). Does $f(ax + b)$ appear to be a good model for these data?

102. **Bird Populations** Repeat Problem 101 using the data in Table 2 on the next page for Magnolia warblers. Change the viewing window in part (C) to $0 \leq y' \leq 1.6$.

Wind farms have become a valuable source of renewable energy. Problems 103 and 104 involve data collected from the electrical generators in two windmills on an offshore wind farm. The electricity generated by the windmills at various wind speeds is given in Table 3 on the next page. The wind speed is given in mph (miles per hour) and the windmill output is given in mwh (megawatt hours).

103. **Wind Farms** Complete parts (A)–(E) to construct a model for the electricity generated by windmill A.

TABLE 1

Henslow's Sparrows

Year	1990	1992	1994	1996	1998	2000	2002	2004	2006
Index	0.657	0.642	0.437	0.481	0.425	0.399	0.374	0.370	0.341

TABLE 2
Magnolia Warblers

Year	1990	1992	1994	1996	1998	2000	2002	2004	2006
Index	0.686	0.731	0.703	0.884	0.974	0.836	0.912	0.940	0.989

TABLE 3

Wind speed	10	15	20	25	30	35	40
Output from windmill A	69	325	836	1,571	1,965	1,993	1,999
Output from windmill B	52	277	748	1,452	1,785	1,952	1,989

(A) Produce a scatter plot for windmill A using $10 \leq x \leq 40$ and $0 \leq y \leq 2{,}000$.

(B) Use the arcsin transformation in Problem 100 with $B = 2{,}000$ to transform the output from windmill A. That is, evaluate the arcsin transformation at 69, 325, and so on.

(C) Produce a scatter plot for the transformed data using $10 \leq x \leq 40$ and $0 \leq y' \leq 1.6$ (where y' represents the transformed data).

(D) Fit the transformed data with a linear regression line and add the graph of this line to the scatter plot in part (C).

(E) If $y = ax + b$ is the regression line and $f(x)$ is the inverse of the arcsin transformation from Problem 100, add the graph of $y = f(ax + b)$ to the scatter plot in part (A). Does $f(ax + b)$ appear to be a good model for these data?

104. Wind Farms Repeat Problem 103 using the data for windmill B.

☆5.2 Inverse Cotangent, Secant, and Cosecant Functions

- Definition of Inverse Cotangent, Secant, and Cosecant Functions
- Calculator Evaluation

■ Definition of Inverse Cotangent, Secant, and Cosecant Functions

In most applications that require the use of inverse trigonometric functions, statements involving cotangent, secant, and cosecant can be rewritten in terms of

---

☆ Sections marked with a star may be omitted without loss of continuity.

tangent, cosine, and sine. For this reason, inverse cotangent, secant, and cosecant are not widely studied. In this optional section, we will provide a brief look at these functions. Our approach will parallel the methods we used in the previous section.

We begin by identifying the standard choices of restricted domains for cotangent, secant, and cosecant that make each one-to-one. This allows us to define their inverses.

STANDARD DOMAIN RESTRICTIONS FOR COTANGENT, SECANT, AND COSECANT FUNCTIONS

$\cot x$: $0 < x < \pi$

$\sec x$: $0 \le x \le \pi, x \ne \pi/2$

$\csc x$: $-\pi/2 \le x \le \pi/2, x \ne 0$

Now we are ready to define the inverses of these three functions.

INVERSE COTANGENT, SECANT, AND COSECANT FUNCTIONS

$y = \cot^{-1} x$ is equivalent to $x = \cot y$ where $0 < y < \pi$ and x is any real number

$y = \sec^{-1} x$ is equivalent to $x = \sec y$ where $0 \le y \le \pi, y \ne \pi/2$, and $x \le -1$ or $x \ge 1$

$y = \csc^{-1} x$ is equivalent to $x = \csc y$ where $-\pi/2 \le y \le \pi/2, y \ne 0$, and $x \le -1$ or $x \ge 1$

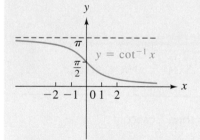

Domain: All real numbers
Range: $0 < y < \pi$

(a)

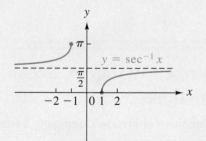

Domain: $x \le -1$ or $x \ge 1$
Range: $0 \le y \le \pi, y \ne \dfrac{\pi}{2}$

(b)

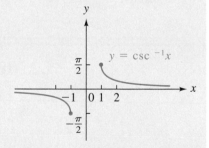

Domain: $x \le -1$ or $x \ge 1$
Range: $-\dfrac{\pi}{2} \le y \le \dfrac{\pi}{2}, y \ne 0$

(c)

[*Note:* The ranges for $\sec^{-1}$ and $\csc^{-1}$ are sometimes selected differently.]

The functions $y = \cot^{-1} x$, $y = \sec^{-1} x$, and $y = \csc^{-1} x$ are also denoted by $y = \text{arccot } x$, $y = \text{arcsec } x$, and $y = \text{arccsc } x$, respectively.

EXPLORE/DISCUSS 1

Which of the following are not defined? Why not?

$\cot^{-1}(-0.5)$ $\sec^{-1}(-0.3)$ $\csc^{-1} 0.9$

$\cot^{-1}(-1{,}000{,}000)$ $\sec^{-1}(3.25 \times 10^8)$ $\csc^{-1}(3.25 \times 10^{-8})$

EXAMPLE 1

Exact Values

Find exact values without using a calculator:

(A) $\operatorname{arccot}(-1)$ (B) $\sec^{-1}(2/\sqrt{3})$

Solution (A) $y = \operatorname{arccot}(-1)$ is equivalent to $\cot y = -1, 0 < y < \pi$. What number between 0 and π has cotangent -1? This y must be positive and in the second quadrant (see Fig. 1):

$$\cot y = -1 = -\frac{1}{1}$$

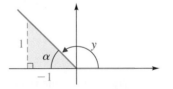

$\alpha = \dfrac{\pi}{4}$

$y = \dfrac{3\pi}{4}$

FIGURE 1

In Figure 1, $y = 3\pi/4$, so

$$\operatorname{arccot}(-1) = \frac{3\pi}{4}$$

(B) $y = \sec^{-1}(2/\sqrt{3})$ is equivalent to $\sec y = 2/\sqrt{3}, 0 \le y \le \pi, y \ne \pi/2$. What number between 0 and π has secant $2/\sqrt{3}$? This y is positive and in the first quadrant. We can draw a reference triangle, as shown in Figure 2.

$$\sec y = \frac{2}{\sqrt{3}}$$

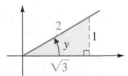

$y = \dfrac{\pi}{6}$

FIGURE 2

In Figure 2, $y = \pi/6$, so

$$\sec^{-1} \frac{2}{\sqrt{3}} = \frac{\pi}{6}$$ ∎

Matched Problem 1 Find exact values without using a calculator:

(A) $\operatorname{arccot}(-\sqrt{3})$ (B) $\csc^{-1}(-2)$ ∎

EXAMPLE 2

Exact Values

Find the exact value of $\tan[\sec^{-1}(-3)]$ without using a calculator.

Solution

Let $y = \sec^{-1}(-3)$; then

$$\sec y = -3 \qquad 0 \le y \le \pi, \quad y \ne \pi/2$$

This y is positive and in the second quadrant. Draw a reference triangle, find the third side, and then determine $\tan y$ from the triangle (see Fig. 3):

$$\sec y = -3 = \frac{3}{-1}$$

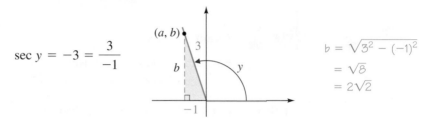

FIGURE 3

Using the values labeled in Figure 3,

$$\tan[\sec^{-1}(-3)] = \tan y = \frac{b}{a} = \frac{2\sqrt{2}}{-1} = -2\sqrt{2} \qquad \blacksquare$$

Matched Problem 2   Find the exact value of $\cot[\csc^{-1}(-\tfrac{5}{3})]$ without using a calculator. $\blacksquare$

■ Calculator Evaluation

Many calculators have keys for sin, cos, tan, $\sin^{-1}$, $\cos^{-1}$, $\tan^{-1}$, or their equivalents. To find sec x, csc x, and cot x using a calculator, we can use the reciprocal identities

$$\sec x = \frac{1}{\cos x} \qquad \csc x = \frac{1}{\sin x} \qquad \cot x = \frac{1}{\tan x}$$

 Caution

It's very tempting to assume that $\sec^{-1}x = 1/(\cos^{-1}x)$, $\csc^{-1}x = 1/(\sin^{-1}x)$, and $\cot^{-1}x = 1/(\tan^{-1}x)$. Unfortunately, these equations are simply not true! The correct reciprocal identities are provided in the box on page 313. These can be used to evaluate $\sec^{-1}x$, $\csc^{-1}x$, and $\cot^{-1}x$ using a calculator that has only $\sin^{-1}$, $\cos^{-1}$, and $\tan^{-1}$ keys. $\square$

We will establish the first part of the inverse cotangent identity stated in the box on page 313. The inverse secant and cosecant identities are left to you to do (see Problems 71 and 72, Exercise 5.2).

Let

$$y = \cot^{-1} x \qquad x > 0$$

Then

$$\cot y = x \qquad 0 < y < \pi/2 \qquad \text{Definition of } \cot^{-1}$$

$$\frac{1}{\tan y} = x \qquad 0 < y < \pi/2 \qquad \text{Reciprocal Identity}$$

$$\tan y = \frac{1}{x} \qquad 0 < y < \pi/2 \qquad \text{Algebra}$$

$$y = \tan^{-1}\frac{1}{x} \quad 0 < y < \pi/2 \qquad \text{Definition of } \tan^{-1}$$

We have shown that

$$\cot^{-1} x = \tan^{-1}\frac{1}{x} \quad \text{for } x > 0$$

INVERSE COTANGENT, SECANT, AND COSECANT IDENTITIES

$$\cot^{-1}x = \begin{cases} \tan^{-1}\dfrac{1}{x} & x > 0 \\[2mm] \pi + \tan^{-1}\dfrac{1}{x} & x < 0 \end{cases}$$

$$\sec^{-1} x = \cos^{-1}\frac{1}{x} \qquad x \geq 1 \quad \text{or} \quad x \leq -1$$

$$\csc^{-1} x = \sin^{-1}\frac{1}{x} \qquad x \geq 1 \quad \text{or} \quad x \leq -1$$

EXAMPLE 3

Calculator Evaluation

Use a calculator to evaluate the following as real numbers to three decimal places.

(A) $\cot^{-1} 4.05$ (B) $\csc^{-1}(-12)$

Solution   (A) With the calculator in radian mode,

$$\cot^{-1} 4.05 = \tan^{-1}\frac{1}{4.05} = 0.242 \qquad \text{Use } \cot^{-1}x = \tan^{-1}\frac{1}{x}.$$

(B) With the calculator in radian mode,

$$\csc^{-1}(-12) = \sin^{-1}(-\tfrac{1}{12}) = -0.083 \qquad \text{Use } \csc^{-1}x = \sin^{-1}\frac{1}{x}. \qquad \blacksquare$$

Matched Problem 3   Use a calculator to evaluate the following as real numbers to three decimal places.

(A) $\cot^{-1} 2.314$ (B) $\sec^{-1}(-1.549)$ $\blacksquare$

**Answers to
Matched Problems**   **1.** (A) $5\pi/6$ (B) $-\pi/6$ **2.** $-\frac{4}{3}$ **3.** (A) 0.408 (B) 2.273

EXERCISE 5.2

A 1. Look at the graph of $y = \sec x$ on the interval $[-2\pi, 2\pi]$ and explain why $[0, \pi]$ is a reasonable choice of restricted domain to use in defining inverse secant.

2. Look at the graph of $y = \csc x$ on the interval $[-2\pi, 2\pi]$ and explain why $[-\pi/2, \pi/2]$ is a reasonable choice of restricted domain to use in defining inverse cosecant.

3. Look at the graph of $y = \cot x$ on the interval $[-2\pi, 2\pi]$ and explain why $[0, \pi]$ is a reasonable choice of restricted domain to use in defining inverse cotangent.

4. Use the graph of $y = \sec x$ to choose an interval of length π, different from the standard interval in this section, so that secant is one-to-one if restricted to that interval. Repeat for $y = \csc x$ and $y = \cot x$.

In Problems 5–32, find the exact real number value of each if possible without using a calculator.

5. $\cot^{-1} \sqrt{3}$

6. $\cot^{-1} 0$

7. $\text{arccsc } 1$

8. $\text{arcsec } 2$

9. $\sec^{-1} \sqrt{2}$

10. $\csc^{-1} 2$

11. $\sin(\cot^{-1} 0)$

12. $\cos(\cot^{-1} 1)$

13. $\tan(\csc^{-1} \frac{5}{4})$

14. $\cot(\sec^{-1} \frac{5}{3})$

B 15. $\cot^{-1}(-1)$

16. $\sec^{-1}(-1)$

17. $\text{arcsec}(-2)$

18. $\text{arccsc}(-\sqrt{2})$

19. $\text{arccsc}(-2)$

20. $\text{arccot}(-\sqrt{3})$

21. $\csc^{-1} \frac{1}{2}$

22. $\sec^{-1}(-\frac{1}{2})$

23. $\cos[\csc^{-1}(-\frac{5}{3})]$

24. $\tan[\cot^{-1}(-1/\sqrt{3})]$

25. $\cot[\sec^{-1}(-\frac{5}{4})]$

26. $\sin[\cot^{-1}(-\frac{3}{4})]$

27. $\cos[\sec^{-1}(-2)]$

28. $\sin[\csc^{-1}(-2)]$

29. $\cot(\cot^{-1} 33.4)$

30. $\sec[\sec^{-1}(-44)]$

31. $\csc[\csc^{-1}(-4)]$

32. $\cot[\cot^{-1}(-7.3)]$

In Problems 33–42, use a calculator to evaluate the following as real numbers if possible to three decimal places.

33. $\sec^{-1} 5.821$

34. $\cot^{-1} 2.094$

35. $\cot^{-1} 0.035$

36. $\csc^{-1}(-1.003)$

37. $\csc^{-1} 0.847$

38. $\sec^{-1}(-0.999)$

39. $\text{arccot}(-3.667)$

40. $\text{arccsc } 8.106$

41. $\text{arcsec}(-15.025)$

42. $\text{arccot}(-0.157)$

In Problems 43–54, find the exact degree measure of θ if possible without using a calculator.

43. $\theta = \text{arcsec}(-2)$

44. $\theta = \text{arccsc}(-\sqrt{2})$

45. $\theta = \cot^{-1}(-1)$

46. $\theta = \text{arccot}(-1/\sqrt{3})$

47. $\theta = \csc^{-1}(-2/\sqrt{3})$

48. $\theta = \sec^{-1}(-1)$

49. $\theta = \text{arcsec}(\sin 60°)$

50. $\theta = \text{arccsc}(\cos 45°)$

51. $\theta = \text{arccsc}(\sec 135°)$

52. $\theta = \text{arccot}[\cot(-30°)]$

53. $\theta = \cot^{-1}[\cot(-15°)]$

54. $\theta = \sec^{-1}(\sec 100°)$

In Problems 55–60, find the degree measure to two decimal places using a calculator.

55. $\theta = \cot^{-1} 0.3288$

56. $\theta = \sec^{-1} 1.3989$

57. $\theta = \text{arccsc}(-1.2336)$

58. $\theta = \text{arcsec}(-1.2939)$

59. $\theta = \text{arccot}(-0.0578)$

60. $\theta = \cot^{-1}(-3.2994)$

C *In Problems 61–64, find exact values for each problem without using a calculator.*

61. $\tan[\csc^{-1}(-\frac{5}{3}) + \tan^{-1}\frac{1}{4}]$

62. $\tan[\tan^{-1} 4 - \sec^{-1}(-\sqrt{5})]$

63. $\tan[2 \cot^{-1}(-\frac{3}{4})]$

64. $\tan[2 \sec^{-1}(-\sqrt{5})]$

In Problems 65–70, write each as an algebraic expression in x free of trigonometric or inverse trigonometric functions.

65. $\sin(\cot^{-1} x)$

66. $\cos(\cot^{-1} x)$

67. $\csc(\sec^{-1} x)$

68. $\tan(\csc^{-1} x)$

69. $\sin(2 \cot^{-1} x)$

70. $\sin(2 \sec^{-1} x)$

71. Show that $\sec^{-1} x = \cos^{-1}(1/x)$ for $x \geq 1$ and $x \leq -1$.

72. Show that $\csc^{-1} x = \sin^{-1}(1/x)$ for $x \geq 1$ and $x \leq -1$.

Problems 73–76 require the use of a graphing calculator. Use an appropriate inverse trigonometric identity to graph each function in the viewing window $-5 \leq x \leq 5$, $-\pi \leq y \leq \pi$.

73. $y = \sec^{-1} x$

74. $y = \csc^{-1} x$

75. $y = \cot^{-1} x$ [Use two viewing windows, one for $-5 \leq x \leq 0$, and the other for $0 \leq x \leq 5$.]

76. $y = \cot^{-1} x$ [Using one viewing window, $-5 \leq x \leq 5$, graph $y_1 = \pi (x < 0) + \tan^{-1}(1/x)$, where $<$ is selected from the TEST menu. The expression $(x < 0)$ assumes a value of 1 for $x < 0$ and 0 for $x \geq 0$.]

For the equations in Problems 77–84, verify those that are identities and give counter examples for those that are not.

77. $\sec^{-1}x + \csc^{-1} x = \pi/2$

78. $\tan^{-1}x + \cot^{-1} x = \pi/2$

79. $\csc(\cot^{-1} x) = \sqrt{x^2 - 1}$

80. $\sec(\tan^{-1} x) = \sqrt{x^2 + 1}$

81. $\cot(\csc^{-1} x) = \sqrt{x^2 - 1}$

82. $\tan(\sec^{-1} x) = \sqrt{x^2 + 1}$

83. $\sec^{-1} x = \csc^{-1} \dfrac{x}{\sqrt{x^2 - 1}}$

84. $\csc^{-1} x = \sec^{-1} \dfrac{x}{\sqrt{x^2 - 1}}$

5.3 Trigonometric Equations: An Algebraic Approach

- Introduction
- Solving Trigonometric Equations Using an Algebraic Approach

■ Introduction

In Chapter 4 we considered trigonometric equations called identities. These equations are true for all replacements of the variable(s) for which both sides are defined. We now consider another class of equations, called **conditional equations,** which may be true for some replacements of the variable(s) but false for others. For example,

$$\sin x = \cos x$$

is a conditional equation, since it is true for $x = \pi/4$ and false for $x = 0$. (Check both values.)

EXPLORE/DISCUSS 1

Consider the simple trigonometric equation

$$\sin x = 0.5$$

Figure 1 shows a partial graph of the left and right sides of the equation.

FIGURE 1
$\sin x = 0.5$

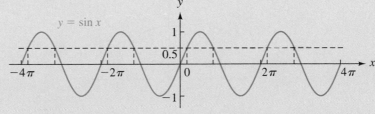

(A) How many solutions does the equation have on the interval $(0, 2\pi)$? What are the solutions?

(B) How many solutions does the equation have on the interval $(-\infty, \infty)$? Discuss a method of writing all solutions to the equation.

Explore/Discuss 1 illustrates the key issue in solving trigonometric equations: In many cases, if there is a solution at all, there are actually infinitely many.

In this section we will solve conditional trigonometric equations using an algebraic approach. In the next section we will use a graphing calculator approach. Solving trigonometric equations using an algebraic approach often requires the use of algebraic manipulation, identities, and some ingenuity. In some cases, an algebraic approach leads to exact solutions. A graphing calculator approach uses graphical methods to approximate solutions to any accuracy desired (or that your calculator will display), which is different from finding exact solutions. The graphing calculator approach can be used to solve trigonometric equations that are either very difficult or impossible to solve algebraically.

■ Solving Trigonometric Equations Using an Algebraic Approach

We will begin developing a general framework for solving trigonometric equations by looking at a simple example. Since we will make liberal use of unit circle diagrams, it would be a good idea to review the unit circle on page 103 before beginning.

EXAMPLE 1

Solving a Simple Sine Equation

Find all solutions to $\sin x = 1/\sqrt{2}$ exactly.

Solution   You may recognize that one solution is $x = \pi/4$. This is a good start. Now we need to find all of the other solutions. We will use a unit circle diagram as an aid (Fig. 2). To simplify notation, the point $(1/\sqrt{2}, 1/\sqrt{2})$ on the unit circle at the angle $\pi/4$ radians is labeled $x = \pi/4$. The dotted horizontal line indicates that the height of that point is $1/\sqrt{2}$.

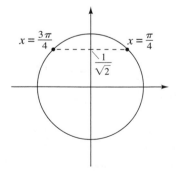

FIGURE 2

Recall that for any x, sin x is defined to be the second coordinate (height) of a point on the unit circle corresponding to x. So there is a second solution directly across the circle from $x = \pi/4$, at the same height: $x = 3\pi/4$. Finally, we need a method of representing all other solutions. Because sine is periodic with period 2π, $\pi/4 + 2\pi$ is also a solution, as are $\pi/4 - 2\pi$, $\pi/4 + 4\pi$, $\pi/4 - 4\pi$, and so on. In general, any number of the form $\pi/4 + 2k\pi$, where k is any integer, is a solution. We will use this idea to write all of the infinitely many solutions as

$$x = \begin{cases} \dfrac{\pi}{4} + 2k\pi \\[2ex] \dfrac{3\pi}{4} + 2k\pi \end{cases} \qquad k \text{ any integer}$$

■

Matched Problem 1   Find all solutions to sin $x = 0.5$ exactly.   ■

 Caution   It's not uncommon for students to think of the "$+ 2k\pi$" part of the answer in Example 1 as an afterthought, which often leads to forgetting it in other solutions. But you could make a case for it being the most important part. The equation sin $x = 1/\sqrt{2}$ has *infinitely many* solutions, but without adding $2k\pi$ to $\pi/4$ and $3\pi/4$, you've only found *two* of them!   □

A similar approach will work for equations of the form cos $x = a$ and tan $x = a$, as illustrated in Example 2.

 **EXAMPLE 2**

Solving Cosine and Tangent Equations

Find all solutions to each equation. Find exact solutions if possible; if not, approximate to two decimal places.

(A) $3 \cos x - 1 = 0$ (B) $\sqrt{3} \tan x = 1$

Solution   (A) We will begin by solving for cos x so that we can use the procedure we developed in Example 1.

$$3 \cos x - 1 = 0 \qquad \text{\small Add 1 to both sides.}$$
$$3 \cos x = 1 \qquad \text{\small Divide both sides by 3.}$$
$$\cos x = \frac{1}{3}$$

Now we need to find all numbers x whose cosine is $1/3$. This is not one of the values we recognize for cosine, so we use a calculator (set in radian mode) to find $\cos^{-1}(1/3)$.

$$x = \cos^{-1}(1/3) = 1.23$$

Since cosine is the first coordinate of a point on the unit circle, a second solution will lie directly below $x = 1.23$ (Fig. 3). This is $x = 2\pi - 1.23$, or 5.05.

FIGURE 3

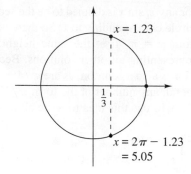

Since the period of cosine is 2π, all solutions are given by

$$x = \begin{cases} 1.23 + 2k\pi \\ 5.05 + 2k\pi \end{cases} \quad k \text{ any integer}$$

(B) Begin by solving for $\tan x$:

$$\sqrt{3} \tan x = 1 \qquad \textit{Divide both sides by } \sqrt{3}.$$
$$\tan x = 1/\sqrt{3}$$

Since $\sin(\pi/6) = 1/2$ and $\cos(\pi/6) = \sqrt{3}/2$, $\tan(\pi/6) = 1/\sqrt{3}$. That is, one solution is $x = \pi/6$. There is one other location on the unit circle whose tangent is $1/\sqrt{3}$ (Fig. 4)—in the third quadrant, where tangent is positive: $x = \pi/6 + \pi = 7\pi/6$.

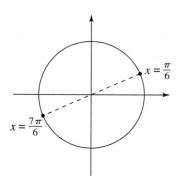

FIGURE 4

But it turns out that we don't need to know this: Recall that the period of tangent is π, so all solutions are given by

$$x = \frac{\pi}{6} + k\pi, \qquad k \text{ any integer.} \qquad \blacksquare$$

Matched Problem 2   Find all solutions to each equation. Find exact solutions if possible; if not, approximate to two decimal places.

(A) $5 \cos x + 3 = 0$ (B) $5 \tan x = -5$ $\blacksquare$

The results of the first two examples suggest a general procedure for solving trigonometric equations using an algebraic approach.

SUGGESTIONS FOR SOLVING TRIGONOMETRIC EQUATIONS ALGEBRAICALLY

1. Solve for a particular trigonometric function first. If it is not apparent how to do so:
   (A) Try using identities.
   (B) Try algebraic manipulations such as factoring, combining fractions, and so on.
2. Find all solutions within one period of the trigonometric function or functions involved.
3. Write all solutions by adding $2k\pi$, where k is any integer, for sine and cosine equations, or $k\pi$ for tangent equations.

Now we will focus on equations where solving for the trigonometric function is more challenging.

EXAMPLE 3

Exact Solutions Using Factoring

Find all solutions exactly for $2 \sin^2 x + \sin x = 0$.

Solution   *Step 1   Solve for* $\sin x$:

$$2 \sin^2 x + \sin x = 0 \qquad \text{Factor out } \sin x.$$
$$\sin x \, (2 \sin x + 1) = 0 \qquad ab = 0 \text{ only if } a = 0 \text{ or } b = 0.$$
$$\sin x = 0 \quad \text{or} \quad 2 \sin x + 1 = 0$$
$$\sin x = -\tfrac{1}{2}$$

Step 2   Solve each equation over one period $[0, 2\pi)$. Using a unit circle diagram (Fig. 5), we see that there are two solutions for $\sin x = 0$ between 0 and 2π, and two for $\sin x = -1/2$ as well. The four solutions are $x = 0, \pi, 7\pi/6$, and $11\pi/6$.

FIGURE 5

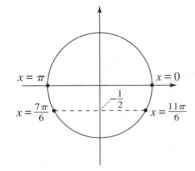

Step 3   *Write an expression for all solutions.* Sine has period 2π, so all solutions are given by

$$x = \begin{cases} 0 + 2k\pi \\ \pi + 2k\pi \\ 7\pi/6 + 2k\pi \\ 11\pi/6 + 2k\pi \end{cases} \quad k \text{ any integer} \qquad ■$$

Matched Problem 3   Find all solutions exactly for $2\cos^2 x - \cos x = 0$.   ■

   **EXAMPLE 4**

Exact Solutions Using Identities and Factoring

Find all solutions exactly for $\sin 2x = \sin x, 0 \le x < 2\pi$.

Solution   *Step 1*   *Solve for* $\sin x$ *and/or* $\cos x$.

$$\sin 2x = \sin x \qquad \text{Use double-angle identity.}$$
$$2\sin x \cos x = \sin x \qquad \text{Algebra}$$
$$2\sin x \cos x - \sin x = 0 \qquad \text{Factor out } \sin x.$$
$$\sin x \,(2\cos x - 1) = 0 \qquad ab = 0 \text{ only if } a = 0 \text{ or } b = 0.$$
$$\sin x = 0 \quad \text{or} \quad 2\cos x - 1 = 0$$
$$\cos x = \frac{1}{2}$$

Step 2   *Solve each equation over one period,* $[0, 2\pi)$.   Note that we are asked only for solutions in the interval $[0, 2\pi)$. Using a unit circle diagram (Fig. 6), we see that there are two solutions for $\sin x = 0$ between 0 and 2π, and two for $\cos x = 1/2$ as well. The four solutions are $x = 0, \pi/3, \pi,$ and $5\pi/3$. (Step 3 is unnecessary.)

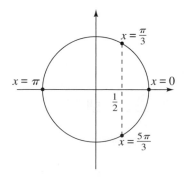

FIGURE 6   ■

Matched Problem 4   Find exact solutions for $\sin^2 x = \dfrac{1}{2} \sin 2x$, $0 \le x < 2\pi$. ∎

 **EXAMPLE 5**

Exact Solutions Using Identities and Factoring

Find all solutions exactly (in degree measure) for $\sin(\theta/2) = \cos\theta - 1$.

Solution   *Step 1   Solve for* $\sin\theta$ *and/or* $\cos\theta$.

$$\sin\frac{\theta}{2} = \cos\theta - 1$$

Use the half-angle identity:
$$\sin\frac{\theta}{2} = \pm\sqrt{\frac{1-\cos\theta}{2}}$$

$$\pm\sqrt{\frac{1-\cos\theta}{2}} = \cos\theta - 1$$

Square both sides.

$$\boxed{\left(\pm\sqrt{\frac{1-\cos\theta}{2}}\right)^2 = (\cos\theta - 1)^2}$$

Recall from algebra, squaring both sides of an equation may introduce **extraneous solutions.**[*]

$$\frac{1-\cos\theta}{2} = \cos^2\theta - 2\cos\theta + 1$$

Algebra

$$1 - \cos\theta = 2\cos^2\theta - 4\cos\theta + 2$$

Algebra

$$2\cos^2\theta - 3\cos\theta + 1 = 0$$

$$(2\cos\theta - 1)(\cos\theta - 1) = 0$$

Factor: $2u^2 - 3u + 1$
$$= (2u - 1)(u - 1)$$

$$2\cos\theta - 1 = 0 \quad \text{or} \quad \cos\theta - 1 = 0$$

$$\cos\theta = \frac{1}{2} \qquad\qquad \cos\theta = 1$$

Step 2   Solve each equation over $[0, 720°)$, *one period of* $\sin(\theta/2)$. Figure 7 shows the locations on the unit circle where $\cos\theta = 1/2$ or $\cos\theta = 1$. These correspond to degree measures of $\theta = 0°$, $60°$, $300°$, $360°$, $420°$, and $660°$ in the first period of $\sin(\theta/2)$.

FIGURE 7

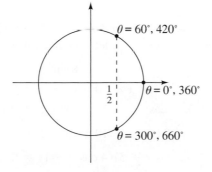

---

[*] In general, an extraneous solution is a solution introduced during the solution process that does not satisfy the original equation.

In the solution process, we squared both sides of an equation. This operation may have introduced extraneous solutions. Checking each solution in the original equation will identify any extraneous solutions:

θ	$0°$	$60°$	$300°$	$360°$	$420°$	$660°$
$\sin \dfrac{\theta}{2}$	0	-0.5	-0.5	0	-0.5	-0.5
$\cos \theta - 1$	0	0.5	0.5	0	-0.5	-0.5

From the table, we see that $60°$ and $300°$ are not solutions of the original equation, while $0°$, $360°$, $420°$, and $660°$ are solutions.

Step 3 *Write an expression for all solutions.* Because the period of $\sin(\theta/2)$ is $720°$, all solutions are given by

$$\theta = \begin{cases} 0° + k(720)° \\ 360° + k(720)° \\ 420° + k(720)° \\ 660° + k(720)° \end{cases} \quad k \text{ any integer.}$$

■

Matched Problem 5 Find all solutions exactly (in degree measure) for $\cos(\theta/2) = \cos \theta$. ■

EXAMPLE 6

Approximate Solutions Using Identities and Factoring

Approximate all real solutions for $8 \sin^2 x = 5 + 10 \cos x$ to four decimal places.

Solution *Step 1* *Solve for* $\sin x$ *and/or* $\cos x$: Move all nonzero terms to the left of the equal sign and express the left side in terms of $\cos x$:

$$8 \sin^2 x = 5 + 10 \cos x$$

$$8 \sin^2 x - 10 \cos x - 5 = 0 \qquad \text{Use } \sin^2 x = 1 - \cos^2 x.$$

$$8(1 - \cos^2 x) - 10 \cos x - 5 = 0 \qquad \text{Distribute and simplify.}$$

$$8 \cos^2 x + 10 \cos x - 3 = 0 \qquad \begin{aligned} &\text{Factor: } 8u^2 + 10u - 3 \\ &= (2u + 3)(4u - 1) \end{aligned}$$

$$(2 \cos x + 3)(4 \cos x - 1) = 0$$

$$2 \cos x + 3 = 0 \quad \text{or} \quad 4 \cos x - 1 = 0 \qquad \text{Solve for } \cos x.$$

$$\cos x = -\tfrac{3}{2} \qquad\qquad \cos x = \tfrac{1}{4}$$

Step 2 *Solve each equation over one period* $[0, 2\pi)$. Figure 8 shows that $\cos x = 1/4$ has solutions in the first and fourth quadrants. It also shows that $\cos x = -3/2$ has no solution (there is no point on the unit circle with first coordinate $-3/2$). We will use inverse cosine on a calculator to find solutions for $\cos x = 1/4$:

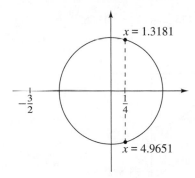

FIGURE 8

$$x = \cos^{-1}\tfrac{1}{4} = 1.3181 \qquad \text{First-quadrant solution}$$

$$x = 2\pi - 1.3181 = 4.9651 \qquad \text{Fourth-quadrant solution}$$

✔ Check $\qquad\qquad \cos 1.3181 = 0.2500 \qquad \cos 4.9651 = 0.2500$

(Checks may not be exact, because of rounding errors.)

Step 3 *Write an expression for all solutions.* Because the cosine function has period 2π, all solutions are given by

$$x = \begin{cases} 1.3181 + 2k\pi \\ 4.9651 + 2k\pi \end{cases} k \text{ any integer} \qquad ■$$

Matched Problem 6 Approximate all real solutions for $3\cos^2 x + 8\sin x = 7$ to four decimal places. ■

EXAMPLE 7

Approximate Solutions Using Identities and the Quadratic Formula

Approximate all real solutions for $\cos 2x = 2(\sin x - 1)$ to four decimal places.

Solution *Step 1* *Solve for* $\sin x$:

$$\cos 2x = 2(\sin x - 1) \qquad \text{Use double-angle identity.}$$

$$1 - 2\sin^2 x = 2\sin x - 2 \qquad \text{Rearrange to get zero on one side.}$$

$$2\sin^2 x + 2\sin x - 3 = 0 \qquad \text{Quadratic in sin x. Left side does}$$

$$\sin x = \frac{-2 \pm \sqrt{4 - 4(2)(-3)}}{4} \qquad \begin{array}{l}\text{not factor using integer coefficients,}\\ \text{so use the quadratic formula.}\end{array}$$

$$= -1.822876 \quad \text{or} \quad 0.822876$$

Step 2   *Solve each equation over one period* $[0, 2\pi)$. Figure 9 shows that $\sin x = 0.822876$ has solutions in the first and second quadrants, while $\sin x = -1.822876$ has no solution. We will use inverse sine on a calculator to find solutions for $\sin x = 0.822876$.

FIGURE 9

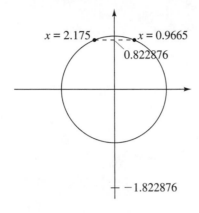

$$x = \sin^{-1} 0.822876 = 0.9665$$
$$x = \pi - 0.9665 = 2.1751$$

✔ Check $\sin 0.9665 = 0.8229$ $\sin 2.1751 = 0.8229$

Step 3   *Write an expression for all solutions.* Because the sine function has period 2π, all solutions are given by

$$x = \begin{cases} 0.9665 + 2k\pi \\ 2.1751 + 2k\pi \end{cases} \quad k \text{ any integer} \qquad ■$$

Matched Problem 7   Approximate all real solutions for $\cos 2x = 4 \cos x - 2$ to four decimal places.   ■

We will close the section with an application problem.

   **EXAMPLE 8**   **Spring–Mass System**

The equation $y = -4 \cos 8t$ represents the motion of a weight hanging on a spring after it has been pulled 4 cm below its equilibrium point and released. (Air resistance and friction are neglected.) The output y tells us how many inches above (positive y values) or below (negative y values) the equilibrium point the weight is after t seconds. Find the first four times when the weight is 2 cm above the equilibrium point.

Solution   The weight is 2 cm above the equilibrium point when $y = 2$, so we are asked to find the first four positive solutions of the equation $-4 \cos 8t = 2$.

$$-4 \cos 8t = 2 \qquad \text{Divide both sides by } -4.$$
$$\cos 8t = -\frac{2}{4} = -\frac{1}{2}$$

Figure 10 shows the first four positive inputs for which the cosine function is $-1/2$. Keep in mind that these are values for $8t$ (the input of cosine in this equation), not t!

So we have $8t = 2\pi/3, 4\pi/3, 8\pi/3$, and $10\pi/3$. Dividing by 8 to find t, we get $t = \pi/12, \pi/6, \pi/3$, and $5\pi/12$ sec. To two decimal places, these times are 0.26, 0.52, 1.05, and 1.31 sec.

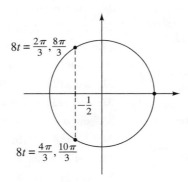

$$8t = \frac{2\pi}{3}, \frac{8\pi}{3}$$

$$-\frac{1}{2}$$

$$8t = \frac{4\pi}{3}, \frac{10\pi}{3}$$

FIGURE 10

Matched Problem 8   Find the first four times when the weight in Example 8 is at its highest point.

Answers to Matched Problems

1. $x = \begin{cases} \pi/6 + 2k\pi \\ 5\pi/6 + 2k\pi \end{cases}$ k any integer

2. (A) $x = \begin{cases} 2.21 + 2k\pi \\ 4.07 + 2k\pi \end{cases}$ k any integer

   (B) $x = \dfrac{3\pi}{4} + k\pi$, k any integer

3. $x = \begin{cases} \pi/3 + 2k\pi \\ \pi/2 + 2k\pi \\ 3\pi/2 + 2k\pi \\ 5\pi/3 + 2k\pi \end{cases}$ k any integer

4. $x = 0, \pi/4, \pi, 5\pi/4$

5. $x = \begin{cases} 0° + k(720°) \\ 120° + k(720°) \\ 360° + k(720°) \\ 480° + k(720°) \end{cases}$ k any integer

6. $x = \begin{cases} 0.7297 + 2k\pi \\ 2.4119 + 2k\pi \end{cases}$ k any integer

7. $x = \begin{cases} 1.2735 + 2k\pi \\ 5.0096 + 2k\pi \end{cases}$ k any integer

8. $\pi/8, 3\pi/8, 5\pi/8$, and $7\pi/8$ sec, or 0.39, 1.18, 1.96, 2.75 sec.

EXERCISE 5.3

A

1. Use a unit circle diagram to explain why the equations $\sin x = 1/2$ and $\cos x = 1/2$ have infinitely many solutions.

2. Use the graphs of $y = \sin x$ and $y = \cos x$ to explain why the equations $\sin x = 1/2$ and $\cos x = 1/2$ have infinitely many solutions.

3. Use a unit circle diagram to explain why the equations $\sin x = 5/4$ and $\cos x = -5/4$ have no solution.

4. Use the graphs of $y = \tan x$ and $y = \sin x$ to explain why $\tan x = 5$ has solutions, while $\sin x = 5$ does not.

In Problems 5–20, find exact solutions over the indicated intervals (x real and θ in degrees).

5. $\cos x = \sqrt{3}/2$, all real x

6. $\sin x = -1/\sqrt{2}$, all real x

7. $\tan x = 1$, all real x

8. $\tan x = 0$, all real x

9. $\sin x = -1$, all real x

10. $\cos x = 1$, all real x

11. $2 \cos x + 1 = 0$, $0 \le x < 2\pi$

12. $2 \sin x + 1 = 0$, $0 \le x < 2\pi$

13. $2 \cos x + 1 = 0$,   all real x

14. $2 \sin x + 1 = 0$,   all real x

15. $\sqrt{2} \sin \theta - 1 = 0$, $0° \le \theta < 360°$

16. $2 \cos \theta - \sqrt{3} = 0$, $0° \le \theta < 360°$

17. $\sqrt{2} \sin \theta - 1 = 0$,   all θ

18. $2 \cos \theta - \sqrt{3} = 0$,   all θ

19. $3 \cos x - 6 = 0$, all real x

20. $\sqrt{5} \sin x + 20 = 0$, all real x

In Problems 21–26, solve each to four decimal places (x real and θ in degrees).

21. $4 \tan \theta + 15 = 0$, $0° \le \theta < 180°$

22. $2 \tan \theta - 7 = 0$, $0° \le \theta < 180°$

23. $5 \cos x - 2 = 0$, $0 \le x < 2\pi$

24. $7 \cos x - 3 = 0$, $0 \le x < 2\pi$

25. $5.0118 \sin x - 3.1105 = 0$,   all real x

26. $1.3224 \sin x + 0.4732 = 0$,   all real x

B *For Problems 27–46, find exact solutions (x real and θ in degrees).*

27. $\cos x = \cot x$, $0 \le x < 2\pi$

28. $\tan x = -2 \sin x$, $0 \le x < 2\pi$

29. $\cos^2 \theta = \frac{1}{2} \sin 2\theta$,   all θ

30. $2 \sin^2 \theta + \sin 2\theta = 0$,   all θ

31. $\tan(x/2) - 1 = 0$, $0 \le x < 2\pi$

32. $\sec(x/2) + 2 = 0$, $0 \le x < 2\pi$

33. $\sin^2 \theta + 2 \cos \theta = -2$, $0° \le \theta < 360°$

34. $2 \cos^2 \theta + 3 \sin \theta = 0$, $0° \le \theta < 360°$

35. $\cos 2\theta + \sin^2 \theta = 0$, $0° \le \theta < 360°$

36. $\cos 2\theta + \cos \theta = 0$, $0° \le \theta < 360°$

37. $\cos 2x = \sin x$, all real x

38. $\sin 2x = 2 \sin^2 x$, all real x

39. $\cos 2x = \cos x - 1$, all real x

40. $\sin 2x = 1 + \cos 2x$, all real x

41. $2 \cos(\theta/2) = \cos \theta + 1$, all θ (in degrees)

42. $\cos(\theta/2) = \cos \theta + 1$, all θ (in degrees)

43. $\sin(\theta/2) = \cos \theta$, $0° \le \theta < 720°$

44. $2 \sin(\theta/2) = \cos \theta - 1$, $0° \le \theta < 720°$

45. $2\sqrt{2} \cos \dfrac{x}{2} = \cos x + 2$, all real x

46. $\tan \dfrac{x}{2} = 1 - \cos x$, all real x

Solve Problems 47–50 (x real and θ in degrees) to four significant digits.

47. $4 \cos^2 \theta = 7 \cos \theta + 2$, $0° \le \theta \le 180°$

48. $6 \sin^2 \theta + 5 \sin \theta = 6$, $0° \le \theta \le 90°$

49. $\cos 2x + 10 \cos x = 5$, $0 \le x < 2\pi$

50. $2 \sin x = \cos 2x$, $0 \le x < 2\pi$

Solve Problems 51–56 for all real solutions to four significant digits.

51. $\cos^2 x = 3 - 5 \cos x$

52. $2 \sin^2 x = 1 - 2 \sin x$

53. $\sin^2 x = -2 + 3 \cos x$

54. $\cos^2 x = 3 + 4 \sin x$

55. $\cos(2x) = 2 \cos x$

56. $\cos(2x) = 2 \sin x$

57. Explain the difference between evaluating the expression $\cos^{-1}(-0.7334)$ and solving the equation $\cos x = -0.7334$.

58. Explain the difference between evaluating the expression $\tan^{-1}(-5.377)$ and solving the equation $\tan x = -5.377$.

C *Find exact solutions to Problems 59–62. [Hint: Square both sides at an appropriate point, solve, then eliminate any extraneous solutions at the end.]*

59. $\sin x + \cos x = 1, \quad 0 \le x < 2\pi$

60. $\cos x - \sin x = 1, \quad 0 \le x < 2\pi$

61. $\sec x + \tan x = 1, \quad 0 \le x < 2\pi$

62. $\tan x - \sec x = 1, \quad 0 \le x < 2\pi$

 Applications

63. Spring–Mass System The equation $y = -10 \cos 3t$ represents the motion of a weight hanging on a spring after it has been pulled 10 in. below its equilibrium point and released. (Air resistance and friction are neglected.) The output y gives the position of the weight in inches above (positive y values) or below (negative y values) the equilibrium point after t seconds. Find the first four times when the weight is at its highest point.

64. Spring–Mass System Refer to Problem 63. Find the first four times when the weight is 5 in. above its equilibrium point.

65. Electric Current An alternating current generator produces a current given by the equation

$$I = 30 \sin 120\pi t$$

where t is time in seconds and I is current in amperes. Find the least positive t (to four significant digits) for which $I = 25$ amperes.

66. Electric Current Find the least positive t in Problem 65 (to four significant digits) for which $I = -10$ amperes.

67. Rotary Motion A ferris wheel with a diameter of 60 ft rotates counterclockwise at 5 revolutions per minute. At its lowest point, each seat is 3 ft from the ground. For a rider starting at the lowest point, his height in feet above the ground after t minutes is given by $y = 63 - 60 \cos(10\pi t)$. Find all times in the first half minute when the rider is 80 ft above the ground. Round to the nearest tenth of a second.

68. Rotary Motion For the ferris wheel in Problem 67, find all times in the first minute when the rider is at the ferris wheel's highest point.

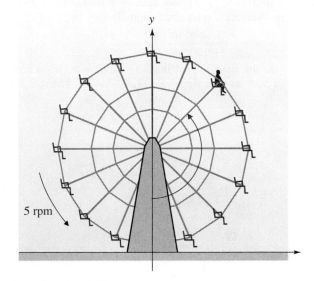

Figure for 67 and 68

69. Photography A polarizing filter for a camera contains two parallel plates of polarizing glass, one fixed and the other able to rotate. If θ is the angle of rotation from the position of maximum light transmission, then the intensity of light leaving the filter is $\cos^2 \theta$ times the intensity entering the filter (see the figure). Find the least positive θ (in decimal degrees, to two decimal places) so that the intensity of light leaving the filter is 70% of that entering. [*Hint:* Solve $I \cos^2 \theta = 0.70I$.]

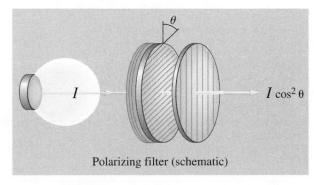

Polarizing filter (schematic)

Figure for 69 and 70

70. Photography Find θ in Problem 69 so that the light leaving the filter is 40% of that entering.

71. Astronomy   The planet Mercury travels around the sun in an elliptical orbit given approximately by

$$r = \frac{3.44 \times 10^7}{1 - 0.206 \cos \theta}$$

(see the figure). Find the least positive θ (in decimal degrees, to three significant digits) for which Mercury is 3.78×10^7 mi from the sun.

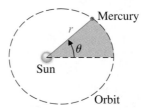

Figure for 71 and 72

72. Astronomy   Find the least positive θ (in decimal degrees, to three significant digits) in Problem 71 for which Mercury is 3.09×10^7 mi from the sun.

Precalculus   *In Problems 73 and 74, find simultaneous solutions for each system of equations for $0° \le \theta \le 360°$. These are polar equations, which will be discussed in Chapter 7.*

73. $r = 2 \sin \theta$
  $r = 2(1 - \sin \theta)$

74. $r = 2 \sin \theta$
  $r = \sin 2\theta$

75. Precalculus   Given the equation $xy = -2$, replace x and y with

$$x = u \cos \theta - v \sin \theta$$
$$y = u \sin \theta + v \cos \theta$$

and simplify the left side of the resulting equation. Find the least positive θ (in degree measure) so that the coefficient of the uv term will be 0.

76. Precalculus   Repeat Problem 75 for the equation $2xy = 1$.

77. Modeling Hours of Daylight   Use the data in Table 1 to construct a model for the hours of daylight for Columbus, OH, of the form $y = k + A \cos(Bx)$, where x is time in months. Let $x = 1$ represent January 15. (Don't forget to convert hours and minutes to decimal hours before constructing the model.) When does Columbus have exactly 12 hours of daylight? Round months to one decimal place.

78. Modeling Hours of Daylight   Use the data in Table 1 to construct a model for the hours of daylight for New Orleans, LA, of the form $y = k + A \cos(Bx)$, where x is time in months. Let $x = 1$ represent January 15. (Don't forget to convert hours and minutes to decimal hours before constructing the model.) When does New Orleans have exactly 12 hours of daylight? Round months to one decimal place.

79. Modeling Hours of Daylight   Refer to Problem 77. Use sinusoidal regression to find a model for the hours of daylight in Columbus. When does Columbus have exactly 12 hours of daylight? Round months to one decimal place.

80. Modeling Hours of Daylight   Refer to Problem 78. Use sinusoidal regression to find a model for the hours of daylight in New Orleans. When does New Orleans have exactly 12 hours of daylight? Round months to one decimal place.

TABLE 1												
Hours of Daylight for Two Cities (in Hours and Minutes on the 15th of Each Month)												
Month	Jan	Feb	Mar	Apr	May	Jun	Jul	Aug	Sep	Oct	Nov	Dec
Columbus	9:37	10:42	11:53	13:17	14:24	15:00	14:49	13:48	12:31	11:10	10:01	9:20
New Orleans	10:24	11:10	11:57	12:53	13:40	14:04	13:56	13:16	12:23	11:28	10:40	10:14

 5.4 Trigonometric Equations and Inequalities: A Graphing Calculator Approach

- Solving Trigonometric Equations Using a Graphing Calculator
- Solving Trigonometric Inequalities Using a Graphing Calculator

All of the trigonometric equations that were solved in the last section using an algebraic approach can also be solved using graphing calculator methods. This is not to say that algebraic methods are not useful. In most cases, graphing calculator methods don't provide exact solutions, and it can require interpretation to find all real solutions when there are infinitely many. However, many trigonometric equations that cannot be solved easily using an algebraic approach can be solved (to any accuracy desired) using graphing calculator methods.

Consider the simple-looking equation

$$2x \cos x = 1$$

Try solving this equation using any of the methods discussed in Section 5.3. You will not be able to isolate x on one side of the equation with a number on the other side. Such equations are easily solved using a graphing calculator, as will be seen in the examples that follow.

Solving trigonometric inequalities, such as

$$\cos x - 2 \sin x > 0.4 - 0.3x$$

using the numerical approximation routines implemented on most graphing calculators is almost as easy as solving trigonometric equations with a graphing calculator. The second part of this section illustrates the process.

■ Solving Trigonometric Equations Using a Graphing Calculator

The best way to proceed is by working through some examples.

EXAMPLE 1

Solutions Using a Graphing Calculator

Find all real solutions (to four decimal places) for $\sin(x/2) = 0.2x - 0.5$.

Solution   We start by graphing $y_1 = \sin(x/2)$ and $y_2 = 0.2x - 0.5$ in the same viewing window for $-4\pi \le x \le 4\pi$ (Fig. 1a on page 380). The figure shows that there is one intersection point between $x = \pi$ and $x = 2\pi$, and a possible intersection near $x = -\pi$. Since y_1 is always between -1 and 1, while y_2 is greater than 1 or less than -1 for x outside the window in Figure 1b ($y_2 = 0.2x - 0.5$ is a line, so the graph can't change direction), the two curves do not intersect anywhere outside this viewing window. Using the $\boxed{\text{INTERSECT}}$ command we find that $x = 5.1609$ is a solution to the original equation (Fig. 1b).

Next, we move the cursor near the suspected intersection in the third quadrant, ZOOM in, and graph (Fig. 1c). Now we can see that there is no intersection point in the third quadrant, and $x = 5.1609$ is the only solution of the original equation.

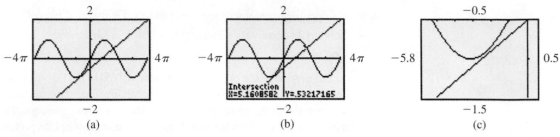

FIGURE 1

Matched Problem 1   Find all real solutions (to four decimal places) for $2 \cos 2x = 1.35x - 2$.

   **EXAMPLE 2**   ### Solutions Using a Graphing Calculator

A rectangle is inscribed under the graph of $y = 2 \cos(\pi x/2)$, $-1 \le x \le 1$, with the base on the x axis, as shown in Figure 2.

FIGURE 2

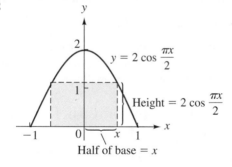

(A) Write an equation for the area A of the rectangle in terms of x.
(B) Graph the equation found in part (A) in a graphing calculator, and use TRACE to describe how the area A changes as x goes from 0 to 1.
(C) Find the value(s) of x (to three decimal places) that produce(s) a rectangle area of 1 square unit.

Solution   (A) $A(x) = (\text{Length of base}) \times (\text{Height})$

$$= (2x) \times \left(2 \cos \frac{\pi x}{2} \right)$$

$$= 4x \cos \frac{\pi x}{2} \qquad 0 \le x \le 1$$

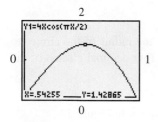

FIGURE 3

(B) The graph of the area function is shown in Figure 3. Referring to the figure, we see that the area A increases from 0 to a maximum of about 1.429 square units, then decreases to 0, as x goes from 0 to 1.

(C) To find x for a rectangle area of 1 square unit, we must solve the equation

$$1 = 4x \cos \frac{\pi x}{2}$$

This equation cannot be solved by methods discussed in Section 5.3, but it is easily solved using a graphing calculator. Graph $y_1 = 1$ and $y_2 = 4x \cos(\pi x/2)$ in the same viewing window, and find the point(s) of intersection of the two graphs using the $\boxed{\text{INTERSECT}}$ command (see Fig. 4). Note that the graph of y_1 intersects the graph of y_2 in two points; therefore, there are two solutions. From Figure 4 we can see that the area is 1 for $x = 0.275$ and for $x = 0.797$.

FIGURE 4

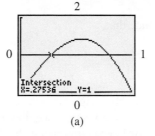

(a)

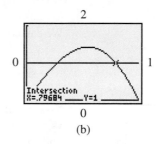

(b)

✔ Check   Substitute $x = 0.275$ and $x = 0.797$ into $A(x) = 4x \cos(\pi x/2)$ to see if the result is 1 (or nearly 1):

```
4*0.275cos(π0.27
5/2)
             .99896
4*0.797cos(π0.79
7/2)
             .99942
```

■

Matched Problem 2   Repeat Example 2 to find the value(s) of x (to three decimal places) that produce(s) a rectangle area of 1.25 square units.   ■

EXPLORE/DISCUSS 1

A 12 cm arc of a circle has a 10 cm chord. A 16 cm arc of a second circle also has a 10 cm chord. For parts (A) and (B), reason geometrically by drawing larger and smaller circles.

(A) Which circle has the larger radius?

(B) Is there a *largest* circle that has a chord of 10 cm? Is there a *smallest* circle that has a chord of 10 cm? Explain.

EXAMPLE 3   **An Application in Geometry**

A 12 cm arc on a circle has a 10 cm chord. What is the radius of the circle (to four decimal places)? What is the radian measure (to four decimal places) of the central angle subtended by the arc?

Solution   Sketch a figure and introduce the auxiliary lines, as shown in Figure 5.

Recall from Section 2.1 that the radian measure of an angle is the length of the arc it subtends divided by the radius. From the figure (θ in radians), we see that

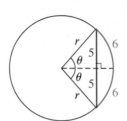

$$\theta = \frac{6}{r} \quad \frac{\text{Arc length}}{\text{Radius}} \qquad \text{and} \qquad \sin\theta = \frac{5}{r} \quad \frac{\text{Opposite}}{\text{Hypotenuse}}$$

We can conclude that

$$\sin\frac{6}{r} = \frac{5}{r}$$

FIGURE 5

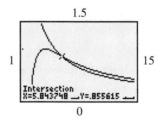

FIGURE 6

Now we'll solve this trigonometric equation for r. Graph $y_1 = \sin(6/x)$ and $y_2 = 5/x$ in the same viewing window for $1 \le x \le 15$ and $0 \le y \le 1.5$; then find the point of intersection using the $\boxed{\text{INTERSECT}}$ command (Fig. 6). From Figure 6, we see that $r = 5.8437$ cm.

The radian measure of the central angle subtended by the 12 cm chord is

$$2\theta = \frac{12}{5.8437} = 2.0535 \text{ rad} \qquad \blacksquare$$

Matched Problem 3   A 10 ft arc on a circle has an 8 ft chord. What is the radius of the circle (to four decimal places)? What is the radian measure (to four decimal places) of the central angle subtended by the arc?   ■

EXPLORE/DISCUSS 2

Solve the equation $\sin x = 0.9x$, $-2 \le x \le 2$, two ways:

(A) Find the points of intersection of $y_1 = \sin x$ and $y_2 = 0.9x$.

(B) Find the zeros of $y_3 = y_1 - y_2$.

(C) Discuss the advantages and disadvantages of each method. Which do you prefer?

EXAMPLE 4

Solution Using a Graphing Calculator

Find all real solutions (to four decimal places) for $\tan(x/2) = 5x - x^2$, $0 \leq x \leq 3\pi$.

Solution   Graph $y_1 = \tan(x/2)$ and $y_2 = 5x - x^2$ in the same viewing window for $0 \leq x \leq 3\pi$ and $-10 \leq y \leq 10$. Solutions occur at the three points of intersection shown in Figure 7.

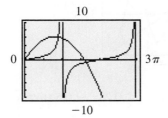

FIGURE 7

Using the $\boxed{\text{INTERSECT}}$ command, the three solutions are found to be $x = 0.0000, 2.8191, 5.1272$. ∎

Matched Problem 4   Find all real solutions (to four decimal places) for $\tan(x/2) = 1/x$, $-\pi < x \leq 3\pi$. ∎

EXPLORE/DISCUSS 3

For each of the following equations, determine the number of real solutions. Is there a largest solution? A smallest solution? Explain.

(A) $\sin x = x/2$ (B) $\tan x = x/2$

■ Solving Trigonometric Inequalities Using a Graphing Calculator

Solving trigonometric inequalities using a graphing calculator is almost as easy as solving trigonometric equations using a graphing calculator. Example 5 illustrates the process.

EXAMPLE 5

Solving a Trigonometric Inequality

Solve $\cos x - 2 \sin x > 0.4 - 0.4x$ (to two decimal places).

Solution Graph $y_1 = \cos x - 2 \sin x$ and $y_2 = 0.4 - 0.4x$ in the same viewing window
(Fig. 8). Finding the three points of intersection using the $\boxed{\text{INTERSECT}}$ command,
we see that the graph of y_1 is above the graph of y_2 for the following two intervals:
$(-2.09, 0.35)$ and $(3.20, \infty)$. Formally, the solution set for the inequality is
$(-2.09, 0.35) \cup (3.20, \infty)$

FIGURE 8

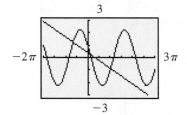

■

Matched Problem 5 Solve $\sin x - \cos x < 0.25x - 0.5$ (to two decimal places). ■

 **EXAMPLE 6** **An Application of Trigonometric Inequalities**

In Example 5 of Section 3.3, we developed a function to model the phases of the
moon: $y = 50 + 50 \cos(0.213x + 0.924)$. In this model, y is the phase of the
moon from 0 (new moon) to 100 (full moon) in terms of x, the number of days
after October 1, 2007. Use this model to find the first stretch of days in 2009 when
more than 70% of the moon is visible.

Solution We are asked to solve the inequality

$$50 + 50 \cos(0.213x + 0.924) > 70$$

First, we need to find an appropriate viewing window for the time period we're interested in. The beginning of 2008 is 91 days after October 1, 2007 [30 days in October
(after October 1) + 30 in November + 31 in December]. There were 366 days
in 2008, so January 1, 2009, is $366 + 91 = 457$ days after October 1, 2007.
  We will graph $y_1 = 50 + 50 \cos(0.213x + 0.924)$ and $y_2 = 70$ on the interval $457 \le x \le 488$, which represents January 2009 (Fig. 9). The graph of y_1 is
above height 70 between $x = 462$ and $x = 473$ (to the nearest day). This corresponds to January 6 through January 17, 2009.

FIGURE 9

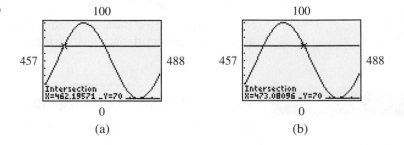

(a)   (b)

■

Matched Problem 6   On how many days in February 2009 is less than 30% of the moon visible?   ∎

Answers to Matched Problems

1. 0.9639
2. 0.376, 0.710
3. $r = 4.4205$ ft, $\theta = 2.2622$ rad
4. $-1.3065, 1.3065, 6.5846$
5. $(-1.65, 0.52) \cup (3.63, \infty)$
6. 11 days

EXERCISE 5.4

Unless stated to the contrary, all the problems in this exercise require the use of a graphing calculator.

A 1. Discuss the advantages and disadvantages of solving trigonometric equations using a graphing calculator as opposed to algebraically.

2. Explain the differences between solving a trigonometric equation and solving a trigonometric inequality using a graphing calculator.

Solve Problems 3–10 (to four decimal places).

3. $2x = \cos x$, all real x
4. $2 \sin x = 1 - x$, all real x
5. $3x + 1 = \tan 2x$, $0 \le x < \pi/4$
6. $8 - x = \tan(x/2)$, $0 \le x < \pi$
7. $\cos x < \sin x$, $0 \le x < 2\pi$
8. $\tan x \ge \cos x$, $0 \le x < 2\pi$
9. $x \ge \cos x$, all real x
10. $\sin x < x^2$, all real x

B *Solve Problems 11–18 (to four decimal places).*

11. $\cos 2x + 10 \cos x = 5$, $0 \le x < 2\pi$
12. $2 \sin x = \cos 2x$, $0 \le x < 2\pi$
13. $\cos^2 x = 3 - 5 \cos x$, all real x
14. $2 \sin^2 x = 1 - 2 \sin x$, all real x
15. $2 \sin(x - 2) < 3 - x^2$, all real x
16. $\cos 2x > x^2 - 2$, all real x

17. $\sin(3 - 2x) \ge 1 - 0.4x$, all real x
18. $\cos(2x + 1) \le 0.5x - 2$, all real x
19. Without graphing, explain why the following inequality is true for all real x:

$$\sin^2 x - 2 \sin x + 1 \ge 0$$

20. Without graphing, explain why the following inequality is false for all real x for which the left side of the statement is defined:

$$\tan^2 x - 4 \tan x + 4 < 0$$

C *Solve Problems 21–28 exactly if possible, or to four significant digits if an exact solution cannot be found.*

21. $2 \cos(1/x) = 950x - 4$, $0.006 < x < 0.007$
22. $\sin(1/x) = 1.5 - 5x$, $0.04 \le x \le 0.2$
23. $\cos(\sin x) > \sin(\cos x)$, $-2\pi \le x \le 2\pi$
24. $1 + \tan^2 x \ge \cos^2(-x)$, $-2\pi \le x \le 2\pi$
25. $\tan^{-1} x = 0.5x$, all real x
26. $x = \sec x$, $-2\pi \le x \le 2\pi$
27. $\sin^2 x > \sin(x^2)$, $0 \le x \le \pi$
28. $\cos(x^2) < 0.5 x$, $-2\pi \le x \le 2\pi$

29. Find subintervals (to four decimal places) of the interval $[0, \pi]$ over which $\sqrt{3 \sin(2x) - 2 \cos(x/2) + x}$ is a real number. (Check the end points of each subinterval.)

30. Find subintervals (to four decimal places) of the interval $[0, 2\pi]$ over which $\sqrt{2 \sin(2x) - \cos^2 x + 0.5x}$ is a real number. (Check the end points of each subinterval.)

Applications

31. Geometry The area of a segment of the circle in the figure is given by

$$A = \tfrac{1}{2}r^2(\theta - \sin\theta)$$

Find the angle θ subtended by the segment if the radius is 10 m and the area is 40 m^2. Compute the solution in radians to two decimal places.

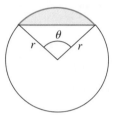

Figure for 31 and 32

32. Geometry Repeat Problem 31 if the radius is 8 m and the area is 48 m^2.

33. Seasonal Business Cycles A producer of ice cream experiences regular seasonal fluctuation in sales but is also experiencing growth in sales overall. An economic consultant uses sales data and regression analysis to model sales with the function $y = 12 + x/8 - 9\cos(\pi x/26)$, where y is weekly sales in thousands of gallons x weeks after January 1 of the current year. When will the company surpass 27,000 gallons in weekly sales for the first time?

34. Seasonal Business Cycles A competitor of the company in Problem 33 experiences similar seasonal fluctuations, but has been losing market share. Its weekly sales in thousands of gallons can be modeled by $y = 26.5 - x/12 - 7\cos(\pi x/26)$, where x is the number of weeks after January 1 of the current year. When will sales for this company drop below sales for the company in Problem 33?

35. Architecture An arched doorway is formed by placing a circular arc on top of a rectangle (see the figure). If the rectangle is 4 ft wide and 8 ft high, and the circular arc is 5 ft long, what is the area of the doorway? Compute the answer to two decimal places.

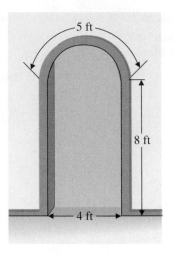

Figure for 35

36. Architecture Repeat Problem 35 if the rectangle is 3 ft wide and 7 ft high, and the circular arc is 4 ft long.

37. Eye Surgery A surgical technique for correcting an astigmatism involves removing small pieces of tissue in order to change the curvature of the cornea.* In the cross section of a cornea shown in part (b) of the figure, the circular arc with radius r and central angle 2θ represents the surface of the cornea.

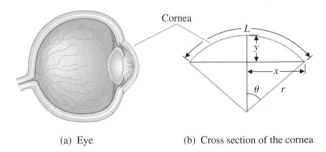

(a) Eye   (b) Cross section of the cornea

Figure for 37 and 38

(A) If $x = 5.5$ mm and $y = 2.5$ mm, find L, the length of the corneal cross section, correct to four decimal places.

* Based on the article "The Surgical Correction of Astigmatism" by Sheldon Rothman and Helen Strassberg in *The UMAP Journal,* Vol. V, No. 2 (1984).

(B) Reducing the width of the cross section without changing its length has the effect of pushing the cornea outward and giving it a rounder, yet still circular, shape. Approximate y to four decimal places if x is reduced to 5.4 mm and L remains the same as in part (A).

38. Eye Surgery Refer to Problem 37. Increasing the width of a cross section of the cornea without changing its length has the effect of pushing the cornea inward and giving it a flatter, yet still circular, shape. Approximate y to four decimal places if x is increased to 5.6 mm and L remains the same as in part (A) of Problem 37.

39. Golf When an object is launched with an initial velocity v_0 (in feet per second) at an angle of α relative to the ground, its horizontal distance traveled (neglecting air resistance) can be modeled by the equation

$$d = \frac{v_0^2}{16} \sin \alpha \cos \alpha$$

Suppose a golfer can hit a ball with an initial velocity of 130 ft per second, and she is aiming for a green that is 160 yd away.

(A) At what angle should she be trying to hit the ball (to the nearest tenth of a degree)?

(B) Would she be able to hit a green that is 200 yd away? Explain.

40. Baseball When making a throw to first base, a third baseman is typically about 130 ft from first base.

(A) According to the equation in Problem 39, find the smallest angle at which a third basemen, who throws with an initial velocity of 120 ft per second (82 mph), should aim to make the ball reach the first baseman at the throwing height. Round to the nearest tenth of a degree.

(B) If the ball were thrown at that angle and not subject to gravity, so that it traveled in a straight line, how far over the first baseman's head would it go?

41. Modeling Climate Use the high-temperature data in Table 1 to construct a model of the form $y = k + A \sin(Bx + C)$, where x is time in months. Graph the points in the table and your model to estimate C. When is the high temperature greater than 70°? Round months to one decimal place.

42. Modeling Climate Use the low-temperature data in Table 1 to construct a model of the form $y = k + A \sin(Bx + C)$, where x is time in months. Graph the points in the table and your model to estimate C. When is the low temperature less than 34°? Round months to one decimal place.

43. Modeling Climate Refer to Problem 41. Use sinusoidal regression to find a model for the monthly high temperatures in Pittsburgh. When is the high temperature less than 50°? Round months to one decimal place.

44. Modeling Climate Refer to Problem 42. Use sinusoidal regression to find a model for the monthly low temperatures in Pittsburgh. When is the low temperature greater than 30°? Round months to one decimal place.

TABLE 1

Monthly Average High and Low Temperatures (in °F) for Pittsburgh, PA

	Jan	Feb	Mar	Apr	May	Jun	Jul	Aug	Sep	Oct	Nov	Dec
High	33.6	36.9	48.9	60.3	70.5	78.8	82.6	80.8	74.3	62.4	50.4	38.5
Low	13.6	17.9	24.8	33.3	42.1	51.3	55.8	54.7	48.0	36.3	30.0	21.0

CHAPTER 5 GROUP ACTIVITY

 $$\sin \frac{1}{x} = 0 \text{ and } \sin^{-1}\frac{1}{x} = 0$$

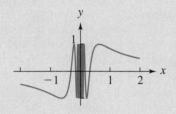

I. EXPLORATION OF SOLUTIONS TO $\sin \dfrac{1}{x} = 0,\ x > 0$

In this activity, we will explore solutions to the equation $\sin(1/x) = 0$, which are the zeros of the function $f(x) = \sin(1/x)$, for $x > 0$. Note that the function f is not defined at $x = 0$. We will restrict our analysis to positive zeros; a similar analysis can be made for negative zeros.

(A) *An overview of the problem.* Graph $f(x) = \sin(1/x)$ for $(0, 1]$. Discuss your first observations regarding the zeros for f.

(B) *Exploring zeros of f on the interval* $[0.1, b],\ b > 0.1$. Graph f over the interval $[0.1, b]$ for various values of b, $b > 0.1$. Does the function f have a largest zero? If so, what is it (to four decimal places)? Explain what happens to the graph of f as x increases without bound. Does the graph appear to have an asymptote? If so, what is its equation?

(C) *Exploring zeros of f near the origin.* Graph f over the intervals $[0.05, 0.1]$, $[0.025, 0.05]$, $[0.0125, 0.025]$, and so on. How many zeros exist between 0 and b, for any positive b, however small? Explain why this happens. Does f have a smallest positive zero? Explain.

II. EXPLORATION OF SOLUTIONS TO $\sin^{-1} \dfrac{1}{x} = 0,\ x > 0$

Next, we will explore solutions to the equation $\sin^{-1}(1/x) = 0$, which are the zeros of the function $g(x) = \sin^{-1}(1/x)$, for $x > 0$. Note that the function g is not defined at $x = 0$.

(A) *An overview of the problem.* Graph $g(x) = \sin^{-1}(1/x)$ over the interval $(0, 5]$. Discuss your first observations regarding the zeros for g.

(B) *The interval* $(0, 1]$. Explain why there is no graph on the interval $(0, 1]$.

(C) *The interval* $[1, \infty)$. Explain why there are no zeros for g on the interval $[1, \infty)$ and, therefore, the equation $\sin^{-1}(1/x) = 0$, $x > 0$, has no solutions. Does the graph have an asymptote? If so, what is its equation?

CHAPTER 5 REVIEW

5.1
INVERSE SINE,
COSINE, AND
TANGENT FUNCTIONS

Definition of Inverse Trigonometric Functions

$y = \sin^{-1} x$ is equivalent to $x = \sin y$
where $-1 \le x \le 1$ and $-\pi/2 \le y \le \pi/2$

$y = \cos^{-1} x$ is equivalent to $x = \cos y$
where $-1 \le x \le 1$ and $0 \le y \le \pi$

$y = \tan^{-1} x$ is equivalent to $x = \tan y$
where x is any real number and $-\pi/2 < y < \pi/2$

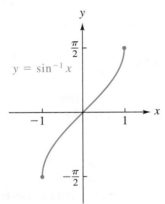

Domain: $-1 \le x \le 1$
Range: $-\frac{\pi}{2} \le y \le \frac{\pi}{2}$

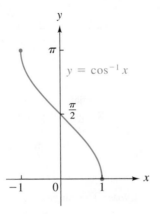

Domain: $-1 \le x \le 1$
Range: $0 \le y \le \pi$

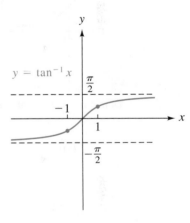

Domain: All real numbers
Range: $-\frac{\pi}{2} < y < \frac{\pi}{2}$

The inverse sine, cosine, and tangent functions are also denoted by arcsin x, arccos x, and arctan x, respectively.

Inverse Sine, Cosine, and Tangent Identities

$$\sin(\sin^{-1} x) = x \qquad -1 \le x \le 1$$

$$\sin^{-1}(\sin x) = x \qquad -\pi/2 \le x \le \pi/2$$

$$\cos(\cos^{-1} x) = x \qquad -1 \le x \le 1$$

$$\cos^{-1}(\cos x) = x \qquad 0 \le x \le \pi$$

$$\tan(\tan^{-1} x) = x \qquad \text{for all } x$$

$$\tan^{-1}(\tan x) = x \qquad -\pi/2 < x < \pi/2$$

☆ **5.2**

INVERSE COTANGENT, SECANT, AND COSECANT FUNCTIONS

Definitions of Inverse Trigonometric Functions

$y = \cot^{-1} x$ is equivalent to $x = \cot y$

where $0 < y < \pi$ and x is any real number

$y = \sec^{-1} x$ is equivalent to $x = \sec y$

where $0 \leq y \leq \pi$, $y \neq \pi/2$, and $x \leq -1$ or $x \geq 1$

$y = \csc^{-1} x$ is equivalent to $x = \csc y$

where $-\pi/2 \leq y \leq \pi/2$, $y \neq 0$, and $x \leq -1$ or $x \geq 1$

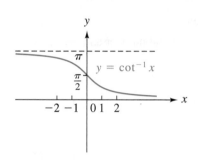

Domain: All real numbers
Range: $0 < y < \pi$

(a)

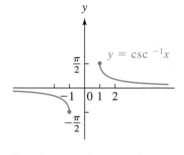

Domain: $x \leq -1$ or $x \geq 1$
Range: $0 \leq y \leq \pi$, $y \neq \dfrac{\pi}{2}$

(b)

Domain: $x \leq -1$ or $x \geq 1$
Range: $-\dfrac{\pi}{2} \leq y \leq \dfrac{\pi}{2}$, $y \neq 0$

(c)

[*Note:* The ranges for $\sec^{-1}$ and $\csc^{-1}$ are sometimes selected differently.]

Inverse Cotangent, Secant, and Cosecant Identities

$$\cot^{-1} x = \begin{cases} \tan^{-1} \dfrac{1}{x} & x > 0 \\[2mm] \pi + \tan^{-1} \dfrac{1}{x} & x < 0 \end{cases}$$

$$\sec^{-1} x = \cos^{-1} \dfrac{1}{x} \qquad x \geq 1 \quad \text{or} \quad x \leq -1$$

$$\csc^{-1} x = \sin^{-1} \dfrac{1}{x} \qquad x \geq 1 \quad \text{or} \quad x \leq -1$$

5.3

TRIGONOMETRIC EQUATIONS: AN ALGEBRAIC APPROACH

An equation that may be true for some replacements of the variable, but is false for others for which both sides are defined, is called a **conditional equation**. An algebraic approach to solving trigonometric equations can yield exact solutions, and the approach may be aided by the following:

Suggestions for Solving Trigonometric Equations Algebraically

1. Solve for a particular trigonometric function first. If it is not apparent how to do so:

   (A) Try using identities.

   (B) Try algebraic manipulations such as factoring, combining fractions, and so on.

2. Find all solutions within one period of the trigonometric function or functions involved.

3. Write all solutions by adding $2k\pi$, where k is any integer, for sine and cosine equations, or $k\pi$ for tangent equations.

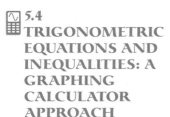
**5.4
TRIGONOMETRIC
EQUATIONS AND
INEQUALITIES: A
GRAPHING
CALCULATOR
APPROACH**

A graphing calculator can be used to solve most trigonometric equations to any accuracy allowed by the calculator but usually will not give exact solutions. Graph each side of the equation in a graphing calculator, then use the INTERSECT command to find any points of intersection.

CHAPTER 5 REVIEW EXERCISE

Work through all the problems in this chapter review and check the answers. Answers to all review problems appear in the back of the book; following each answer is an italic number that indicates the section in which that type of problem is discussed. Where weaknesses show up, review the appropriate sections in the text. Review problems flagged with a star (☆) are from optional section.

A *Evaluate exactly as real numbers.*

   1. $\tan^{-1}(-1)$ **2.** $\sin^{-1}(\sqrt{3}/2)$

   3. $\arccos 1$ **4.** $\arctan(-1/\sqrt{3})$

   5. $\arcsin \sqrt{2}$ **6.** $\cos^{-1}(-1/\sqrt{2})$

 ☆ **7.** $\csc^{-1}(-2/\sqrt{3})$ ☆ **8.** $\cot^{-1} 0$

 ☆ **9.** $\sec^{-1}(-2)$ ☆**10.** $\csc^{-1}\frac{1}{2}$

   Evaluate as a real number to four significant digits.

   11. $\sin^{-1} 0.6298$ **12.** $\arccos(-0.9704)$

   13. $\tan^{-1} 23.55$ ☆**14.** $\cot^{-1}(-1.414)$

   Find the degree measure of each to two decimal places.

   15. $\theta = \cos^{-1}(-1.025)$ **16.** $\theta = \arctan 8.333$

   17. $\theta = \sin^{-1}(-0.1010)$

In Problems 18–21, find all solutions exactly.

   18. $\sin x = -\sqrt{3}/2$ **19.** $\tan x = -\sqrt{3}$

   20. $\cos x = 1/\sqrt{2}$ **21.** $-3\cos x + 10 = 0$

In Problems 22–27, find exact solutions over the indicated interval.

   22. $2\cos x - \sqrt{3} = 0$, $0 \le x < 2\pi$

   23. $2\sin^2\theta = \sin\theta$, $0° \le \theta < 360°$

   24. $4\cos^2 x - 3 = 0$, $0 \le x < 2\pi$

   25. $2\cos^2\theta + 3\cos\theta + 1 = 0$, $0° \le \theta < 360°$

   26. $\sqrt{2}\sin 4x - 1 = 0$, $0 \le x < \pi/2$

   27. $\tan(\theta/2) + \sqrt{3} = 0$, $-180° < \theta < 180°$

   28. Explain how to find the value of x that produces the result shown in the graphing calculator display and find it. The calculator is in degree mode. Give the answer to six decimal places.

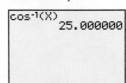
```
cos⁻¹(X)
        25.000000
```

B *In Problems 29–34, find exact values.*

29. $\cos(\cos^{-1} 0.315)$ **30.** $\tan^{-1}[\tan(-1.5)]$

31. $\sin[\tan^{-1}(-\frac{3}{4})]$ **32.** $\cot[\arccos(-\frac{2}{3})]$

☆**33.** $\csc[\cot^{-1}(-\frac{1}{3})]$ ☆**34.** $\cos(\text{arccsc } 5)$

In Problems 35–40, evaluate to four significant digits.

35. $\sin^{-1}(\cos 22.37)$ **36.** $\sin^{-1}(\tan 1.345)$

37. $\sin[\tan^{-1}(-14.00)]$ **38.** $\csc[\cos^{-1}(-0.4081)]$

☆**39.** $\cos(\cot^{-1} 6.823)$ ☆**40.** $\sec[\text{arccsc}(-25.52)]$

41. Referring to the two displays from a graphing calculator below, explain why one of the displays illustrates a sine–inverse sine identity and the other does not.

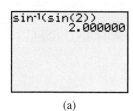
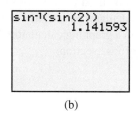

       (a)   (b)

In Problems 42–48, find exact solutions over the indicated interval.

42. $\sin^2 \theta = -\cos 2\theta, \quad 0° \le \theta \le 360°$

43. $\sin 2x = \frac{1}{2}, \quad 0 \le x < \pi$

44. $2 \cos x + 2 = -\sin^2 x, \quad -\pi \le x < \pi$

45. $2 \sin^2 \theta - \sin \theta = 0, \quad \text{all } \theta$

46. $\sin 2x = \sqrt{3} \sin x, \quad \text{all real } x$

47. $2 \sin^2 \theta + 5 \cos \theta + 1 = 0, \quad 0° \le \theta < 360°$

48. $3 \sin 2x = -2 \cos^2 2x, \quad 0 \le x \le \pi$

In Problems 49–52, solve for all real x to four significant digits.

49. $\sin x = 0.7088$ **50.** $\tan x = -4.318$

51. $\sin^2 x + 2 = 4 \sin x$ **52.** $\tan^2 x = 2 \tan x + 1$

In Problems 53–60, find all solutions over the indicated interval to three decimal places using a graphing calculator.

53. $\sin x = 0.25, \quad -\pi \le x \le \pi$

54. $\cot x = -4, \quad -\pi \le x \le \pi$

55. $\sec x = 2, \quad -\pi \le x \le \pi$

56. $\cos x = x^2, \quad \text{all real } x$

57. $\sin x = \sqrt{x}, \quad x \ge 0$

58. $2 \sin x \cos 2x = 1, \quad 0 \le x \le 2\pi$

59. $\sin \dfrac{x}{2} + 3 \sin x = 2, \quad 0 \le x \le 4\pi$

60. $\sin x + 2 \sin 2x + 3 \sin 3x = 3, \quad 0 \le x \le 2\pi$

61. Graph $y_1 = \tan(\sin^{-1} x)$ in the window $-2 \le x \le 2$, $-10 \le y \le 10$. What is the domain for y_1? Explain.

62. Given $h(x) = \sin^{-1}\left(\dfrac{x-2}{2}\right)$:

  (A) Explain how you would find the domain of h, and find it.

  (B) Graph h over the interval $-5 \le x \le 5$ and explain the result.

63. Does $\tan^{-1} 23.255$ represent all the solutions to the equation $\tan x = 23.255$? Explain.

C *In Problems 64 and 65, find exact solutions over the indicated interval.*

64. $\cos x = 1 - \sin x, \quad 0 \le x < 2\pi$

65. $\cos^2 2x = \cos 2x + \sin^2 2x, \quad 0 \le x < \pi$

66. Solve to three significant digits:

$$2 + 2 \sin x = 1 + 2 \cos^2 x \qquad 0 \le x \le 2\pi$$

In Problems 67 and 68, find the exact value.

67. $\sin[2 \tan^{-1}(-\frac{3}{4})]$ **68.** $\sin(\sin^{-1} \frac{3}{5} + \cos^{-1} \frac{4}{5})$

Write Problems 69 and 70 each as an algebraic expression in x free of trigonometric or inverse trigonometric functions.

69. $\tan(\sin^{-1} x)$ **70.** $\cos(\tan^{-1} x)$

71. The identity $\tan^{-1}(\tan x) = x$ is valid over the interval $-\pi/2 \le x \le \pi/2$.

  (A) Graph $y = \tan^{-1}(\tan x)$ for $-\pi/2 \le x \le \pi/2$. (Use dot mode.)

  (B) What happens if you graph $y = \tan^{-1}(\tan x)$ over a larger interval, say $-2\pi \le x \le 2\pi$? Explain. (Use dot mode.)

 Applications

72. **Music** The note A above middle C has a frequency of 440 Hz. If the intensity I of the sound at a certain point t seconds after the sound is made can be described by the equation

$$I = 0.08 \sin(880\pi t)$$

find the smallest positive t for which $I = 0.05$. Compute the answer to two significant digits.

73. Electric Current An alternating current generator produces a current given by the equation

$$I = 30 \sin(120\pi t)$$

where t is time in seconds and I is current in amperes. Find the least positive t (to four significant digits) for which $I = 20$ amperes.

74. Navigation A small craft is approaching a large vessel on the course shown in the figure.

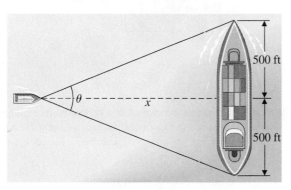

Figure for 74

(A) Express the angle θ subtended by the large vessel in terms of the distance x between the two ships.

(B) Find θ in decimal degrees to one decimal place for $x = 1,200$ ft.

75. Viewing Angle A sightseeing helicopter is approaching the Statue of Liberty at an altitude of 3,500 ft above ground level. The staue is 305 ft tall from ground level to the torch.

(A) Show that the viewing angle below the horizontal from a viewer in the plane to the torch is given by

$$\theta = \tan^{-1}\left(\frac{3,195}{x}\right)$$

where x is the horizontal distance in feet between the plane and the statue.

(B) At what angle (to the nearest tenth of a degree) should a viewer 1 mi away look in order to be looking at the torch?

76. Precalculus: Viewing Angle For advertising purposes, a large brokerage house has a 1.5 ft by 12 ft ticker screen mounted 20 ft above the floor on a high wall at an

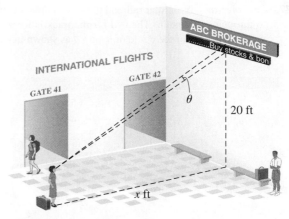

Figure for 76

airport terminal (see the figure). A woman's eyes are 5 ft above the floor. If the best view of the tape is when θ is maximum, how far from the wall should she stand? Parts (A)–(F) explore this problem.

(A) Describe what you think happens to θ as x increases from 0 ft to 100 ft.

(B) Show that

$$\theta = \tan^{-1}\frac{1.5x}{x^2 + 247.5} \qquad x \geq 0$$

(C) Complete Table 1 (to two decimal places, θ in degrees), and from the table select the maximum θ and the distance x that produces it. (Use a table generator if your calculator has one.)

TABLE 1

x (ft)	0	5	10	15	20	25	30
θ (deg)		1.58					

(D) In a graphing calculator, graph the equation in part (B) for $0 \leq x \leq 30$, and describe what the graph shows.

(E) Use the MAXIMUM command in your graphing calculator to find the maximum θ and the x that produces it. Find both to two decimal places.

(F) How far away from the wall should the woman stand to have a viewing angle of 2.5°? Solve graphically to two decimal places using a graphing calculator.

77. Engineering   A circular railroad tunnel of radius r is to go through a mountain. The bed for the track is formed by using a chord of the circle of length d as shown in the figure.

Figure for 77

(A) Show that the cross-sectional area of the tunnel is given by

$$A = \pi r^2 - r^2 \sin^{-1} \frac{d}{2r} + \frac{d}{4} \sqrt{4r^2 - d^2}$$

(B) Complete Table 2 (to the nearest square foot) for $r = 15$ ft and $10 \le d \le 20$. (Use a table generator if your calculator has one.)

TABLE 2

d (ft)	8	10	12	14	16	18	20
A(ft^2)	704						

(C) Find the value of d (to one decimal place) that will produce a cross-sectional area of 675 ft^2. Solve graphically using a graphing calculator and the INTERSECT command.

78. Architecture   The roof for a 10 ft wide storage shed is formed by bending a 12 ft wide steel panel into a circular arc (see the figure). Approximate (to two decimal places) the height h of the arc.

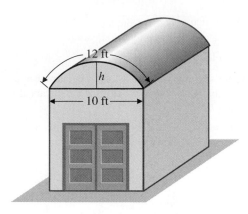

Figure for 78

79. The height above the ground of a certain bungee jumper after he reaches the low point of his jump and is being pulled back up by the cord can be modeled by the function

$$y = 50 - \frac{40}{x + 2} \cos(0.815x)$$

where x is the number of seconds after the jumper reaches the low point. Find the total amount of time (to the nearest tenth of a second) after reaching the low point that the jumper is at least 52 feet above the ground.

80. Physics   The equation of motion for a weight suspended from a spring is given by

$$y = -1.8 \sin 4t - 2.4 \cos 4t$$

where y is displacement of the weight from its equilibrium position (positive direction upward) and t is time in seconds.

(A) Graph y for $0 \le t \le \pi/2$.

(B) Approximate (to two decimal places) the time(s) t, $0 \le t \le \pi/2$, when the weight is 2 in. above the equilibrium position.

(C) Approximate (to two decimal places) the time(s) t, $0 \le t \le \pi/2$, when the weight is 2 in. below the equilibrium position.

CUMULATIVE REVIEW EXERCISE CHAPTERS 1–5

Work through all the problems in this cumulative review and check the answers. Answers to all review problems appear in the back of the book; following each answer is an italic number that indicates the section in which that type of problem is discussed. Where weaknesses show up, review the appropriate sections in the text.

 A

1. Find the degree measure of 4.21 rad.

2. Find the radian measure of 505°42′.

3. Explain what is meant by an angle of radian measure 0.5.

4. The hypotenuse of a right triangle is 7.6 m and one of the sides is 4.5 m. Find the acute angles and the other side.

5. Find the value of $\cos\theta$ and $\cot\theta$ if the terminal side of θ contains $P = (-5, -12)$.

6. Evaluate to four significant digits using a calculator:

   (A) $\cos 67°45'$ (B) $\csc 176.2°$ (C) $\cot 2.05$

7. Sketch a graph of each function for $-2\pi \le x \le 2\pi$.

   (A) $y = \cos x$ (B) $y = \csc x$ (C) $y = \cot x$

8. Is it possible to find an angle θ such that $\sin\theta$ is negative and $\csc\theta$ is positive? Explain.

Verify each identity in Problems 9–12 without looking at a table of identities.

9. $\tan x \csc x = \sec x$

10. $\csc\theta - \sin\theta = \cos\theta \cot\theta$

11. $(\sin^2 u)(\tan^2 u + 1) = \sec^2 u - 1$

12. $\dfrac{\sin^2\alpha - \cos^2\alpha}{\sin\alpha\cos\alpha} = \dfrac{\tan\alpha - \cot\alpha}{\tan\alpha\cot\alpha}$

13. Is the following equation an identity? Explain.

   $\cos x = \sin x$ for $x = \pi/4 + 2k\pi, k$ an integer

Evaluate Problems 14–20 exactly as real numbers.

14. $\cos\dfrac{-7\pi}{4}$

15. $\tan\dfrac{7\pi}{3}$

16. $\sec\dfrac{3\pi}{2}$

17. $\arctan 0$

18. $\cos^{-1}(-\sqrt{3}/2)$

19. $\arcsin 3$

☆20. $\text{arccot}(-\sqrt{3})$

Evaluate Problems 21–25 as real numbers to four significant digits using a calculator.

21. $\sin^{-1} 0.0505$

22. $\cos^{-1}(-0.7228)$

23. $\arctan(-9)$

☆24. $\text{arccot } 3$

☆25. $\sec^{-1} 2.6$

26. Explain how to find the value of x that produces the result shown in the graphing calculator display that follows, and find it. The calculator is in radian mode. Give the answer to six decimal places.

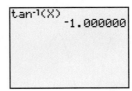

Find exact solutions for Problems 27–29 over the indicated interval.

27. $2\sin\theta - 1 = 0,\quad 0° \le \theta < 360°$

28. $3\tan x + \sqrt{3} = 0,\quad -\pi/2 < x < \pi/2$

29. $2\cos x + 2 = 0,\quad -\pi \le x < \pi$

☆30. Write $\sin 7u \cos 3u$ as a sum or difference.

☆31. Write $\cos 5w - \cos w$ as a product.

B 32. If the radian measure of an angle is halved, is the degree measure of the angle also halved? Explain.

33. Convert 92.462° to degree-minute-second form.

34. Find the degree measure of a central angle subtended by an arc of 12 in. in a circle with a circumference of 30 in.

35. Find x exactly and θ to the nearest 0.1° in the figure.

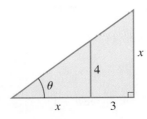

Figure for 35

36. Convert 72° to radian measure in terms of π.

37. Find the exact value of each of the other five trigonometric functions if $\tan\theta = \frac{1}{2}$ and $\sin\theta < 0$.

In Problems 38–40, sketch a graph of each function for the indicated interval. State the period and, if applicable, the amplitude and phase shift.

38. $y = 2 - 2 \sin \dfrac{x}{2}, \quad -\pi \le x \le 5\pi$

39. $y = 3 \cos(2x - \pi), \quad -\pi \le x \le 2\pi$

40. $y = 2 \tan(\pi x - \pi/4), \quad 0 \le x \le 3$

41. Find an equation of the form $y = k + A \cos Bx$ whose graph is shown in the figure.

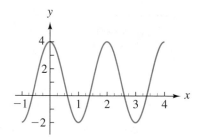

42. Find an equation of the form $y = k + A \sin Bx$ that produces the graph shown in the following graphing calculator display:

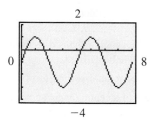

43. If the sides of a right triangle are 19.4 cm and 41.7 cm, find the hypotenuse and find the acute angles to the nearest 10′.

Verify the identities in Problems 44–48.

44. $\dfrac{\cos x}{1 + \sin x} + \tan x = \sec x$

45. $\dfrac{1 + \cos \theta}{1 + \sin \theta} = (\sec \theta - \tan \theta)(\sec \theta + 1)$

46. $\cot \dfrac{u}{2} = \csc u + \cot u$

47. $\dfrac{2}{1 + \sec 2\theta} = 1 - \tan^2 \theta$

48. $\dfrac{\sin x - \sin y}{\cos x + \cos y} = \tan \dfrac{x - y}{2}$

49. For the following graphing calculator displays, find the value of the final expression without finding x or using a calculator:

(A)   (B)

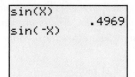

50. Write $y = \sin 3x \cos x - \cos 3x \sin x$ in terms of a single trigonometric function. Check the result by entering the original equation in a graphing calculator as y_1 and the converted form as y_2. Then graph y_1 and y_2 in the same viewing window. Use $\boxed{\text{TRACE}}$ to compare the two graphs.

51. Find the exact values of $\sin(x/2)$ and $\cos 2x$, given $\tan x = \dfrac{8}{15}$ and $\pi < x < 3\pi/2$.

52. Graph $y_1 = \cos(x/2)$ and $y_2 = -\sqrt{(1 + \cos x)/2}$ in the same viewing window for $-2\pi \le x \le 2\pi$, and indicate the subintervals for which y_1 and y_2 are equal.

53. Use a graphing calculator to test whether each equation is an identity. If the equation appears to be an identity, verify it. If the equation does not appear to be an identity, find a value of x for which both sides are defined but are not equal.

(A) $\dfrac{\sin x}{1 + \cos x} = \dfrac{1 + \cos x}{\sin x}$

(B) $\dfrac{\sin x}{1 - \cos x} = \dfrac{1 + \cos x}{\sin x}$

54. Find exact values for each of the following:

(A) $\sin(\cos^{-1} 0.4)$ (B) $\sec[\arctan(-\sqrt{5})]$

(C) $\csc(\sin^{-1} \frac{1}{3})$ ☆ (D) $\tan(\sec^{-1} 4)$

55. Find the exact degree value of $\theta = \tan^{-1} \sqrt{3}$ without using a calculator.

56. Find the degree measure of $\theta = \sin^{-1} 0.8989$ to two decimal places using a calculator.

Find exact solutions to Problems 57–59 over the indicated interval.

57. $2 + 3 \sin x = \cos 2x, \quad 0 \le x \le 2\pi$

58. $\sin 2\theta = 2 \cos \theta, \quad$ all θ

59. $4 \tan^2 x - 3 \sec^2 x = 0, \quad$ all real x

Find all real solutions to Problems 60–62 to four significant digits.

60. $\sin x = -0.5678$

61. $\sec x = 2.345$

62. $2 \cos 2x = 7 \cos x$

63. Referring to the two displays from a graphing calculator below, explain why one of the displays illustrates a cosine–inverse cosine identity and the other does not.

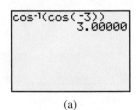

(a)

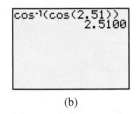

(b)

In Problems 64–66, use a graphing calculator to approximate all solutions (to three decimal places) over the indicated interval.

64. $\tan x = 3, \quad -\pi \le x \le \pi$

65. $\cos x = \sqrt{x}, \quad x > 0$

66. $\cos \dfrac{x}{2} - 2 \sin x = 1, \quad 0 \le x \le 4\pi$

C 67. A point P moves counterclockwise around a unit circle starting at $(1, 0)$ for a distance of 28.703 units. Explain how you would find the coordinates of the point P at its final position and how you would determine which quadrant P is in. Find the coordinates (to four decimal places) and the quadrant in which P lies.

68. The following graphing calculator display shows the coordinates of a point on a unit circle. Let s be the length of the least positive arc from $(1, 0)$ to the point. Find s to four decimal places.

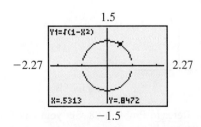

69. If θ is an angle in the fourth quadrant and $\cos \theta = a$, $0 < a < 1$, express the other five trigonometric functions of θ in terms of a.

70. Show that $\tan 3x = \tan x \, \dfrac{2 \cos 2x + 1}{2 \cos 2x - 1}$ is an identity.

71. Find the exact value of $\cos\left(2 \sin^{-1} \frac{1}{3}\right)$.

72. Write $\sin\left(\cos^{-1} x - \tan^{-1} x\right)$ as an algebraic expression in x free of trigonometric or inverse trigonometric functions.

73. Find exact solutions for all real x:

$$\sin x = 1 + \cos x$$

74. Find solutions to four significant digits:

$$\sin x = \cos^2 x \qquad 0 \le x \le 2\pi$$

In Problems 75–78, graph $f(x)$, find a simpler function $g(x)$ that has the same graph as $f(x)$, and verify the identity $f(x) = g(x)$. [Assume $g(x) = k + A \cdot t(Bx)$, where $t(x)$ is one of the six trigonometric functions.]

75. $f(x) = \dfrac{\sin^2 x}{1 - \cos x} + \dfrac{2 \tan^2 x \cos^2 x}{1 + \cos x}$

76. $f(x) = 2 \sin^2 x + 6 \cos^2 x$

77. $f(x) = \dfrac{2 - 2 \sin^2 x}{2 \cos^2 x - 1}$

78. $f(x) = \dfrac{3 \cos x + \sin x - 3}{\cos x - 1}$

79. Find the subinterval(s) of $[0, 4]$ over which

$$2 \sin \dfrac{\pi x}{2} - 3 \cos \dfrac{\pi x}{2} \ge 2x - 1$$

Find the end points to three decimal places.

Applications

80. Navigation An airplane flying on a level course directly toward a beacon on the ground records two angles of depression as indicated in the figure. Find the altitude h of the plane to the nearest meter.

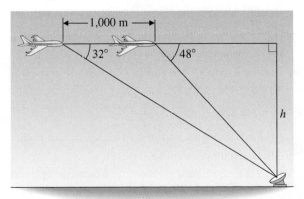

Figure for 80

81. Precalculus: Trigonometric Substitution In the expression $\sqrt{u^2 - a^2}, a > 0$, let $u = a \csc x$, $0 < x < \pi/2$, simplify, and write in a form that is free of radicals.

82. Engineering Find the exact values of x and θ in the figure.

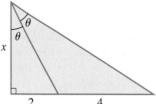

2   4   Figure for 82

83. Physics The equation of motion for a weight suspended from a spring is given by

$$y = -7.2 \sin 5t - 9.6 \cos 5t$$

where y is the displacement of the weight from its equilibrium position in centimeters (positive direction upward) and t is time in seconds. Find the smallest positive t (to three decimal places) for which the weight is in the equilibrium position.

84. Physics Refer to Problem 83. Graph y for $0 \le t \le \pi/2$ and approximate (to three decimal places) the time(s) t in this interval for which the weight is 5 cm above the equilibrium position.

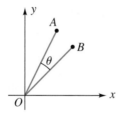

Figure for 85

85. Analytic Geometry Find the radian measure (to three decimal places) of the angle θ in the figure, if A has coordinates $(2, 5)$ and B has coordinates $(5, 3)$. [*Hint:* Label the angle between OB and the x axis as α; then use an appropriate sum identity.]

86. Electric Circuits The current I (in amperes) in an electrical circuit is given by

$$I = 60 \sin\left(90 \pi t - \frac{\pi}{2}\right)$$

(A) Find the amplitude, period, frequency, and phase shift.

(B) Graph the equation for $0 \le t \le 0.02$.

(C) Find the smallest positive t (to four decimal places) for which $I = 35$ amperes.

87. Boat Safety A boat is approaching a 200 ft high vertical cliff (see the figure that follows).

(A) Write an equation for the distance d from the boat to the base of the cliff in terms of the angle of elevation θ from the boat to the top of the cliff.

(B) Sketch a graph of this equation for $0 < \theta \le \pi/2$.

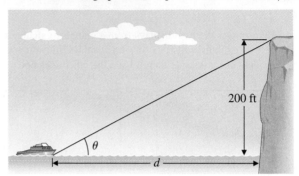

200 ft

d

Figure for 87

88. Precalculus Two guy wires are attached to a radio tower as shown in the figure.

(A) Show that $\theta = \arctan \dfrac{100x}{x^2 + 20{,}000}$

(B) Find θ in decimal degrees to one decimal place for $x = 50$ ft.

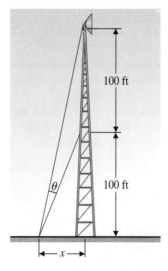

100 ft

100 ft

θ

x

Figure for 88

89. Precalculus Refer to Problem 88. Set your calculator in degree mode, graph

$$y_1 = \tan^{-1} \frac{100x}{x^2 + 20{,}000} \quad \text{and} \quad y_2 = 15$$

in the same viewing window, and use approximation techniques to find x to one decimal place when $\theta = 15°$.

90. Engineering   The chain on a bicycle goes around the pedal sprocket and the rear wheel sprocket (see the figure). The radius of the pedal sprocket is 11.0 cm, the radius of the rear sprocket is 4.00 cm, and the diameter of the rear wheel is 70.0 cm. If the pedal sprocket rotates through an angle of 18π radians, through how many radians does the rear wheel rotate?

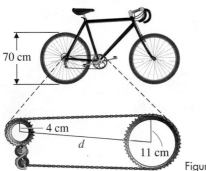

70 cm

4 cm

d

11 cm

Figure for 90

☆**91. Engineering**   Refer to Problem 90. If the pedal sprocket rotates at 60.0 rpm, how fast is the bicycle traveling (in centimeters per minute)?

92. Precalculus: Surveying   A log is to be floated around the right angle corner of a canal as shown in the figure.

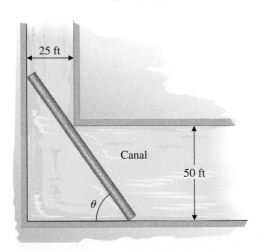

25 ft

Canal

50 ft

θ

Figure for 92

We are interested in finding the length of the longest log that will make it around the corner. (Assume the log is represented by a straight line segment with no diameter.)

(A) From the figure, and assuming that the log touches the inside corner of the canal, explain how L, the length of the segment, varies as θ varies from 0° to 90°.

(B) how that the length of the segment, L, is given by
$$L = 50 \csc \theta + 25 \sec \theta \quad 0° < \theta < 90°$$

(C) Complete Table 1 (to one decimal place). (If you have a table-generating calculator, use it.)

TABLE 1							
θ (deg)	35	40	45	50	55	60	65
L (ft)	117.7						

(D) From the table, select the minimum length L and the angle θ that produces it. Is this the length of the longest log that will go around the corner? Explain.

(E) Graph the equation from part (B) in a graphing calculator for $0° < \theta < 90°$ and $80 \le L \le 200$. Use the $\boxed{\text{MINIMUM}}$ command to find the minimum length L and the value of θ that produces it.

93. Modeling Daylight Duration   Table 2 gives the duration of daylight on the fifteenth day of each month for 1 year at Anchorage, Alaska.

(A) Convert the data in Table 2 from hours and minutes to two-place decimal hours. Enter the data for a 2 year period in your graphing calculator and produce a scatter plot in the following viewing window: $1 \le x \le 24, 6 \le y \le 20$.

(B) A function of the form $y = k + A \sin(Bx + C)$ can be used to model the data. Use the converted data from Table 2 to determine k and A (to two decimal places), and B (exact value). Use the graph from part (A) to visually estimate C (to one decimal place).

(C) Plot the data from part (A) and the equation from part (B) in the same viewing window. If necessary, adjust your value of C to produce a better fit.

TABLE 2												
x (months)	1	2	3	4	5	6	7	8	9	10	11	12
y (daylight duration)	6:31	9:10	11:50	14:48	17:33	19:16	18:27	15:51	12:57	10:09	7:19	5:36

Additional Topics: Triangles and Vectors

6

6.1   Law of Sines

6.2   Law of Cosines

☆**6.3**   Areas of Triangles

6.4   Vectors: Geometrically Defined

6.5   Vectors: Algebraically Defined

☆**6.6**   The Dot Product

Chapter 6 Group Activity: The SSA Case and the Law of Cosines

Chapter 6 Review

─────────

☆ Sections marked with a star may be omitted without loss of continuity.

n chapter 1 we introduced the trigonometric ratios to solve right triangles. We also saw that this could be very useful in solving applied problems, particularly those involving measurement. But it's unreasonable to assume that every situation that involves triangles will involve right triangles. So we will now return to solving triangles, but without being restricted to right triangles. We will establish the *law of sines* and *law of cosines*; they are the principal tools we'll need to solve general triangles. We will then use these tools to solve a wide variety of applied problems. We will also develop several formulas for calculating the area of a triangle.

Velocity and force are examples of vector quantities. In Section 6.3, we will introduce the concept of *vector*, first in geometric form and then in algebraic form. We will also define algebraic operations on vectors, including the dot product. Finally, we will illustrate how vectors and trigonometry can be used together to solve problems in physics and engineering.

6.1   Law of Sines

- Deriving the Law of Sines
- Solving ASA and AAS Cases
- Solving the Ambiguous SSA Case

Until now, we have considered only triangle problems that involved right triangles. We now turn to **oblique triangles,** that is, triangles that contain no right angle. Every oblique triangle is either **acute** (all angles are between 0° and 90°) or **obtuse** (one angle is between 90° and 180°; because the three angles always sum to 180°, there can't be more than one greater than 90°). Figure 1 illustrates an acute triangle and an obtuse triangle.

FIGURE 1

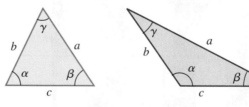

(a) Acute triangle   (b) Obtuse triangle

Notice how we labeled the sides and angles of the oblique triangles shown in Figure 1: Side a is opposite angle α, side b is opposite angle β, and side c is opposite angle γ. Note also that the largest side of a triangle is opposite the largest angle. Given any three of the six quantities indicated in Figure 1, we will be trying to find the remaining three, if possible. This process is called **solving the triangle.**

If only the three angles α, β, and γ of a triangle are known, it is impossible to solve for the sides. (Why?) But if the given information consists of two angles and a side, two sides and an angle, or all three sides, then it is possible to determine whether a triangle having the given quantities exists and, if so, to solve for the remaining quantities.

The two basic tools for solving oblique triangles are the *law of sines,* developed in this section, and the *law of cosines,* developed in the next section.

Before we proceed with specifics, recall the rules governing angle measure and significant digits for side measure (listed in Table 1 and inside the front cover for easy reference).

TABLE 1	
Angle to nearest	Significant digits for side measure
1°	2
10′ or 0.1°	3
1′ or 0.01°	4
10″ or 0.001°	5

Remark on Calculations   When you solve for a particular side or angle, you should always carry out all operations within the calculator and then round to the appropriate number of significant digits (following the rules in Table 1) at the very end of the calculation. Note that your answers still may differ slightly from those in the book, depending on the order in which you solve for the sides and angles.   □

■ Deriving the Law of Sines

The law of sines is relatively easy to derive using the right triangle properties we studied earlier. We also use the fact that

$$\sin(180° - x) = \sin x$$

which can be obtained from the difference identity for sine. Referring to the triangles in Figure 2, we will proceed as follows: For each triangle,

$$\sin \alpha = \frac{h}{b} \quad \text{and} \quad \sin \beta = \frac{h}{a}$$

Therefore,

$$h = b \sin \alpha \quad \text{and} \quad h = a \sin \beta$$

FIGURE 2

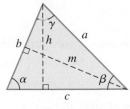
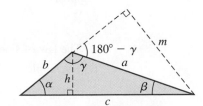

(a) Acute triangle   (b) Obtuse triangle

Since $b \sin \alpha$ and $a \sin \beta$ are both equal to h,

$$b \sin \alpha = a \sin \beta$$

and

$$\frac{\sin \alpha}{a} = \frac{\sin \beta}{b} \tag{1}$$

Similarly, for each triangle in Figure 2,

$$\sin \alpha = \frac{m}{c} \qquad \text{and} \qquad \sin \gamma = \sin(180° - \gamma)$$
$$= \frac{m}{a}$$

Therefore,

$$m = c \sin \alpha \qquad \text{and} \qquad m = a \sin \gamma$$

Since $c \sin \alpha$ and $a \sin \gamma$ are both equal to m,

$$c \sin \alpha = a \sin \gamma$$

and

$$\frac{\sin \alpha}{a} = \frac{\sin \gamma}{c} \tag{2}$$

If we combine equations (1) and (2), we obtain the **law of sines.**

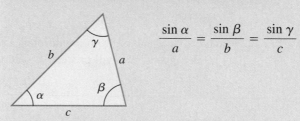

LAW OF SINES

$$\frac{\sin \alpha}{a} = \frac{\sin \beta}{b} = \frac{\sin \gamma}{c}$$

In words, in any triangle the ratio of the sine of an angle to its opposite side is the same as the ratio of the sine of either of the other angles to its opposite side.

The law of sines is used to solve triangles, given:

1. Two angles and any side (ASA or AAS), or
2. Two sides and an angle opposite one of them (SSA)

We will apply the law of sines to the easier ASA and AAS cases first; then we will turn to the more difficult SSA case.

EXPLORE/DISCUSS 1

(A) Suppose that in a triangle $\alpha = 35°$, $\beta = 79°$, and $a = 13$ cm. Using only these three pieces of information and the law of sines, which side or angle can you find? Why?

(B) If instead we know that $\alpha = 35°$, $b = 10$ cm, and $c = 21$ cm, explain why we can't use the law of sines to find any other side or angle.

Explore/Discuss 1 illustrates the key point about when you can and cannot use the law of sines. In order to set up an equation with only one unknown, you need to know at least one side opposite a known angle. In the next section, we will develop a formula for dealing with situations the law of sines cannot handle.

■ Solving ASA and AAS Cases

If we are given two angles and the included side (ASA case) or two angles and the side opposite one of the angles (AAS case), the simplest approach is to first find the measure of the third angle (remember that the sum of the measures of all three angles in any triangle is 180°). We can then use the law of sines to find the other two sides.

⚠ Caution   Note that for the ASA or AAS case to determine a unique triangle, the sum of the two given angles must be between 0° and 180° (see Fig. 3).

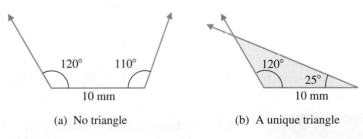

(a) No triangle   (b) A unique triangle

FIGURE 3   □

Examples 1 and 2 illustrate the use of the law of sines in solving the ASA and AAS cases.

Using the Law of Sines (ASA)

Solve the triangle shown in Figure 4.

Solution   This is an example of the ASA case.

Solve for γ:

$$\alpha + \beta + \gamma = 180°$$

$$\gamma = 180° - (45.1° + 75.8°)$$

$$= 59.1°$$

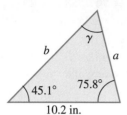

FIGURE 4

Since the only side we know is c (the side opposite angle γ), we have no choice other than to use $(\sin \gamma)/c$. We choose to first find side a:

$$\frac{\sin \alpha}{a} = \frac{\sin \gamma}{c}$$

$$\frac{\sin 45.1°}{a} = \frac{\sin 59.1°}{10.2} \qquad \text{Cross-multiply.}$$

$$a \sin 59.1° = 10.2 \sin 45.1° \qquad \text{Divide both sides by } \sin 59.1°.$$

$$a = \frac{10.2 \sin 45.1°}{\sin 59.1°} \qquad \text{Use calculator.}$$

$$= 8.42 \text{ in.}$$

Solve for b:

$$\frac{\sin \beta}{b} = \frac{\sin \gamma}{c}$$

$$\frac{\sin 75.8°}{b} = \frac{\sin 59.1°}{10.2} \qquad \text{Cross-multiply.}$$

$$b \sin 59.1° = 10.2 \sin 75.8° \qquad \text{Divide both sides by } \sin 59.1°.$$

$$b = \frac{10.2 \sin 75.8°}{\sin 59.1°} \qquad \text{Use calculator.}$$

$$= 11.5 \text{ in.} \qquad ■$$

Matched Problem 1   Solve the triangle with $\alpha = 28.0°$, $\beta = 45.3°$, and $c = 122$ m.   ■

Using the Law of Sines (AAS): Sundials

A sundial can be used to trace a moving shadow during the day to measure the time. The raised part that casts the shadow is called the *gnomon*. In order to measure time accurately, the angle the gnomon makes with the face of the dial (labeled β in Fig. 5) must be the same as the latitude where the sundial is used. If the latitude of San Francisco is 38°N, and the angle of elevation of the sun, α, is 63° at noon, how long will the shadow of a 12 in. gnomon be on the face of the sundial (see Fig. 5)?

FIGURE 5
Sundial

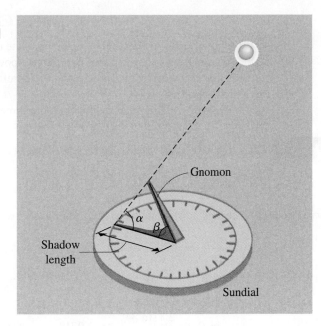

Solution   We are given two angles and the side opposite one of the angles (AAS). First we find the third angle; then we find the length of the shadow using the law of sines. It is helpful to make a simple drawing (Fig. 6) showing the given and unknown parts.

Solve for γ:

$$\gamma = 180° - (63° + 38°) = 79°$$

Solve for c:

$$\frac{\sin \alpha}{a} = \frac{\sin \gamma}{c}$$

$$\frac{\sin 63°}{12} = \frac{\sin 79°}{c} \qquad \text{Cross-multiply.}$$

$$c \sin 63° = 12 \sin 79° \qquad \text{Divide both sides by sin 63°.}$$

$$c = \frac{12 \sin 79°}{\sin 63°} = 13 \text{ in.} \qquad \text{Shadow length} \qquad ■$$

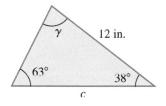

FIGURE 6

Matched Problem 2   Repeat Example 2 for Columbus, Ohio, with a latitude of 40°N, a gnomon of length 15 in., and a 52° angle of elevation for the sun.   ■

Remark   Note that the AAS case can always be converted to the ASA case by first solving for the third angle, so we can think of the two situations where we know two angles as a single case. For this case to determine a unique triangle, the sum of the two given angles must be between 0° and 180°.   □

■ Solving the Ambiguous SSA Case

If we are given two sides and an angle opposite one of the sides—the SSA case—then it is possible to have zero, one, or two possible triangles, depending on the measures of the two sides and the angle. Therefore, we will refer to the SSA case as the ambiguous case.

We will use examples to illustrate the three possibilities.

EXAMPLE 3

Using the Law of Sines (SSA): No Triangle

Find β in the triangle with $\alpha = 34°$, $b = 2.3$ mm, and $a = 1.2$ mm.

Solution

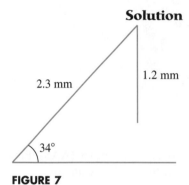

FIGURE 7

Since we know an angle and the side opposite (α and a), we can use the law of sines:

$$\frac{\sin \beta}{b} = \frac{\sin \alpha}{a} \qquad \text{Law of sines}$$

$$\frac{\sin \beta}{2.3} = \frac{\sin 34°}{1.2} \qquad \text{Multiply both sides by 2.3.}$$

$$\sin \beta = \frac{2.3 \sin 34°}{1.2} \approx 1.0718 \qquad \text{No solution since 1.0718 > 1}$$

Since the equation $\sin \beta = 1.0718$ has no solution, there is no triangle having the given measurements. Figure 7 illustrates geometrically what went wrong: Side a is not long enough to reach the bottom side of a 34° angle. ■

Matched Problem 3

Use the law of sines to find β in the triangle with $\alpha = 130°$, $b = 1.3$ m, and $a = 1.2$ m. ■

EXAMPLE 4

Using the Law of Sines (SSA): One Triangle

Solve the triangle with $\alpha = 47°$, $a = 3.7$ ft, and $b = 3.5$ ft.

Solution *Solve for β:*

$$\frac{\sin \beta}{b} = \frac{\sin \alpha}{a} \qquad \text{Law of sines}$$

$$\frac{\sin \beta}{3.5} = \frac{\sin 47°}{3.7} \qquad \text{Multiply both sides by 3.5.}$$

$$\sin \beta = \frac{3.5 \sin 47°}{3.7} \qquad \text{Apply inverse sine to solve for } \beta.$$

$$\beta = \sin^{-1}\left(\frac{3.5 \sin 47°}{3.7}\right) \qquad \text{Use the inverse sine key on a calculator.}$$

$$= 44°$$

The calculator gives the measure of the acute angle whose sine is 3.5 sin 47°/3.7. But there is another possible choice for β (see Fig. 8): the supplement of 44°, that is, 180° − 44° = 136°. But if we add 136° to α = 47°, we get 183°, which is greater than 180° (the sum of the measures of the three angles in a triangle). Because it is not possible to form a triangle having two angles with measures 47° and 136°, the supplement 136° must be rejected, and we are left with only one triangle using β = 44°. We complete the problem by solving for γ and c.

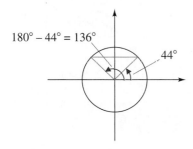

Solve for γ:

$$\gamma = 180° - (\alpha + \beta)$$
$$= 180° - (47° + 44°) = 89°$$

180° − 44° = 136°

44°

FIGURE 8

Solve for c:

$$\frac{\sin \alpha}{a} = \frac{\sin \gamma}{c}$$ Law of sines

$$\frac{\sin 47°}{3.7} = \frac{\sin 89°}{c}$$ Cross-multiply.

$$c \sin 47° = 3.7 \sin 89°$$ Divide both sides by sin 47°.

$$c = \frac{3.7 \sin 89°}{\sin 47°} = 5.1 \text{ ft}$$

Matched Problem 4   Solve the triangle with α = 139°, a = 42 yd, and b = 27 yd.

EXAMPLE 5   **Using the Law of Sines (SSA): Two Triangles**

Solve the triangle(s) with α = 26°, a = 11 cm, and b = 18 cm.

Solution   *Solve for β:*

$$\frac{\sin \beta}{b} = \frac{\sin \alpha}{a}$$ Law of sines

$$\frac{\sin \beta}{18} = \frac{\sin 26°}{11}$$ Solve for sin β.

$$\sin \beta = \frac{18 \sin 26°}{11}$$ Apply inverse sine to solve for β.

$$\beta = \sin^{-1}\left(\frac{18 \sin 26°}{11}\right)$$ Use inverse sine key on a calculator.

$$= 46°$$

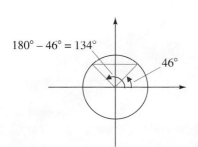

180° − 46° = 134°

46°

FIGURE 9

But β could also be the supplement of 46°; that is, 180° − 46° = 134° (see Fig. 9). To see if the supplement, 134°, can be an angle in a second triangle, we add 134° to α = 26° to obtain 160°. Since 160° is less than 180° (the sum of the measures of all angles of a triangle), there are two possible triangles meeting the original conditions, one with $\beta = 46°$ and the other with $\beta = 134°$. Figure 10 illustrates the results: $\beta = 46°$ and $\beta' = 134°$.

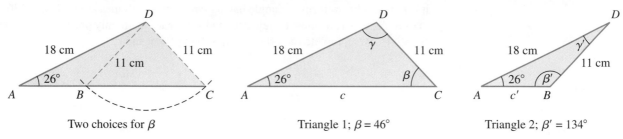

Two choices for β Triangle 1; $\beta = 46°$ Triangle 2; $\beta' = 134°$

FIGURE 10

Solve for γ and γ':

$$\gamma = 180° - (26° + 46°) = 108°$$

$$\gamma' = 180° - (26° + 134°) = 20°$$

Solve for c and c':

$$\frac{\sin 26°}{11} = \frac{\sin 108°}{c} \qquad \frac{\sin 26°}{11} = \frac{\sin 20°}{c'}$$

$$c \sin 26° = 11 \sin 108° \qquad c' \sin 26° = 11 \sin 20°$$

$$c = \frac{11 \sin 108°}{\sin 26°} \qquad c' = \frac{11 \sin 20°}{\sin 26°}$$

$$c = 24 \text{ cm} \qquad c' = 8.6 \text{ cm} \qquad ■$$

Matched Problem 5   Solve the triangle(s) with $a = 8.0$ mm, $b = 11$ mm, and $\alpha = 35°$.   ■

For the ambiguous SSA case, it is probably easiest to begin solving the triangle and see if zero, one, or two triangles result, as in Examples 3–5. It is possible, however, to summarize the various cases, as shown in Table 2 on page 361. With this information, you can determine in advance how many triangles exist.

EXPLORE/DISCUSS 2

If it is found that $\sin \beta > 1$ in the process of solving an SSA triangle with α acute, indicate which case(s) in Table 2 apply and why. Repeat the problem for $\sin \beta = 1$ and for $0 < \sin \beta < 1$.

TABLE 2
SSA Variations

	a ($h = b \sin \alpha$)	Number of triangles	Figure	
α acute	$0 < a < h$	0		(a)
	$a = h$	1		(b)
	$h < a < b$	2		(c)
	$a \geq b$	1		(d)
α obtuse	$0 < a \leq b$	0		(e)
	$a > b$	1		(f)

We close the section with another application.

EXAMPLE 6

Distances Between Cities

The cities of Jacksonville and Miami are 335 mi apart on Florida's east coast, with Miami southeast of Jacksonville. Tampa is 181 mi southwest of Jacksonville, on the west coast. If a triangle is formed by lines connecting the cities, the angle at Tampa is 120.3°. How far is it from Tampa to Miami?

Solution We should begin by drawing a diagram matching the described situation, labeling the angles with the cities (Fig. 11).

FIGURE 11

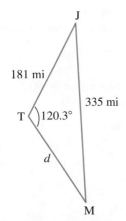

Based on the given information, our only choice is to find angle M first:

$$\frac{\sin 120.3°}{335} = \frac{\sin M}{181} \qquad \text{Multiply both sides by 181.}$$

$$\sin M = \frac{181 \sin 120.3°}{335} \qquad \text{Apply inverse sine to find M.}$$

$$M = \sin^{-1}\left(\frac{181 \sin 120.3°}{335}\right) \qquad \text{Use calculator.}$$

$$= 27.8°$$

(There is only one possible triangle because the angle at Tampa is greater than 90°). Now we can find angle J, then use the law of sines again to find the distance d.

$$J = 180° - (120.3° + 27.8°) = 31.9°$$

$$\frac{\sin 120.3°}{335} = \frac{\sin 31.9°}{d} \qquad \text{Cross-multiply.}$$

$$d \sin 120.3° = 335 \sin 31.9° \qquad \text{Divide both sides by sin 120.3°.}$$

$$d = \frac{335 \sin 31.9°}{\sin 120.3°} \qquad \text{Use calculator.}$$

$$= 205 \text{ mi}$$

Miami is 205 miles from Tampa.   ■

Matched Problem 6   San Francisco is 446 mi northwest of San Diego, and Las Vegas is 259 mi northeast of San Diego. If a triangle is drawn from the lines connecting these cities, the angle at Las Vegas is 79.6°. How far is it from San Francisco to Las Vegas?   ■

Remark   The cities in Example 6 are relatively close together, so it's reasonable to assume that the actual distances between them are the lengths of the line segments connecting the points J, T, and M in Figure 11. For cities like New York, Tokyo, and São Paulo, which are much farther apart, the shortest distances between them must take the curvature of the earth into account. Spherical trigonometry, a branch of trigonometry that is treated in more advanced books, can be used to calculate such distances.   □

Answers to Matched Problems

1. $\gamma = 106.7°$; $a = 59.8$ m; $b = 90.5$ m
2. Shadow length $= 19$ in.
3. No triangle: α is obtuse and $a \leq b$.
4. $\beta = 25°$; $\gamma = 16°$; $c = 18$ yd
5. $\beta = 128°$, $\beta' = 52°$; $\gamma = 17°$, $\gamma' = 93°$; $c = 4.1$ mm, $c' = 14$ mm
6. 413 mi

EXERCISE 6.1

Assume all triangles are labeled as in the figure unless stated to the contrary. Your answers may differ slightly from those in the back of the book, depending on the order in which you solve for the sides and angles.

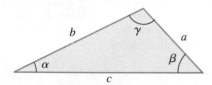

A 1. What is meant by "solving a triangle"?

2. Use illustrations to explain why there are sometimes two possible triangles when we are given two sides and an angle not between the sides.

3. Explain why it's always possible to use the law of sines in solving a triangle when we are given two angles and any side.

4. Discuss a simple way to decide if the law of sines can be used when we are given three parts of a triangle.

In Problems 5–10, decide whether or not the law of sines can be used to solve the triangle with the given information. Do not solve.

5. $a = 5$ in, $b = 7$ in, $\alpha = 31°$

6. $b = 10.1$ ft, $\beta = 116°$, $\gamma = 9°$

7. $a = 9$ mm, $b = 10$ mm, $c = 11$ mm

8. $a = 110$ yd, $\beta = 81°$, $c = 91$ yd

9. $b = 2.4$ mi, $\alpha = 78.4°$, $c = 1.6$ mi

10. $a = 6$ m, $b = 2$ m, $c = 90$ m

In Problems 11–18, solve each triangle given the indicated measures of angles and sides.

11. $\beta = 43°$, $\gamma = 36°$, $a = 92$ cm

12. $\alpha = 122°$, $\gamma = 18°$, $b = 12$ mm

13. $\beta = 27.5°$, $\gamma = 54.5°$, $a = 9.27$ mm

14. $\alpha = 118.3°$, $\gamma = 12.2°$, $b = 17.3$ km

15. $\alpha = 122.7°$, $\beta = 34.4°$, $b = 18.3$ cm

16. $\alpha = 67.7°$, $\beta = 54.2°$, $b = 123$ ft

17. $\beta = 12°40'$, $\gamma = 100°0'$, $b = 13.1$ km

18. $\alpha = 73°50'$, $\beta = 51°40'$, $a = 36.6$ mm

B *In Problems 19–30, determine whether the information in each problem allows you to construct zero, one, or two triangles. Do not solve the triangle. Explain which case in Table 2 applies.*

19. $a = 5$ ft, $b = 4$ ft, $\alpha = 60°$

20. $a = 1$ ft, $b = 2$ ft, $\alpha = 30°$

21. $a = 3$ in., $b = 8$ in., $\alpha = 30°$

22. $a = 6$ in., $b = 7$ in., $\alpha = 135°$

23. $a = 5$ cm, $b = 6$ cm, $\alpha = 45°$

24. $a = 9$ cm, $b = 8$ cm, $\alpha = 75°$

25. $a = 3$ mm, $b = 2$ mm, $\alpha = 150°$

26. $a = 5$ mm, $b = 7$ mm, $\alpha = 30°$

27. $a = 4$ ft, $b = 8$ ft, $\alpha = 30°$

28. $a = 7$ ft, $b = 1$ ft, $\alpha = 108°$

29. $a = 2$ in., $b = 3$ in., $\alpha = 120°$

30. $a = 4$ in., $b = 5$ in., $\alpha = 60°$

In Problems 31–48, solve each triangle. If a problem has no solution, say so. If a problem involves two triangles, solve both.

31. $\alpha = 134°$, $a = 38$ cm, $b = 24$ cm

32. $\alpha = 15°$, $a = 53$ cm, $b = 48$ cm

33. $\alpha = 69°$, $a = 86$ ft, $b = 91$ ft

34. $\alpha = 30°$, $a = 19$ ft, $b = 38$ ft

35. $\alpha = 21°$, $a = 4.7$ in., $b = 6.2$ in.

36. $\alpha = 145°$, $a = 8.6$ in., $b = 9.1$ in.

37. $\alpha = 52°$, $\beta = 65°$, $a = 57$ m

38. $\alpha = 45°$, $a = 58$ m, $b = 72$ m

39. $\beta = 30°$, $a = 54$ cm, $b = 27$ cm

40. $\alpha = 108°$, $\beta = 36°$, $b = 66$ cm

41. $\beta = 57°$, $a = 47$ ft, $b = 62$ ft

42. $\beta = 95°$, $a = 7.9$ ft, $b = 8.7$ ft

43. $\beta = 150.7°$, $a = 26.4$ in., $b = 19.5$ in.

44. $\beta = 70.5°$, $a = 85.3$ in., $b = 61.4$ in.

45. $\alpha = 130°20'$, $\beta = 31°30'$, $c = 37.2$ cm

46. $\beta = 45°50'$, $a = 8.25$ cm, $b = 5.83$ cm

47. $\beta = 22°10'$, $a = 56.3$ mm, $b = 25.1$ mm

48. $\alpha = 65°40'$, $\beta = 13°50'$, $b = 42.3$ mm

49. Mollweide's equation,

$$(a - b) \cos\frac{\gamma}{2} = c \sin\frac{\alpha - \beta}{2}$$

is often used to check the final solution of a triangle since all six parts of a triangle are involved in the equation. If, after substitution, the left side does not equal the right side, then an error has been made in solving a triangle. Use this equation to check Problem 11 to two decimal places. (Remember that rounding may not produce exact equality, but the left and right sides of the equation should be close.)

50. Use Mollweide's equation (see Problem 49) to check Problem 13 to two decimal places.

51. Use the law of sines and suitable identities to show that for any triangle,

$$\frac{a - b}{a + b} = \frac{\tan\dfrac{\alpha - \beta}{2}}{\tan\dfrac{\alpha + \beta}{2}}$$

52. Verify (to three decimal places) the formula in Problem 51 with values from Problem 11 and its solution.

53. Let $\beta = 46.8°$ and $a = 66.8$ yd. Determine a value k so that if $0 < b < k$, there is no solution; if $b = k$, there is one solution; and if $k < b < a$, there are two solutions.

54. Let $\beta = 36.6°$ and $b = 12.2$ m. Determine a value k so that if $0 < b < k$, there is no solution; if $b = k$, there is one solution; and if $k < b < a$, there are two solutions.

Applications

55. Surveying To determine the distance across the Grand Canyon in Arizona, a 1.00 mi baseline, *AB*, is established along the southern rim of the canyon. Sightings are then

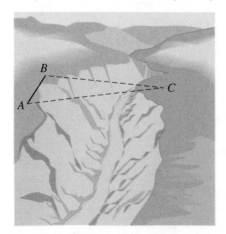

Figure for 55 and 56

made from the ends (*A* and *B*) of the baseline to a point *C* across the canyon (see the figure). Find the distance from *A* to *C* if $\angle BAC = 118.1°$ and $\angle ABC = 58.1°$.

56. Surveying Refer to Problem 55. The Grand Canyon was surveyed at a narrower part of the canyon using a similar 1.00 mi baseline along the southern rim. Find the length of *AC* if $\angle BAC = 28.5°$ and $\angle ABC = 144.6°$.

57. Fire Spotting A fire at *F* is spotted from two fire lookout stations, *A* and *B*, which are located 10.3 mi apart. If station *B* reports the fire at angle $ABF = 52.6°$, and station *A* reports the fire at angle $BAF = 25.3°$, how far is the fire from station *A*? From station *B*?

Figure for 57

58. Coast Patrol Two lookout posts, *A* and *B*, which are located 12.4 mi apart, are established along a coast to watch for illegal foreign fishing boats coming within the 3 mi limit. If post *A* reports a ship *S* at angle $BAS = 37.5°$, and post *B* reports the same ship at angle $ABS = 19.7°$, how far is the ship from post *A*? How far is the ship from the shore (assuming the shore is along the line joining the two observation posts)?

59. Surveying An underwater telephone cable is to cross a shallow lake from point *A* to point *B* (see the figure).

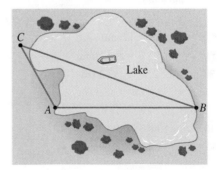

Figure for 59

Stakes are located at *A, B,* and *C.* Distance *AC* is measured to be 112 m, ∠*CAB* to be 118.4°, and ∠*ABC* to be 19.2°. Find the distance *AB.*

60.   Surveying   A suspension bridge is to cross a river from point *B* to point *C* (see the figure). Distance *AB* is measured to be 0.652 mi, ∠*ABC* to be 81.3°, and ∠*BCA* to be 41.4°. Compute the distance *BC.*

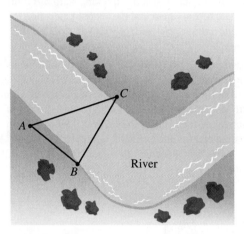

Figure for 60

61. Cleveland, Columbus, and Toledo form a triangle in Ohio with Columbus 120 mi southeast of Toledo, and Cleveland 112 mi northeast of Columbus. If the angle at Toledo is 59.7°, how far is it from Toledo to Cleveland?

62.   Coastal Piloting   A boat is traveling along a coast at night. A flashing buoy marks a reef. While proceeding on the same course, the navigator of the boat sights the buoy twice, 4.6 nautical mi apart, and forms triangles as

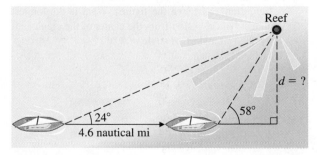

Figure for 62

shown in the figure. If the boat continues on course, by how far will it miss the reef?

63.   Surveying   Find the height of the mountain above the valley in the figure.

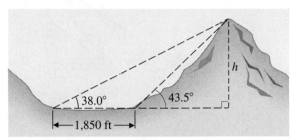

How high is the mountain?

Figure for 63

64.   Tree Height   A tree growing on a hillside casts a 157 ft shadow straight down the hill (see the figure). Find the vertical height of the tree if relative to the horizontal, the hill slopes 11.0° and the angle of elevation of the sun is 42.0°.

Figure for 64 and 65

65.   Tree Height   Find the height of the tree in Problem 64 if the shadow length is 102 ft and relative to the horizontal, the hill slopes 15.0° and the angle of elevation of the sun is 62.0°.

66. Aircraft Design   Find the measures of β and c for the sweptback wing of a supersonic jet, given the information in the figure.

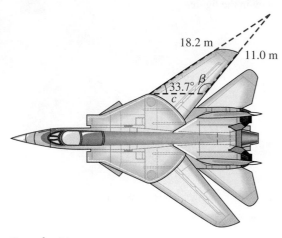

Figure for 66

67. Aircraft Design   Find the measures of β and c in Problem 66 if the extended leading and trailing edges of the wing are 20.0 m and 15.5 m, respectively, and the measure of the given angle is 35.3° instead of 33.7°.

68. Eye   A cross section of the cornea of an eye, a circular arc, is shown in the figure. With the information given in the figure, find the radius of the arc, r, and the length of the arc, s.

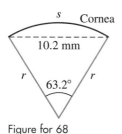

Figure for 68

69. Eye   Repeat Problem 68 using a central angle of 98.9° and a chord of length 11.8 mm.

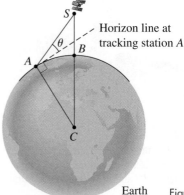

Earth   Figure for 70 and 71

70. Space Science   When a satellite is directly over tracking station B (see the figure), tracking station A measures the angle of elevation θ of the satellite (from the horizon line) to be 24.9°. If the tracking stations are 504 mi apart (that is, the arc AB is 504 mi) and the radius of the earth is 3,964 mi, how high is the satellite above B? [*Hint:* Find all angles for the triangle ACS first.]

71. Space Science   Refer to Problem 70. Compute the height of the satellite above tracking station B if the stations are 632 mi apart and the angle of elevation θ of the satellite (above the horizon line) is 26.2° at tracking station A.

72. Astronomy   The orbits of the earth and Venus are approximately circular, with the sun at the center (see the figure). A sighting of Venus is made from the earth and the angle α is found to be 18°40′. If the diameter of the orbit of the earth is 2.99×10^8 km and the diameter of the orbit of Venus is 2.17×10^8 km, what are the possible distances from the earth to Venus?

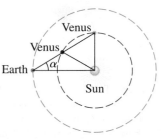

Figure for 72 and 73

73. Astronomy   In Problem 72 find the maximum value of α. [*Hint:* The value of α is maximum when a straight line joining the earth and Venus in the figure is tangent to Venus's orbit.]

74. Engineering   A 12 cm piston rod joins a piston to a 4.2 cm crankshaft (see the figure). What is the longest distance d of the piston from the center of the crankshaft when the rod makes an angle of 8.0°?

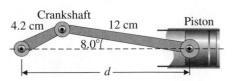

Figure for 74 and 75

75. Engineering Refer to Problem 74. What is the shortest distance d of the piston from the center of the crankshaft when the rod makes an angle of 8.0°?

76. Surveying The scheme illustrated in the figure is used to determine inaccessible heights when d, α, β, and γ can be measured. Show that

$$h = d \sin \alpha \csc(\alpha + \beta) \tan \gamma$$

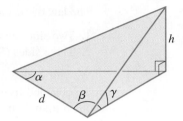

Figure for 76

6.2 Law of Cosines

- Deriving the Law of Cosines
- Solving the SAS Case
- Solving the SSS Case

In Figure 1, we are given two types of triangles that we didn't study in Section 6.1. In the first, we know two sides and the included angle (SAS), and in the second we are given three sides (SSS). Neither case can be solved using the law of sines because we don't know both an angle *and* the opposite side. In this section we will develop the *law of cosines,* which can be used for these two cases.

FIGURE 1

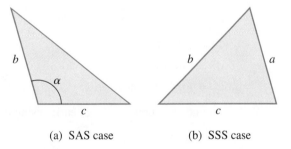

(a) SAS case   (b) SSS case

■ Deriving the Law of Cosines

The following three formulas are known collectively as the **law of cosines.**

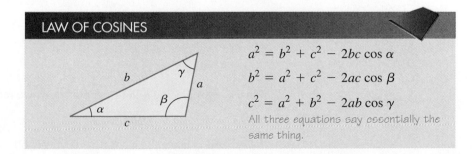

LAW OF COSINES

$$a^2 = b^2 + c^2 - 2bc \cos \alpha$$

$$b^2 = a^2 + c^2 - 2ac \cos \beta$$

$$c^2 = a^2 + b^2 - 2ab \cos \gamma$$

All three equations say essentially the same thing.

The law of cosines is used to solve a triangle if we are given:

1. Two sides and the included angle (SAS), or
2. Three sides (SSS)

We will derive only the first equation in the box. (The other equations can then be obtained from the first case simply by relabeling the figure.) We start by locating a triangle in a rectangular coordinate system. Figure 2 shows three typical triangles.

FIGURE 2

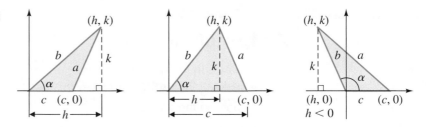

Since the left side of the formula we are trying to develop is a^2, we will begin by finding an expression for a^2.

For an arbitrary triangle located as in Figure 2, we use the formula for the distance between two points to obtain

$$a = \sqrt{(h - c)^2 + (k - 0)^2}$$

or, squaring both sides,

$$a^2 = (h - c)^2 + k^2$$
$$= h^2 - 2hc + c^2 + k^2 \qquad (1)$$

We can find an expression for b^2 easily using the Pythagorean theorem in Figure 2:

$$b^2 = h^2 + k^2$$

Notice that $h^2 + k^2$ is part of the right side in equation (1). Substituting b^2 for $h^2 + k^2$, we get

$$a^2 = b^2 + c^2 - 2hc \qquad (2)$$

But using a right triangle ratio,

$$\cos \alpha = \frac{h}{b} \quad \text{and}$$
$$h = b \cos \alpha$$

Finally, by replacing h in (2) with $b \cos \alpha$, we reach our objective:

$$a^2 = b^2 + c^2 - 2bc \cos \alpha$$

[*Note:* If α is acute, then $\cos \alpha > 0$; if α is obtuse, then $\cos \alpha < 0$.]

■ Solving the SAS Case

In this case we start by using the law of cosines to find the side opposite the given angle. We can then use either the law of cosines or the law of sines to find a second angle. Because of the simpler computation, we will generally use the law of sines to find a second angle.

EXPLORE/DISCUSS 1

After using the law of cosines to find the side opposite the angle for the SAS case, the law of sines is used to find a second angle. There are two choices for the second angle, and the following discussion shows that one choice may be better than the other.

(A) If the given angle between the two given sides is obtuse, explain why neither of the remaining angles can be obtuse.

(B) If the given angle between the two given sides is acute, explain why choosing the angle opposite the shorter given side guarantees the selection of an acute angle.

(C) Starting with $(\sin \beta)/b = (\sin \alpha)/a$, show that

$$\beta = \sin^{-1}\left(\frac{b \sin \alpha}{a} \right) \tag{3}$$

(D) Explain why equation (3) gives us the correct angle β only if β is acute.

This discussion leads to the following strategy for solving the SAS case:

STRATEGY FOR SOLVING THE SAS CASE		
Step	**Find**	**Method**
1.	Side opposite given angle	Law of cosines
2.	Second angle (Find the angle opposite the shorter of the two given sides—this angle will always be acute.)	Law of sines or law of cosines
3.	Third angle	Subtract the sum of the measures of the given angle and the angle found in step 2 from 180°.

EXAMPLE 1

Using the Law of Cosines (SAS)

Solve the triangle shown in Figure 3.

Solution   *Solve for b:*   With the given information, we have only one choice: the second equation in the law of cosines.

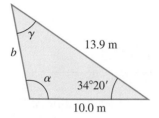

$$b^2 = a^2 + c^2 - 2ac \cos \beta \qquad \text{Solve for b.}$$

$$b = \sqrt{a^2 + c^2 - 2ac \cos \beta}$$

$$= \sqrt{(13.9)^2 + (10.0)^2 - 2(13.9)(10.0) \cos 34°20'}$$

$$= 7.98 \text{ m}$$

(We choose only the positive root because b is a length.)

Solve for γ: Since side c is shorter than side a, γ must be acute, so we use the law of sines to solve for γ.

$$\frac{\sin \gamma}{c} = \frac{\sin \beta}{b} \qquad \text{Law of sines}$$

$$\frac{\sin \gamma}{10.0} = \frac{\sin 34°20'}{7.98} \qquad \text{Multiply both sides by 10.0.}$$

$$\sin \gamma = \frac{10.0 \sin 34°20'}{7.98} \qquad \text{Use inverse sine to solve for } \gamma.$$

$$\gamma = \sin^{-1}\left(\frac{10.0 \sin 34°20'}{7.98}\right) \qquad \begin{array}{l}\text{Since } \gamma \text{ is acute, we can use the} \\ \text{inverse sine key on a calculator to} \\ \text{find the measure of } \gamma.\end{array}$$

$$= 45.0° \quad or \quad 45°0'$$

Solve for α:

$$\alpha = 180° - (\beta + \gamma)$$

$$= 180° - (34°20' + 45°0') = 100°40' \qquad ■$$

Matched Problem 1   Solve the triangle with $\alpha = 86.0°$, $b = 15.0$ m, and $c = 24.0$ m (angles in decimal degrees).   ■

EXPLORE DISCUSS 2

After finding side b in Example 1, we next used the law of sines to find angle γ. Rework that part of Example 1 by using the law of sines to find α instead of γ. Did you get the right answer? Explain why or why not.

Explore/Discuss 2 illustrates why it's a good idea to find the smaller of the remaining angles when using the law of sines: If you try to find the larger, and it happens to be obtuse, you will get the incorrect angle using inverse sine.

■ Solving the SSS Case

When we start with three sides of a triangle, our problem is to find the three angles. Although the law of cosines can be used to find any of the three angles, we will always start with the angle opposite the longest side to get an obtuse angle, if present, out of the way at the start. This will ensure that if we use the law of sines to find a second angle, we won't have to worry about two potential solutions. Then we will switch to the law of sines to find a second angle.

EXPLORE/DISCUSS 3

(A) Starting with $a^2 = b^2 + c^2 - 2bc \cos \alpha$, show that

$$\alpha = \cos^{-1}\left(\frac{a^2 - b^2 - c^2}{-2bc}\right) \tag{4}$$

(B) Explain why equation (4) gives us the correct angle α regardless of whether α is obtuse or acute.

This discussion leads to the following strategy for solving the SSS case:

STRATEGY FOR SOLVING THE SSS CASE		
Step	Find	Method
1.	Angle opposite longest side (This will take care of an obtuse angle, if present.)	Law of cosines
2.	Either of the remaining angles (Always acute, since a triangle cannot have more than one obtuse angle.)	Law of sines or law of cosines
3.	Third angle	Subtract the sum of the measures of the angles found in steps 1 and 2 from 180°.

EXAMPLE 2

Solving the SSS Case: Surveying

A triangular plot of land has sides $a = 21.2$ m, $b = 24.6$ m, and $c = 12.0$ m. Find the measures of all three angles in decimal degrees.

Solution   Find the measure of the angle opposite the longest side, which in this problem is angle β, first using the law of cosines. We make a rough sketch to keep the various parts of the triangle straight (Fig. 4).

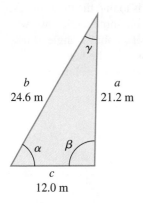

FIGURE 4

Solve for β: It is not clear whether β is acute, obtuse, or 90°, but we do not need to know beforehand, because the law of cosines will automatically tell us in the solution process.

$$b^2 = a^2 + c^2 - 2ac \cos \beta \qquad \text{Law of cosines}$$

$$(24.6)^2 = (21.2)^2 + (12.0)^2 - 2(21.2)(12.0) \cos \beta \quad \text{Solve for } \cos \beta.$$

$$\cos \beta = \frac{(24.6)^2 - (21.2)^2 - (12.0)^2}{-2(21.2)(12.0)} \qquad \text{Use inverse cosine to solve for } \beta.$$

$$\beta = \cos^{-1}\left(\frac{(24.6)^2 - (21.2)^2 - (12.0)^2}{-2(21.2)(12.0)}\right) \qquad \text{Use a calculator.}$$

$$= 91.3°$$

Solve for α: Both α and γ must be acute, since β is obtuse. We arbitrarily choose to find α first using the law of sines. Since α is acute, the inverse sine function in a calculator will give us the measure of this angle directly.

$$\frac{\sin \alpha}{a} = \frac{\sin \beta}{b} \qquad \text{Law of sines.}$$

$$\frac{\sin \alpha}{21.2} = \frac{\sin 91.3°}{24.6} \qquad \text{Multiply both sides by 21.2.}$$

$$\sin \alpha = \frac{21.2 \sin 91.3°}{24.6} \qquad \text{Use inverse sine to solve for } \alpha.$$

$$\alpha = \sin^{-1}\left(\frac{21.2 \sin 91.3}{24.6}\right) \qquad \text{Use a calculator.}$$

$$= 59.5°$$

Solve for γ:

$$\gamma = 180° - (\alpha + \beta)$$

$$= 180° - (59.5° + 91.3°) = 29.2° \qquad ■$$

Matched Problem 2   A triangular plot of land has sides $a = 217$ ft, $b = 362$ ft, and $c = 345$ ft. Find the measures of all three angles to the nearest $10'$.   ■

EXAMPLE 3

Navigation

A small plane is flying from Chicago to St. Louis, a distance of 258 mi. After flying at 130 mi/hr for 45 min, the pilot finds that she is 12° off course. If she corrects the course and maintains speed and direction for the remainder of the flight, how much longer will the flight take?

Solution We first need to know how far the plane has flown after 45 min, or 3/4 hr.

$$130 \, \frac{mi}{hr} \cdot \frac{3}{4} \, hr = 97.5 \, mi$$

Now we can draw a diagram (Fig. 5). The distance remaining is labeled d, and we can find it using the law of cosines.

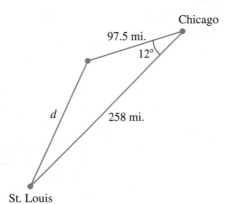

FIGURE 5

$$d^2 = (97.5)^2 + (258)^2 - 2(97.5)(258) \cos 12°$$

$$d = \sqrt{(97.5)^2 + (258)^2 - 2(97.5)(258) \cos 12°}$$

$$= 164 \, mi$$

Finally, we can find the remaining time:

$$\frac{164 \, mi}{130 \, \frac{mi}{hr}} \approx 1.3 \, hr, \text{ or about 1 hr, 18 min} \qquad ■$$

Matched Problem 3 Repeat Example 3 if the pilot was 21° off course after flying 110 mi/hr for 90 min. ■

We conclude our study of solving triangles with a brief summary describing a simple overall strategy.

STRATEGY FOR SOLVING OBLIQUE (NON-RIGHT) TRIANGLES

Given information	Overall strategy
Two angles and any side	Subtract from 180° to find the third angle, then use the law of sines.
All three sides	Use the law of cosines to find the largest angle.
Two sides and an angle	If one of the sides is opposite the angle, use the law of sines. If not, use the law of cosines.

Answers to Matched Problems

1. $a = 27.4$ m, $\beta = 33.1°$, $\gamma = 60.9°$
2. $\alpha = 35°40'$, $\beta = 76°30'$, $\gamma = 67°50'$
3. 1 hr, 5min

EXERCISE 6.2

All triangles in this exercise are labeled as in the figure unless stated to the contrary. Your answers may differ slightly from those in the back of the book, depending on the order in which you solve for the sides and angles.

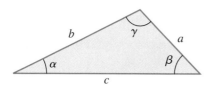

A   1. Explain why it's a good idea to first find the largest angle when solving a triangle, given all three sides.

2. Use drawings to decide if you can always find a solution to a triangle when given two sides and an angle between them that is between 0° and 180°.

In Problems 3–10, decide whether you should use the law of sines or the law of cosines to begin solving the triangle. Do not solve.

3. $a = 5$ ft, $b = 6$ ft, $c = 7$ ft
4. $\alpha = 10°$, $\gamma = 119°$, $c = 10$ mi
5. $\beta = 45°$, $a = 8$ in., $b = 11$ in.
6. $\gamma = 9.1°$, $a = 14$ km, $c = 20$ km
7. $a = 9$ ft, $b = 12$ ft, $\gamma = 101°$
8. $a = 1.3$ m, $b = 7$ m, $c = 8.1$ m

9. $\beta = 81°$, $\gamma = 4°$, $b = 21$ yd
10. $\alpha = 5.6°$, $b = 3$ mm, $c = 2.1$ mm

11. Referring to the figure at the beginning of the exercise, if $\beta = 38.7°$, $a = 25.3$ ft, and $c = 19.6$ ft, which of the two angles, α or γ, can you say for certain is acute? Why?

12. Referring to the figure at the beginning of the exercise, if $\alpha = 92.6°$, $b = 33.8$ cm, and $c = 49.1$ cm, which of the two angles, β or γ, can you say for certain is acute? Why?

Solve each triangle in Problems 13–16.

13. $\alpha = 50°40'$, $b = 7.03$ mm, $c = 7.00$ mm
14. $\alpha = 71°0'$, $b = 5.32$ cm, $c = 5.00$ cm
15. $\gamma = 134.0°$, $a = 20.0$ m, $b = 8.00$ m
16. $\alpha = 120.0°$, $b = 5.00$ km, $c = 10.0$ km

B   17. Referring to the figure at the beginning of the exercise, if $a = 36.5$ mm, $b = 22.7$ mm, and $c = 19.1$ mm, then, if the triangle has an obtuse angle, which angle must it be? Why?

18. You are told that a triangle has sides $a = 29.4$ ft, $b = 12.3$ ft, and $c = 16.7$ ft. Explain why the triangle has no solution.

Solve each triangle in Problems 19–22.

19. $a = 9.00$ yd, $b = 6.00$ yd, $c = 10.0$ yd
    (decimal degrees)

20. $a = 5.00$ km, $b = 5.50$ km, $c = 6.00$ km (decimal degrees)

21. $a = 420.0$ km, $b = 770.0$ km, $c = 860.0$ km (degrees and minutes)

22. $a = 15.0$ cm, $b = 12.0$ cm, $c = 10.0$ cm (degrees and minutes)

Problems 23–40 represent a variety of problems involving the first two sections of this chapter. Solve each triangle using the law of sines or the law of cosines (or both). If a problem does not have a solution, say so.

23. $\beta = 132.4°$, $\gamma = 17.3°$, $b = 67.6$ ft

24. $\alpha = 57.2°$, $\gamma = 112.0°$, $c = 24.8$ ft

25. $\beta = 66.5°$, $a = 13.7$ m, $c = 20.1$ m

26. $\gamma = 54.2°$, $a = 112$ ft, $b = 87.2$ ft

27. $\beta = 84.4°$, $\gamma = 97.8°$, $a = 12.3$ cm

28. $\alpha = 95.6°$, $\gamma = 86.3°$, $b = 43.5$ cm

29. $a = 10.5$ in., $b = 5.23$ in., $c = 9.66$ in. (decimal degrees)

30. $a = 15.0$ ft, $b = 18.0$ ft, $c = 22.0$ ft (decimal degrees)

31. $\gamma = 80.3°$, $a = 14.5$ mm, $c = 10.0$ mm

32. $\beta = 63.4°$, $b = 50.5$ in., $c = 64.4$ in.

33. $\alpha = 46.3°$, $\gamma = 105.5°$, $b = 643$ m

34. $\beta = 123.6°$, $\gamma = 21.9°$, $a = 108$ cm

35. $a = 12.2$ m, $b = 16.7$ m, $c = 30.0$ m

36. $a = 28.2$ yd, $b = 52.3$ yd, $c = 22.0$ yd

37. $\alpha = 46.7°$, $a = 18.1$ yd, $b = 22.6$ yd

38. $\gamma = 58.4°$, $b = 7.23$ cm, $c = 6.54$ cm

39. $\alpha = 36.5°$, $\beta = 72.4°$, $\gamma = 71.1°$

40. $\alpha = 29°20'$, $\beta = 32°50'$, $\gamma = 117°50'$

C **41.** Using the law of cosines, show that if $\beta = 90°$, then $b^2 = c^2 + a^2$ (the Pythagorean theorem).

42. Using the law of cosines, show that if $b^2 = c^2 + a^2$, then $\beta = 90°$.

43. Check Problem 13 using Mollweide's equation (see Problem 49, Exercise 6.1),

$$(a - b)\cos\frac{\gamma}{2} = c\sin\frac{\alpha - \beta}{2}$$

44. Check Problem 5 using Mollweide's equation (see Problem 43).

45. Show that for any triangle with standard labeling (see the figure at the beginning of the exercise),

$$c = b\cos\alpha + a\cos\beta$$

46. Use Problem 45 to show that the sum of the lengths of any two sides of a triangle is greater than the length of the third side.

47. The base of an isosceles triangle is 1.5 times the length of the congruent sides. Use the law of cosines to find the base angles.

48. Show that for any triangle with standard labeling (see the figure at the beginning of the exercise),

$$\frac{a^2 + b^2 + c^2}{2abc} = \frac{\cos\alpha}{a} + \frac{\cos\beta}{b} + \frac{\cos\gamma}{c}$$

Applications

49. **Surveying**   A geologist wishes to determine the distance CB across the base of a volcanic cinder cone (see the figure). Distances AB and AC are measured to be 425 m and 384 m, respectively, and $\angle CAB$ is 98.3°. Find the approximate distance across the base of the cinder cone.

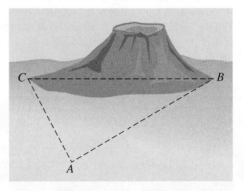

Figure for 49

50. **Surveying**   To estimate the length CB of the lake in the figure on the next page, a surveyor measures AB and AC to be 89 m and 74 m, respectively, and $\angle CAB$ to be 95°. Find the approximate length of the lake.

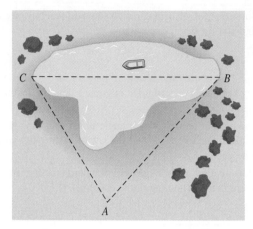

Figure for 50

51. Geometry: Engineering Find the measure in decimal degrees of a central angle subtended by a chord of length 13.8 cm in a circle of radius 8.26 cm.

52. Geometry: Engineering Find the measure in decimal degrees of a central angle subtended by a chord of length 112 ft in a circle of radius 72.8 ft.

53. Search and Rescue At midnight, two Coast Guard helicopters set out from San Francisco to find a sail-boat in distress. Helicopter *A* flies due west over the Pacific Ocean at 250 km/hr, and helicopter *B* flies northwest at 210 km/hr. At 1 A.M., helicopter *A* spots a flare from the boat and radios helicopter *B* to come and assist in the rescue. How far is helicopter *B* from heli-copter *A* at this time? Compute the answer to two sig-nificant digits.

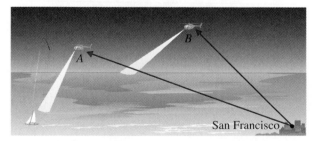

Figure for 53

54. Navigation Los Angeles and San Francisco are approximately 600 km apart. A pilot flying from Los Angeles to San Francisco notices when she is 200 km from Los Angeles that the plane is 20° off course. How far is the plane from San Francisco at this time (to two significant digits)?

55. Navigation A cruise ship is sailing from St. Thomas to St. Lucia, a distance of 398 mi. After 6 hr at a cruising speed of 26 mi/hr in heavy seas, the captain is informed that the ship is 14.1° off course. If the course is correct-ed and speed and direction are maintained, how much longer will the trip take?

56. Geometry: Engineering A 58.3 cm chord of a circle subtends a central angle of 27.8°. Find the radius of the circle to three significant digits using the law of cosines.

57. Geometry: Engineering Find the perimeter (to the nearest centimeter) of a regular pentagon inscribed in a circle with radius 5 cm (see the figure).

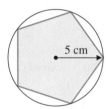

5 cm

Figure for 57

58. Geometry: Engineering Three circles of radius 2 cm, 3 cm, and 8 cm are tangent to each other (see the figure). Find (to the nearest 10′) the three angles formed by the lines joining their centers.

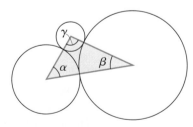

Figure for 58

59. Engineering A tunnel for hydroelectric power is to be constructed through a mountain from one reservoir to another at a lower level (see the figure on the next page). The distance from the top of the mountain to the lower end of the tunnel is 5.32 mi, and that from the top of the mountain to the upper end of the tunnel is 2.63 mi. The angles of depression of the two slopes of the mountain are 42.7° and 48.8°, respectively.

(A) What is the length of the tunnel?

(B) What angle does the tunnel make with the horizontal?

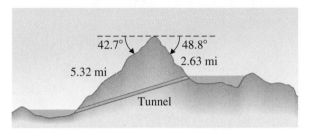

Figure for 59

60. **Engineering: Construction**   A fire lookout is to be constructed as indicated in the figure. Support poles *AD* and *BC* are each 18.0 ft long and tilt 8.0° inward from the vertical. The distance between the tops of the poles, *DC,* is 12.0 ft. Find the length of a brace *AC* and the distance *AB* between the supporting poles at ground level.

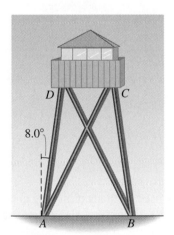

Figure for 60

61. **Engineering: Construction**   Repeat Problem 60 with support poles *AD* and *BC* making an angle of 11.0° with the vertical instead of 8.0°.

62. **Space Science**   A satellite, *S,* in circular orbit around the earth, is sighted by a tracking station *T* (see the figure). The distance *TS* is determined by radar to be 1,034 mi, and the angle of elevation above the horizon is 32.4°. How high is the satellite above the earth at the time of the sighting? The radius of the earth is *r* = 3,964 mi.

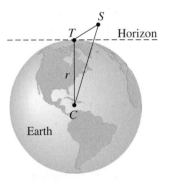

Figure for 62

63. **Space Science**   For communications between a space shuttle and the White Sands Missile Range in southern New Mexico, two satellites are placed in geostationary orbit, 130° apart. Each is 22,300 mi above the surface of the earth (see the figure). (When a satellite is in **geostationary orbit** it remains stationary above a fixed point on the earth's surface.) Radio signals are sent from an orbiting shuttle by way of the satellites to the White Sands facility, and vice versa. This arrangement enables the White Sands facility to maintain radio contact with the shuttle over most of the earth's surface. How far (to the nearest 100 mi) is one of these geostationary satellites from the White Sands facility (*W*)? The radius of the earth is 3,964 mi.

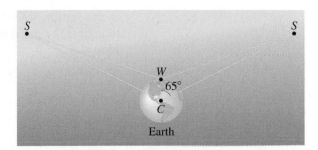

Figure for 63

64. Geometry: Engineering A rectangular solid has sides of 6.0 cm, 3.0 cm, and 4.0 cm (see the figure). Find $\angle ABC$.

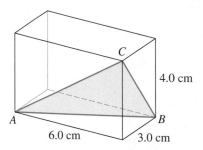

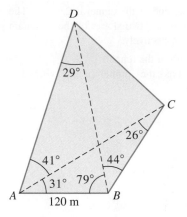

Figure for 66

Figure for 64 and 65

65. Geometry: Engineering Refer to Problem 64. Find $\angle ACB$.

66. Surveying A plot of land was surveyed, with the resulting information shown in the figure. Find the length of DC.

67. Surveying A plot of land was surveyed, with the resulting information shown in the figure. Find the length of BC.

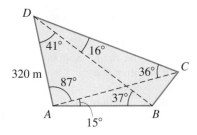

Figure for 67

☆6.3 Areas of Triangles

- Base and Height Given
- Two Sides and Included Angle Given
- Three Sides Given (Heron's Formula)
- Arbitrary Triangles

In this section, we will discuss three frequently used methods of finding areas of triangles. The derivation of *Heron's formula* also illustrates a significant use of identities.

■ Base and Height Given

If the base b and height h of a triangle are given (see Fig. 1), then the area A is one-half the area of a parallelogram with the same base and height.

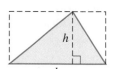

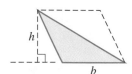

FIGURE 1

$A = \frac{1}{2}bh$

☆ Sections marked with a star may be omitted without loss of continuity.

$A = \frac{1}{2}bh$

Notice that in each case in Figure 1, the shaded triangle fills exactly half of the parallelogram.

EXPLORE/DISCUSS 1

Given two parallel lines m and n, explain why all three triangles (I, II, and III) sharing the same base b have the same area.

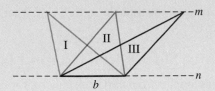

■ Two Sides and Included Angle Given

The area formula, $A = \frac{1}{2}bh$, can be rewritten into a form that doesn't require knowing the height of the triangle. Suppose that for the triangles in Figure 2, all we know are two sides, a and b, and the included angle, θ. According to the figure,

FIGURE 2

$$A = \frac{ab}{2}\sin\theta$$

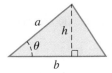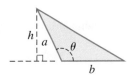

$\sin\theta = h/a$, so $h = a\sin\theta$. [This is true for both cases in Fig. 2 because $\sin(180° - \theta) = \sin\theta$.] Our area formula then becomes

$$A = \frac{1}{2}bh = \frac{1}{2}b(a\sin\theta) \quad \text{or}$$

$$A = \frac{ab}{2}\sin\theta \qquad \text{This is the area of the triangle.}$$

EXAMPLE 1

Find the Area of a Triangle Given Two Sides and the Included Angle

Find the area of the triangle shown in Figure 3 on page 380.

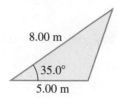

FIGURE 3

Solution

$$A = \frac{ab}{2} \sin \theta = \frac{1}{2}(8.00)(5.00) \sin 35.0°$$

$$= 11.5 \, \text{m}^2$$

∎

Matched Problem 1   Find the area of the triangle with $a = 12.0$ cm, $b = 7.00$ cm, and included angle $\theta = 125.0°$.

∎

■ Three Sides Given (Heron's Formula)

A famous formula from the Greek philosopher-mathematician Heron of Alexandria (75 AD) enables us to compute the area of a triangle directly when given only the lengths of the three sides of the triangle.

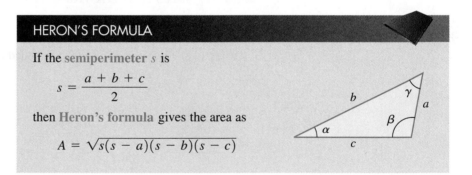

HERON'S FORMULA

If the **semiperimeter** s is

$$s = \frac{a + b + c}{2}$$

then **Heron's formula** gives the area as

$$A = \sqrt{s(s - a)(s - b)(s - c)}$$

Heron's formula is obtained from

$$A = \frac{bc}{2} \sin \alpha \qquad\qquad (1)$$

by expressing $\sin \alpha$ in terms of the sides a, b, and c. Several identities and the law of cosines play a central role in the derivation of the formula. We will first get $\sin(\alpha/2)$ and $\cos(\alpha/2)$ in terms of a, b, and c. Then we will use a double-angle identity in the form

$$\sin \alpha = 2 \sin \frac{\alpha}{2} \cos \frac{\alpha}{2} \qquad\qquad (2)$$

to write $\sin \alpha$ in terms of a, b, and c. We start with a half-angle identity for sine in the form

$$\sin^2 \frac{\alpha}{2} = \frac{1 - \cos \alpha}{2} \qquad (3)$$

The following version of the law of cosines involves $\cos \alpha$ and all three sides of the triangle a, b, and c:

$$a^2 = b^2 + c^2 - 2bc \cos \alpha$$

Solving for $\cos \alpha$, we obtain

$$\cos \alpha = \frac{b^2 + c^2 - a^2}{2bc} \qquad (4)$$

Substituting (4) into (3), we can write $\sin^2(\alpha/2)$ in terms of a, b, and c:

$$\sin^2 \frac{\alpha}{2} = \frac{1 - \dfrac{b^2 + c^2 - a^2}{2bc}}{2}$$

Multiply numerator and denominator by $2bc$.

$$= \frac{2bc \left(1 - \dfrac{b^2 + c^2 - a^2}{2bc} \right)}{2bc(2)}$$

Simplify.

$$= \frac{2bc - b^2 - c^2 + a^2}{4bc}$$

Numerator factors (not obvious).

$$= \frac{(a - b + c)(a + b - c)}{4bc} \qquad (5)$$

To bring the semiperimeter

$$s = \frac{a + b + c}{2} \qquad (6)$$

into the picture, we write (5) in the form

$$\sin^2 \frac{\alpha}{2} = \frac{(a + b + c - 2b)(a + b + c - 2c)}{4bc}$$

$$= \frac{2 \left(\dfrac{a + b + c}{2} - b \right) 2 \left(\dfrac{a + b + c}{2} - c \right)}{4bc}$$

$$= \frac{\left(\dfrac{a + b + c}{2} - b \right) \left(\dfrac{a + b + c}{2} - c \right)}{bc} \qquad (7)$$

Substituting (6) into (7) and solving for $\sin(\alpha/2)$, we obtain

$$\sin \frac{\alpha}{2} = \sqrt{\frac{(s - b)(s - c)}{bc}} \qquad (8)$$

If we repeat this reasoning starting with a half-angle identity for cosine in the form

$$\cos^2 \frac{\alpha}{2} = \frac{1 + \cos \alpha}{2} \tag{9}$$

we obtain

$$\cos \frac{\alpha}{2} = \sqrt{\frac{s(s-a)}{bc}} \tag{10}$$

Substituting (8) and (10) into identity (2) produces

$$\sin \alpha = 2 \sqrt{\frac{(s-b)(s-c)}{bc}} \sqrt{\frac{s(s-a)}{bc}} = \frac{2}{bc} \sqrt{s(s-a)(s-b)(s-c)} \tag{11}$$

And we are almost there. We now substitute (11) into formula (1) to obtain Heron's formula:

$$
\begin{aligned}
A &= \frac{bc}{2} \left[\frac{2}{bc} \sqrt{s(s-a)(s-b)(s-c)} \right] \\
&= \sqrt{s(s-a)(s-b)(s-c)}
\end{aligned}
$$

The derivation of Heron's formula provides a good illustration of the importance of identities. Try to imagine a derivation of this formula without identities!

EXPLORE/DISCUSS 2

Derive equation (10) above following the same type of reasoning that was used to derive equation (8).

EXAMPLE 2

Finding the Area of a Triangle Given Three Sides

Find the area of a triangle with sides $a = 12.0$ cm, $b = 8.0$ cm, and $c = 6.0$ cm.

Solution First, find the semiperimeter s:

$$s = \frac{a+b+c}{2} = \frac{12.0 + 8.0 + 6.0}{2} = 13.0 \text{ cm}$$

Then

$$
\begin{aligned}
s - a &= 13.0 - 12.0 = 1.0 \text{ cm} \\
s - b &= 13.0 - 8.0 = 5.0 \text{ cm} \\
s - c &= 13.0 - 6.0 = 7.0 \text{ cm}
\end{aligned}
$$

Using Heron's formula, we get

$$A = \sqrt{s(s-a)(s-b)(s-c)}$$
$$= \sqrt{13.0(1.0)(5.0)(7.0)}$$
$$= 21 \text{ cm}^2 \qquad \text{To two significant digits}$$

[*Note:* The computed area has the same number of significant digits as the side with the least number of significant digits.] ■

Matched Problem 2   Find the area of a triangle with $a = 6.0$ m, $b = 10.0$ m, and $c = 8.0$ m. ■

■ **Arbitrary Triangles**

To find the area of a triangle when the given information does not include (i) two sides and the included angle or (ii) all three sides, we partially solve the triangle to reduce to (i) or (ii). The law of sines may be useful, as illustrated by Example 3.

EXAMPLE 3

Finding the Area of a Triangle

Find the area of a triangle with $\alpha = 39°$, $a = 9.4$ ft, and $b = 7.2$ ft.

Solution   Angle γ is the included angle for sides a and b, so we will find γ. But first, we have to use the law of sines to find β.

$$\frac{\sin \beta}{b} = \frac{\sin \alpha}{a} \qquad \text{Law of sines}$$

$$\frac{\sin \beta}{7.2} = \frac{\sin 39°}{9.4} \qquad \text{Multiply both sides by 7.2.}$$

$$\sin \beta = \frac{7.2 \sin 39°}{9.4} \qquad \text{Use inverse sine to solve for } \beta.$$

$$\beta = \sin^{-1}\left(\frac{7.2 \sin 39°}{9.4}\right) = 29°$$

The third angle γ is given by

$$\gamma = 180° - (\alpha + \beta) = 112°$$

Since γ is the included angle for sides a and b,

$$A = \frac{ab}{2} \sin \gamma = \frac{1}{2}(9.4)(7.2) \sin 112°$$
$$= 31 \text{ ft}^2 \qquad\qquad\qquad\qquad\qquad ■$$

Matched Problem 3   Find the area of a triangle with $\alpha = 104°$, $a = 51$ in., and $b = 42$ in. ■

Answers to
Matched Problems

1. 34.4 cm^2
2. 24 m^2
3. 420 in.^2

EXERCISE 6.3

A **1.** Explain, using illustrations, why the area of a triangle is half the area of a parallelogram with the same base and height.

2. Discuss a method you could use to find the area of the triangle below if you didn't know Heron's formula.

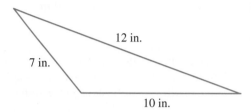

Find the area of the triangle matching the information given in Problems 3–22.

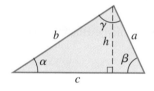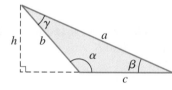

Figure for 3–22

3. $h = 12.0$ m, $c = 17.0$ m

4. $h = 7.0$ ft, $c = 10.0$ ft

5. $c = 3\sqrt{5}$ m, $h = \dfrac{\sqrt{5}}{2}$ m

6. $c = 2\sqrt{6}$ yd, $h = \dfrac{\sqrt{3}}{2}$ yd

7. $\alpha = 30.0°$, $b = 6.0$ cm, $c = 8.0$ cm

8. $\alpha = 45°$, $b = 5.0$ m, $c = 6.0$ m

9. $a = 4.00$ in., $b = 6.00$ in., $c = 8.00$ in.

10. $a = 4.00$ ft, $b = 10.00$ ft, $c = 12.00$ ft

11. $\alpha = 23°20'$, $b = 403$ ft, $c = 512$ ft

12. $\alpha = 58°40'$, $b = 28.2$ in., $c = 6.40$ in.

13. $\alpha = 132.67°$, $b = 12.1$ cm, $c = 10.2$ cm

14. $\alpha = 147.5°$, $b = 125$ mm, $c = 67.0$ mm

15. $a = 12.7$ m, $b = 20.3$ m, $c = 24.4$ m

16. $a = 5.24$ cm, $b = 3.48$ cm, $c = 6.04$ cm

B **17.** $\alpha = 15°$, $\beta = 35°$, $c = 4.5$ ft

18. $\alpha = 25°$, $\beta = 115°$, $c = 14.2$ ft

19. $\alpha = 72°$, $a = 38$ in., $b = 29$ in.

20. $\alpha = 115°$, $a = 93$ in., $b = 82$ in.

21. $\beta = 175°$, $a = 3.2$ cm, $b = 4.5$ cm

22. $\beta = 12°$, $a = 7.8$ cm, $b = 6.4$ cm, α acute

C *In Problems 23–26, determine whether the statement is true or false. If true, explain. If false, give a specific counterexample.*

23. If two triangles have the same side lengths, then they have the same area.

24. If two quadrilaterals have the same side lengths, then they have the same area.

25. If two triangles have the same semiperimeter, then they have the same area.

26. If two triangles have the same area, then they have the same side lengths.

27. Show that the diagonals of a parallelogram divide the figure into four triangles, all having the same area.

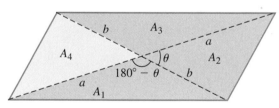

Figure for 27

28. If $s = (a + b + c)/2$ is the semiperimeter of a triangle with sides a, b, and c (see the figure), show that the radius r of an inscribed circle is given by

$$r = \sqrt{\frac{(s - a)(s - b)(s - c)}{s}}$$

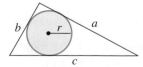

Figure for 28

29. Find the area of a regular pentagon that has a perimeter of 35 ft.

30. Find the area of a regular decagon that has a perimeter of 35 ft.

31. Find the area of a regular heptagon (seven-sided polygon) if the length of a side is 25 m.

32. Find the area of a regular nonagon (nine-sided polygon) if the length of a side is 25 m.

 Applications

33. Construction Cost   A building owner contracts with a local construction company to build a brick patio in a triangular courtyard. The three sides of the courtyard measure 42 ft, 51 ft 4 in., and 37 ft 9 in. The construction company charges $15.50 per square foot for materials and installation. Find the cost of the patio to the nearest hundred dollars.

34. Gardening   A gardener is building four triangular wildflower beds along one wall of a museum. The triangles will be equilateral with side length 14 ft 3 in. The wildflower seeds are to be spread at a rate of one packet for each 20 sq ft. How many packets will the gardener need?

35. Land Cost   A four-sided plot of land, shown in the figure, occupies the cul-de-sac in a new development.

The land in the rest of the development has sold for $4.50 per square foot. Find the price of this plot to the nearest thousand dollars. [*Hint:* Draw a diagonal that divides the plot into two triangles.]

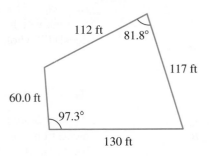

36. Land Cost   Refer to Problem 35. The plot drawn below is in a different development nearby, where land costs $5.20 per square foot. Find the price of the plot to the nearest thousand dollars.

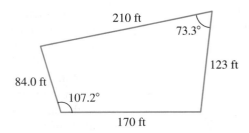

6.4 Vectors: Geometrically Defined

- Geometric Vectors; Vector Addition
- Velocity Vectors
- Force Vectors
- Resolution of a Vector into Components
- Centrifugal Force

Scalar quantities are physical quantities that can be completely specified by a single real number, such as length, area, or volume. But many physical quantities—for example, displacements, velocities, and forces—can't be described by just a size: They require both a size (magnitude) *and* a direction for a complete description. We call these **vector quantities.**

Vector quantities are widely used in both pure and applied mathematics. They are used extensively in the physical sciences and engineering and are seeing increased use in the social and life sciences.

In this section we will limit our development to intuitive notions of *geometric vectors in a plane*. In Section 6.5 we will present an algebraic treatment of vectors, which is just the first step in developing an area of study that can (and does!) fill whole books.

Geometric Vectors; Vector Addition

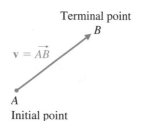

Terminal point

$\mathbf{v} = \vec{AB}$

B

A
Initial point

FIGURE 1
Vector $\vec{AB} = v$

If *A* and *B* are two points in the plane, then the **directed line segment** (displacement) from *A* to *B*, denoted by $\vec{AB}$, is the line segment joining *A* to *B* with an arrowhead placed at *B* to indicate the direction is from *A* to *B* (see Fig. 1). Point *A* is called the **initial point**, and point *B* is called the **terminal point.**

A **geometric vector in the plane** is a quantity that possesses both a length and a direction and can be represented by a directed line segment. The directed line segment in Figure 1 represents a vector that is denoted by $\vec{AB}$. A vector also may be denoted by a boldface letter, such as **v** or **F**, or by a letter with an arrow over it, such as $\vec{v}$. Letters with arrows over them are easier to write by hand, but boldface letters are easier to recognize in print. We will generally use boldface letters for vector quantities in this book, but when you write a vector by hand you should include the arrow above it as a reminder that the quantity is a vector.

The **magnitude** of the vector $\vec{AB}$, denoted by $|\vec{AB}|$, $|\vec{v}|$, or $|\boldsymbol{v}|$, is the length of the directed line segment. Two vectors have the **same direction** if they are parallel and point in the same direction. Two vectors have **opposite direction** if they are parallel and point in opposite directions. The **zero vector,** denoted by $\vec{0}$ or $\mathbf{0}$, has a magnitude of zero and an arbitrary direction. Two vectors are **equal** if they have the same magnitude and direction. That is, as long as the length and direction are the same, we consider two vectors to be the same no matter where they begin. So a vector can be **translated** from one location to another as long as the magnitude and direction do not change.

The **sum of two vectors u and v** can be defined using the **tail-to-tip rule:** Translate **v** so that its tail end (initial point) is at the tip end (terminal point) of **u.** Then the vector from the tail end of **u** to the tip end of **v** is the sum, denoted by **u + v,** of the vectors **u** and **v** (see Fig. 2). The sum of two nonparallel vectors also can be defined using the **parallelogram rule:** The **sum of two nonparallel vectors u and v** is the diagonal of the parallelogram formed using **u** and **v** as adjacent sides (see Fig. 3). (If **u** and **v** are parallel, use the tail-to-tip rule.) Of course, both rules give the same sum, as they should. The choice of which rule to use depends on the situation.

The sum vector **u + v** is also called the **resultant** of the two vectors **u** and **v,** and **u** and **v** are called **components** of **u + v.** As we will show in the exercise, vector addition is **commutative** and **associative.** That is, **u + v = v + u** and **u + (v + w) = (u + v) + w.**

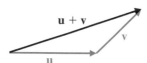

u + v

v

u

FIGURE 2
Tail-to-tip rule for addition

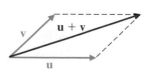

v   **u + v**

u

FIGURE 3
Parallelogram rule for addition

EXAMPLE 1

Finding the Sum of Two Vectors

For the vectors **u** and **v** drawn in Figure 4, draw **u** + **v** using the parallelogram rule, then find the magnitude of **u** + **v**.

FIGURE 4

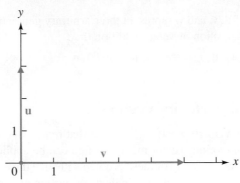

Solution The sum is drawn in Figure 5. Because the vectors are perpendicular, we can use the Pythagorean theorem to find $|\mathbf{u} + \mathbf{v}|$. Note that $|\mathbf{u}| = 3$ and $|\mathbf{v}| = 5$.

$$|\mathbf{u} + \mathbf{v}|^2 = 3^2 + 5^2 = 34$$

$$|\mathbf{u} + \mathbf{v}| = \sqrt{34}$$

FIGURE 5

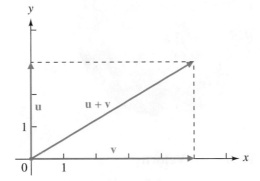

Matched Problem 1 Repeat Example 1 for vectors **u** and **v** in Figure 6.

FIGURE 6

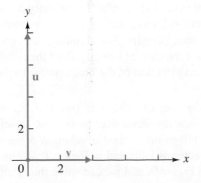

EXPLORE/DISCUSS 1

If **u, v,** and **w** represent three arbitrary geometric vectors, illustrate using either definition of vector addition that:

(A) $\mathbf{u} + \mathbf{v} = \mathbf{v} + \mathbf{u}$ (B) $\mathbf{u} + (\mathbf{v} + \mathbf{w}) = (\mathbf{u} + \mathbf{v}) + \mathbf{w}$

■ Velocity Vectors

A **velocity vector** is a vector that represents the speed and direction of an object in motion. Vector methods often can be applied to problems involving objects in motion. Many of these problems involve the use of a *navigational compass,* which is marked clockwise in degrees starting at north (see Fig. 7).

FIGURE 7

EXAMPLE 2

Resultant Velocity

A power boat traveling at 24 km/hr relative to the water has a compass heading (the direction the boat is pointing) of 95°. A strong tidal current, with a heading of 35°, is flowing at 12 km/hr. The velocity of the boat relative to the water is called its **apparent velocity,** and the velocity relative to the ground is called the **resultant** or **actual velocity.** The resultant velocity is the vector sum of the apparent velocity and the current velocity. Find the resultant velocity; that is, find the actual speed and direction of the boat relative to the ground.

Solution   We can use geometric vectors (see Fig. 8a on the next page) to represent the apparent velocity vector and the current velocity vector. We add the two vectors using the tail-to-tip method of addition of vectors to obtain the resultant (actual) velocity vector as indicated in Figure 8b. From this vector diagram we obtain the triangle in Figure 9, and we can solve this triangle for β, b, and α.

FIGURE 8

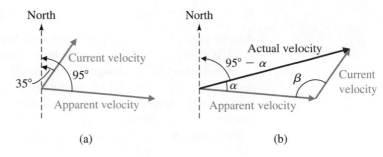

(a)   (b)

FIGURE 9

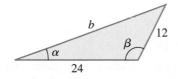

Boat's actual heading: $95° - \alpha$
Boat's actual speed: b

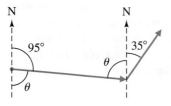

FIGURE 10

Solve for β: Using the apparent velocity heading (95°) and the current velocity heading (35°), we can find β using Figure 10. Note that the two angles labeled θ are equal because they are alternate interior angles relative to the parallel dotted lines.

$$\theta = 180° - 95° = 85°$$
$$\beta = \theta + 35° = 85° + 35° = 120°$$

Solve for b: Use the law of cosines:

$$b^2 = a^2 + c^2 - 2ac \cos \beta \qquad\qquad a = 12, c = 24, \beta = 120°$$
$$= 12^2 + 24^2 - 2(12)(24) \cos 120°$$
$$b = \sqrt{12^2 + 24^2 - 2(12)(24) \cos 120°}$$
$$= 32 \text{ km/hr} \qquad\qquad \text{Actual speed relative to the ground}$$

Solve for α: Use the law of sines:

$$\frac{\sin \alpha}{a} = \frac{\sin \beta}{b} \qquad\qquad \text{Law of sines}$$
$$\frac{\sin \alpha}{12} = \frac{\sin 120°}{32} \qquad\qquad \text{Multiply both sides by 12.}$$
$$\sin \alpha = \frac{12 \sin 120°}{32} \qquad\qquad \text{Use inverse sine to solve for } \alpha$$
$$\alpha = \sin^{-1}\left[\frac{12 \sin 120°}{32}\right] \qquad\qquad \text{Use calculator.}$$
$$= 19°$$

Actual heading $= 95° - \alpha = 95° - 19° = 76°$ ■

Matched Problem 2   Repeat Example 2 with a current of 8.0 km/hr at 22° and the speedometer on the boat reading 35 km/hr with a compass heading of 85°.   ■

■ Force Vectors

A **force vector** is a vector that represents the direction and magnitude of an applied force. If an object is subjected to two forces, then the sum of these two forces (the **resultant force**) can be thought of as a single force. In other words, if the resultant force replaced the original two forces, it would act on the object in the same way as the two original forces taken together. In physics it is shown that the resultant force vector can be obtained using vector addition to add the two individual force vectors. It seems natural to use the parallelogram rule for adding force vectors, as illustrated in the next example.

Resultant Force

Figure 11 shows a man and a horse pulling on a large piece of granite. The man is pulling with a force of 200 lb, and the horse is pulling with a force of 1,000 lb 50° away from the direction in which the man is pulling. The length of the main diagonal of the parallelogram will be the actual magnitude of the resultant force on the stone, and its direction will be the direction of motion of the stone (if it moves). Find the magnitude and direction α of the resultant force. (The magnitudes of the forces are measured to two significant digits and the angle to the nearest degree.)

FIGURE 11

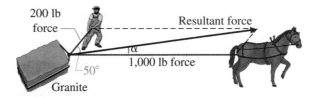

Solution   From Figure 11 we obtain the triangle in Figure 12 and then solve for α and b.

FIGURE 12

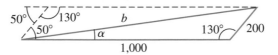

Solve for b:

$$b^2 = 1{,}000^2 + 200^2 - 2(1{,}000)(200)\cos 130° \quad \text{Law of cosines}$$
$$= 1{,}297{,}115.04$$
$$b = \sqrt{1{,}297{,}115.04} = 1{,}100 \text{ lb}$$

Solve for α:

$$\frac{\sin \alpha}{200} = \frac{\sin 130°}{1{,}100} \qquad\qquad \text{Law of sines}$$

$$\sin \alpha = \frac{200}{1,100} \sin 130°$$

$$\alpha = \sin^{-1}\left(\frac{200}{1,100} \sin 130°\right) = 8°$$ ■

Matched Problem 3 Repeat Example 3, but change the angle between the force vector from the horse and the force vector from the man to 45° instead of 50°, and change the magnitude of the force vector from the horse to 800 lb instead of 1,000 lb. ■

■ Resolution of a Vector into Components

Instead of adding vectors, many problems require the opposite: writing a single vector as a sum. Whenever a vector is expressed as a resultant of two vectors, these two vectors are called **components** of the given vector. In many cases, we will find it helpful to find the horizontal and vertical components of a vector, as in Example 4.

EXAMPLE 4 **Finding Components of a Vector**

Find the horizontal and vertical components of the vector **u** in Figure 13.

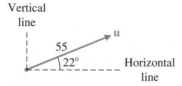

FIGURE 13

Solution We will find the magnitudes of these component vectors, labeled **H** and **V** in Figure 14, using the sine and cosine functions. The directions are already known since the vectors are horizontal and vertical.

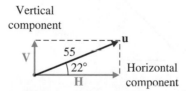

FIGURE 14

Magnitude of horizontal component:

$$\cos 22° = \frac{|\mathbf{H}|}{55}$$

$$|\mathbf{H}| = 55 \cos 22°$$

$$= 51$$

Magnitude of vertical component:

$$\sin 22° = \frac{|\mathbf{V}|}{55}$$

$$|\mathbf{V}| = 55 \sin 22°$$

$$= 21 \qquad \blacksquare$$

Matched Problem 4   Find the horizontal and vertical components **H** and **V** of the vector **u** in Figure 15.

FIGURE 15

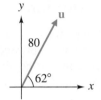

■ Centrifugal Force

Ideally, a race car going around a curve will not slide sideways if the track is banked appropriately. How much should a track be banked for a given speed v and a given curve of radius r to eliminate sideways forces? Figure 16 shows the relevant forces.

FIGURE 16

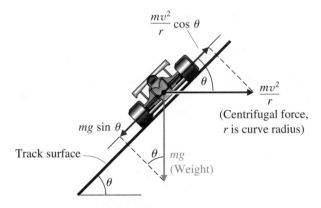

To avoid sideways forces, the components of the forces parallel to the track surface due to the mass of the car and its movement around the curve (centrifugal force) must be equal. That is,

$$mg \sin \theta = \frac{mv^2}{r} \cos \theta \qquad \text{Divide both sides by } mg \cos \theta.$$

$$\frac{\sin \theta}{\cos \theta} = \frac{mv^2}{mgr} \qquad \text{Simplify.}$$

$$\tan \theta = \frac{v^2}{gr} \qquad \text{Use inverse tangent to solve for } \theta.$$

$$\theta = \tan^{-1} \frac{v^2}{gr} \qquad v \text{ in meters per second, } r \text{ in meters, } g \approx 9.81 \text{ m/sec}^2$$

EXAMPLE 5

Racetrack Design

At what angle θ must a track be banked at a curve to eliminate sideways forces parallel to the track, given that the curve has a radius of 525 m and the racing car is moving at 85.0 m/sec (about 190 mph)?

Solution

$$\theta = \tan^{-1} \frac{v^2}{gr}$$

$$= \tan^{-1} \frac{(85.0)^2}{(9.81)(525)}$$

$$= 54.5°$$ ∎

Matched Problem 5

Repeat Example 5 with a car moving at 25.0 m/sec (about 56 mph) around a curve with a radius of 125 m. ∎

Answers to Matched Problems

1. $|\mathbf{u} + \mathbf{v}| = \sqrt{80}$ or $4\sqrt{5}$

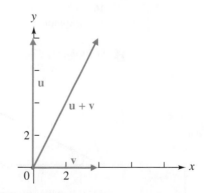

2. Actual speed: 39 km/hr; actual heading: 75°
3. Resultant force: $b = 950$ lb, $\alpha = 9°$
4. $|H| = 38$, $|V| = 71$
5. $\theta = 27.0°$

EXERCISE 6.4

A *Express all angles in decimal degrees.*

1. Explain the two methods of geometrically adding vectors.

2. Explain how addition of vectors can be applied to the motion of boats and airplanes.

In Problems 3 and 4, for the vectors **u** *and* **v** *drawn in the figure, draw* **u** + **v**, *then find the magnitude of* **u** + **v**.

3.
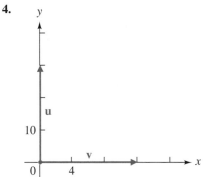

4.
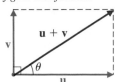

In Problems 5–10, find |**u** + **v**| *and θ, given* |**u**| *and* |**v**| *in the figure that follows.*

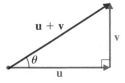

(a) Parallelogram rule   (b) Tail-to-tip rule

* A knot is 1 nautical mile per hour.

5. |**u**| = 44 km/hr, |**v**| = 32 km/hr
6. |**u**| = 65 km/hr, |**v**| = 17 km/hr
7. |**u**| = 23 lb, |**v**| = 98 lb
8. |**u**| = 37 lb, |**v**| = 39 lb
9. |**u**| = 15.6 knots*, |**v**| = 28.3 knots
10. |**u**| = 11.5 knots, |**v**| = 61.4 knots

In Problems 11–14, find the magnitudes of the horizontal and vertical components, |**H**| *and* |**V**|, *respectively, of the vector* **u** *given* |**u**| *and θ in the figure.*

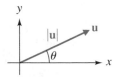

11. |**u**| = 95 lb, θ = 23°
12. |**u**| = 72 lb, θ = 68°
13. |**u**| = 325 km/hr, θ = 85.9°
14. |**u**| = 82.6 km/hr, θ = 10.2°
15. Can the magnitude of a vector ever be negative? Explain.
16. If two vectors have the same magnitude, are they equal? Explain.

B *In Problems 17–20, find* |**u** + **v**| *and α, given* |**u**|, |**v**|, *and θ in the figure that follows.*

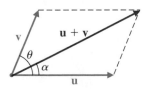   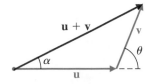

(a) Parallelogram rule   (b) Tail-to-tip rule

17. |**u**| = 125 lb, |**v**| = 84 lb, θ = 44°
18. |**u**| = 66 lb, |**v**| = 22 lb, θ = 68°
19. |**u**| = 655 mi/hr, |**v**| = 97.3 mi/hr, θ = 66.8°
20. |**u**| = 487 mi/hr, |**v**| = 74.2 mi/hr, θ = 37.4°

In Problems 21–28, determine whether the vectors $\overrightarrow{AB}$ and $\overrightarrow{CD}$ have the same direction, opposite direction, or neither.

21. $A=(-1,4)$; $B=(5,4)$; $C=(3,-2)$; $D=(12,-2)$

22. $A=(3,2)$; $B=(3,7)$; $C=(1,-1)$; $D=(1,-9)$

23. $A=(0,1)$; $B=(3,4)$; $C=(5,0)$; $D=(1,-4)$

24. $A=(6,3)$; $B=(-2,11)$; $C=(2,7)$; $D=(9,0)$

25. $A=(5,-1)$; $B=(6,5)$; $C=(-4,3)$; $D=(2,4)$

26. $A=(12,4)$; $B=(0,12)$; $C=(-10,15)$; $D=(11,1)$

27. $A=(1,3)$; $B=(4,9)$; $C=(-1,-4)$; $D=(4,6)$

28. $A=(2,0)$; $B=(-1,4)$; $C=(0,5)$; $D=(-4,8)$

29. Explain why it is or is not correct to say that the zero vector is parallel to every vector.

30. Explain why it is or is not correct to say that the zero vector is perpendicular to every vector.

31. Draw two arbitrary vectors **u** and **v**, then use the parallelogram law to show that $\mathbf{u} + \mathbf{v} = \mathbf{v} + \mathbf{u}$.

32. Draw three vectors as follows: **u** on the positive *y* axis, **w** on the positive *x* axis, and **v** in the first quadrant. Then use the tail-to-tip rule to illustrate that vector addition is associative. That is, show that $\mathbf{u} + (\mathbf{v} + \mathbf{w}) = (\mathbf{u} + \mathbf{v}) + \mathbf{w}$.

Applications

Navigation   *In Problems 33–36, assume the north, east, south, and west directions are exact. Remember that in navigational problems a compass is divided clockwise into 360° starting at north (see Fig. 7, page 388).*

33. A river is flowing east (90°) at 3.0 km/hr. A boat crosses the river with a compass heading of 180° (south). If the speedometer on the boat reads 4.0 km/hr, speed relative to the water, what is the boat's actual speed and direction (resultant velocity) relative to the river bottom?

34. A boat capable of traveling 12 knots on still water maintains a westward compass heading (270°) while crossing a river. If the river is flowing southward (180°) at 4.0 knots, what is the velocity (magnitude and direction) of the boat relative to the river bottom?

35. An airplane can cruise at 255 mi/hr in still air. If a steady wind of 46 mi/hr is blowing from the west, what compass heading should the pilot fly in order for the true course of the plane to be north (0°)? Compute the ground speed for this course.

36. Two docks are directly opposite each other on a southward flowing river. A boat pilot wishes to go in a straight line from the east dock to the west dock in a ferryboat with a cruising speed of 8.0 knots. If the river's current is 2.5 knots, what compass heading should the pilot maintain while crossing the river? What is the actual speed of the boat relative to land?

37. **Resultant Force**   Two tugs are trying to pull a barge off a shoal as indicated in the figure. Find the magnitude of the resulting force and its direction relative to $\mathbf{F_1}$.

$$|\mathbf{F_1}| = 1,500 \text{ lb}$$
$$|\mathbf{F_2}| = 1,100 \text{ lb}$$

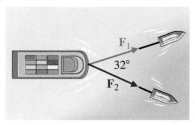

Figure for 37

38. **Resultant Force**   Repeat Problem 37 with $|\mathbf{F_1}| = 1,300$ lb and the angle between the force vectors 45°.

39. **Centrifugal Force**   At what angle must a freeway be banked at a curve to eliminate sideways forces parallel to the road, given that the curve has a radius of 138 m and cars are expected to move at 29 m/sec (about 65 mph)?

40. **Centrifugal Force**   Repeat Problem 39 with an expected car speed of 24.6 m/sec (about 55 mph) around a curve with a radius of 105 m.

41. **Resolution of Forces**   A car weighing 2,500 lb is parked on a hill inclined 15° to the horizontal (see the figure). Neglecting friction, what magnitude of force parallel to the hill will keep the car from rolling down the hill? What is the force magnitude perpendicular to the hill?

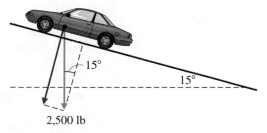

Figure for 41

42. Resolution of Forces Repeat Problem 41 with the car weighing 4,200 lb and the hill inclined 12°.

43. Biomechanics A person in the hospital has severe leg and hip injuries from a motorcycle accident. His leg is put under traction as indicated in the figure. The upper leg is pulled with a force of 52 lb at the angle indicated, and the lower leg is pulled with a force of

37 lb at the angle indicated. Compute the horizontal and vertical scalar components of the force vector for the upper leg.

44. Biomechanics Refer to Problem 43. Compute the horizontal and vertical scalar components of the force vector for the lower leg.

45. Resolution of Forces If two weights are fastened together and placed on inclined planes as indicated in the figure, neglecting friction, which way will they slide? (Weights are accurate to two significant digits, and angles are measured to the nearest degree) [*Hint:* The forces of the weights act perpendicularly to the ground, not the incline.]

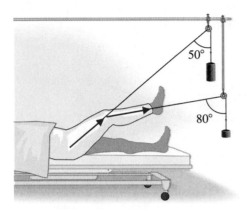

Figure for 43 and 44

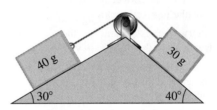

Figure for 45

6.5 Vectors: Algebraically Defined

- From Geometric Vectors to Algebraic Vectors
- Vector Addition and Scalar Multiplication
- Unit Vectors
- Algebraic Properties
- Application—Static Equilibrium

In Section 6.4 we introduced vectors in a geometric setting and considered their applications to velocity and force. We discussed geometric vectors in a plane, but they are readily extended to three-dimensional space. However, for the generalization of vector concepts to "higher-dimensional" abstract spaces, it's difficult, or impossible, to work with them geometrically. Instead, we'll study vectors algebraically. This process is done in such a way that geometric vectors become special cases of the more general algebraic vector. It also turns out that for the solution of certain types of problems in two- and three-dimensional space, algebraic

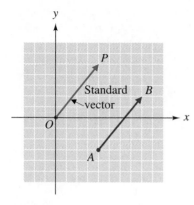

FIGURE 1

$\overrightarrow{OP}$ is the standard vector for $\overrightarrow{AB}$ ($\overrightarrow{OP} = \overrightarrow{AB}$)

vectors have an advantage over geometric vectors. Static equilibrium problems, which are discussed at the end of this section, provide an example of this type of problem.

■ From Geometric Vectors to Algebraic Vectors

We will start with an arbitrary geometric vector $\overrightarrow{AB}$ in a rectangular coordinate system. A geometric vector $\overrightarrow{AB}$ translated so that its initial point is at the origin is said to be in **standard position,** and the vector $\overrightarrow{OP}$ (as shown in Fig. 1) such that $\overrightarrow{OP} = \overrightarrow{AB}$ is said to be the **standard vector** for $\overrightarrow{AB}$. (Recall that two vectors are equal if they have the same magnitude and direction.) The standard vector is also called the **position vector** or the **radius vector.** Infinitely many geometric vectors have $\overrightarrow{OP}$ in Figure 1 as a standard vector—all vectors with the same magnitude and direction as $\overrightarrow{OP}$.

EXPLORE/DISCUSS 1

(A) In Figure 1 (or a copy) draw two other vectors having $\overrightarrow{OP}$ as their standard vector.

(B) If the tail of a vector is at point $A = (1, -2)$ and the tip is at $B = (3, 5)$, explain how you would find the coordinates of P such that $\overrightarrow{OP}$ is the standard vector for $\overrightarrow{AB}$.

How do we find the standard vector $\overrightarrow{OP}$ for a geometric vector $\overrightarrow{AB}$? The process is not difficult. We need only find the coordinates of P, since the coordinates of O, the origin, are known. If the coordinates of A are (x_a, y_a) and the coordinates of B are (x_b, y_b), then the coordinates of $P = (x, y)$ are given by

$$x = x_b - x_a \qquad y = y_b - y_a$$

The process is illustrated geometrically in Example 1.

EXAMPLE 1

Finding a Standard Vector for a Given Geometric Vector

Given the geometric vector $\overrightarrow{AB}$, with initial point $A = (8, -3)$ and terminal point $B = (4, 5)$, find the standard vector $\overrightarrow{OP}$ for $\overrightarrow{AB}$; that is, find the coordinates of the point P so that $\overrightarrow{OP} = \overrightarrow{AB}$.

Solution The coordinates (x, y) of P are given by

$$x = x_b - x_a = 4 - 8 = -4$$
$$y = y_b - y_a = 5 - (-3) = 8$$

Figure 2 illustrates these results geometrically.

FIGURE 2

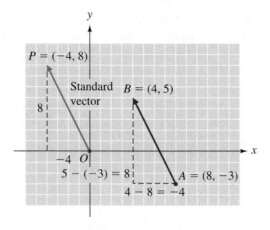

Matched Problem 1 Given the geometric vector $\vec{AB}$, with initial point $A = (4, 2)$ and terminal point $B = (2, -3)$, find the standard vector $\vec{OP}$ for $\vec{AB}$; that is, find the coordinates of the point P so that $\vec{OP} = \vec{AB}$ ∎

Let's extend our discussion to develop another way of looking at vectors. Since, given any geometric vector $\vec{AB}$, there always exists a point $P = (x, y)$ such that $\vec{OP} = \vec{AB}$, the point P completely determines the vector $\vec{AB}$ (except for its position, which we are not concerned about because we are free to translate $\vec{AB}$ anywhere we please). Conversely, given any point $P = (x, y)$ in the plane, the directed line segment joining O to P forms the geometric vector $\vec{OP}$. We have made the key observation in dealing with vectors algebraically:

> Every geometric vector in a plane corresponds to an ordered pair of real numbers, and every ordered pair of real numbers corresponds to a geometric vector.

This leads us to a definition of an **algebraic vector** as an ordered pair of real numbers. To avoid confusing a point (a, b) with a vector (a, b), we use special brackets to write the symbol $\langle a, b \rangle$, which represents an algebraic vector with initial point the origin and terminal point $\langle a, b \rangle$, as indicated in Figure 3.

The real numbers a and b are called **scalar components** of the vector $\langle a, b \rangle$. Two vectors $\mathbf{u} = \langle a, b \rangle$ and $\mathbf{v} = \langle c, d \rangle$ are said to be **equal** if their components are equal—that is, if $a = c$ and $b = d$. The **zero vector** is denoted by $\mathbf{0} = \langle 0, 0 \rangle$ and has arbitrary direction.

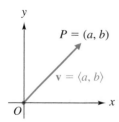

FIGURE 3
Algebraic vector $\langle a, b \rangle$;
geometric vector $\vec{OP}$

Geometric vectors are limited to two- and three-dimensional spaces, which we can visualize. Algebraic vectors do not have such restrictions. The following are algebraic vectors from two-, three-, four-, and five-dimensional spaces:

$$\langle 4, -2 \rangle \qquad \langle 8, 1, 3 \rangle \qquad \langle 12, 2, -6, 1 \rangle \qquad \langle 5, 11, 3, -8, 4 \rangle$$

Our development in this book is limited to algebraic vectors in two-dimensional space (a plane). Higher-dimensional vector spaces form the subject matter for more advanced treatments of the subject.

The *magnitude* of an algebraic vector is defined as follows:

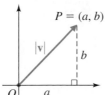

MAGNITUDE* OF A VECTOR $\mathbf{v} = \langle a, b \rangle$

The **magnitude** (or **norm**) of a vector $\mathbf{v} = \langle a, b \rangle$ denoted by $|\mathbf{v}|$, is given by

$$|\mathbf{v}| = \sqrt{a^2 + b^2}$$

FIGURE 4
$|v| = \sqrt{a^2 + b^2}$

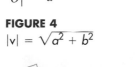

Geometrically, $\sqrt{a^2 + b^2}$ is the length of the standard geometric vector $\overrightarrow{OP}$ associated with the algebraic vector $\langle a, b \rangle$ (see Fig. 4).

EXAMPLE 2

The Magnitude of a Vector

Find the magnitude of the vector $\mathbf{v} = \langle -2, 4 \rangle$.

Solution $\quad |\mathbf{v}| = \sqrt{(-2)^2 + 4^2} = \sqrt{20} \quad$ or $\quad 2\sqrt{5}$ ■

Matched Problem 2 $\quad$ Find the magnitude of the vector $\mathbf{v} = \langle 4, -5 \rangle$. ■

■ Vector Addition and Scalar Multiplication

Adding two algebraic vectors is very easy: We simply add the corresponding components of the two vectors.

VECTOR ADDITION

If $\mathbf{u} = \langle a, b \rangle$ and $\mathbf{v} = \langle c, d \rangle$, then

$$\mathbf{u} + \mathbf{v} = \langle a + c, b + d \rangle$$

---

* The definition of magnitude is readily generalized to higher-dimensional vector spaces. For example, if $\mathbf{v} = \langle a, b, c, d, e \rangle$, then the norm of $\mathbf{v}$ is given by $|\mathbf{v}| = \sqrt{a^2 + b^2 + c^2 + d^2 + e^2}$. But in this case we lose the geometric interpretation of length.

Since the algebraic vectors in this definition of addition can be associated with geometric vectors in a plane, we can interpret the definition geometrically. As you will see in Explore/Discuss 2, the definition is consistent with the parallelogram and tail-to-tip rules for adding geometric vectors given in Section 6.4.

EXPLORE/DISCUSS 2

If $\mathbf{u} = \langle -1, 4 \rangle$ and $\mathbf{v} = \langle 6, -2 \rangle$, then $\mathbf{u} + \mathbf{v} = \langle (-1) + 6, 4 + (-2) \rangle = \langle 5, 2 \rangle$ Locate $\mathbf{u}$, $\mathbf{v}$, and $\mathbf{u} + \mathbf{v}$ in a rectangular coordinate system, and interpret geometrically in terms of the parallelogram and tail-to-tip rules discussed in the last section.

To multiply a vector by a scalar (a real number), simply multiply each component by the scalar.

SCALAR MULTIPLICATION

If $\mathbf{u} = \langle a, b \rangle$ and k is a scalar, then

$$k\mathbf{u} = k\langle a, b \rangle = \langle ka, kb \rangle$$

FIGURE 5
Scalar Multiplication

Geometrically, if a vector $\mathbf{v}$ is multiplied by a scalar k, the magnitude of the vector $\mathbf{v}$ is multiplied by $|k|$. If k is positive, then $k\mathbf{v}$ has the same direction as $\mathbf{v}$; if k is negative, then $k\mathbf{v}$ has the opposite direction as $\mathbf{v}$. Figure 5 illustrates several cases. We can use scalar multiplication to define subtraction for vectors: The difference $\mathbf{u} - \mathbf{v}$ of vectors $\mathbf{u}$ and $\mathbf{v}$ is equal to $\mathbf{u} + (-1)\mathbf{v}$.

EXAMPLE 3

Arithmetic of Algebraic Vectors

Let $\mathbf{u} = \langle -5, 3 \rangle$, $\mathbf{v} = \langle 4, -6 \rangle$, and $\mathbf{w} = \langle -2, 0 \rangle$. Find:

(A) $\mathbf{u} + \mathbf{v}$ (B) $-3\mathbf{u}$ (C) $3\mathbf{u} - 2\mathbf{v}$ (D) $2\mathbf{u} - \mathbf{v} + 3\mathbf{w}$

Solution   (A) $\mathbf{u} + \mathbf{v} = \langle -5, 3 \rangle + \langle 4, -6 \rangle = \langle -1, -3 \rangle$ Add components.

(B) $-3\mathbf{u} = -3\langle -5, 3 \rangle = \langle 15, -9 \rangle$ Multiply each component by -3.

(C) $3\mathbf{u} - 2\mathbf{v} = 3\langle -5, 3 \rangle - 2\langle 4, -6 \rangle$ First scalar multiply, then add.

$= \langle -15, 9 \rangle + \langle -8, 12 \rangle = \langle -23, 21 \rangle$

(D) $2\mathbf{u} - \mathbf{v} + 3\mathbf{w} = 2\langle -5, 3 \rangle - \langle 4, -6 \rangle + 3\langle -2, 0 \rangle$ *First scalar-multiply, then add.*

$\qquad\qquad\qquad = \langle -10, 6 \rangle + \langle -4, 6 \rangle + \langle -6, 0 \rangle = \langle -20, 12 \rangle$ ■

Matched Problem 3 Let $\mathbf{u} = \langle 4, -3 \rangle$, $\mathbf{v} = \langle 2, 3 \rangle$, and $\mathbf{w} = \langle 0, -5 \rangle$. Find:

(A) $\mathbf{u} + \mathbf{v}$ (B) $-2\mathbf{u}$ (C) $2\mathbf{u} - 3\mathbf{v}$ (D) $3\mathbf{u} + 2\mathbf{v} - \mathbf{w}$ ■

■ Unit Vectors

A vector $\mathbf{v}$ is called a **unit vector** if its magnitude is 1, that is, if $|\mathbf{v}| = 1$. Special unit vectors play an important role in certain applications of vectors. There is a simple way to turn any nonzero vector $\mathbf{v}$ into a unit vector with the same direction as $\mathbf{v}$.

FORMING A UNIT VECTOR

If $\mathbf{v}$ is a nonzero vector, then

$$\mathbf{u} = \frac{1}{|\mathbf{v}|}\mathbf{v}$$

is a unit vector with the same direction as $\mathbf{v}$.

EXAMPLE 4 **Unit Vector**

Find a unit vector $\mathbf{u}$ with the same direction as the vector $\mathbf{v} = \langle 3, 1 \rangle$.

Solution $|\mathbf{v}| = \sqrt{3^2 + 1^2} = \sqrt{10}$ *Find the magnitude of* $\mathbf{v}$.

$\qquad \mathbf{u} = \dfrac{1}{|\mathbf{v}|}\mathbf{v} = \dfrac{1}{\sqrt{10}}\langle 3, 1 \rangle = \left\langle \dfrac{3}{\sqrt{10}}, \dfrac{1}{\sqrt{10}} \right\rangle$ *Scalar-multiply by* $\dfrac{1}{\sqrt{10}}$.

✔ **Check** $|\mathbf{u}| = \sqrt{\left(\dfrac{3}{\sqrt{10}}\right)^2 + \left(\dfrac{1}{\sqrt{10}}\right)^2}$

$\qquad\qquad = \sqrt{\dfrac{9}{10} + \dfrac{1}{10}} = \sqrt{1} = 1$

So $\mathbf{u}$ is a unit vector. Since it was obtained from $\mathbf{v}$ by scalar multiplication, it has the same direction as $\mathbf{v}$. ■

Matched Problem 4 Find a unit vector $\mathbf{u}$ with the same direction as the vector $\mathbf{v} = \langle 1, -2 \rangle$. ■

Next, we'll define two very important unit vectors: the $\mathbf{i}$ and $\mathbf{j}$ unit vectors.

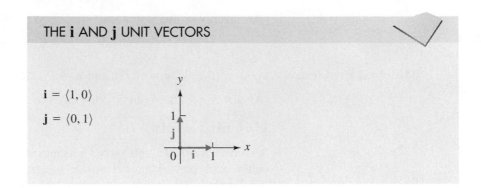

THE **i** AND **j** UNIT VECTORS

$\mathbf{i} = \langle 1, 0 \rangle$

$\mathbf{j} = \langle 0, 1 \rangle$

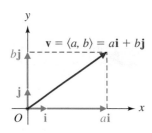

FIGURE 6

One of the reasons the **i** and **j** unit vectors are important is that any vector $\mathbf{v} = \langle a, b \rangle$ can be expressed as a linear combination of these two vectors, that is, as $a\mathbf{i} + b\mathbf{j}$. For any vector $\langle a, b \rangle$,

$$\mathbf{v} = \langle a, b \rangle = \langle a, 0 \rangle + \langle 0, b \rangle$$
$$= a\langle 1, 0 \rangle + b\langle 0, 1 \rangle$$
$$= a\mathbf{i} + b\mathbf{j}$$

Conclusion: Any vector $\mathbf{v} = \langle a, b \rangle$ can be written as $a\mathbf{i} + b\mathbf{j}$. This result is illustrated geometrically in Figure 6.

EXAMPLE 5

Expressing a Vector in Terms of **i** and **j** Unit Vectors

Express each vector as a linear combination of the **i** and **j** unit vectors.

(A) $\langle 3, -4 \rangle$ (B) $\langle 7, 0 \rangle$ (C) $\langle 0, -2 \rangle$

Solution

(A) $\langle 3, -4 \rangle = 3\mathbf{i} - 4\mathbf{j}$ (B) $\langle 7, 0 \rangle = 7\mathbf{i} + 0\mathbf{j} = 7\mathbf{i}$

(C) $\langle 0, -2 \rangle = 0\mathbf{i} - 2\mathbf{j} = -2\mathbf{j}$ ■

Matched Problem 5

Express each vector as a linear combination of the **i** and **j** unit vectors.

(A) $\langle -8, -1 \rangle$ (B) $\langle 0, -3 \rangle$ (C) $\langle 6, 0 \rangle$ ■

EXAMPLE 6

Expressing a Geometric Vector in Terms of *i* and *j* Unit Vectors

Express the vector **w** in Figure 7 as a linear combination of **i** and **j** unit vectors.

FIGURE 7

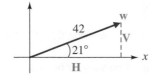

Solution   First, we find the horizontal and vertical components **H** and **V** for vector **w**:

$$\cos 21° = \frac{|\mathbf{H}|}{42}$$

$$|\mathbf{H}| = 42 \cos 21° = 39$$

$$\sin 21° = \frac{|\mathbf{V}|}{42}$$

$$|\mathbf{V}| = 42 \sin 21° = 15$$

We can write **H** as 39**i** and **V** as 15**j**, so **w** = 39**i** + 15**j**.   ■

Matched Problem 6   If **w** is a vector with $|\mathbf{w}| = 37$ that forms a 51° angle with the positive x axis, express **w** as a linear combination of the **i** and **j** unit vectors.   ■

■ Algebraic Properties

Vector addition and scalar multiplication possess algebraic properties similar to those of the real numbers. These properties enable us to manipulate symbols representing vectors and scalars in much the same way we manipulate symbols that represent real numbers in algebra. These properties are listed in the following box for convenient reference.

ALGEBRAIC PROPERTIES OF VECTORS

(A) The following **addition properties** are satisfied for all vectors **u**, **v**, and **w**:
1. $\mathbf{u} + \mathbf{v} = \mathbf{v} + \mathbf{u}$ Commutative property
2. $\mathbf{u} + (\mathbf{v} + \mathbf{w}) = (\mathbf{u} + \mathbf{v}) + \mathbf{w}$ Associative property
3. $\mathbf{u} + \mathbf{0} = \mathbf{0} + \mathbf{u} = \mathbf{u}$ Additive identity
4. $\mathbf{u} + (-\mathbf{u}) = (-\mathbf{u}) + \mathbf{u} = \mathbf{0}$ Additive inverse

(B) The following **scalar multiplication properties** are satisfied for all vectors **u** and **v** and all scalars m and n:
1. $m(n\mathbf{u}) = (mn)\mathbf{u}$ Associative property
2. $m(\mathbf{u} + \mathbf{v}) = m\mathbf{u} + m\mathbf{v}$ Distributive property
3. $(m + n)\mathbf{u} = m\mathbf{u} + n\mathbf{u}$ Distributive property
4. $1\mathbf{u} = \mathbf{u}$ Multiplication identity

When vectors are represented in terms of the **i** and **j** unit vectors, the algebraic properties listed in the preceding box provide us with efficient procedures for performing algebraic operations with vectors.

EXAMPLE 7

Algebraic Operations Involving Unit Vectors

If **u** = 2**i** − **j** and **v** = 4**i** + 5**j**, compute each of the following:

(A) **u** + **v**   (B) **u** − **v**   (C) 3**u** − 2**v**

Solution In essence, we will treat **i** and **j** as variables and combine like terms.

(A) $\mathbf{u} + \mathbf{v} = (2\mathbf{i} - \mathbf{j}) + (4\mathbf{i} + 5\mathbf{j})$ *Clear parentheses.*

$= 2\mathbf{i} - \mathbf{j} + 4\mathbf{i} + 5\mathbf{j}$ *Combine like terms.*

$= 2\mathbf{i} + 4\mathbf{i} - \mathbf{j} + 5\mathbf{j}$

$= (2 + 4)\mathbf{i} + (-1 + 5)\mathbf{j}$

$= 6\mathbf{i} + 4\mathbf{j}$

(B) $\mathbf{u} - \mathbf{v} = (2\mathbf{i} - \mathbf{j}) - (4\mathbf{i} + 5\mathbf{j})$ *Clear parentheses.*

$= 2\mathbf{i} - \mathbf{j} - 4\mathbf{i} - 5\mathbf{j}$ *Combine like terms.*

$= 2\mathbf{i} - 4\mathbf{i} - \mathbf{j} - 5\mathbf{j}$

$= (2 - 4)\mathbf{i} + (-1 - 5)\mathbf{j}$

$= -2\mathbf{i} - 6\mathbf{j}$

(C) $3\mathbf{u} - 2\mathbf{v} = 3(2\mathbf{i} - \mathbf{j}) - 2(4\mathbf{i} + 5\mathbf{j})$ *Distribute twice.*

$= 6\mathbf{i} - 3\mathbf{j} - 8\mathbf{i} - 10\mathbf{j}$ *Combine like terms.*

$= 6\mathbf{i} - 8\mathbf{i} - 3\mathbf{j} - 10\mathbf{j}$

$= (6 - 8)\mathbf{i} + (-3 - 10)\mathbf{j}$

$= -2\mathbf{i} - 13\mathbf{j}$ ∎

Matched Problem 7 If $\mathbf{u} = \mathbf{i} - 2\mathbf{j}$ and $\mathbf{v} = 5\mathbf{i} + 2\mathbf{j}$, compute each of the following:

(A) $\mathbf{u} + \mathbf{v}$ (B) $\mathbf{u} - \mathbf{v}$ (C) $2\mathbf{u} + 3\mathbf{v}$ ∎

■ Application–Static Equilibrium

We will now show how algebraic vectors can be used to solve certain physics and engineering problems. Basic to the process of solving such problems is the following principle from physics: An object at rest is in **static equilibrium.** (Even though the earth is moving in space, it is convenient to consider an object at rest if it is not moving relative to the earth.) Two conditions (from *Newton's second law of motion*) are required for an object to be in static equilibrium: The sum of all forces acting on the object must be zero, and the sum of all *torques* (rotational forces) acting on the object must be zero. A discussion of torque will be left to a more advanced treatment of the subject, and we will limit our attention to problems involving forces in a plane, where the lines of all forces on an object intersect in a single point. Such forces are called **coplanar concurrent forces.**

For example, the three forces acting on the ski chair in Figure 8 on the next page are coplanar concurrent. In contrast, if a chandelier is hung by four chains attached to the four corners of a room, the forces are not coplanar.

CONDITIONS FOR STATIC EQUILIBRIUM

An object subject only to coplanar concurrent forces is in static equilibrium if and only if the sum of all these forces is zero.

Examples 8 and 9 illustrate how important physics and engineering problems can be solved using vector methods and the conditions just stated.

EXAMPLE 8

Cable Tension

Two skiers on a stalled chair lift deflect the cable relative to the horizontal as indicated in Figure 8. If the skiers and chair weigh 325 lb (neglect cable weight), what is the tension in each cable running to each tower?

FIGURE 8

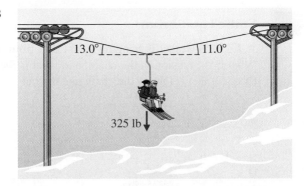

Solution

Step 1   Form a force diagram with all force vectors in standard position at the origin (Fig. 9). Our objective is to find $|\mathbf{u}|$ and $|\mathbf{v}|$.

Step 2   Write each force vector in terms of **i** and **j** unit vectors (see Example 6):

$$\mathbf{u} = |\mathbf{u}|(\cos 11.0°)\mathbf{i} + |\mathbf{u}|(\sin 11.0°)\mathbf{j}$$
$$\mathbf{v} = |\mathbf{v}|(-\cos 13.0°)\mathbf{i} + |\mathbf{v}|(\sin 13.0°)\mathbf{j}$$
$$\mathbf{w} = -325\mathbf{j}$$

FIGURE 9
Force diagram

Step 3   Set the sum of all force vectors equal to the zero vector, and solve for any unknown quantities. For the system to be in static equilibrium, the sum of the force vectors must be zero. That is,

$$\mathbf{u} + \mathbf{v} + \mathbf{w} = \mathbf{0}$$

Writing the vectors as in step 2, we have

$$[|\mathbf{u}|(\cos 11.0°) + |\mathbf{v}|(-\cos 13.0°)]\mathbf{i} + [|\mathbf{u}|(\sin 11.0°) + |\mathbf{v}|(\sin 13.0°) - 325]\mathbf{j} = 0\mathbf{i} + 0\mathbf{j}$$

Now, since two vectors are equal if and only if their corresponding components are equal (that is, $a\mathbf{i} + b\mathbf{j} = 0\mathbf{i} + 0\mathbf{j}$ if and only if $a = 0$ and $b = 0$), we are led to the following system of two equations in the two variables $|\mathbf{u}|$ and $|\mathbf{v}|$:

$$(\cos 11.0°)|\mathbf{u}| + (-\cos 13.0°)|\mathbf{v}| = 0$$
$$(\sin 11.0°)|\mathbf{u}| + (\sin 13.0°)|\mathbf{v}| - 325 = 0$$

Solving this system by standard methods from algebra (details omitted), we find that

$$|\mathbf{u}| = 779 \text{ lb} \qquad \text{and} \qquad |\mathbf{v}| = 784 \text{ lb}$$

You probably did not guess that the tension in each part of the cable is more than the weight hanging from the cable!   ■

Matched Problem 8   Repeat Example 8 with 13.0° replaced by 14.5°, 11.0° replaced by 9.5°, and 325 lb replaced by 435 lb.   ■

   **EXAMPLE 9**   **Static Equilibrium**

A 1,250 lb weight is hanging by a cable from a hoist as indicated in Figure 10. What is the tension in the chain, and what is the compression (force) on the bar? (Neglect the weight of the chain and the bar.)

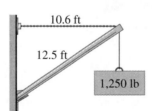

10.6 ft

12.5 ft

1,250 lb

FIGURE 10

Solution   ***Step 1***   Form a force diagram with all force vectors in standard position at the origin (Fig. 11). The 1,250 lb weight (**w**) is acting straight down; the tension in the chain (**v**) acts directly toward the wall; the force exerted by the bar (**u**) acts at the angle θ. Then

$$\theta = \cos^{-1} \frac{10.6}{12.5} = 32.0°$$

Our objective is to find the compression $|\mathbf{u}|$ and the tension $|\mathbf{v}|$.

Step 2   Write each force vector in terms of $\mathbf{i}$ and $\mathbf{j}$ unit vectors:

$$\mathbf{u} = |\mathbf{u}|(\cos 32.0°)\mathbf{i} + |\mathbf{u}|(\sin 32.0°)\mathbf{j}$$
$$\mathbf{v} = -|\mathbf{v}|\mathbf{i}$$
$$\mathbf{w} = -1,250\mathbf{j}$$

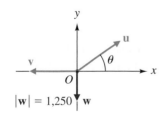

$|\mathbf{w}| = 1,250$ **w**

FIGURE 11

Step 3 Set the sum of all force vectors equal to the zero vector and solve for any unknown quantities:

$$\mathbf{u} + \mathbf{v} + \mathbf{w} = \mathbf{0}$$

Writing the vectors as in step 2, we have

$$[\,|\mathbf{u}|(\cos 32.0°) - |\mathbf{v}|\,]\mathbf{i} + [\,|\mathbf{u}|(\sin 32.0°) - 1{,}250\,]\mathbf{j} = 0\mathbf{i} + 0\mathbf{j}$$

This leads to the following system of two equations in the variables $|\mathbf{u}|$ and $|\mathbf{v}|$:

$$(\cos 32.0°)|\mathbf{u}| - |\mathbf{v}| = 0$$
$$(\sin 32.0°)|\mathbf{u}| - 1{,}250 = 0$$

Solving this system by standard methods, we find that

$$|\mathbf{u}| = 2{,}360 \text{ lb} \qquad \text{and} \qquad |\mathbf{v}| = 2{,}000 \text{ lb} \qquad\blacksquare$$

Matched Problem 9 A 450 lb sign is suspended as shown in Figure 12. Find the compression force on the 2.0 ft rod and the tension on the 4.0 ft rod. (Neglect the weight of the rods.)

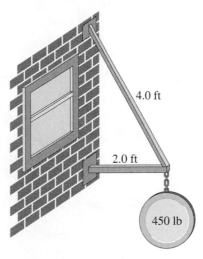

4.0 ft

2.0 ft

450 lb

FIGURE 12 ■

Answers to Matched Problems

1. $\langle -2, -5 \rangle$
2. $\sqrt{41}$
3. (A) $\langle 6, 0 \rangle$ (B) $\langle -8, 6 \rangle$ (C) $\langle 2, -15 \rangle$ (D) $\langle 16, 2 \rangle$
4. $\langle 1/\sqrt{5}, -2/\sqrt{5} \rangle$
5. (A) $-8\mathbf{i} - \mathbf{j}$ (B) $0\mathbf{i} - 3\mathbf{j} = -3\mathbf{j}$ (C) $6\mathbf{i} + 0\mathbf{j} = 6\mathbf{i}$
6. $23\mathbf{i} + 29\mathbf{j}$
7. (A) $6\mathbf{i}$ (B) $-4\mathbf{i} - 4\mathbf{j}$ (C) $17\mathbf{i} + 2\mathbf{j}$
8. $|\mathbf{u}| = 1{,}040$ lb; $|\mathbf{v}| = 1{,}050$ lb
9. $|\mathbf{u}| = 260$ lb (compression); $|\mathbf{v}| = 520$ lb (tension)

EXERCISE 6.5

A 1. Explain the difference between (a, b) and $\langle a, b \rangle$.

   2. Describe a procedure for finding the algebraic representation of a vector $\mathbf{v} = \langle a, b \rangle$ when you are given the magnitude and direction.

   3. Given any vector, how can you decide if it is a unit vector?

   4. What is the significance of the vectors $\mathbf{i}$ and $\mathbf{j}$?

In Problems 5–8, draw the vector $\overrightarrow{AB}$ in a rectangular coordinate system. Draw the standard vector $\overrightarrow{OP}$ so that $\overrightarrow{OP} = \overrightarrow{AB}$. Indicate the coordinates of P.

   5. $A = (2, -3)$; $B = (5, 1)$

   6. $A = (1, -2)$; $B = (4, 1)$

   7. $A = (-1, 3)$; $B = (-3, -1)$

   8. $A = (-1, -2)$; $B = (-3, 2)$

In Problems 9–12, represent each geometric vector $\overrightarrow{AB}$ as an algebraic vector $\langle a, b \rangle$:

   9. $A = (-1, -2)$; $B = (3, 0)$

  10. $A = (-1, 2)$; $B = (2, 3)$

  11. $A = (0, 2)$; $B = (4, -2)$

  12. $A = (-4, 0)$; $B = (-2, -4)$

In Problems 13–16, find the magnitude of each vector $\langle a, b \rangle$.

  13. $\langle -3, 4 \rangle$ 14. $\langle 4, -3 \rangle$

  15. $\langle -5, -2 \rangle$ 16. $\langle 3, -5 \rangle$

  17. When are two geometric vectors equal?

  18. When are two algebraic vectors equal?

B *In Problems 19–22, find:*

   (A) $\mathbf{u} + \mathbf{v}$ (B) $\mathbf{u} - \mathbf{v}$

   (C) $2\mathbf{u} - 3\mathbf{v}$ (D) $3\mathbf{u} - \mathbf{v} + 2\mathbf{w}$

  19. $\mathbf{u} = \langle 1, 4 \rangle$; $\mathbf{v} = \langle -3, 2 \rangle$; $\mathbf{w} = \langle 0, 4 \rangle$

  20. $\mathbf{u} = \langle -1, 3 \rangle$; $\mathbf{v} = \langle 2, -3 \rangle$; $\mathbf{w} = \langle -2, 0 \rangle$

  21. $\mathbf{u} = \langle 2, -3 \rangle$; $\mathbf{v} = \langle -1, -3 \rangle$; $\mathbf{w} = \langle -2, 0 \rangle$

  22. $\mathbf{u} = \langle -1, -4 \rangle$; $\mathbf{v} = \langle 3, 3 \rangle$; $\mathbf{w} = \langle 0, -3 \rangle$

In Problems 23–26, form a unit vector $\mathbf{u}$ with the same direction as $\mathbf{v}$.

  23. $\mathbf{v} = \langle 1, -1 \rangle$ 24. $\mathbf{v} = \langle 5, -12 \rangle$

  25. $\mathbf{v} = \langle -4, 3 \rangle$ 26. $\mathbf{v} = \langle 6, 1 \rangle$

In Problems 27–32, express $\mathbf{v}$ in terms of the $\mathbf{i}$ and $\mathbf{j}$ unit vectors.

  27. $\mathbf{v} = \langle -10, 7 \rangle$ 28. $\mathbf{v} = \langle 0, 9 \rangle$

  29. $\mathbf{v} = \langle -12, 0 \rangle$ 30. $\mathbf{v} = \langle 8, -3 \rangle$

  31. $\mathbf{v} = \overrightarrow{AB}$; $A = (3, 8)$; $B = (13, -6)$

  32. $\mathbf{v} = \overrightarrow{AB}$; $A = (-5, 2)$; $B = (4, 9)$

In Problems 33–36, use the vectors shown in the figure below. Write each listed vector as a linear combination of the unit vectors $\mathbf{i}$ and $\mathbf{j}$.

  33. Vector $\mathbf{u}$.   34. Vector $\mathbf{v}$.

  35. Vector $\mathbf{w}$.   36. Vector $\mathbf{t}$.

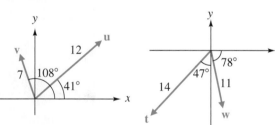

Figure for 33 and 34   Figure for 35 and 36

In Problems 37–42, let $\mathbf{u} = 2\mathbf{i} - 3\mathbf{j}$, $\mathbf{v} = 3\mathbf{i} + 4\mathbf{j}$, and $\mathbf{w} = 5\mathbf{j}$, and perform the indicated operations.

  37. $5\mathbf{u} + \mathbf{w}$ 38. $\mathbf{v} - 5\mathbf{w}$

  39. $\mathbf{u} - \mathbf{v} + \mathbf{w}$ 40. $\mathbf{u} + \mathbf{v} - \mathbf{w}$

  41. $2\mathbf{u} + 4\mathbf{v} - 6\mathbf{w}$ 42. $-7\mathbf{u} - 2\mathbf{v} + 10\mathbf{w}$

  43. An object that is free to move has three nonzero coplanar concurrent forces acting on it, and it remains at rest. How is any one of these forces related to the other two?

  44. An object that is free to move has two nonzero coplanar concurrent forces acting on it, and it remains at rest. How are the two force vectors related?

  45. Find a vector of length 5 that has the same direction as $\langle 3, 2 \rangle$.

46. Find a vector of length 0.3 that has the same direction as $\langle 12, 5 \rangle$.

47. Find a vector of length 2 that has direction opposite to $\langle -8, 6 \rangle$.

48. Find a vector of length 12 that has direction opposite to $\langle 1, 1 \rangle$.

C *A vector* $\mathbf{u}$ *is said to be a linear combination of vectors* $\mathbf{v}$ *and* $\mathbf{w}$ *if there exist real numbers* c_1 *and* c_2 *such that* $\mathbf{u} = c_1\mathbf{v} + c_2\mathbf{w}$. *In Problems 49–52, show that:*

49. $\langle -4, -5 \rangle$ is a linear combination of $\langle 1, 0 \rangle$ and $\langle 3, 1 \rangle$

50. $\langle -1, 3 \rangle$ is not a linear combination of $\langle 6, 2 \rangle$ and $\langle 9, 3 \rangle$

51. $\langle 1, 2 \rangle$ is not a linear combination of $\langle 8, 4 \rangle$ and $\langle 10, 5 \rangle$

52. $\langle 0, 1 \rangle$ is a linear combination of $\langle 2, 0 \rangle$ and $\langle 7, 1 \rangle$

In Problems 53–60, let $\mathbf{u} = \langle a, b \rangle, \mathbf{v} = \langle c, d \rangle$, *and* $\mathbf{w} = \langle e, f \rangle$ *be vectors, and let m and n be scalars. Prove each of the following vector properties using appropriate properties of real numbers.*

53. $\mathbf{u} + \mathbf{v} = \mathbf{v} + \mathbf{u}$

54. $\mathbf{u} + (\mathbf{v} + \mathbf{w}) = (\mathbf{u} + \mathbf{v}) + \mathbf{w}$

55. $\mathbf{v} + (-\mathbf{v}) = \mathbf{0}$

56. $\mathbf{v} + \mathbf{0} = \mathbf{v}$

57. $m(\mathbf{u} + \mathbf{v}) = m\mathbf{u} + m\mathbf{v}$

58. $(m + n)\mathbf{v} = m\mathbf{v} + n\mathbf{v}$

59. $1\mathbf{v} = \mathbf{v}$

60. $(mn)\mathbf{v} = m(n\mathbf{v})$

Applications

61. Static Equilibrium A tightrope walker weighing 112 lb deflects a rope as indicated in the figure. How much tension is in each part of the rope?

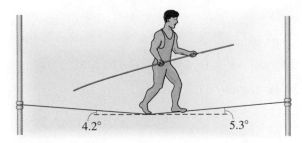

Figure for 61

62. Static Equilibrium Repeat Problem 61 but change the left angle to 5.5°, the right angle to 6.2°, and the weight of the person to 155 lb.

63. Static Equilibrium Two movers lift a piano using a thick strap, which runs under the bottom of the piano. It extends upward on each side to wrap around the neck of each mover, with one mover on each side of the piano. The left side of the strap has a tension of 230 lb and makes an angle of 49° with the ground, and the right side has a tension of 197 lb and makes an angle of 40° with the ground.

(A) Show that the piano is not moving horizontally, given these forces.

(B) Assuming that the piano is not moving vertically, find the weight of the piano.

64. Static Equilibrium Repeat Problem 63 for a freezer: The left side of the strap has a tension of 127 lb and makes a 52° angle with the ground, and the right side has a tension of 150 lb and makes an angle of 56° with the ground.

65. Static Equilibrium A weight of 5,000 lb is supported as indicated in the figure. What are the magnitudes of the forces on the members AB and BC (to two significant digits)?

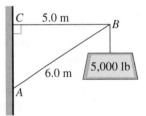

Figure for 65

66. Static Equilibrium A weight of 1,000 lb is supported as indicated in the figure. What are the magnitudes of the forces on the members AB and BC (to two significant digits)?

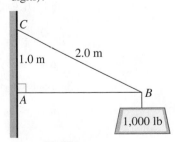

Figure for 66

67. Biomechanics   A woman slipped on ice on the sidewalk and fractured her arm. The severity of the accident required her arm to be put in traction as indicated in the figure. (The weight of the arm exerts a downward force of 6 lb.) Find angle θ and the tension on the line fastened to the overhead bar. Compute each to one decimal place.

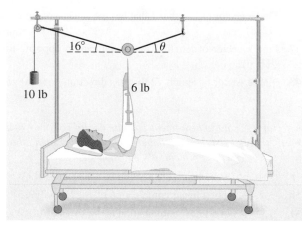

Figure for 67

✩6.6   The Dot Product

- The Dot Product of Two Vectors
- Angle Between Two Vectors
- Scalar Component of One Vector on Another
- Work

In Section 6.5 we saw that the sum of two vectors is a vector and the scalar multiple of a vector is a vector. In this section we will introduce another operation on vectors, called the *dot product*. This product of two vectors, however, is a scalar (a real number), not a vector. The dot product is also called the *scalar product* or the *inner product*. Dot products are used to find angles between vectors, to define concepts and solve problems in physics and engineering, and in the solution of geometric problems.

■ The Dot Product of Two Vectors

We define the dot product in the following box.

THE DOT PRODUCT

The **dot product** of the two vectors

$$\mathbf{u} = \langle a, b \rangle = a\mathbf{i} + b\mathbf{j} \quad \text{and} \quad \mathbf{v} = \langle c, d \rangle = c\mathbf{i} + d\mathbf{j}$$

denoted by $\mathbf{u} \cdot \mathbf{v}$, is the scalar given by

$$\mathbf{u} \cdot \mathbf{v} = ac + bd$$

✩ Sections marked with a star may be omitted without loss of continuity.

For now, we will deal with the dot product algebraically. In subsequent examples we'll consider geometric interpretations.

EXAMPLE 1

Finding Dot Products

Find each dot product.

(A) $\langle 4, 2 \rangle \cdot \langle 1, -3 \rangle$ (B) $(6\mathbf{i} + 2\mathbf{j}) \cdot (\mathbf{i} - 4\mathbf{j})$

(C) $\langle 5, 2 \rangle \cdot \langle 2, -5 \rangle$ (D) $\mathbf{j} \cdot \mathbf{i}$

Solution

(A) $\langle 4, 2 \rangle \cdot \langle 1, -3 \rangle = 4 \cdot 1 + 2 \cdot (-3) = -2$

(B) $(6\mathbf{i} + 2\mathbf{j}) \cdot (\mathbf{i} - 4\mathbf{j}) = 6 \cdot 1 + 2 \cdot (-4) = -2$

(C) $\langle 5, 2 \rangle \cdot \langle 2, -5 \rangle = 5 \cdot 2 + 2 \cdot (-5) = 0$

(D) $\mathbf{j} \cdot \mathbf{i} = (0\mathbf{i} + 1\mathbf{j}) \cdot (1\mathbf{i} + 0\mathbf{j}) = 0 \cdot 1 + 1 \cdot 0 = 0$ ∎

Matched Problem 1

Find each dot product.

(A) $\langle 3, -2 \rangle \cdot \langle 4, 1 \rangle$ (B) $(5\mathbf{i} - 2\mathbf{j}) \cdot (3\mathbf{i} + 4\mathbf{j})$

(C) $\langle 4, -3 \rangle \cdot \langle 3, 4 \rangle$ (D) $\mathbf{i} \cdot \mathbf{j}$ ∎

Some important properties of the dot product are listed in the following box. All these properties follow directly from the definitions of the vector operations involved and the properties of real numbers.

PROPERTIES OF THE DOT PRODUCT

For all vectors $\mathbf{u}$, $\mathbf{v}$, and $\mathbf{w}$, and k a real number:

1. $\mathbf{u} \cdot \mathbf{u} = |\mathbf{u}|^2$ 4. $(\mathbf{u} + \mathbf{v}) \cdot \mathbf{w} = \mathbf{u} \cdot \mathbf{w} + \mathbf{v} \cdot \mathbf{w}$

2. $\mathbf{u} \cdot \mathbf{v} = \mathbf{v} \cdot \mathbf{u}$ 5. $k(\mathbf{u} \cdot \mathbf{v}) = (k\mathbf{u}) \cdot \mathbf{v} = \mathbf{u} \cdot (k\mathbf{v})$

3. $\mathbf{u} \cdot (\mathbf{v} + \mathbf{w}) = \mathbf{u} \cdot \mathbf{v} + \mathbf{u} \cdot \mathbf{w}$ 6. $\mathbf{u} \cdot \mathbf{0} = 0$

The proofs of the dot product properties are left to Problems 33–38 in Exercise 6.6.

EXPLORE/DISCUSS 1

Verify each of the dot product properties for $\mathbf{u} = \langle -1, 2 \rangle$, $\mathbf{v} = \langle 3, 1 \rangle$, $\mathbf{w} = \langle 1, \ 1 \rangle$, and $k = 2$.

■ Angle Between Two Vectors

Our first geometric application of the dot product will be finding the angle between two vectors. If **u** and **v** are two nonzero vectors, we can position **u** and **v** so that their initial points coincide. The **angle between the vectors u and v** is defined to be the angle θ, $0 \leq \theta \leq \pi$, formed by the vectors (see Fig. 1).

FIGURE 1
Angle between two vectors,
$0 \leq \theta \leq \pi$

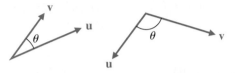

If $\theta = \pi/2$, the vectors **u** and **v** are said to be **orthogonal**, or **perpendicular.** If $\theta = 0$ or $\theta = \pi$, the vectors are said to be **parallel.**

We will now show how the dot product can be used to find the angle between two vectors.

ANGLE BETWEEN TWO VECTORS

If **u** and **v** are two nonzero vectors and θ is the angle between **u** and **v**, then

$$\cos \theta = \frac{\mathbf{u} \cdot \mathbf{v}}{|\mathbf{u}||\mathbf{v}|}$$

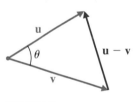

FIGURE 2

We will sketch a proof of this result for nonparallel vectors. Given two non-zero, nonparallel vectors **u** and **v**, we can position **u**, **v**, and **u** − **v** so that they form a triangle where θ is the angle between **u** and **v** (see Fig. 2). Applying the law of cosines, we obtain

$$|\mathbf{u} - \mathbf{v}|^2 = |\mathbf{u}|^2 + |\mathbf{v}|^2 - 2|\mathbf{u}||\mathbf{v}| \cos \theta \qquad (1)$$

Next, we apply some of the properties of the dot product listed earlier to $|\mathbf{u} - \mathbf{v}|^2$.

$$
\begin{aligned}
|\mathbf{u} - \mathbf{v}|^2 &= (\mathbf{u} - \mathbf{v}) \cdot (\mathbf{u} - \mathbf{v}) && \text{Property 1}\\
&= (\mathbf{u} - \mathbf{v}) \cdot \mathbf{u} - (\mathbf{u} - \mathbf{v}) \cdot \mathbf{v} && \text{Property 3}\\
&= \mathbf{u} \cdot (\mathbf{u} - \mathbf{v}) - \mathbf{v} \cdot (\mathbf{u} - \mathbf{v}) && \text{Property 2}\\
&= \mathbf{u} \cdot \mathbf{u} - \mathbf{u} \cdot \mathbf{v} - \mathbf{v} \cdot \mathbf{u} + \mathbf{v} \cdot \mathbf{v} && \text{Property 3}\\
&= |\mathbf{u}|^2 - 2(\mathbf{u} \cdot \mathbf{v}) + |\mathbf{v}|^2 && \text{Properties 1 and 2}
\end{aligned}
$$

Now substitute this last expression in the left side of equation (1):

$$|\mathbf{u}|^2 - 2(\mathbf{u} \cdot \mathbf{v}) + |\mathbf{v}|^2 = |\mathbf{u}|^2 + |\mathbf{v}|^2 - 2|\mathbf{u}||\mathbf{v}| \cos \theta$$

This equation simplifies to

$$2|\mathbf{u}||\mathbf{v}| \cos \theta = 2(\mathbf{u} \cdot \mathbf{v})$$

Since **u** and **v** are nonzero, we can solve for $\cos\theta$:

$$\cos\theta = \frac{\mathbf{u}\cdot\mathbf{v}}{|\mathbf{u}||\mathbf{v}|}$$

EXAMPLE 2

Angle Between Two Vectors

Find the angle (in decimal degrees, to one decimal place) between each pair of vectors.

(A) $\mathbf{u} = \langle 2,3\rangle;\quad \mathbf{v} = \langle 4,1\rangle$ (B) $\mathbf{u} = \langle 3,1\rangle;\quad \mathbf{v} = \langle -3,-3\rangle$

(C) $\mathbf{u} = -\mathbf{i} + 3\mathbf{j};\quad \mathbf{v} = 4\mathbf{i} - \mathbf{j}$

Solution

(A) $\cos\theta = \dfrac{\mathbf{u}\cdot\mathbf{v}}{|\mathbf{u}||\mathbf{v}|} = \dfrac{11}{\sqrt{13}\sqrt{17}}$

$\theta = \cos^{-1}\left(\dfrac{11}{\sqrt{13}\sqrt{17}}\right) = 42.3°$

(B) $\cos\theta = \dfrac{\mathbf{u}\cdot\mathbf{v}}{|\mathbf{u}||\mathbf{v}|} = \dfrac{-12}{\sqrt{10}\sqrt{18}}$

$\theta = \cos^{-1}\left(\dfrac{-12}{\sqrt{10}\sqrt{18}}\right) = 153.4°$

(C) $\cos\theta = \dfrac{\mathbf{u}\cdot\mathbf{v}}{|\mathbf{u}||\mathbf{v}|} = \dfrac{-7}{\sqrt{10}\sqrt{17}}$

$\theta = \cos^{-1}\left(\dfrac{-7}{\sqrt{10}\sqrt{17}}\right) = 122.5°$ ■

Matched Problem 2

Find the angle (in decimal degrees, to one decimal place) between each pair of vectors.

(A) $\mathbf{u} = \langle 4,2\rangle;\quad \mathbf{v} = \langle 3,-2\rangle$ (B) $\mathbf{u} = \langle -3,-1\rangle;\quad \mathbf{v} = \langle 2,4\rangle$

(C) $\mathbf{u} = -4\mathbf{i} - 2\mathbf{j};\quad \mathbf{v} = -2\mathbf{i} + \mathbf{j}$ ■

EXPLORE/DISCUSS 2

You will need the formula for the angle between two vectors to answer the following questions.

(A) Graph the vectors from parts (C) and (D) of Example 1 in standard position. What do you notice about the relationship between the vectors in each case?

(B) Given that **u** and **v** are orthogonal vectors, what can you say about $\mathbf{u}\cdot\mathbf{v}$?

(C) Given $\mathbf{u}\cdot\mathbf{v} = 0$, what can you say about the angle between **u** and **v**?

Explore/Discuss 2 leads to the following very useful test for orthogonality:

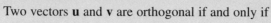

TEST FOR ORTHOGONAL VECTORS

Two vectors **u** and **v** are orthogonal if and only if

$$\mathbf{u} \cdot \mathbf{v} = 0$$

Note: The zero vector is orthogonal to every vector.

EXAMPLE 3   **Determining Orthogonality**

Determine which of the following pairs of vectors are orthogonal.

(A) $\mathbf{u} = \langle 3, -2 \rangle$; $\mathbf{v} = \langle 4, 6 \rangle$ (B) $\mathbf{u} = 2\mathbf{i} - \mathbf{j}$; $\mathbf{v} = \mathbf{i} - 2\mathbf{j}$

Solution   (A) $\mathbf{u} \cdot \mathbf{v} = \langle 3, -2 \rangle \cdot \langle 4, 6 \rangle = 12 + (-12) = 0$

The dot product is zero, so **u** and **v** are orthogonal.

(B) $\mathbf{u} \cdot \mathbf{v} = (2\mathbf{i} - \mathbf{j}) \cdot (\mathbf{i} - 2\mathbf{j}) = 2 + 2 = 4 \neq 0$

The dot product is not zero, so **u** and **v** are not orthogonal. ■

Matched Problem 3   Determine which of the following pairs of vectors are orthogonal.

(A) $\mathbf{u} = \langle -2, 4 \rangle$; $\mathbf{v} = \langle -2, 1 \rangle$ (B) $\mathbf{u} = 2\mathbf{i} - \mathbf{j}$; $\mathbf{v} = \mathbf{i} + 2\mathbf{j}$ ■

■ **Scalar Component of One Vector on Another**

Our second geometric application of the dot product will be the determination of the *projection* of a vector **u** onto a vector **v**. To project a vector **u** onto an arbitrary nonzero vector **v**, locate **u** and **v** so they have the same initial point *O;* then drop a perpendicular line from the terminal point of **u** to the line that contains **v**. The **vector projection of u onto v** is the vector **p** that lies on the line containing **v**, as shown in Figure 3. The vector **p** has the same direction as **v** if θ is acute and the opposite direction as **v** if θ is obtuse. The vector **p** is the zero vector if **u** and **v** are orthogonal. Problems 39–42 in Exercise 6.6 ask you to compute some projections.

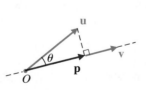
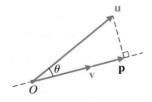
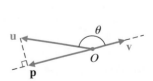

FIGURE 3
Projection of **u** onto **v**

The **scalar component of u on v**, denoted by Comp$_v$ **u**, is given by $|\mathbf{u}| \cos \theta$. This quantity is positive if θ is acute, 0 if $\theta = \pi/2$, and negative if θ is obtuse. Geometrically, Comp$_v$ **u** is the length of the projection of **u** onto **v**.

The scalar component of **u** on **v** can be expressed in terms of the dot product as follows:

$$\text{Comp}_v \mathbf{u} = |\mathbf{u}| \cos \theta \qquad \text{Multiply numerator and denominator by } |\mathbf{v}|.$$

$$= \frac{|\mathbf{u}||\mathbf{v}| \cos \theta}{|\mathbf{v}|}$$

$$= \frac{\mathbf{u} \cdot \mathbf{v}}{|\mathbf{v}|} \qquad \text{Since } \cos \theta = \frac{\mathbf{u} \cdot \mathbf{v}}{|\mathbf{u}||\mathbf{v}|}$$

THE SCALAR COMPONENT OF **u** ON **v**

Let $\theta, 0 \leq \theta \leq \pi$, be the angle between the two vectors **u** and **v**. The **scalar component of u on v**, denoted by Comp$_v$ **u**, is given by

$$\text{Comp}_v \mathbf{u} = |\mathbf{u}| \cos \theta$$

or, in terms of the dot product, by

$$\text{Comp}_v \mathbf{u} = \frac{\mathbf{u} \cdot \mathbf{v}}{|\mathbf{v}|}$$

EXAMPLE 4

Finding the Scalar Component of u on v

Find Comp$_v$ **u** for each pair of vectors **u** and **v**. Compute answers to three significant digits.

(A) $\mathbf{u} = \langle 2, 3 \rangle; \quad \mathbf{v} = \langle 4, 1 \rangle$ (B) $\mathbf{u} = \langle 3, 2 \rangle; \quad \mathbf{v} = \langle 4, -6 \rangle$

(C) $\mathbf{u} = -3\mathbf{i} + 3\mathbf{j}; \quad \mathbf{v} = 3\mathbf{i} + \mathbf{j}$

Solution

(A) $\text{Comp}_v \mathbf{u} = \dfrac{\mathbf{u} \cdot \mathbf{v}}{|\mathbf{v}|} = \dfrac{11}{\sqrt{17}} = 2.67$

(B) $\text{Comp}_v \mathbf{u} = \dfrac{\mathbf{u} \cdot \mathbf{v}}{|\mathbf{v}|} = \dfrac{0}{\sqrt{52}} = 0$

(C) $\text{Comp}_v \mathbf{u} = \dfrac{\mathbf{u} \cdot \mathbf{v}}{|\mathbf{v}|} = \dfrac{-6}{\sqrt{10}} = -1.90$ ∎

Matched Problem 4

Find Comp$_v$ **u** for each pair of vectors **u** and **v**. Compute answers to three significant digits.

(A) $\mathbf{u} = \langle 3, 3 \rangle; \quad \mathbf{v} = \langle 6, 2 \rangle$ (B) $\mathbf{u} = \langle -4, 2 \rangle; \quad \mathbf{v} = \langle 1, 2 \rangle$

(C) $\mathbf{u} = \mathbf{i} + 4\mathbf{j}; \quad \mathbf{v} = 4\mathbf{i} - 4\mathbf{j}$ ∎

■ Work

To close this section, we will look at an application of the scalar component.

Intuitively, we know work is done when a force causes an object to be moved a certain distance. One example is illustrated in Figure 4, where work is done by the force that the harness exerts on a dogsled to move the sled a certain distance (although the real work is being done by the dogs, not the harness!).

FIGURE 4

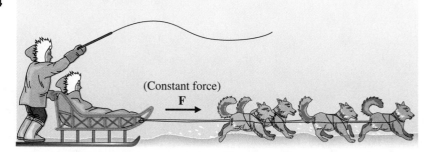

(Constant force)

F

The concept of work is very important in science studies dealing with energy and requires a definition that is quantitative: If the motion of an object takes place in a straight line in the direction of a constant force **F** causing the motion, then we define work W done by the force **F** to be the product of the magnitude of the force $|\mathbf{F}|$ and the distance d through which the object moves. In symbols,

$$W = |\mathbf{F}|d$$

For example, suppose the dogsled in Figure 4 is moved a distance of 200 ft under a constant force of 100 lb. Then the work done by the force is

$$W = (100 \text{ lb})(200 \text{ ft}) = 20,000 \text{ ft-lb*}$$

Although work is easy to quantify when the force acts in the same direction as the motion, it is more common that the motion of an object is caused by a constant force acting in a direction different from the line of motion. For example, a person pushing a lawn mower across a level lawn exerts a force downward along the handle (see Fig. 5 on the next page). In this case, only the component of force in the direction of motion (the horizontal component) is responsible for the motion. Suppose the lawn mower is pushed 52 ft across a level lawn with a constant force of 61 lb directed down the handle. If the handle makes a constant angle of 38° relative to the horizontal, then the work done by the force is

$$W = \left(\begin{array}{c}\text{Component of force in} \\ \text{the direction of motion}\end{array}\right)(\text{Displacement})$$
$$= (61 \cos 38°)(52)$$
$$= 2,500 \text{ ft-lb}$$

----

* In scientific work, the basic unit of force is the newton, the basic unit of linear measure is the meter, and the basic unit of work is the newton-meter or joule. In the British engineering system, which we will use, the basic unit of work is the foot-pound (ft-lb).

FIGURE 5

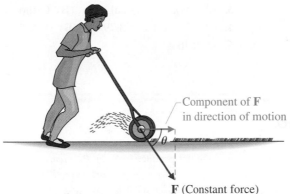

Component of **F** in direction of motion

θ

F (Constant force)

A vector **d** is a **displacement vector** for an object that is moved in a straight line if **d** points in the direction of motion and if $|\mathbf{d}|$ is the distance moved. If the displacement vector of an object is **0**, then no matter how much force is applied, no work is done. Also, for an object to move in the direction of the displacement vector, the angle between the force vector and the displacement vector must be acute. We now give a more general definition of work.

> **WORK**
>
>
> If the displacement vector is **d** for an object moved by a force **F**, then the **work** done, W, is the product of the component of force in the direction of motion and the actual displacement. In symbols,
>
> $$W = (\text{Comp}_{\mathbf{d}}\,\mathbf{F})|\mathbf{d}| = \frac{\mathbf{F} \cdot \mathbf{d}}{|\mathbf{d}|}\,|\mathbf{d}| = \mathbf{F} \cdot \mathbf{d}$$

EXAMPLE 5

Work Done by a Force

How much work is done by a force $\mathbf{F} = \langle 6, 4 \rangle$ that moves an object from the origin to the point $P = (8, 2)$? (Force is in pounds and displacement is in feet.)

Solution

$$
\begin{aligned}
W &= \mathbf{F} \cdot \mathbf{d} \\
&= \langle 6, 4 \rangle \cdot \langle 8, 2 \rangle \qquad d - \langle 8, 2 \rangle \\
&= 56 \text{ ft-lb}
\end{aligned}
$$

Matched Problem 5

How much work is done by a force $\mathbf{F} = -8\mathbf{i} + 4\mathbf{j}$ that moves an object from the origin to the point $P = (-12, -4)$? (Force is in pounds and displacement is in feet.)

Answers to Matched Problems

1. (A) 10   (B) 7   (C) 0   (D) 0
2. (A) 60.3°   (B) 135.0°   (C) 53.1°

3. (A) Not orthogonal   (B) Orthogonal
4. (A) $24/\sqrt{10} = 3.79$ (B) $0/\sqrt{5} = 0$ (C) $-12/\sqrt{32} = -2.12$
5. 80 ft-lb

EXERCISE 6.6

A **1.** Explain in your own words how to find the dot product of two vectors.

2. How can you tell if two vectors are orthogonal? Parallel?

In Problems 3–10, find each dot product.

3. $\langle 0, 5 \rangle \cdot \langle 6, 2 \rangle$ **4.** $\langle 3, -7 \rangle \cdot \langle -4, 0 \rangle$

5. $\langle 9, 6 \rangle \cdot \langle -3, 8 \rangle$ **6.** $\langle -4, -2 \rangle \cdot \langle 7, -5 \rangle$

7. $2\mathbf{i} \cdot 5\mathbf{j}$ **8.** $-9\mathbf{j} \cdot 8\mathbf{i}$

9. $(3\mathbf{i} + 4\mathbf{j}) \cdot (5\mathbf{i} + 4\mathbf{j})$ **10.** $(8\mathbf{i} - \mathbf{j}) \cdot (10\mathbf{i} + \mathbf{j})$

In Problems 11–16, find the angle (in decimal degrees, to one decimal place) between each pair of vectors.

11. $\mathbf{u} = \langle 0, 1 \rangle$; $\mathbf{v} = \langle 5, 5 \rangle$

12. $\mathbf{u} = \langle -3, -3 \rangle$; $\mathbf{v} = \langle 1, 0 \rangle$

13. $\mathbf{u} = \langle 2, 9 \rangle$; $\mathbf{v} = \langle 2, 10 \rangle$

14. $\mathbf{u} = \langle -5, 4 \rangle$; $\mathbf{v} = \langle -5, 6 \rangle$

15. $\mathbf{u} = 8\mathbf{i} + \mathbf{j}$; $\mathbf{v} = -8\mathbf{i} + \mathbf{j}$

16. $\mathbf{u} = \mathbf{i} - 12\mathbf{j}$; $\mathbf{v} = 2\mathbf{i} + 7\mathbf{j}$

B *In Problems 17 and 18, use the following alternative form for the dot product (from the formula for the angle between two vectors):*

$$\mathbf{u} \cdot \mathbf{v} = |\mathbf{u}||\mathbf{v}| \cos \theta$$

Determine the sign of the dot product for $\mathbf{u}$ and $\mathbf{v}$ without calculating it, and explain how you found it.

17.   **18.**

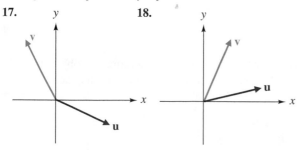

In Problems 19–22, determine which pairs of vectors are orthogonal.

19. $\mathbf{u} = 2\mathbf{i} + \mathbf{j}$; $\mathbf{v} = \mathbf{i} - 2\mathbf{j}$

20. $\mathbf{u} = 5\mathbf{i} - 4\mathbf{j}$; $\mathbf{v} = -4\mathbf{i} - 5\mathbf{j}$

21. $\mathbf{u} = \langle 1, 3 \rangle$; $\mathbf{v} = \langle -3, -1 \rangle$

22. $\mathbf{u} = \langle 4, -3 \rangle$; $\mathbf{v} = \langle 6, -8 \rangle$

In Problems 23–26, determine which pairs of vectors are parallel.

23. $\mathbf{u} = \langle -2, 7 \rangle$; $\mathbf{v} = \langle 2, 7 \rangle$

24. $\mathbf{u} = \langle 1, -4 \rangle$; $\mathbf{v} = \langle -3, 12 \rangle$

25. $\mathbf{u} = 4\mathbf{i} - 3\mathbf{j}$; $\mathbf{v} = -20\mathbf{i} + 15\mathbf{j}$

26. $\mathbf{u} = \mathbf{i} + 5\mathbf{j}$; $\mathbf{v} = 2\mathbf{i} - 10\mathbf{j}$

In Problems 27–32, find $\text{Comp}_{\mathbf{v}}\, \mathbf{u}$, the scalar component of $\mathbf{u}$ on $\mathbf{v}$. Compute answers to three significant digits.

27. $\mathbf{u} = \langle 5, 12 \rangle$; $\mathbf{v} = \langle 0, 1 \rangle$

28. $\mathbf{u} = \langle 14, 9 \rangle$; $\mathbf{v} = \langle -1, 0 \rangle$

29. $\mathbf{u} = \langle -6, 4 \rangle$; $\mathbf{v} = \langle 2, 3 \rangle$

30. $\mathbf{u} = \langle 8, -6 \rangle$; $\mathbf{v} = \langle -15, -20 \rangle$

31. $\mathbf{u} = -3\mathbf{i} + \mathbf{j}$; $\mathbf{v} = 5\mathbf{i} + 14\mathbf{j}$

32. $\mathbf{u} = 2\mathbf{i} - \mathbf{j}$; $\mathbf{v} = 7\mathbf{i} - 3\mathbf{j}$

C *Given $\mathbf{u} = \langle a, b \rangle$, $\mathbf{v} = \langle c, d \rangle$, $\mathbf{w} = \langle e, f \rangle$, and k a scalar, prove Problems 33–38 using the definitions of the given operations and the properties of real numbers.*

33. $\mathbf{u} \cdot \mathbf{u} = |\mathbf{u}|^2$

34. $\mathbf{u} \cdot \mathbf{v} = \mathbf{v} \cdot \mathbf{u}$

35. $\mathbf{u} \cdot (\mathbf{v} + \mathbf{w}) = \mathbf{u} \cdot \mathbf{v} + \mathbf{u} \cdot \mathbf{w}$

36. $(\mathbf{u} + \mathbf{v}) \cdot \mathbf{w} = \mathbf{u} \cdot \mathbf{w} + \mathbf{v} \cdot \mathbf{w}$

37. $k(\mathbf{u} \cdot \mathbf{v}) = (k\mathbf{u}) \cdot \mathbf{v} = \mathbf{u} \cdot (k\mathbf{v})$

38. $\mathbf{u} \cdot \mathbf{0} = 0$

The vector projection of **u** *onto* **v**, *denoted by* Proj$_\mathbf{v}$ **u**, *is given by*

$$\text{Proj}_\mathbf{v}\,\mathbf{u} = (\text{Comp}_\mathbf{v}\,\mathbf{u})\frac{\mathbf{v}}{|\mathbf{v}|} = \frac{\mathbf{u}\cdot\mathbf{v}}{\mathbf{v}\cdot\mathbf{v}}\,\mathbf{v}$$

In Problems 39–42, find Proj$_\mathbf{v}$ **u**.

39. $\mathbf{u} = \langle 3, 4\rangle$; $\mathbf{v} = \langle 4, 0\rangle$

40. $\mathbf{u} = \langle 3, -4\rangle$; $\mathbf{v} = \langle 0, -3\rangle$

41. $\mathbf{u} = -6\mathbf{i} + 3\mathbf{j}$; $\mathbf{v} = -3\mathbf{i} - 2\mathbf{j}$

42. $\mathbf{u} = -2\mathbf{i} - 3\mathbf{j}$; $\mathbf{v} = \mathbf{i} - 6\mathbf{j}$

43. Show that **u** and **v** have the same direction if $\mathbf{u}\cdot\mathbf{v} = |\mathbf{u}||\mathbf{v}|$.

44. Show that **u** and **v** have opposite directions if $\mathbf{u}\cdot\mathbf{v} = -|\mathbf{u}||\mathbf{v}|$.

45. Show that the angle between **u** and **v** is acute if $0 < \mathbf{u}\cdot\mathbf{v} < |\mathbf{u}||\mathbf{v}|$.

46. Show that the angle between **u** and **v** is obtuse if $-|\mathbf{u}||\mathbf{v}| < \mathbf{u}\cdot\mathbf{v} < 0$.

In Problems 47–50, determine whether the statement is true or false. If true, explain. If false, give a specific counterexample.

47. If $\mathbf{u}\cdot\mathbf{v} = 0$, then $\mathbf{u} = 0$ or $\mathbf{v} = 0$.

48. If $\mathbf{u}\cdot\mathbf{v} = 1$, then **u** and **v** have the same direction.

49. $|\mathbf{u}\cdot\mathbf{v}| \le |\mathbf{u}||\mathbf{v}|$ [*Note:* $|\mathbf{u}\cdot\mathbf{v}|$ denotes the absolute value of the real number $\mathbf{u}\cdot\mathbf{v}$.]

50. If $\mathbf{u}\cdot\mathbf{v} < 0$, then the angle between **u** and **v** is obtuse.

 Applications

51. **Work** A parent pulls a child in a wagon (see the figure) for one block (440 ft). If a constant force of 15 lb is exerted along the handle, how much work is done?

Figure for 51

52. **Work** Repeat Problem 51 using a constant force of 12 lb, a distance of 5,300 ft (approximately a mile), and an angle of 35° relative to the horizontal.

53. **Work** Refer to Problem 51. If the angle is reduced to 30°, is more or less work done? How much?

54. **Work** Refer to Problem 52. If the angle is increased to 45°, is more or less work done? How much?

Work *In Problems 55–60, determine how much work is done by a force* **F** *moving an object from the origin to the point P. (Force is in pounds and displacement is in feet.)*

55. $\mathbf{F} = \langle 10, 5\rangle$; $P = (8, 1)$

56. $\mathbf{F} = \langle 5, 1\rangle$; $P = (8, 5)$

57. $\mathbf{F} = -2\mathbf{i} + 3\mathbf{j}$; $P = (-3, 1)$

58. $\mathbf{F} = 3\mathbf{i} - 4\mathbf{j}$; $P = (5, 0)$

59. $\mathbf{F} = 10\mathbf{i} + 10\mathbf{j}$; $P = (1, 1)$

60. $\mathbf{F} = 8\mathbf{i} - 4\mathbf{j}$; $P = (2, -1)$

61. **Geometry** Use the accompanying figure with the indicated vector assignments and appropriate properties of the dot product to prove that an angle inscribed in a semicircle is a right angle. [*Note:* $|\mathbf{a}| = |\mathbf{c}| = Radius$]

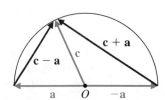

Figure for 61

62. **Geometry** Use the accompanying figure with the indicated vector assignments and appropriate properties of the dot product to prove that the diagonals of a rhombus are perpendicular. [*Note:* $|\mathbf{a}| = |\mathbf{b}|$]

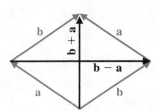

Figure for 62

CHAPTER 6 GROUP ACTIVITY

The SSA Case and the Law of Cosines

In Section 6.1 we spent quite a bit of time discussing SSA triangles and their solution using the law of sines. Table 1 is the part of Table 2 in Section 6.1 limited to α acute and $a < b$.

TABLE 1
For α Acute and $a < b$

a	Number of triangles	Figure
$0 < a < h$	0	
$a = h$	1	
$h < a < b$	2	

The law of cosines can be used on the SSA cases in Table 1, and it has the advantage of automatically sorting out the situations illustrated. The computations are more complicated using the law of cosines, but using a calculator helps overcome this difficulty.

Given a, b, and α, with α acute and $a < b$, we want to find c. Start with the law of cosines in the form

$$a^2 = b^2 + c^2 - 2bc \cos \alpha$$

and show that c is given by

$$c = \frac{2b \cos \alpha \pm \sqrt{(2b \cos \alpha)^2 - 4(b^2 - a^2)}}{2} \qquad (1)$$

Now, denote the discriminant by

$$D = (2b \cos \alpha)^2 - 4(b^2 - a^2)$$

and complete Table 2. Explain your reasoning.

TABLE 2
For α Acute and $a < b$

D	Number of solutions
$D < 0$	?
$D = 0$	?
$D > 0$	?

Problem 1 For the indicated values, determine whether the triangle has zero, one, or two solutions without computing the solutions.

(A) $a = 5.1, b = 7.6, \alpha = 32.7°$

(B) $a = 2.1, b = 8.2, \alpha = 21.5°$

Problem 2   Use equation (1) to find all values of c for the triangle(s) in Problem 1 that have at least one solution.   ■

Problem 3   If $a > b$ and α is obtuse, how many solutions will equation (1) give? Explain your reasoning.   ■

Problem 4   Use equation (1) to find all values of c for a triangle with $a = 8.8$, $b = 6.4$, and $\alpha = 123.4°$.   ■

CHAPTER 6 REVIEW

**6.1
LAW OF SINES**

An **oblique triangle** is a triangle without a right angle. An oblique triangle is **acute** if all angles are between 0° and 90° and **obtuse** if one angle is between 90° and 180°. See Figure 1.

    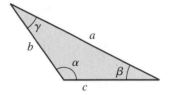

(a) Acute triangle   (b) Obtuse triangle

FIGURE 1

Solving a triangle involves finding three of the six quantities indicated in Figure 1 when given the other three. Table 1 is used to determine the accuracy of these computations.

TABLE 1	
Angle to nearest	Significant digits for side measure
1°	2
10′ or 0.1°	3
1′ or 0.01°	4
10″ or 0.001°	5

If the given quantities include an angle and the opposite side (ASA, AAS, or SSA), the **law of sines** is used to solve the triangle in Figure 2 on the next page. In the case where we are given two angles, we can first find the remaining angle by subtracting from 180°.

FIGURE 2

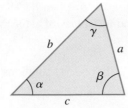

$$\frac{\sin \alpha}{a} = \frac{\sin \beta}{b} = \frac{\sin \gamma}{c}$$

The SSA case, often called the **ambiguous case,** has a number of variations (see Table 2). It is usually simplest to solve for one angle using the law of sines and consider both the acute and obtuse solutions to the resulting sine equations.

TABLE 2
SSA Variations

	a ($h = b \sin \alpha$)	Number of triangles	Figure	
α acute	$0 < a < h$	0		(a)
	$a = h$	1		(b)
	$h < a < b$	2		(c)
	$a \geq b$	1		(d)
α obtuse	$0 < a \leq b$	0		(e)
	$a > b$	1		(f)

6.2
LAW OF COSINES

We stated the **law of cosines** for the triangle in Figure 3 as follows:

FIGURE 3

$$a^2 = b^2 + c^2 - 2bc \cos \alpha$$
$$b^2 = a^2 + c^2 - 2ac \cos \beta$$
$$c^2 = a^2 + b^2 - 2ab \cos \gamma$$

The law of cosines is generally used as the first step in solving the SAS and SSS cases for oblique triangles. After a side or angle is found using the law of cosines, it is usually easier to use the law of sines to find a second angle.

*6.3
AREAS OF TRIANGLES

The area of a triangle (Fig. 4) can be determined from the base and altitude, two sides and the included angle, or all three sides (**Heron's formula**).

FIGURE 4

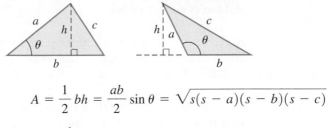

$$A = \frac{1}{2}bh = \frac{ab}{2}\sin\theta = \sqrt{s(s - a)(s - b)(s - c)}$$

where $s = \frac{1}{2}(a + b + c)$ is the **semiperimeter.**

6.4
VECTORS: GEOMETRICALLY DEFINED

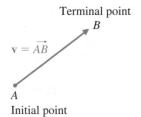

Terminal point

B

$\mathbf{v} = \vec{AB}$

A

Initial point

FIGURE 5
Vector $\vec{AB} = \mathbf{v}$

A **scalar** is a real number. The **directed line segment** from point A to point B in the plane, denoted $\vec{AB}$, is the line segment from A to B with the arrowhead placed at B to indicate the direction (see Fig. 5). Point A is called the **initial point,** and point B is called the **terminal point.**

A **geometric vector** in a plane, denoted $\mathbf{v}$, is a quantity that possesses both a length and a direction and can be represented by a directed line segment as indicated in Figure 5. The **magnitude** of the vector $\vec{AB}$, denoted by $|\vec{AB}|$, $|\vec{v}|$, or $|\mathbf{v}|$, is the length of the directed line segment. Two vectors have the **same direction** if they are parallel and point in the same direction. Two vectors have **opposite direction** if they are parallel and point in opposite directions. The **zero vector,** denoted by $\vec{0}$ or $\mathbf{0}$, has a magnitude of zero and an arbitrary direction. Two vectors are **equal** if they have the same magnitude and direction. Thus, a vector may be **translated** from one location to another as long as the magnitude and direction do not change.

The **sum of two vectors u and v** can be defined using the **tail-to-tip rule.** The **sum of two nonparallel vectors** also can be defined using the **parallelogram rule.** Both forms are shown in Figure 6.

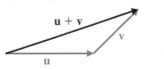

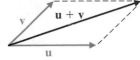

Tail-to-tip rule for addition   Parallelogram rule for addition

FIGURE 6
Vector addition

The sum vector $\mathbf{u} + \mathbf{v}$ is also called the **resultant** of the two vectors $\mathbf{u}$ and $\mathbf{v}$, and $\mathbf{u}$ and $\mathbf{v}$ are called **components** of $\mathbf{u} + \mathbf{v}$. Vector addition is **commutative** and **associative.** That is, $\mathbf{u} + \mathbf{v} = \mathbf{v} + \mathbf{u}$ and $\mathbf{u} + (\mathbf{v} + \mathbf{w}) = (\mathbf{u} + \mathbf{v}) + \mathbf{w}.$

A **velocity vector** represents the direction and speed of an object in motion. The velocity of a boat relative to the water is called the **apparent velocity,** and the velocity relative to the ground is called the **resultant** or **actual velocity.** The resultant velocity is the vector sum of the apparent velocity and the current velocity. Similar statements apply to objects in air subject to winds.

A **force vector** represents the direction and magnitude of an applied force. If an object is subjected to two forces, then the sum of these two forces, the **resultant force,** is a single force acting on the object in the same way as the two original forces taken together.

6.5
VECTORS:
ALGEBRAICALLY
DEFINED

A geometric vector $\overrightarrow{AB}$ in a rectangular coordinate system translated so that its initial point is at the origin is said to be in **standard position.** The vector $\overrightarrow{OP}$ such that $\overrightarrow{OP} = \overrightarrow{AB}$ is said to be the **standard vector** for $\overrightarrow{AB}$. This is shown in Figure 7.

Note that the vector $\overrightarrow{OP}$ in Figure 7 is the standard vector for infinitely many vectors: all vectors with the same magnitude and direction as $\overrightarrow{OP}$. If the coordinates of A in Figure 7 are (x_a, y_a) and the coordinates of B are (x_b, y_b), then the coordinates of P are given by

$$(x_p, y_p) = (x_b - x_a, y_b - y_a)$$

Every geometric vector in a plane corresponds to an ordered pair of real numbers, and every ordered pair of real numbers corresponds to a geometric vector. This leads to the definition of an **algebraic vector** as an ordered pair of real numbers, denoted by $\langle a, b \rangle$. The real numbers a and b are **scalar components** of the vector $\langle a, b \rangle$.

Two vectors $\mathbf{u} = \langle a, b \rangle$ and $\mathbf{v} = \langle c, d \rangle$ are said to be **equal** if their components are equal. The **zero vector** is denoted by $\mathbf{0} = \langle 0, 0 \rangle$ and has arbitrary direction.

The **magnitude,** or **norm,** of a vector $\mathbf{v} = \langle a, b \rangle$, denoted by $|\mathbf{v}|$, is given by

$$|\mathbf{v}| = \sqrt{a^2 + b^2}$$

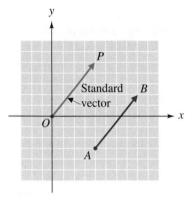

FIGURE 7
$\overrightarrow{OP}$ is the standard vector for $\overrightarrow{AB}(\overrightarrow{OP} = \overrightarrow{AB})$

Geometrically, $\sqrt{a^2 + b^2}$ is the length of the standard geometric vector $\overrightarrow{OP}$ associated with the algebraic vector $\langle a, b \rangle$.

If $\mathbf{u} = \langle a, b \rangle$, $\mathbf{v} = \langle c, d \rangle$, and k is a scalar, then the **sum** of $\mathbf{u}$ and $\mathbf{v}$ is given by

$$\mathbf{u} + \mathbf{v} = \langle a + c, b + d \rangle$$

and **scalar multiplication** of $\mathbf{u}$ by k is given by

$$k\mathbf{u} = k\langle a, b \rangle = \langle ka, kb \rangle$$

If $\mathbf{v}$ is a nonzero vector, then

$$\mathbf{u} = \frac{1}{|\mathbf{v}|} \mathbf{v}$$

is a **unit vector** with the same direction as $\mathbf{v}$. The $\mathbf{i}$ and $\mathbf{j}$ unit vectors are defined as follows (Fig. 8):

$$\mathbf{i} = \langle 1, 0 \rangle$$
$$\mathbf{j} = \langle 0, 1 \rangle$$

Any vector $\mathbf{v}$ can be expressed in terms of $\mathbf{i}$ and $\mathbf{j}$:

$$\mathbf{v} = \langle a, b \rangle = a\mathbf{i} + b\mathbf{j}$$

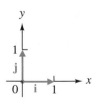

FIGURE 8

The following algebraic properties of vector addition and scalar multiplication enable us to manipulate symbols representing vectors and scalars in much the same way we manipulate symbols that represent real numbers in algebra.

(A) The following **addition properties** are satisfied for all vectors **u**, **v**, and **w**:

1. $\mathbf{u} + \mathbf{v} = \mathbf{v} + \mathbf{u}$ Commutative property
2. $\mathbf{u} + (\mathbf{v} + \mathbf{w}) = (\mathbf{u} + \mathbf{v}) + \mathbf{w}$ Associative property
3. $\mathbf{u} + \mathbf{0} = \mathbf{0} + \mathbf{u} = \mathbf{u}$ Additive identity
4. $\mathbf{u} + (-\mathbf{u}) = (-\mathbf{u}) + \mathbf{u} = \mathbf{0}$ Additive inverse

(B) The following **scalar multiplication properties** are satisfied for all vectors **u** and **v** and all scalars m and n:

1. $m(n\mathbf{u}) = (mn)\mathbf{u}$ Associative property
2. $m(\mathbf{u} + \mathbf{v}) = m\mathbf{u} + m\mathbf{v}$ Distributive property
3. $(m + n)\mathbf{u} = m\mathbf{u} + n\mathbf{u}$ Distributive property
4. $1\mathbf{u} = \mathbf{u}$ Multiplication identity

Algebraic vectors can be used to solve **static equilibrium** problems. An object at rest is said to be in static equilibrium. An object subject only to **coplanar concurrent forces** is in static equilibrium if and only if the sum of all these forces is zero.

*6.6
THE DOT PRODUCT

The **dot product** of the two vectors

$$\mathbf{u} = \langle a, b \rangle = a\mathbf{i} + b\mathbf{j} \quad \text{and} \quad \mathbf{v} = \langle c, d \rangle = c\mathbf{i} + d\mathbf{j}$$

denoted by $\mathbf{u} \cdot \mathbf{v}$, is the scalar given by

$$\mathbf{u} \cdot \mathbf{v} = ac + bd$$

For all vectors **u**, **v**, and **w**, and k a real number:

1. $\mathbf{u} \cdot \mathbf{u} = |\mathbf{u}|^2$ 4. $(\mathbf{u} + \mathbf{v}) \cdot \mathbf{w} = \mathbf{u} \cdot \mathbf{w} + \mathbf{v} \cdot \mathbf{w}$
2. $\mathbf{u} \cdot \mathbf{v} = \mathbf{v} \cdot \mathbf{u}$ 5. $k(\mathbf{u} \cdot \mathbf{v}) = (k\mathbf{u}) \cdot \mathbf{v} = \mathbf{u} \cdot (k\mathbf{v})$
3. $\mathbf{u} \cdot (\mathbf{v} + \mathbf{w}) = \mathbf{u} \cdot \mathbf{v} + \mathbf{u} \cdot \mathbf{w}$ 6. $\mathbf{u} \cdot \mathbf{0} = 0$

The **angle between the vectors u and v** is the angle θ, $0 \leq \theta \leq \pi$, formed by positioning the vectors so that their initial points coincide (Fig. 9). The vectors

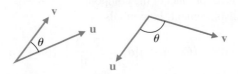

FIGURE 9
Angle between two vectors

u and **v** are **orthogonal,** or **perpendicular,** if $\theta = \pi/2$, and **parallel** if $\theta = 0$ or $\theta = \pi$. If **u** and **v** are nonzero, then

$$\cos\theta = \frac{\mathbf{u}\cdot\mathbf{v}}{|\mathbf{u}||\mathbf{v}|}$$

Two vectors **u** and **v** are orthogonal if and only if

$$\mathbf{u}\cdot\mathbf{v} = 0$$

[*Note:* The zero vector is orthogonal to every vector.]

To project a vector **u** onto an arbitrary nonzero vector **v**, locate **u** and **v** so that they have the same initial point O and drop a perpendicular from the terminal point of **u** to the line that contains **v**. The **vector projection of u onto v** is the vector **p** that lies on the line containing **v**, as shown in Figure 10.

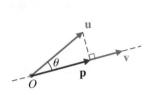
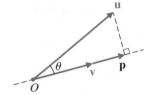
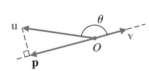

Projection of **u** onto **v**

FIGURE 10
Projection of **u** onto **v**

The **scalar component of u on v** is given by

$$\text{Comp}_\mathbf{v}\,\mathbf{u} = |\mathbf{u}|\cos\theta$$

or, in terms of the dot product, by

$$\text{Comp}_\mathbf{v}\,\mathbf{u} = \frac{\mathbf{u}\cdot\mathbf{v}}{|\mathbf{v}|}$$

A vector **d** is a **displacement vector** for an object that is moved in a straight line if **d** points in the direction of motion and if $|\mathbf{d}|$ is the distance moved. If the displacement vector is **d** for an object moved by a force **F**, then the **work** done, W, is the product of the component of force in the direction of motion and the actual displacement. In symbols,

$$W = (\text{Comp}_\mathbf{d}\mathbf{F})|\mathbf{d}| = \frac{\mathbf{F}\cdot\mathbf{d}}{|\mathbf{d}|}\,|\mathbf{d}| = \mathbf{F}\cdot\mathbf{d}$$

CHAPTER 6 REVIEW EXERCISE

Work through all the problems in this chapter review and check the answers. Answers to all review problems appear in the back of the book; following each answer is an italic number that indicates the section in which that type of problem is discussed. Where weaknesses show up, review the appropriate sections in the text. Review problems flagged with a star (☆) are from optional section.

Where applicable, quantities in the problems refer to a triangle labeled as in the figure.

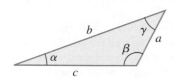

A **1.** If two sides and the included angle are given for a triangle, explain why the law of sines cannot be used to find another side or angle.

2. If two forces acting on a point have equal magnitudes and the resultant is **0**, what can you say about the two forces?

3. If two forces acting on a point have equal magnitudes and the magnitude of the resultant is the sum of the magnitudes of the individual forces, what can you say about the two forces?

4. Without solving each triangle, determine whether the given information allows you to construct zero, one, or two triangles. Explain your reasoning.
(A) $a = 4$ cm, $b = 8$ cm, $\alpha = 30°$
(B) $a = 5$ m, $b = 7$ m, $\alpha = 30°$
(C) $a = 3$ in., $b = 8$ in., $\alpha = 30°$

In Problems 5–8, solve each triangle given the indicated measures of angles and sides.

5. $\alpha = 53°$, $\gamma = 105°$, $b = 42$ cm

6. $\alpha = 66°$, $\beta = 32°$, $b = 12$ m

7. $\alpha = 49°$, $b = 22$ in., $c = 27$ in.

8. $\alpha = 62°$, $a = 14$ cm, $b = 12$ cm

☆ **9.** Find the area of the triangle in Problem 7.

☆ **10.** Find the area of the triangle in Problem 8.

11. Two vectors **u** and **v** are located in a coordinate system, as indicated in the figure. Find the direction (relative to the x axis) and magnitude of **u** + **v** if $|\mathbf{u}| = 8.0$ and $|\mathbf{v}| = 5.0$.

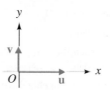

Figure for 11

12. Find the magnitude of the horizontal and vertical components of the vector **v** located in a coordinate system as indicated in the figure, then write **v** in terms of the unit vectors **i** and **j**.

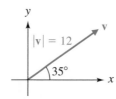

Figure for 12

13. Given $A = (-3,\ 2)$ and $B = (-1,\ -3)$ represent the geometric vector $\overrightarrow{AB}$ as an algebraic vector $\langle a, b \rangle$.

14. Find the magnitude of the vector $\langle -5, 12 \rangle$.

☆ **15.** $\langle 2, -1 \rangle \cdot \langle -3, 2 \rangle = $ **?**

☆ **16.** $(2\mathbf{i} + \mathbf{j}) \cdot (3\mathbf{i} - 2\mathbf{j}) = $ **?**

☆ **17.** Find the angle (in decimal degrees, to one decimal place) between $\mathbf{u} = \langle 4, 3 \rangle$ and $\mathbf{v} = \langle 3, 0 \rangle$.

☆ **18.** Find the angle (in decimal degrees, to one decimal place) between $\mathbf{u} = 5\mathbf{i} + \mathbf{j}$ and $\mathbf{v} = -2\mathbf{i} + 2\mathbf{j}$.

19. For each pair of vectors, are they orthogonal, parallel, or neither?
(A) $\langle 4, -3 \rangle$ and $\langle 8, 6 \rangle$
(B) $\left\langle \dfrac{5}{2}, -\dfrac{1}{2} \right\rangle$ and $\langle -10, 2 \rangle$
(C) $\langle 10, -6 \rangle$ and $\langle 3, 5 \rangle$

B **20.** In the process of solving an SSA triangle, it is found that $\sin \beta > 1$. How many triangles are possible? Explain.

In Problems 21–24, solve each triangle given the indicated measures of angles and sides.

21. $\alpha = 65.0°$, $b = 103$ m, $c = 72.4$ m

22. $\alpha = 35°20'$, $a = 13.2$ in., $b = 15.7$ in., β acute

23. $\alpha = 35°20'$, $a = 13.2$ in., $b = 15.7$ in., β obtuse

24. $a = 43$ mm, $b = 48$ mm, $c = 53$ mm

☆**25.** Find the area of the triangle in Problem 21.

☆**26.** Find the area of the triangle in Problem 24.

27. There are two vectors **u** and **v** such that $|\mathbf{u}| = |\mathbf{v}|$ and $\mathbf{u} \neq \mathbf{v}$. Explain how this can happen.

28. Under what conditions are the two algebraic vectors $\langle a, b \rangle$ and $\langle c, d \rangle$ equal?

29. Given the vector diagram, find $|\mathbf{u} + \mathbf{v}|$ and θ.

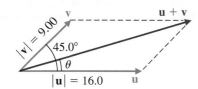

Figure for 29

In Problems 30 and 31, find:

(A) $\mathbf{u} + \mathbf{v}$ (B) $\mathbf{u} - \mathbf{v}$

(C) $3\mathbf{u} - 2\mathbf{v}$ (D) $2\mathbf{u} - 3\mathbf{v} + \mathbf{w}$

30. $\mathbf{u} = \langle 4, 0 \rangle$; $\mathbf{v} = \langle -2, -3 \rangle$; $\mathbf{w} = \langle 1, -1 \rangle$

31. $\mathbf{u} = 3\mathbf{i} - \mathbf{j}$; $\mathbf{v} = 2\mathbf{i} - 3\mathbf{j}$; $\mathbf{w} = -2\mathbf{j}$

32. Find a unit vector **u** with the same direction as $\mathbf{v} = \langle -8, 15 \rangle$.

33. Express **v** in terms of **i** and **j** unit vectors.

(A) $\mathbf{v} = \langle -5, 7 \rangle$ (B) $\mathbf{v} = \langle 0, -3 \rangle$

(C) $\mathbf{v} = \overrightarrow{AB}$; $A = (4, -2)$; $B = (0, -3)$

☆**34.** Determine which vector pairs are orthogonal using properties of the dot product.

(A) $\mathbf{u} = \langle -12, 3 \rangle$; $\mathbf{v} = \langle 2, 8 \rangle$

(B) $\mathbf{u} = -4\mathbf{i} + \mathbf{j}$; $\mathbf{v} = -\mathbf{i} + 4\mathbf{j}$

☆**35.** Find Comp$_\mathbf{v}$ **u**, the scalar component of **u** on **v**. Compute answers to three significant digits.

(A) $\mathbf{u} = \langle 4, 5 \rangle$; $\mathbf{v} = \langle 3, 1 \rangle$

(B) $\mathbf{u} = -\mathbf{i} + 4\mathbf{j}$; $\mathbf{v} = 3\mathbf{i} - \mathbf{j}$

36. An object that is free to move has three nonzero coplanar concurrent forces acting on it, and it remains at rest. How is any one of the three forces related to the other two?

C 37. An oblique triangle is solved, with the following results:

$$\alpha = 45.1° \qquad a = 8.42 \text{ cm}$$
$$\beta = 75.8° \qquad b = 11.5 \text{ cm}$$
$$\gamma = 59.1° \qquad c = 10.2 \text{ cm}$$

Allowing for rounding, use Mollweide's equation (below) to check these results.

$$(a - b) \cos \frac{\gamma}{2} = c \sin \frac{\alpha - \beta}{2}$$

38. Given an oblique triangle with $\alpha = 52.3°$ and $b = 12.7$ cm, determine a value k so that if $0 < a < k$, there is no solution; if $a = k$, there is one solution; and if $k < a < b$, there are two solutions.

In Problems 39–44, let $\mathbf{u} = \langle a, b \rangle$, $\mathbf{v} = \langle c, d \rangle$, and $\mathbf{w} = \langle e, f \rangle$ be vectors, and let m and n be scalars. Prove each property using the definitions of the operations involved and properties of real numbers. (Although some of these problems appeared in Exercises 6.5 and 6.6, it is useful to consider them again as part of this review.)

39. $\mathbf{u} + \mathbf{v} = \mathbf{v} + \mathbf{u}$

☆**40.** $\mathbf{u} \cdot \mathbf{v} = \mathbf{v} \cdot \mathbf{u}$

41. $(mn)\mathbf{v} = m(n\mathbf{v})$

☆**42.** $m(\mathbf{u} \cdot \mathbf{v}) = (m\mathbf{u}) \cdot \mathbf{v} = \mathbf{u} \cdot (m\mathbf{v})$

☆**43.** $\mathbf{u} \cdot (\mathbf{v} + \mathbf{w}) = \mathbf{u} \cdot \mathbf{v} + \mathbf{u} \cdot \mathbf{w}$

☆**44.** $\mathbf{u} \cdot \mathbf{u} = |\mathbf{u}|^2$

Applications

45. **Geometry**   Find the lengths of the sides of a parallelogram with diagonals 20.0 cm and 16.0 cm long intersecting at 36.4°.

46. **Geometry**   Find h to three significant digits in the triangle:

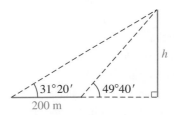

Figure for 46

47. **Geometry** A chord of length 34 cm subtends a central angle of 85° in a circle of radius r. Find the radius of the circle.

48. **Surveying** A plot of land has been surveyed, with the resulting information shown in the figure. Find the length of CD.

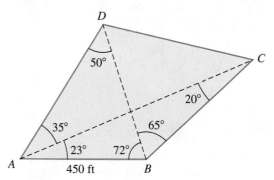

Figure for 48

☆ 49. **Surveying** Refer to Problem 48. Find the area of the plot.

50. **Engineering** A tunnel for a highway is to be constructed through a mountain, as indicated in the figure. How long is the tunnel?

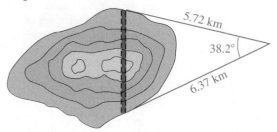

Figure for 50

51. **Space Science** A satellite S, in circular orbit around the earth, is sighted by a tracking station T (see the figure).

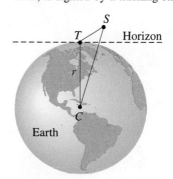

Figure for 51

The distance TS determined by radar is 1,147 mi, and the angle of elevation above the horizon is 28.6°. How high is the satellite above the earth at the time of the sighting? The radius of the earth is $r = 3,964$ mi.

52. **Space Science** When a satellite is directly over tracking station B (see the figure), tracking station A measures the angle of elevation θ of the satellite (from the horizon line) to be 21.7°. If the tracking stations are 632 mi apart (that is, the arc AB is 632 mi long) and the radius of the earth is 3,964 mi, how high is the satellite above B? [*Hint:* Find all angles for the triangle ACS first.]

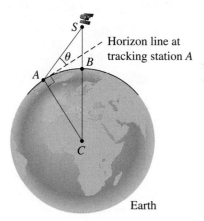

Figure for 52

53. **Navigation** An airplane flies with a speed of 230 km/hr and a compass heading of 68°. If a 55 km/hr wind is blowing in the direction of 5°, what is the plane's actual direction (relative to north) and ground speed?

54. **Navigation** A pilot in a small plane is flying from New Orleans to Houston, a distance of 302 mi. After flying 1 hr at an average of 125 mi/hr, he finds that he is 17° off course. If he corrects his course and maintains that speed for the remainder of the trip, how much longer will it take?

55. **Resultant Force** Two forces act on an object as indicated in the figure:

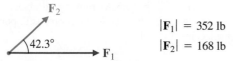

$|\mathbf{F}_1| = 352$ lb

$|\mathbf{F}_2| = 168$ lb

Find the magnitude of the resultant force and its direction relative to the horizontal force $\mathbf{F}_1$.

☆ **56. Work**   Refer to Problem 55. How much work is done by the resultant force if the object is moved 22 ft in the direction of the resultant force?

57. Engineering   A 260 lb sign is hung as shown in the figure. Determine the compression force on *AB* and the tension on *CB*.

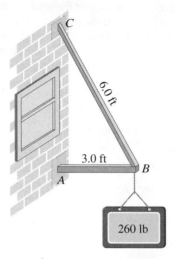

Figure for 57

58. Space Science   To simulate a reduced-gravity environment such as that found on the surface of the moon, NASA constructed an inclined plane and a harness suspended on a cable fastened to a movable track, as indicated in the figure. A person suspended in the harness walks at a right angle to the inclined plane.

(A) Using the accompanying force diagram, find $|\mathbf{u}|$, the force on the astronaut's feet, in terms of $|\mathbf{w}|$ and θ. Also find $|\mathbf{v}|$, the tension on the cable, in terms of $|\mathbf{w}|$ and θ.

(B) What is the force against the feet of a woman astronaut weighing 130 lb ($|\mathbf{w}| = 130$ lb) if $\theta = 72°$? What is the tension on the cable? Compute answers to the nearest pound.

(C) Find θ, to the nearest degree, so that the force on an astronaut's feet will be the same as the force on the moon (about one-sixth of that on earth).

☆ **59. Engineering: Physics**   Determine how much work is done by the force $\mathbf{F} = \langle -5, 8 \rangle$ moving an object from the origin to the point $P = (-8, 2)$. (Force is in pounds and displacement is in feet.)

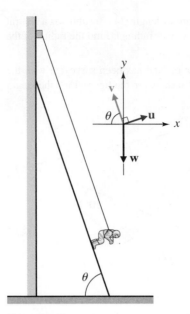

Figure for 58

60. Biomechanics   A person with a broken leg has it in a cast and must keep the leg elevated, as shown in the figure. (The weight of the leg with cast exerts a downward force of 12 lb.) Find angle θ and the tension on the line fastened to the overhead bar. Compute each to one decimal place.

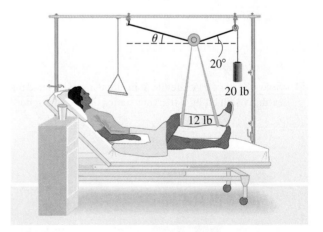

Figure for 60

Polar Coordinates; Complex Numbers

7.1 Polar and Rectangular Coordinates

7.2 Sketching Polar Equations

7.3 The Complex Plane

7.4 De Moivre's Theorem and the nth-Root Theorem

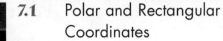 *Chapter 7 Group Activity: Orbits of Planets*

Chapter 7 Review

Cumulative Review Exercise, Chapters 1–7

W e specify the position of an object—a ship, a robotic hand, the eye of a hurricane—by giving it coordinates with respect to some coordinate system. In previous chapters we used the rectangular coordinate system. In this chapter we will introduce the polar coordinate system. We will graph polar equations and use polar coordinates to represent complex numbers. Representing complex numbers in polar form gives geometric insight into their multiplication and division. With this insight we will obtain formulas for the nth power and nth roots of a complex number, for n a positive integer.

7.1 Polar and Rectangular Coordinates

- Polar Coordinate System
- From Polar Form to Rectangular Form and Vice Versa

In previous chapters we used the rectangular coordinate system to associate ordered pairs of numbers with points in a plane. The polar coordinate system is an alternative to the rectangular coordinate system. When distance from a central location is of primary concern—for example, the distance from a city to a storm front, or the distance from a control tower to an airplane—then polar coordinates have advantages over rectangular coordinates.

■ Polar Coordinate System

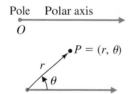

FIGURE 1
Polar coordinate system

To form a **polar coordinate system** in a plane (see Fig. 1), we start with a fixed point O and call it the **pole,** or **origin.** From this point we draw a half-line (usually horizontal and to the right), and we call this line the **polar axis.**

If P is an arbitrary point in a plane, then we associate **polar coordinates** (r, θ) with it as follows: Starting with the polar axis as the initial side of an angle, we rotate the terminal side until it, or the extension of it through the pole, passes through the point. The θ coordinate in (r, θ) is this angle, in degree or radian measure. The angle θ is positive if the rotation is counterclockwise and negative if the rotation is clockwise. The r coordinate in (r, θ) is the directed distance from the pole to the point P, which is positive if measured from the pole along the terminal side of θ and negative if measured along the terminal side extended through the pole. Figure 2 on the next page illustrates how one point in a polar coordinate system can have many—in fact, an unlimited number of—polar coordinates.

We can now see a major difference between the rectangular coordinate system and the polar coordinate system. In the former, each point has exactly one set of rectangular coordinates; in the latter, a point may have infinitely many polar coordinates.

The pole itself has polar coordinates of the form $(0, \theta)$, where θ is arbitrary. For example, $(0, 37°)$ and $(0, -\pi/4)$ are both polar coordinates of the pole, and there are infinitely many others.

FIGURE 2

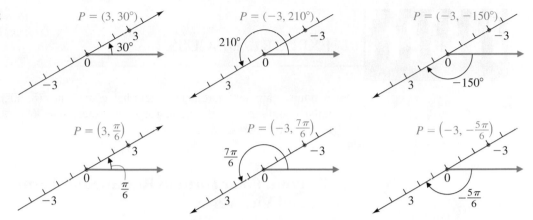

Just as graph paper can be used for work related to rectangular coordinate systems, polar graph paper can be used for work related to polar coordinates. The following examples illustrate its use.

EXAMPLE 1

Plotting Points in a Polar Coordinate System

Plot the following points in a polar coordinate system:

(A) $A = (4, 45°)$; $B = (-6, 45°)$; $C = (7, 240°)$; $D = (3, -75°)$

(B) $A = (8, \pi/6)$; $B = (5, -3\pi/4)$; $C = (-7, 2\pi/3)$; $D = (-9, -\pi/6)$

Solution

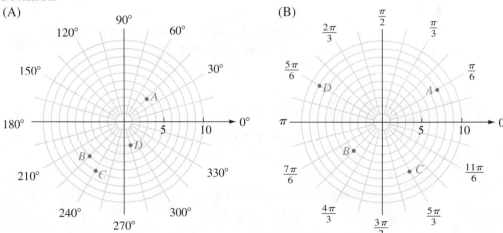

Matched Problem 1   Plot the following points in a polar coordinate system:

(A) $A = (7, 30°)$; $B = (-6, 165°)$; $C = (-9, -90°)$

(B) $A = (10, \pi/3)$; $B = (8, -7\pi/6)$; $C = (-5, -5\pi/4)$

EXPLORE/DISCUSS 1

A point in a polar coordinate system has coordinates (8, 60°). Find the other polar coordinates of the point for θ restricted to $-360° \leq \theta \leq 360°$, and explain how they are found.

■ From Polar Form to Rectangular Form and Vice Versa

It is often convenient to be able to transform coordinates or equations in rectangular form into polar form, or vice versa. The following polar–rectangular relationships are useful in this regard.

POLAR–RECTANGULAR RELATIONSHIPS

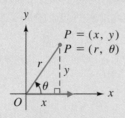

$$r^2 = x^2 + y^2$$

$$\sin \theta = \frac{y}{r} \qquad y = r \sin \theta$$

$$\cos \theta = \frac{x}{r} \qquad x = r \cos \theta$$

$$\tan \theta = \frac{y}{x}$$

[*Note:* The signs of x and y determine the quadrant for θ. The angle θ is usually chosen so that $-\pi < \theta \leq \pi$ or $-180° < \theta \leq 180°$.]

EXAMPLE 2

Polar Coordinates to Rectangular Coordinates

Change $A = (5, \pi/6)$, $B = (-3, 3\pi/4)$, and $C = (-2, -5\pi/6)$ to exact rectangular coordinates.

Solution   Use $x = r \cos \theta$ and $y = r \sin \theta$.

For A (see Fig. 3):

$$x = 5 \cos \frac{\pi}{6} = 5\left(\frac{\sqrt{3}}{2}\right) = \frac{5\sqrt{3}}{2}$$

$$y = 5 \sin \frac{\pi}{6} = 5\left(\frac{1}{2}\right) = \frac{5}{2}$$

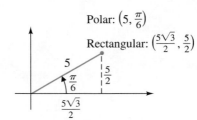

FIGURE 3

Rectangular coordinates: $\left(\dfrac{5\sqrt{3}}{2}, \dfrac{5}{2}\right)$

For B:

$$x = -3\cos\frac{3\pi}{4} = (-3)\left(\frac{-\sqrt{2}}{2}\right) = \frac{3\sqrt{2}}{2}$$

$$y = -3\sin\frac{3\pi}{4} = (-3)\left(\frac{\sqrt{2}}{2}\right) = \frac{-3\sqrt{2}}{2}$$

Rectangular coordinates: $\left(\dfrac{3\sqrt{2}}{2}, \dfrac{-3\sqrt{2}}{2}\right)$

For C:

$$x = -2\cos\left(\frac{-5\pi}{6}\right) = (-2)\left(\frac{-\sqrt{3}}{2}\right) = \sqrt{3}$$

$$y = -2\sin\left(\frac{-5\pi}{6}\right) = (-2)\left(\frac{-1}{2}\right) = 1$$

Rectangular coordinates: $(\sqrt{3}, 1)$ ■

Matched Problem 2   Change $A = (8, \pi/3)$, $B = (-6, 5\pi/4)$, and $C = (-4, -7\pi/6)$ to exact rectangular coordinates.   ■

  EXAMPLE 3   **Rectangular Coordinates to Polar Coordinates**

Change $A = (1, \sqrt{3})$ and $B = (-\sqrt{3}, -1)$ into exact polar form with $r \geq 0$ and $-\pi < \theta \leq \pi$.

Solution   Use $r^2 = x^2 + y^2$ and $\tan\theta = y/x$.

For A:

$$r^2 = 1^2 + (\sqrt{3})^2 = 4$$

$$r = 2$$

$$\tan\theta = \frac{\sqrt{3}}{1}$$

$$\theta = \frac{\pi}{3} \qquad \textit{Since A is in the first quadrant}$$

Polar coordinates: $\left(2, \dfrac{\pi}{3}\right)$

For *B:*

$$r^2 = (-\sqrt{3})^2 + (-1)^2 = 4$$
$$r = 2$$
$$\tan \theta = \frac{-1}{-\sqrt{3}} = \frac{1}{\sqrt{3}}$$
$$\theta = -\frac{5\pi}{6} \qquad \textit{Since B is in the third quadrant}$$

Polar coordinates: $\left(2, -\dfrac{5\pi}{6}\right)$ ■

Matched Problem 3   Change $A = (\sqrt{3}, 1)$ and $B = (1, -\sqrt{3})$ to polar form with $r \geq 0$ and $-\pi < \theta \leq 2\pi$. ■

Many calculators can automatically convert rectangular coordinates to polar form and vice versa; read the manual for your particular calculator. Example 4 illustrates conversions using the method shown in Examples 2 and 3 followed by the conversions on a graphing calculator with a built-in conversion routine.

   **EXAMPLE 4**   **Calculator Conversion**

Perform the conversions following the methods illustrated in Examples 2 and 3; then use a calculator with a built-in conversion routine (if you have one).

(A) Convert the rectangular coordinates $(-6.434, 4.023)$ to polar coordinates (to three decimal places), with θ in degree measure, $-180° < \theta \leq 180°$, and $r \geq 0$.

(B) Convert the polar coordinates $(8.677, -1.385)$ to rectangular coordinates (to three decimal places).

Solution   (A) Use a calculator set in degree mode.

$$(x, y) = (-6.434, 4.023)$$
$$r = \sqrt{x^2 + y^2} = \sqrt{(-6.434)^2 + 4.023^2} = 7.588$$
$$\tan \theta = \frac{y}{x} = \frac{4.023}{-6.434}$$

The angle θ is in the second quadrant, and it is to be chosen so that $-180° < \theta \leq 180°$. Therefore,

$$\theta = 180° + \tan^{-1}\left(\frac{4.023}{-6.434}\right) = 147.983°$$

Polar coordinates: $(7.588, 147.983°)$

Figure 4 shows the same conversion done on a graphing calculator with a built-in conversion routine.

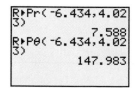

FIGURE 4

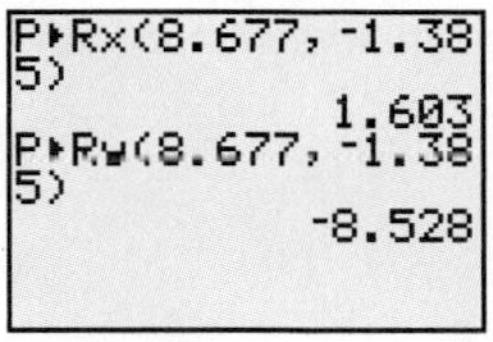

```
P▸Rx(8.677,-1.38
5)
           1.603
P▸Ry(8.677,-1.38
5)
          -8.528
```

FIGURE 5

(B) Use a calculator set in radian mode.

$$(r, \theta) = (8.677, -1.385)$$
$$x = r \cos \theta = 8.677 \cos(-1.385) = 1.603$$
$$y = r \sin \theta = 8.677 \sin(-1.385) = -8.528$$

Rectangular coordinates: $(1.603, -8.528)$

Figure 5 shows the same conversion done on a graphing calculator with a built-in conversion routine. ■

Matched Problem 4   Perform the conversions following the methods illustrated in Examples 2 and 3; then use a calculator with a built-in conversion routine (if you have one).

(A) Convert the rectangular coordinates $(-4.305, -5.117)$ to polar coordinates (to three decimal places), with θ in radian measure, $-\pi < \theta \le \pi$, and $r \ge 0$.

(B) Convert the polar coordinates $(-10.314, -37.232°)$ to rectangular coordinates (to three decimal places). ■

EXAMPLE 5

From Rectangular to Polar Form

Change $x^2 + y^2 - 2x = 0$ to polar form.

Solution   Use $r^2 = x^2 + y^2$ and $x = r \cos \theta$:

$$x^2 + y^2 - 2x = 0 \qquad \text{Replace } x^2 + y^2 \text{ by } r^2, \text{ and } x \text{ by } r \cos \theta.$$
$$r^2 - 2r \cos \theta = 0 \qquad \text{Factor.}$$
$$r(r - 2 \cos \theta) = 0 \qquad \text{If } ab = 0, \text{ then } a = 0 \text{ or } b = 0.$$
$$r = 0 \qquad \text{or} \qquad r - 2 \cos \theta = 0$$

The graph of $r = 0$ is the pole, and since the pole is included as a solution of $r - 2 \cos \theta = 0$ (let $\theta = \pi/2$), we can discard $r = 0$ and keep only

$$r - 2 \cos \theta = 0$$

or

$$r = 2 \cos \theta \qquad ■$$

Matched Problem 5   Change $x^2 + y^2 - 2y = 0$ to polar form. ■

EXAMPLE 6

From Polar to Rectangular Form

Change $r + 3 \sin \theta = 0$ to rectangular form.

Solution   The conversion of this equation, as it stands, to rectangular form gets messy. A simple trick, however, makes the conversion easy: We multiply both sides by r, which simply adds the pole to the graph. In this case, the pole is already included as a solution of $r + 3 \sin \theta = 0$ (let $\theta = 0$), so we have not actually changed anything by doing this. Therefore,

$$r + 3 \sin \theta = 0 \qquad \text{Multiply both sides by } r.$$
$$r^2 + 3r \sin \theta = 0 \qquad r^2 = x^2 + y^2, \ \ y = r \sin \theta$$
$$x^2 + y^2 + 3y = 0 \qquad \blacksquare$$

Matched Problem 6   Change $r = -8 \cos \theta$ to rectangular form.   ■

   **EXAMPLE 7**

From Polar to Rectangular Form

Change $r^2 \cos 2\theta = 9$ to rectangular form.

Solution   We first use a trigonometric identity to write $\cos 2\theta$ in terms of $\cos \theta$ and $\sin \theta$.

$$r^2 \cos 2\theta = 9 \qquad \text{Use a double-angle identity.}$$
$$r^2 \left(\cos^2 \theta - \sin^2 \theta\right) = 9 \qquad \text{Use the distributive law.}$$
$$r^2 \cos^2 \theta - r^2 \sin^2 \theta = 9 \qquad \text{Substitute } x = r \cos \theta, y = r \sin \theta.$$
$$x^2 - y^2 = 9 \qquad \blacksquare$$

Matched Problem 7   Change $r^2 \sin 2\theta = 8$ to rectangular form.   ■

Answers to Matched Problems   **1.** (A)

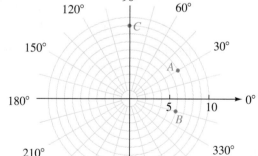

(B)

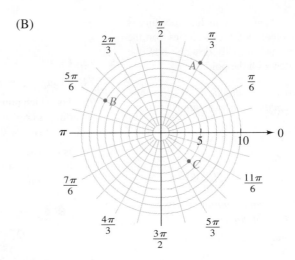

2. $A = (4, 4\sqrt{3})$, $B = (3\sqrt{2}, 3\sqrt{2})$, $C = (2\sqrt{3}, -2)$

3. $A = (2, \pi/6)$, $B = (2, -\pi/3)$

4. (A) $(r, \theta) = (6.687, -2.270)$ (B) $(x, y) = (-8.212, 6.240)$

5. $r = 2 \sin \theta$ **6.** $x^2 + y^2 + 8x = 0$ **7.** $xy = 4$

EXERCISE 7.1

A *In Problems 1–14, plot in a polar coordinate system.*

1. $A = (8, 0°)$; $B = (5, 90°)$; $C = (6, 30°)$

2. $A = (4, 0°)$; $B = (7, 180°)$; $C = (9, 45°)$

3. $A = (-8, 0°)$; $B = (-5, 90°)$; $C = (-6, 30°)$

4. $A = (-4, 0°)$; $B = (-7, 180°)$; $C = (-9, 45°)$

5. $A = (5, -30°)$; $B = (4, -45°)$; $C = (9, -90°)$

6. $A = (8, -45°)$; $B = (6, -60°)$; $C = (4, -30°)$

7. $A = (-5, -30°)$; $B = (-4, -45°)$; $C = (-9, -90°)$

8. $A = (-8, -45°)$; $B = (-6, -60°)$; $C = (-4, -30°)$

9. $A = (6, \pi/6)$; $B = (5, \pi/2)$; $C = (8, \pi/4)$

10. $A = (8, \pi/3)$; $B = (4, \pi/4)$; $C = (10, 0)$

11. $A = (-6, \pi/6)$; $B = (-5, \pi/2)$; $C = (-8, \pi/4)$

12. $A = (-8, \pi/3)$; $B = (-4, \pi/4)$; $C = (-10, 0)$

13. $A = (6, -\pi/6)$; $B = (5, -\pi/2)$; $C = (8, -\pi/4)$

14. $A = (8, -\pi/3)$; $B = (4, -\pi/4)$; $C = (10, -\pi/6)$

15. Is it possible for a point in the plane to have two different ordered pairs of polar coordinates? Explain.

16. Is it possible for one ordered pair of polar coordinates to correspond to two different points in the plane? Explain.

17. Explain how the polar coordinates of a point P in the plane can be obtained from the rectangular coordinates (x, y) of P.

18. Explain how the rectangular coordinates of a point P in the plane can be obtained from the polar coordinates (r, θ) of P.

In Problems 19–24, change to exact rectangular coordinates.

19. $(2, -\pi/2)$ **20.** $(5, \pi/2)$

21. $(10, \pi/3)$ **22.** $(6, \pi/4)$

23. $(-3\sqrt{2}, 3\pi/4)$ **24.** $(4, -5\pi/6)$

In Problems 25–30, change to exact polar coordinates with $r \geq 0$ and $-\pi < \theta \leq \pi$.

25. $(0, 6)$ **26.** $(-3, 0)$

27. $(-4, 4)$ **28.** $(2, -2\sqrt{3})$

29. $(-\sqrt{3}, -3)$ **30.** $(-\sqrt{5}, -\sqrt{5})$

31. A point in a polar coordinate system has coordinates $(6, -30°)$. Find all other polar coordinates for the point, $-360° < \theta \leq 360°$, and verbally describe how the coordinates are associated with the point.

32. A point in a polar coordinate system has coordinates $(-5, 3\pi/4)$. Find all other polar coordinates for the point, $-2\pi < \theta \le 2\pi$, and verbally describe how the coordinates are associated with the point.

B *In Problems 33–38, convert the rectangular coordinates to polar coordinates (to three decimal places), with $r \ge 0$ and $-180° < \theta \le 180°$.*

33. $(1.625, 3.545)$ **34.** $(8.721, 2.067)$

35. $(9.984, -1.102)$ **36.** $(-4.033, -4.614)$

37. $(-3.217, -8.397)$ **38.** $(-2.175, 5.005)$

In Problems 39–44, convert the polar coordinates to rectangular coordinates (to three decimal places).

39. $(2.718, -31.635°)$ **40.** $(1.824, 98.484°)$

41. $(-4.256, 3.085)$ **42.** $(6.518, -0.016)$

43. $(0.903, 1.514)$ **44.** $(-5.999, -2.450)$

In Problems 45–58, change to polar form.

45. $6x - x^2 = y^2$ **46.** $y^2 = 5y - x^2$

47. $2x + 3y = 5$ **48.** $3x - 5y = -2$

49. $x^2 + y^2 = 9$ **50.** $y = x$

51. $2xy = 1$ **52.** $y^2 = 4x$

53. $4x^2 - y^2 = 4$ **54.** $x^2 + 9y^2 = 9$

55. $x = 3$ **56.** $y = 8$

57. $y = -7$ **58.** $x = -10$

In Problems 59–72, change to rectangular form.

59. $r(2 \cos \theta + \sin \theta) = 4$

60. $r(3 \cos \theta - 4 \sin \theta) = -1$

61. $r = 8 \cos \theta$ **62.** $r = -2 \sin \theta$

63. $r = 2 \cos \theta + 3 \sin \theta$ **64.** $r = 5 \sin \theta - 4 \cos \theta$

65. $r^2 \cos 2\theta = 4$ **66.** $r^2 \sin 2\theta = 2$

67. $r^2 = 3 \cos 2\theta$ **68.** $r^2 = 5 \sin 2\theta$

69. $r = 4$ **70.** $r = -5$

71. $\theta = 30°$ **72.** $\theta = \pi/4$

C **73.** Change $r = 3/(\sin \theta - 2)$ into rectangular form.

74. Change $(y - 3)^2 = 4(x^2 + y^2)$ into polar form.

75. For which points P in the plane are the rectangular coordinates (x, y) of P identical to the polar coordinates (r, θ) of P (assume θ is measured in radians)? Explain.

76. Is it possible for every point in the plane to be described by polar coordinates (r, θ), where r and θ are real numbers with $0 \le \theta < \pi$? Explain.

Applications

77. **Precalculus: Analytic Geometry** A distance d (see the figure) between two points in a polar coordinate system is given by a formula that follows directly from the law of cosines:

$$d^2 = r_1^2 + r_2^2 - 2r_1r_2 \cos(\theta_2 - \theta_1)$$
$$d = \sqrt{r_1^2 + r_2^2 - 2r_1r_2 \cos(\theta_2 - \theta_1)}$$

Find the distance between the two points $P_1 = (2, 30°)$ and $P_2 = (3, 60°)$. Compute your answer to four significant digits.

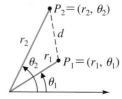

Figure for 77 and 78

78. **Precalculus: Analytic Geometry** Refer to Problem 77 and find the distance between the two points $P_1 = (4, \pi/4)$ and $P_2 = (1, \pi/2)$. Compute your answer to four significant digits.

7.2 Sketching Polar Equations

- Point-by-Point Sketching
- Rapid Sketching
- Graphing Polar Equations on a Graphing Calculator
- A Table of Standard Polar Curves
- Application

In rectangular coordinate systems we sketch graphs of equations involving the variables *x* and *y*. In polar coordinate systems we sketch graphs of equations involving the variables *r* and θ. We will see that certain curves have simpler representations in polar coordinates and other curves have simpler representations in rectangular coordinates. Likewise, some applications have simpler solutions in polar coordinates while other applications have simpler solutions in rectangular coordinates. In this section you will gain experience in recognizing and sketching some standard polar graphs.

■ Point-by-Point Sketching

To graph a polar equation such as $r = 2\theta$ or $r = 4\sin 2\theta$ in a polar coordinate system, we locate all points with coordinates that satisfy the equation. A sketch of a graph can be obtained (just as in rectangular coordinates) by making a table of values that satisfy the equation, plotting these points, and then joining them with a smooth curve. A calculator can be used to generate the table.

The graph of the polar equation

$$r = a\theta \quad a > 0$$

is called **Archimedes' spiral.** Example 1 illustrates how to obtain a particular spiral by point-by-point plotting.

EXAMPLE 1

Spiral

Graph $r = 2\theta, 0 \le \theta \le 5\pi/3$, in a polar coordinate system using point-by-point plotting (θ in radians).

Solution We form a table using multiples of π/6; then we plot the points and join them with a smooth curve (see Fig. 1).

θ	r
0	0.00
π/6	1.05
π/3	2.09
π/2	3.14
2π/3	4.19
5π/6	5.24
π	6.28
7π/6	7.33
4π/3	8.38
3π/2	9.42
5π/3	10.47

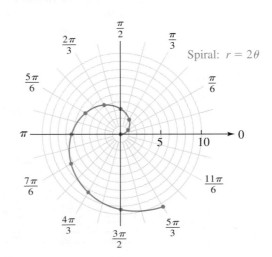

FIGURE 1

Matched Problem 1 Graph $r = 3\theta$, $0 \leq \theta \leq 7\pi/6$, in a polar coordinate system using point-by-point plotting (θ in radians). ∎

EXAMPLE 2 **Circle**

Graph $r = 4 \cos \theta$ (θ in radians).

Solution We start with multiples of $\pi/6$ and continue until the graph begins to repeat (see Fig. 2). (In intervals of uncertainty, add more points.) The graph is a circle with radius 2 and center at (2, 0).

θ	r
0	4.00
$\pi/6$	3.46
$\pi/3$	2.00
$\pi/2$	0.00
$2\pi/3$	−2.00
$5\pi/6$	−3.46
π	−4.00
$7\pi/6$	−3.46

Graph is repeating.

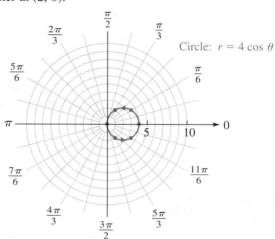

Circle: $r = 4 \cos \theta$

FIGURE 2 ∎

Matched Problem 2 Graph $r = 4 \sin \theta$ (θ in radians). ∎

If the equation in Example 2 is graphed using degrees instead of radians, we get the same graph except that degrees are marked around the polar coordinate system instead of radians (see Fig. 3).

FIGURE 3

θ	r
0°	4.00
30°	3.46
60°	2.00
90°	0.00
120°	−2.00
150°	−3.46
180°	−4.00
210°	−3.46

Graph is repeating.

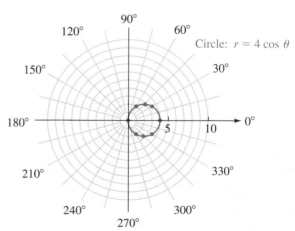

Circle: $r = 4 \cos \theta$

■ Rapid Sketching

If all that is desired is a rough sketch of a polar equation involving sin θ or cos θ, you can speed up the point-by-point process by taking advantage of the uniform variation of sin θ and cos θ as θ moves around a unit circle. We refer to this process as **rapid polar sketching.** It is useful to visualize the unit circle definition of sine and cosine in Figure 4 in the process.

FIGURE 4

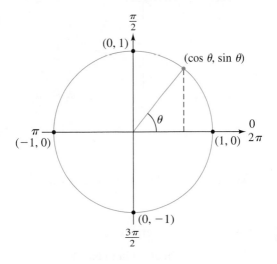

EXAMPLE 3

Rapid Polar Sketching

Sketch $r = 5 + 5 \cos \theta$ (θ in radians).

Solution We set up Table 1, which shows how r varies as θ varies through each set of quadrant values in Figure 4.

Notice that as θ increases from 0 to $\pi/2$, cos θ decreases from 1 to 0, 5 cos θ decreases from 5 to 0, and $r = 5 + 5 \cos \theta$ decreases from 10 to 5, and so on. Sketching the values in Table 1, we obtain the heart-shaped graph, called a **cardioid,** in Figure 5 on page 444.

TABLE 1

θ varies from	cos θ varies from	5 cos θ varies from	$r = 5 + 5\cos\theta$ varies from
0 to $\pi/2$	1 to 0	5 to 0	10 to 5
$\pi/2$ to π	0 to -1	0 to -5	5 to 0
π to $3\pi/2$	-1 to 0	-5 to 0	0 to 5
$3\pi/2$ to 2π	0 to 1	0 to 5	5 to 10

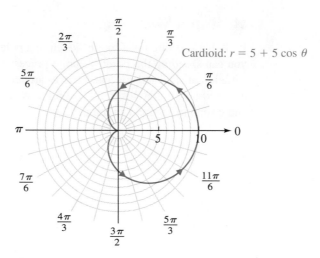

FIGURE 5
Cardioid

Matched Problem 3   Sketch $r = 4 + 4 \sin \theta$ (θ in radians).

Rapid Polar Sketching

Sketch $r = 6 \sin 2\theta$ (θ in radians).

Solution   We start by letting 2θ (instead of θ) range through each set of quadrant values. That is, we start with values of 2θ in the second column of Table 2, fill in the table to the right, and then fill in the first column for θ.

TABLE 2			
θ varies from	2θ varies from	$\sin 2\theta$ varies from	$r = 6 \sin 2\theta$ varies from
0 to $\pi/4$	0 to $\pi/2$	0 to 1	0 to 6
$\pi/4$ to $\pi/2$	$\pi/2$ to π	1 to 0	6 to 0
$\pi/2$ to $3\pi/4$	π to $3\pi/2$	0 to -1	0 to -6
$3\pi/4$ to π	$3\pi/2$ to 2π	-1 to 0	-6 to 0
π to $5\pi/4$	2π to $5\pi/2$	0 to 1	0 to 6
$5\pi/4$ to $3\pi/2$	$5\pi/2$ to 3π	1 to 0	6 to 0
$3\pi/2$ to $7\pi/4$	3π to $7\pi/2$	0 to -1	0 to -6
$7\pi/4$ to 2π	$7\pi/2$ to 4π	-1 to 0	-6 to 0

Using the information in Table 2, we complete the graph shown in Figure 6 on the next page, which is called a **four-leaved rose.**

FIGURE 6
Four-leaved rose

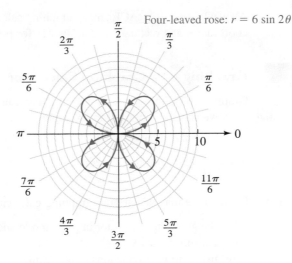

Four-leaved rose: $r = 6 \sin 2\theta$

Matched Problem 4 Sketch $r = 4 \cos 3\theta$ (θ in radians). ■

In a rectangular coordinate system the simplest equations to graph are found by setting the variables x and y equal to constants:

$$x = a \qquad \text{and} \qquad y = b$$

The graphs are straight lines: $x = a$ is a vertical line and $y = b$ is a horizontal line. A look ahead to Table 3 (page 447) will show you that horizontal and vertical lines do not have such simple expressions in polar coordinates.

The simplest equations to graph in a polar coordinate system are found by setting the polar variables r and θ equal to constants:

$$r = a \qquad \text{and} \qquad \theta = b$$

Figure 7 illustrates two particular cases. Notice that a circle centered at the origin has a simpler expression in polar coordinates than in rectangular coordinates.

FIGURE 7

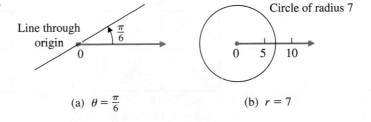

(a) $\theta = \frac{\pi}{6}$ (b) $r = 7$

Graphing Polar Equations on a Graphing Calculator

We now turn to the graphing of polar equations on a graphing calculator. Note that many graphing calculators, including the one used here, do not show a polar grid.

Also, when using TRACE , many graphing calculators offer a choice between polar coordinates and rectangular coordinates for points on the polar curve.

EXAMPLE 5

Graphing on a Graphing Calculator

 Graph each polar equation on a graphing calculator. (These are the same equations we sketched by hand in Examples 1–4.)

(A) $r = 2\theta, 0 \leq \theta \leq 5\pi/3$ (B) $r = 4 \cos \theta, \theta$ in radians

(C) $r = 5 + 5 \cos \theta, \theta$ in radians   (D) $r = 6 \sin 2\theta, \theta$ in radians

Solution   Refer to the manual for your graphing calculator as necessary. In particular:

- Set the graphing calculator in polar mode and select polar coordinates and radian measure.
- Adjust window dimensions to accommodate the whole graph (a little trial and error may be necessary).
- To show the true shape of a curve, choose window dimensions so that unit distances on the x and y axes have the same length (by using a ZOOM SQUARE command, for example).

Figure 8 shows the graphs in square graphing calculator viewing windows.

FIGURE 8

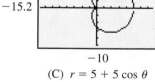

(A) $r = 2\theta, 0 \leq \theta \leq 5\pi/3$

(B) $r = 4 \cos \theta$

(C) $r = 5 + 5 \cos \theta$

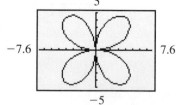

(D) $r = 6 \sin 2\theta$

Matched Problem 5   Graph each polar equation on a graphing calculator. (These are the same equations you sketched by hand in Matched Problems 1–4.) Use a square viewing window.

(A) $r = 3\theta, 0 \leq \theta \leq 7\pi/6$ (B) $r = 4 \sin \theta, \theta$ in radians

(C) $r = 4 + 4 \sin \theta, \theta$ in radians   (D) $r = 4 \cos 3\theta, \theta$ in radians

EXPLORE/DISCUSS 1

 (A) Graph $r_1 = 8 \cos \theta$ and $r_2 = 8 \sin \theta$ in the same viewing window. Use $\boxed{\text{TRACE}}$ on r_1 and estimate the polar coordinates where the two graphs intersect. Repeat for r_2.

(B) Which intersection point appears to have the same polar coordinates on each curve and consequently represents a simultaneous solution to both equations? Which intersection point appears to have different polar coordinates on each curve and consequently does not represent a simultaneous solution?

(C) Solve the system of equations above for r and θ. Conclusions?

(D) Could the situation illustrated in parts (A)–(C) happen with rectangular equations in two variables in a rectangular coordinate system? Discuss.

◼ A Table of Standard Polar Curves

Table 3 presents a few standard polar curves. Graphing polar equations is often made easier if you have an idea of what the shape of the curve will be.

TABLE 3
Some Standard Polar Curves

(a) Vertical line:
$$r = \frac{a}{\cos \theta}$$

(b) Horizontal line:
$$r = \frac{a}{\sin \theta}$$

(c) Radial line:
$$\theta = a$$

(d) Circle:
$$r = a$$

(e) Circle:
$$r = a \cos \theta$$

(f) Circle:
$$r = a \sin \theta$$

(g) Cardioid:
$$r = a + a \cos \theta$$

(h) Cardioid:
$$r = a + a \sin \theta$$

(Continued)

TABLE 3 (*Continued*)

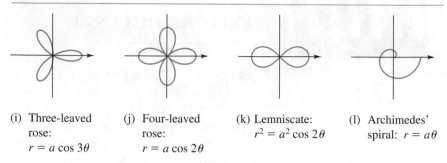

(i) Three-leaved
rose:
$r = a \cos 3\theta$

(j) Four-leaved
rose:
$r = a \cos 2\theta$

(k) Lemniscate:
$r^2 = a^2 \cos 2\theta$

(l) Archimedes'
spiral: $r = a\theta$

◼ Application

Polar coordinate systems are useful in many types of applications. Exercise 7.2 includes applications from sailboat racing and astronomy; Figure 9, which was supplied by the U.S. Coast and Geodetic Survey, illustrates an application from oceanography. In Figure 9 each arrow represents the direction and magnitude of the tide at a particular time of the day at the San Francisco Light Station (a navigational light platform in the Pacific Ocean about 12 nautical miles west of the Golden Gate Bridge).

FIGURE 9
Tidal current curve, San Francisco Light Station [Note: LL + 3^h means 3 hr after low low tide, HL − 1^h means 1 hr before high low tide, and so on.]

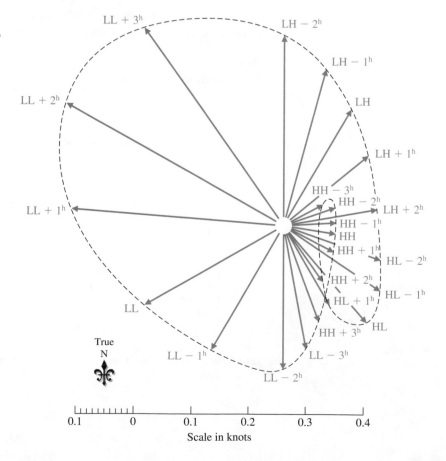

Answers to
Matched Problems

1.

θ	r
0	0.00
$\pi/6$	1.57
$\pi/3$	3.14
$\pi/2$	4.71
$2\pi/3$	6.28
$5\pi/6$	7.85
π	9.42
$7\pi/6$	11.00

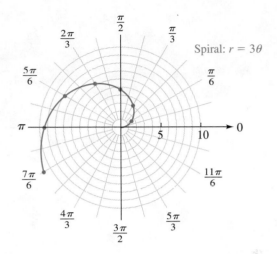

Spiral: $r = 3\theta$

2.

θ	r
0	0.00
$\pi/6$	2.00
$\pi/3$	3.46
$\pi/2$	4.00
$2\pi/3$	3.46
$5\pi/6$	2.00
π	0.00
$7\pi/6$	-2.00

Graph is repeating.

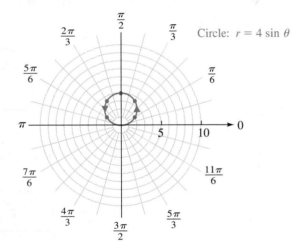

Circle: $r = 4 \sin \theta$

3.

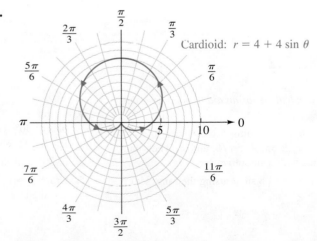

Cardioid: $r = 4 + 4 \sin \theta$

4.

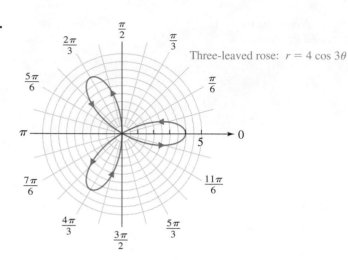

Three-leaved rose: $r = 4 \cos 3\theta$

5. (A) $r = 3\theta$ (B) $r = 4 \sin \theta$

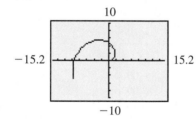

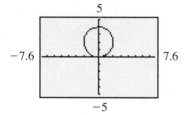

(C) $r = 4 + 4 \sin \theta$ (D) $r = 4 \cos 3\theta$

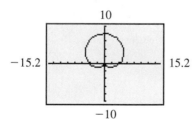

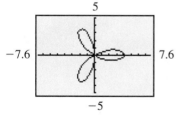

EXERCISE 7.2

A *In Problems 1–6, use a calculator as an aid to point-by-point plotting.*

1. Graph $r = 10 \cos \theta$ by assigning θ the values 0, $\pi/6$, $\pi/4$, $\pi/3$, $\pi/2$, $2\pi/3$, $3\pi/4$, $5\pi/6$, and π. Then join the resulting points with a smooth curve.

2. Repeat Problem 1 for $r = 8 \sin \theta$, using the same set of values for θ.

3. Graph $r = 3 + 3 \cos \theta, 0° \le \theta \le 360°$, using multiples of 30° starting at 0.

4. Graph $r = 8 + 8 \sin \theta, 0° \le \theta \le 360°$, using multiples of 30° starting at 0.

5. Graph $r = \theta, 0 \le \theta \le 2\pi$, using multiples of $\pi/6$ for θ starting at $\theta = 0$.

6. Graph $r = \theta/2$, $0 \leq \theta \leq 2\pi$, using multiples of $\pi/2$ for θ starting at $\theta = 0$.

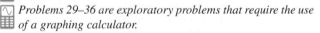 *Verify the graphs for Problems 1–6 on a graphing calculator.*

In Problems 7–14, graph each polar equation.

7. $r = 5$ **8.** $r = 8$ **9.** $\theta = \pi/4$ **10.** $\theta = \pi/3$

11. $\theta^2 = \dfrac{\pi^2}{9}$ **12.** $\theta^2 = \dfrac{\pi^2}{16}$

13. $r^2 = 16$ **14.** $r^2 = 9$

B *In Problems 15–28, sketch each polar equation using rapid sketching techniques.*

15. $r = 4 \cos \theta$ **16.** $r = 4 \sin \theta$

17. $r = 8 \cos 2\theta$ **18.** $r = 10 \sin 2\theta$

19. $r = 6 \sin 3\theta$ **20.** $r = 5 \cos 3\theta$

21. $r = 3 + 3 \cos \theta$ **22.** $r = 2 + 2 \sin \theta$

23. $r = 2 + 4 \cos \theta$ **24.** $r = 2 + 4 \sin \theta$

25. $r = \dfrac{7}{\cos \theta}$ **26.** $r = \dfrac{6}{\sin \theta}$

27. $r = \dfrac{-5}{\sin \theta}$ **28.** $r = \dfrac{-4}{\cos \theta}$

Verify the graphs for Problems 15–28 on a graphing calculator.

Problems 29–36 are exploratory problems that require the use of a graphing calculator.

29. Graph each polar equation in its own viewing window:
(A) $r = 5 + 5 \cos \theta$
(B) $r = 5 + 4 \cos \theta$
(C) $r = 4 + 5 \cos \theta$
(D) Verbally describe the effect of the relative size of a and b on the graph of $r = a + b \cos \theta$, $a, b > 0$.

30. Graph each polar equation in its own viewing window:
(A) $r = 5 + 5 \sin \theta$
(B) $r = 5 + 4 \sin \theta$
(C) $r = 4 + 5 \sin \theta$
(D) Verbally describe the effect of the relative size of a and b on the graph of $r = a + b \sin \theta$, $a, b > 0$.

31. (A) Graph each polar equation in its own viewing window: $r = 9 \cos \theta$, $r = 9 \cos 3\theta$, $r = 9 \cos 5\theta$.
(B) What would you guess to be the number of leaves for the graph of $r = 9 \cos 7\theta$?
(C) What would you guess to be the number of leaves for the graph of $r = a \cos n\theta$, $a > 0$ and n odd?

32. (A) Graph each polar equation in its own viewing window: $r = 9 \sin \theta$, $r = 9 \sin 3\theta$, $r = 9 \sin 5\theta$.
(B) What would you guess to be the number of leaves for the graph of $r = 9 \sin 7\theta$?
(C) What would you guess to be the number of leaves for the graph of $r = a \sin n\theta$, $a > 0$ and n odd?

33. (A) Graph each polar equation in its own viewing window: $r = 9 \cos 2\theta$, $r = 9 \cos 4\theta$, $r = 9 \cos 6\theta$.
(B) What would you guess to be the number of leaves for the graph of $r = 9 \cos 8\theta$?
(C) What would you guess to be the number of leaves for the graph of $r = a \cos n\theta$, $a > 0$ and n even?

34. (A) Graph each polar equation in its own viewing window: $r = 9 \sin 2\theta$, $r = 9 \sin 4\theta$, $r = 9 \sin 6\theta$.
(B) What would you guess to be the number of leaves for the graph of $r = 9 \sin 8\theta$?
(C) What would you guess to be the number of leaves for the graph of $r = a \sin n\theta$, $a > 0$ and n even?

35. Graph each polar equation in its own viewing window:
(A) $r = 9 \cos(\theta/2)$, $0 \leq \theta \leq 4\pi$
(B) $r = 9 \cos(\theta/4)$, $0 \leq \theta \leq 8\pi$
(C) What would you guess to be the maximum number of times a ray from the origin intersects the graph of $r = 9 \cos(\theta/n)$, $0 \leq \theta \leq 2\pi n$, n even?

36. Graph each polar equation in its own viewing window:
(A) $r = 9 \sin(\theta/2)$, $0 \leq \theta \leq 4\pi$
(B) $r = 9 \sin(\theta/4)$, $0 \leq \theta \leq 8\pi$
(C) What would you guess to be the maximum number of times a ray from the origin intersects the graph of $r = 9 \sin(\theta/n)$, $0 \leq \theta \leq 2\pi n$, n even?

C *In Problems 37 and 38, sketch each polar equation using rapid sketching techniques.*

37. $r^2 = 64 \cos 2\theta$ **38.** $r^2 = 64 \sin 2\theta$

Problems 39 and 40 are exploratory problems that require the use of a graphing calculator.

39. Graph $r = 1 + 2 \cos n\theta$ for various values of n, n a natural number. Describe how n is related to the number of large petals and the number of small petals on the graph and how the large and small petals are related to each other relative to n.

40. Graph $r = 1 + 2 \sin n\theta$ for various values of n, n a natural number. Describe how n is related to the number of large petals and the number of small petals on the graph and how the large and small petals are related to each other relative to n.

Precalculus *In Problems 41–44, graph and solve each system of equations in the same polar coordinate system. [Note: Any solution (r_1, θ_1) to the system must satisfy each equation in the system and thus identifies a point of intersection of the two graphs. However, there may be other points of intersection of the two graphs that do not have any coordinates that satisfy both equations. This represents a major difference between the rectangular coordinate system and the polar coordinate system.]*

41. $r = 2 \cos \theta$
$r = 2 \sin \theta$
$0 \le \theta \le \pi$

42. $r = 4 \cos \theta$
$r = -4 \sin \theta$
$0 \le \theta \le \pi$

43. $r = 8 \sin \theta$
$r = 8 \cos 2\theta$
$0° \le \theta \le 360°$

44. $r = 6 \cos \theta$
$r = 6 \sin 2\theta$
$0° \le \theta \le 360°$

In Problems 45–48, refer to Table 3 on pages 447 and 448. A curve is said to be **symmetric with respect to the x axis** *if, when the plane is folded in half along the x axis, the part of the curve in the lower-half plane coincides with the part of the curve in the upper-half plane. Similarly, a curve is said to be* **symmetric with respect to the y axis** *if, when the plane is folded in half along the y axis, the part of the curve in the left-half plane coincides with the part of the curve in the right-half plane.*

45. Which of the polar curves in Table 3 are symmetric with respect to the *x* axis?

46. Which of the polar curves in Table 3 are symmetric with respect to the *y* axis?

47. To test a polar curve for symmetry with respect to the *x* axis, substitute $-\theta$ for θ in its equation and simplify. If the same equation is obtained, the curve is symmetric with respect to the *x* axis.
(A) Use this test to confirm your answers to Problem 45.
(B) Explain why the test works.

48. To test a polar curve for symmetry with respect to the *y* axis, substitute $-r$ for r and $-\theta$ for θ in its equation and simplify. If the same equation is obtained, the curve is symmetric with respect to the *y* axis.
(A) Use this test to confirm your answers to Problem 46.
(B) Explain why the test works.

Applications

Sailboat Racing *Polar diagrams are used extensively by serious sailboat racers. The polar diagram in the figure shows the theoretical speeds that boats in the America's Cup competition (1991) and the older 12 m boats should have been*

able to achieve at different points of sail relative to a 16 knot wind. Problems 49 and 50 refer to this polar diagram.

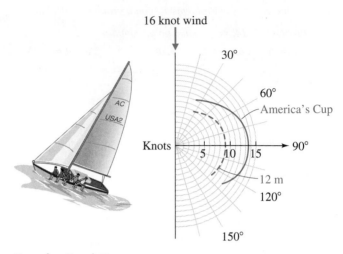

Figure for 49 and 50

49. Refer to the figure. How fast, to the nearest half knot, should the 1991 America's Cup boats have been able to sail going in the following directions relative to the wind?
(A) 30°   (B) 60°   (C) 90°   (D) 120°

50. Refer to the figure. How fast, to the nearest half knot, should the older 12 m boats have been able to sail going in the following directions relative to the wind?
(A) 30°   (B) 60°   (C) 90°   (D) 120°

51. Conic Sections   Using a graphing calculator, graph the equation

$$r = \frac{8}{1 - e \cos \theta}$$

for the following values of *e*, and identify each curve as a hyperbola, ellipse, or parabola.
(A) $e = 0.5$ (B) $e = 1$ (C) $e = 2$

52. Conic Sections   Using a graphing calculator, graph the equation

$$r = \frac{4}{1 - e \cos \theta}$$

for the following values of *e*, and identify each curve as a hyperbola, ellipse, or parabola.
(A) $e = 0.7$ (B) $e = 1$ (C) $e = 1.3$

53. Astronomy
(A) The planet Mercury travels around the sun in an elliptical orbit given approximately by

$$r = \frac{3.44 \times 10^7}{1 - 0.206 \cos \theta}$$

where r is in miles. Graph the orbit with the sun at the pole. Find the distance from Mercury to the sun at **aphelion** (greatest distance from the sun) and at **perihelion** (shortest distance from the sun).

(B) Johannes Kepler (1571–1630) showed that a line joining a planet to the sun swept out equal areas in space in equal intervals in time (see the figure). Use this information to determine whether a planet travels faster or slower at aphelion than at perihelion.

Figure for 53.

7.3 The Complex Plane

- Complex Numbers in Rectangular and Polar Form
- Products and Quotients in Polar Form
- Historical Note

A brief review of Appendix A.2 on complex numbers should prove helpful before we proceed further. Making use of the polar concepts studied in Sections 7.1 and 7.2, we will show how complex numbers can be written in polar form. The polar form is very useful in many applications, including the process of finding all nth roots of any number, real or complex.

■ Complex Numbers in Rectangular and Polar Form

A **complex number** is any number z that can be written in the form

$$z = x + yi$$

where x and y are real numbers and i is the imaginary unit (see Appendix A.2, where addition and multiplication of complex numbers are defined); the **real part** of $x + yi$ is x, and the **imaginary part** of $x + yi$ is yi. Any complex number can be associated with a unique ordered pair of real numbers and vice versa. For example,

$$2 + 2\sqrt{3}\, i \text{ corresponds to } (2, 2\sqrt{3})$$

In general,

$$x + yi \text{ corresponds to } (x, y)$$

With this correspondence, the set of all complex numbers forms a plane called the **complex plane**. The x axis is called the **real axis** and the y axis is called the **imaginary axis** (see Fig. 1).

We can specify any point in the plane using either rectangular coordinates (x, y) or polar coordinates (r, θ). The complex number z associated with the point (x, y) is written $z = x + yi$; $x + yi$ is called the **rectangular form** of z.

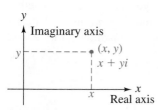

FIGURE 1
Complex plane

The complex number z associated with (r, θ) is written $z = re^{i\theta}$; $re^{i\theta}$ is called the **polar form** of z.

	Rectangular form	Polar form
Points	(x, y)	(r, θ)
Complex numbers	$x + yi$	$re^{i\theta}$

The point with rectangular coordinates $(2, 2\sqrt{3})$ has polar coordinates $(4, \pi/3)$ (why?). The complex number $2 + 2\sqrt{3}\, i$, therefore, has polar form $4e^{i(\pi/3)}$, so

$$2 + 2\sqrt{3}\, i = 4e^{i(\pi/3)}$$

In other words, the left-hand side and right-hand side of the preceding equation represent the same complex number (see Fig. 2).

FIGURE 2
$2 + 2\sqrt{3}\, i = 4e^{i(\pi/3)}$

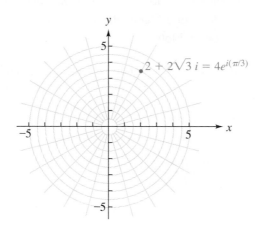

Remark on Notation   In this book, $re^{i\theta}$ is used only as a convenient and advantageous notation for the complex number with polar coordinates (r, θ). Just as with polar coordinates, θ may be given in either radians or degrees. (In advanced mathematics, $e^{i\theta}$ represents an element in the range of a generalized version of the natural exponential function $f(x) = e^x$ with base $e \approx 2.718$, but our use of $re^{i\theta}$ does not depend on that interpretation.) Here is one advantage of complex number notation over ordered pair notation: Given the ordered pair $(3, \pi)$, you must specify whether the coordinates are rectangular or polar; on the other hand, $3 + \pi i$ and $3e^{i\pi}$ determine complex numbers without ambiguity (the first lies in quadrant I, the second on the negative real axis).   □

EXAMPLE 1

Plotting Complex Numbers

Plot each complex number in the complex plane.
(A) $3 + 4i$ (B) $5e^{i(3\pi/4)}$ (C) -2 (D) $4e^{i(-\pi/2)}$

Solution

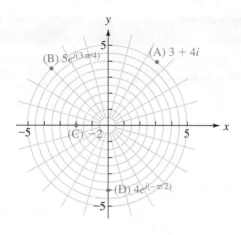

Matched Problem 1   Plot each number in a complex plane.

(A) $4e^{i(-\pi/6)}$ (B) $-4 - 3i$ (C) $5e^{i(\pi/2)}$ (D) 4   ■

If k is any integer, the ordered pairs (r, θ) and $(r, \theta + 2k\pi)$ describe the same point in polar coordinates. Therefore,

$$re^{i\theta} = re^{i(\theta + 2k\pi)} \qquad k \text{ any integer}$$

If θ is given in degrees, then

$$re^{i\theta} = re^{i(\theta + k360°)} \qquad k \text{ any integer}$$

Unless otherwise specified, we will choose $r \geq 0$ and $-\pi < \theta \leq \pi$ (or $-180° < \theta \leq 180°$) when we write the polar form of a complex number.

EXPLORE/DISCUSS 1

There is a one-to-one correspondence between the set of real numbers and the set of points on a *real number line* (see Appendix A.1). Each real number is associated with exactly one point on the line, and each point on the line is associated with exactly one real number. Does a one-to-one correspondence exist between the set of complex numbers and the set of points in a plane? If so, explain how it can be established.

For some purposes (addition and subtraction, for example) it is best to work with complex numbers in rectangular form. For others (multiplication and division, for example) polar form gives greater insight. So it is good to become adept at moving from rectangular form to polar form and vice versa. The polar–rectangular relationships of Section 7.1 imply the following connections between the rectangular and polar forms of a complex number.

POLAR–RECTANGULAR EQUATIONS FOR COMPLEX NUMBERS

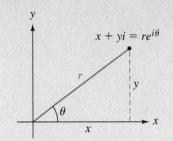

If $x + yi = re^{i\theta}$, then

$$x = r\cos\theta \qquad\qquad r = \sqrt{x^2 + y^2}$$

$$y = r\sin\theta \qquad\qquad \tan\theta = \frac{y}{x},\ x \neq 0$$

If $z = x + yi = re^{i\theta}$, the number r is called the **modulus**, or **absolute value**, of z and is denoted by $|z|$. The modulus of z is the distance from z to the origin. The number θ is called the **argument**, or **angle**, of z, denoted arg z. The argument of z is the angle that the line joining z to the origin makes with the positive real axis (we usually choose the argument θ so that $-\pi < \theta \leq \pi$ or $-180° < \theta \leq 180°$).

Note that if $z = x + yi = re^{i\theta}$ is a complex number, then

$$z = x + yi \qquad\qquad \textit{Use polar–rectangular equations.}$$
$$= (r\cos\theta) + (r\sin\theta)i \qquad \textit{Factor out the real number } r.$$
$$= r(\cos\theta + i\sin\theta)$$

Therefore, any complex number z can be expressed in the **trigonometric form** $z = r(\cos\theta + i\sin\theta)$. (As with the polar form, we usually choose $r \geq 0$ and $-\pi < \theta \leq \pi$ or $-180° < \theta \leq 180°$.) The trigonometric form $z = r(\cos\theta + i\sin\theta)$ makes it easy to see that the real part of z is $r\cos\theta$, and the imaginary part of z is $(r\sin\theta)i$. The notation r cis θ is sometimes used as an abbreviation for $r(\cos\theta + i\sin\theta)$.

⚠ **Caution**   To simplify notation or to avoid confusion, we sometimes prefer to write $x + iy$ instead of $x + yi$; you should keep in mind that both represent the same complex number. For example, the parentheses in $\cos\theta + (\sin\theta)i$ can be safely eliminated by writing $\cos\theta + i\sin\theta$. Similarly, we prefer to write $2 + i\sqrt{3}$ rather than $2 + \sqrt{3}i$, so that i will not be mistakenly included under the radical.   □

    EXAMPLE 2   **From Rectangular to Polar Form**

Write the following in polar form (θ in radians, $-\pi < \theta \leq \pi$). Compute the modulus and arguments for (A) and (B) exactly. Compute the modulus and argument for (C) to two decimal places.

(A) $z_1 = -1 + i$ (B) $z_2 = 1 + i\sqrt{3}$ (C) $z_3 = -3 - 7i$

Solution First, locate the complex number in a complex plane. If x and y are associated with special reference triangles, r and θ can be determined by inspection.

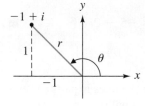

FIGURE 3

(A) A sketch (Fig. 3) shows that z_1 is associated with a special $45°-45°-90°$ reference triangle in the second quadrant. So $\mathrm{mod}\, z_1 = r = \sqrt{2}$, $\arg z_1 = \theta = 3\pi/4$, and the polar form for z_1 is

$$z_1 = \sqrt{2}\, e^{(3\pi/4)i}$$

(B) A sketch (Fig. 4) shows that z_2 is associated with a special $30°-60°-90°$ reference triangle. So $\mathrm{mod}\, z_2 = r = 2$, $\arg z_2 = \theta = \pi/3$, and the polar form for z_2 is

$$z_2 = 2e^{(\pi/3)i}$$

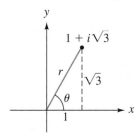

FIGURE 4

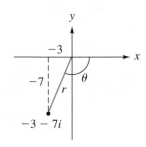

FIGURE 5

(C) A sketch (Fig. 5) shows that z_3 is not associated with a special reference triangle. Consequently, we proceed as follows:

$$\mathrm{mod}\, z_3 = r = \sqrt{(-3)^2 + (-7)^2} = 7.62 \qquad \text{To two decimal places}$$

$$\arg z_3 = \theta = -\pi + \tan^{-1}\tfrac{7}{3} = -1.98 \qquad \text{To two decimal places}$$

Therefore, the polar form for z_3 is

$$z_3 = 7.62 e^{(-1.98)i}$$

Figure 6 shows the same conversion done by a graphing calculator with a built-in conversion routine. ■

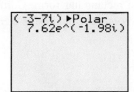

FIGURE 6

Matched Problem 2   Write the following in polar form (θ in radians, $-\pi < \theta \le \pi$). Compute the modulus and arguments for (A) and (B) exactly. Compute the modulus and argument for (C) to two decimal places.

(A) $z_1 = 1 - i$ (B) $z_2 = -\sqrt{3} + i$ (C) $z_3 = -5 - 2i$ ■

   **EXAMPLE 3**

From Polar to Rectangular Form

Write the following in rectangular form. Compute the exact values in parts (A) and (B). For part (C), compute a and b for $a + bi$ to two decimal places.

(A) $z_1 = \sqrt{2}e^{(-\pi/4)i}$ (B) $z_2 = 3e^{(120°)i}$ (C) $z_3 = 6.49e^{(-2.08)i}$

Solution   (A) $x + iy = \sqrt{2}e^{(-\pi/4)i}$

$$= \sqrt{2}\left[\cos(-\pi/4) + i\sin(-\pi/4)\right]$$

$$= \sqrt{2}\left(\frac{1}{\sqrt{2}} + \frac{-1}{\sqrt{2}}i\right)$$

$$= 1 - i$$

(B) $x + iy = 3e^{(120°)i}$

$$= 3(\cos 120° + i\sin 120°)$$

$$= 3\left(\frac{-1}{2} + \frac{\sqrt{3}}{2}i\right)$$

$$= -\frac{3}{2} + \frac{3\sqrt{3}}{2}i$$

```
6.49e^(-2.08i)▶R
ect
       -3.16-5.67i
```

FIGURE 7

(C) $x + iy = 6.49e^{(-2.08)i}$

$$= 6.49\left[\cos(-2.08) + i\sin(-2.08)\right]$$

$$= -3.16 - 5.67i$$

Figure 7 shows the same conversion done by a graphing calculator with a built-in conversion routine.   ■

Matched Problem 3   Write the following in rectangular form. Compute the exact values in parts (A) and (B). For part (C), compute a and b for $a + bi$ two decimal places.

(A) $z_1 = 2e^{(5\pi/6)i}$ (B) $z_2 = 3e^{(-60°)i}$ (C) $z_3 = 7.19e^{(-2.13)i}$ ■

 Caution   For complex numbers in polar form, some calculators require θ to be in radian mode. Check your user's manual.   □

EXPLORE/DISCUSS 2

Let $z_1 = 1 + i$ and $z_2 = -1 + i$.

(A) Find $z_1 z_2$ and z_1/z_2 using the rectangular forms of z_1 and z_2.

(B) Find $z_1 z_2$ and z_1/z_2 using the polar forms of z_1 and z_2, θ in degrees. (Assume the product and quotient exponent laws hold for $e^{i\theta}$.)

(C) Convert the results from part (B) back to rectangular form and compare with the results in part (A).

■ Products and Quotients in Polar Form

We will now see an advantage of representing complex numbers in polar form: Multiplication and division of complex numbers become very easy. Theorem 1 provides the reason: The polar form of a complex number obeys the product and quotient rules for exponents: $b^m b^n = b^{m+n}$ and $b^m/b^n = b^{m-n}$.

Theorem 1

PRODUCTS AND QUOTIENTS IN POLAR FORM

If $z_1 = r_1 e^{i\theta_1}$ and $z_2 = r_2 e^{i\theta_2}$, then:

1. $z_1 z_2 = r_1 e^{i\theta_1} r_2 e^{i\theta_2} = r_1 r_2 e^{i(\theta_1 + \theta_2)}$

2. $\dfrac{z_1}{z_2} = \dfrac{r_1 e^{i\theta_1}}{r_2 e^{i\theta_2}} = \dfrac{r_1}{r_2} e^{i(\theta_1 - \theta_2)}$

We establish the multiplication property and leave the quotient property to Problem 40 in Exercise 7.3.

$$
\begin{aligned}
z_1 z_2 &= r_1 e^{i\theta_1} r_2 e^{i\theta_2} \\
&= [r_1(\cos\theta_1 + i\sin\theta_1)][r_2(\cos\theta_2 + i\sin\theta_2)] \\
&= r_1 r_2(\cos\theta_1 + i\sin\theta_1)(\cos\theta_2 + i\sin\theta_2) \\
&= r_1 r_2(\cos\theta_1\cos\theta_2 + i\cos\theta_1\sin\theta_2 + i\sin\theta_1\cos\theta_2 - \sin\theta_1\sin\theta_2) \\
&= r_1 r_2[(\cos\theta_1\cos\theta_2 - \sin\theta_1\sin\theta_2) + i(\cos\theta_1\sin\theta_2 + \sin\theta_1\cos\theta_2)] \\
&= r_1 r_2[\cos(\theta_1 + \theta_2) + i\sin(\theta_1 + \theta_2)] \qquad \text{Sum identities.} \\
&= r_1 r_2 e^{i(\theta_1 + \theta_2)}
\end{aligned}
$$

In other words, to multiply two complex numbers in polar form, multiply r_1 and r_2, and add θ_1 and θ_2. To divide the complex number z_1 by z_2, divide r_1 by r_2, and subtract θ_2 from θ_1. The process is illustrated in Example 4.

EXAMPLE 4

Products and Quotients

If $z_1 = 8e^{(50°)i}$ and $z_2 = 4e^{(30°)i}$, find:

(A) $z_1 z_2$ (B) z_1/z_2

Solution   (A) $z_1 z_2 = 8e^{(50°)i} \cdot 4e^{(30°)i}$

$$\boxed{= (8 \cdot 4)e^{i(50° + 30°)}} = 32e^{(80°)i}$$

(B) $\dfrac{z_1}{z_2} = \dfrac{8e^{(50°)i}}{4e^{(30°)i}}$

$$\boxed{= \frac{8}{4}\, e^{i(50° - 30°)}} = 2e^{(20°)i}$$ ∎

Matched Problem 4   If $z_1 = 21e^{(140°)i}$ and $z_2 = 3e^{(105°)i}$, find:

(A) $z_1 z_2$ (B) z_1/z_2 ∎

▉ Historical Note

There is hardly an area in mathematics that does not have some imprint of the famous Swiss mathematician Leonhard Euler (1707–1783), who spent most of his productive life in the New St. Petersburg Academy in Russia and the Prussian Academy in Berlin. One of the most prolific writers in the history of mathematics, he is credited with making the following familiar notations standard:

$f(x)$	Function notation
e	Natural logarithmic base
i	Imaginary unit

Of immediate interest, he is also responsible for the remarkable relationship

$$e^{i\theta} = \cos\theta + i\sin\theta$$

If $\theta = \pi$, then an equation results that relates five of the most important numbers in the history of mathematics:

$$e^{i\pi} + 1 = 0$$

Answers to Matched Problems

1.

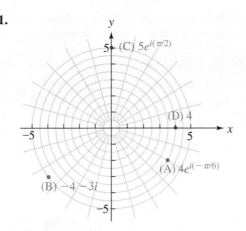

2. (A) $z_1 = \sqrt{2}e^{(-\pi/4)i}$
   (B) $z_2 = 2e^{(5\pi/6)i}$
   (C) $z_3 = 5.39e^{(-2.76)i}$

3. (A) $-\sqrt{3} + i$ (B) $\dfrac{3}{2} - \dfrac{3\sqrt{3}}{2}i$ (C) $-3.81 - 6.09i$

4. (A) $z_1z_2 = 63e^{(245°)i}$ (B) $z_1/z_2 = 7e^{(35°)i}$
      or $63e^{(-115°)i}$

EXERCISE 7.3

A *Plot each set of complex numbers in a complex plane.*

1. $A = 4 + 5i$; $B = -3 + 4i$; $C = -3i$
2. $A = 3 - 2i$; $B = -4 - 2i$; $C = 5i$
3. $A = 4 + i$; $B = -2 + 3i$; $C = -4$
4. $A = -3 - i$; $B = 5 - 4i$; $C = 4$
5. $A = 8e^{(\pi/4)i}$; $B = 6e^{(\pi/2)i}$; $C = 3e^{(\pi/6)i}$
6. $A = 5e^{(5\pi/6)i}$; $B = 3e^{(3\pi/2)i}$; $C = 4e^{(7\pi/4)i}$
7. $A = 5e^{(270°)i}$; $B = 4e^{(60°)i}$; $C = 8e^{(150°)i}$
8. $A = 3e^{(310°)i}$; $B = 4e^{(180°)i}$; $C = 5e^{(210°)i}$

B *Change Problems 9–16 to polar form. For Problems 9–12, choose θ in radians, $-\pi < \theta \leq \pi$; for Problems 13–16, choose θ in degrees, $-180° < \theta \leq 180°$. Compute the modulus and arguments for (A) and (B) exactly; compute the modulus and argument for (C) to two decimal places.*

9. (A) $\sqrt{3} - i$ (B) $-2 + 2i$ (C) $6 - 5i$
10. (A) $-i\sqrt{3}$ (B) $-\sqrt{3} - i$ (C) $-8 + 5i$
11. (A) $21\left(\cos\dfrac{\pi}{4} + i\sin\dfrac{\pi}{4}\right)$
    (B) $12\left[\cos\left(-\dfrac{\pi}{6}\right) + i\sin\left(-\dfrac{\pi}{6}\right)\right]$
    (C) $\sqrt{40}\,(\cos 0.83 + i\sin 0.83)$

12. (A) $25\left(\cos\dfrac{\pi}{2} + i\sin\dfrac{\pi}{2}\right)$
    (B) $30\left[\cos\left(-\dfrac{\pi}{3}\right) + i\sin\left(-\dfrac{\pi}{3}\right)\right]$
    (C) $\sqrt{15}\,(\cos 2.17 + i\sin 2.17)$

13. (A) $-1 + i\sqrt{3}$ (B) $-3i$ (C) $-7 - 4i$
14. (A) $\sqrt{3} + i$ (B) $-1 - i$ (C) $5 - 6i$
15. (A) $5(\cos 30° + i\sin 30°)$
    (B) $4[\cos(-45°) + i\sin(-45°)]$
    (C) $\sqrt{8}\,(\cos 54° + i\sin 54°)$
16. (A) $10(\cos 135° + i\sin 135°)$
    (B) $9(\cos 60° + i\sin 60°)$
    (C) $\sqrt{3}\,(\cos 7° + i\sin 7°)$

Change Problems 17–20 to rectangular form. Compute the exact values for (A) and (B); for (C) compute a and b in a + bi to two decimal places.

17. (A) $2e^{(30°)i}$ (B) $\sqrt{2}e^{(-3\pi/4)i}$ (C) $5.71e^{(-0.48)i}$

18. (A) $2e^{(\pi/3)i}$ (B) $\sqrt{2}e^{(-45°)i}$ (C) $3.08e^{2.44i}$

19. (A) $\sqrt{3}e^{(-\pi/2)i}$ (B) $\sqrt{2}e^{(135°)i}$ (C) $6.83e^{(-108.82°)i}$

20. (A) $6e^{(\pi/6)i}$ (B) $\sqrt{7}e^{(-90°)i}$ (C) $4.09e^{(-122.88°)i}$

In Problems 21–28, find z_1z_2 and z_1/z_2. Leave answers in polar form.

21. $z_1 = 4e^{(25°)i}$; $z_2 = 8e^{(12°)i}$

22. $z_1 = 3e^{(97°)i}$; $z_2 = 5e^{(68°)i}$

23. $z_1 = 6e^{(108°)i}$; $z_2 = 2e^{(-120°)i}$

24. $z_1 = 12e^{(-55°)i}$; $z_2 = 3e^{(165°)i}$

25. $z_1 = 4.36e^{1.27i}$; $z_2 = 1.69e^{0.91i}$

26. $z_1 = 1.75e^{(-0.88)i}$; $z_2 = 2.44e^{(-1.01)i}$

27. $z_1 = 4e^{(\pi/3)i}$, $z_2 = 5e^{(-\pi/4)i}$

28. $z_1 = 1.5e^{(\pi/5)i}$, $z_2 = 2.9e^{(\pi/6)i}$

In Problems 29–34, find each product or power directly and by using polar forms. Write answers in rectangular and polar forms with θ in degrees.

29. $(1 + \sqrt{3}i)(\sqrt{3} + i)$ **30.** $(-1 + i)(1 + i)$

31. $(1 + i)^2$ **32.** $(-1 + i)^2$

33. $(1 + i)^3$ **34.** $(1 - i)^3$

C **35.** If $z = re^{i\theta}$, show that $z^2 = r^2e^{2\theta i}$.

36. If $z = re^{i\theta}$, show that $z^3 = r^3e^{3\theta i}$.

37. Show that $r^{1/2}e^{i\theta/2}$ is a square root of $re^{i\theta}$.

38. Show that $r^{1/3}e^{i\theta/3}$ is a cube root of $re^{i\theta}$.

39. Based on Problems 35 and 36, what do you think z^n will be for n a natural number?

40. Prove: $\dfrac{z_1}{z_2} = \dfrac{r_1e^{i\theta_1}}{r_2e^{i\theta_2}} = \dfrac{r_1}{r_2}e^{i(\theta_1-\theta_2)}$

*In Problems 41–46, recall that the **conjugate** of a complex number $z = x + yi$ is denoted by $\bar{z}$ and is defined by $\bar{z} = x - yi$.*

41. Show that $z + \bar{z}$ lies on the real axis.

42. Show that $z - \bar{z}$ lies on the imaginary axis.

43. Show that z and $\bar{z}$ have the same modulus.

44. Show that $z\bar{z}$ is the square of the modulus of z.

45. If the polar form of z is $re^{i\theta}$, find the polar form of $\bar{z}$.

46. If z lies on the circle of radius 1 with center the origin, show that $\dfrac{1}{z} = \bar{z}$.

Applications

47. **Resultant Force** An object is located at the pole, and two forces $\mathbf{F}_1$ and $\mathbf{F}_2$ act upon the object. Let the forces be vectors going from the pole to the complex numbers $8(\cos 0° + i \sin 0°)$ and $6(\cos 30° + i \sin 30°)$ respectively. (Force $\mathbf{F}_1$ has a magnitude of 8 lb at a direction of $0°$, and force $\mathbf{F}_2$ has a magnitude of 6 lb at a direction of $30°$.)

(A) Convert the polar forms of these complex numbers to rectangular form and add.

(B) Convert the sum from part (A) back to polar form (to three significant digits).

(C) The vector going from the pole to the complex number in part (B) is the resultant of the two original forces. What is its magnitude and direction?

48. **Resultant Force** Repeat Problem 47 with forces $\mathbf{F}_1$ and $\mathbf{F}_2$ associated with the complex numbers

$$20(\cos 0° + i \sin 0°) \text{ and } 10(\cos 60° + i \sin 60°)$$

7.4 De Moivre's Theorem and the *n*th-Root Theorem

- De Moivre's Theorem
- The *n*th-Root Theorem

Abraham De Moivre (1667–1754), of French birth, spent most of his life in London doing private tutoring, writing, and publishing mathematics. He became a close friend of Isaac Newton and belonged to many prestigious professional societies in England, France, and Germany.

De Moivre's theorem enables us to find any power of a complex number in polar form very easily. More important, the theorem is the basis for the *nth-root theorem,* which enables us to find all *n*th roots of any complex number. How many roots does the equation $x^3 = 1$ have? You may think that 1 is the only root. The equation actually has three roots that are complex numbers, only one of which is real (see Example 3).

■ De Moivre's Theorem

We start with Explore/Discuss 1 and encourage you to generalize from this exploration.

EXPLORE/DISCUSS 1

Establish the following by repeated use of the product formula for the polar form discussed in Section 7.3:

(A) $(x + iy)^2 = (re^{\theta i})^2 = r^2 e^{2\theta i}$ (B) $(x + iy)^3 = (re^{\theta i})^3 = r^3 e^{3\theta i}$

(C) $(x + iy)^4 = (re^{\theta i})^4 = r^4 e^{4\theta i}$

For *n* a natural number, what do you think the polar form of $(x + iy)^n$ would be?

If you guessed $(x + iy)^n = r^n e^{n\theta i}$, you have discovered De Moivre's famous theorem, which we now state without proof. (A general proof of De Moivre's theorem requires a technique called *mathematical induction,* which is discussed in more advanced courses.)

DE MOIVRE'S THEOREM

If $z = x + iy = re^{i\theta}$ and *n* is a natural number, then

$$z^n = (x + iy)^n = (re^{i\theta})^n = r^n e^{n\theta i}$$

EXAMPLE 1

Finding a Power of a Complex Number

Find $(\sqrt{3} + i)^{13}$ and write the answer in exact polar and rectangular forms.

Solution

First we convert $\sqrt{3} + i$ to polar form by noting that its modulus is 2 and its argument is 30°. Therefore,

$$(\sqrt{3} + i)^{13} = (2e^{(30°)i})^{13} \quad \text{Use De Moivre's theorem.}$$

$$= 2^{13} e^{(13 \cdot 30°)i} \quad \text{Simplify.}$$

$$= 8{,}192 e^{(390°)i} \qquad \text{\small 390° = 30° + 360°}$$

$$= 8{,}192 e^{(30°+360°)i} \qquad \text{\small $e^{i\theta}$ is periodic with period 360°.}$$

$$= 8{,}192 e^{(30°)i} \qquad \text{\small Write in trigonometric form.}$$

$$= 8{,}192(\cos 30° + i \sin 30°) \qquad \text{\small $\cos 30° = \frac{\sqrt{3}}{2}$ and $\sin 30° = \frac{1}{2}$}$$

$$= 8{,}192\left(\frac{\sqrt{3}}{2} + \frac{1}{2}i \right) \qquad \text{\small Write in rectangular form.}$$

$$= 4{,}096\sqrt{3} + 4{,}096i \qquad \blacksquare$$

Instead of using the method of Example 1, try to evaluate $(\sqrt{3} + i)^{13}$ by repeated multiplication of rectangular forms. You will develop a healthy appreciation for the power of De Moivre's theorem!

Matched Problem 1   Find $(-1 + i)^6$ and write the answer in exact polar and rectangular forms.   ■

■ The *n*th-Root Theorem

Now let us take a look at roots of complex numbers. We say w is an *n*th root of z, where n is a natural number, if

$$w^n = z$$

For example, if $w^2 = z$, then w is a square root of z; if $w^3 = z$, then w is a cube root of z; and so on.

EXPLORE/DISCUSS 2

For $z = re^{i\theta}$ show that $r^{1/2}e^{(\theta/2)i}$ is a square root of z and $r^{1/3}e^{(\theta/3)i}$ is a cube root of z. (Use De Moivre's theorem.)

Proceeding in the same way as in Explore/Discuss 2, it is easy to show that for n a natural number, $r^{1/n}e^{(\theta/n)i}$ is an *n*th root of $re^{i\theta}$:

$$(r^{1/n}e^{(\theta/n)i})^n = (r^{1/n})^n e^{n(\theta/n)i}$$

$$= re^{i\theta}$$

However, we can do better than this. The *n*th-root theorem shows how to find *all* the *n*th roots of a complex number. The proof of the theorem is left to Problems 37 and 38 in Exercise 7.4.

*n*TH ROOT THEOREM

For *n* a positive integer greater than 1 and θ in degrees,

$$r^{1/n}e^{(\theta/n + k\,360°/n)i} \qquad k = 0, 1, \ldots, n-1$$

are the *n* distinct *n*th roots of $re^{i\theta}$, and there are no others.

Remark   If θ is given in radians, then, converting 360° to 2π radians, the *n* distinct roots of $re^{i\theta}$ are given by

$$r^{1/n}\,e^{(\theta/n + k\,2\pi/n)i} \qquad k = 0, 1, \ldots, n-1 \qquad \square$$

EXAMPLE 2

Roots of a Complex Number

Find the six distinct sixth roots of $z = 1 + i\sqrt{3}$.

Solution   First, write $1 + i\sqrt{3}$ in polar form:

$$1 + i\sqrt{3} = 2e^{(60°)i}$$

From the *n*th-root theorem, all six roots are given by

$$2^{1/6}e^{(60°/6 + k360°/6)i} \qquad k = 0, 1, 2, 3, 4, 5$$

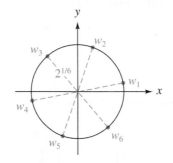

FIGURE 1

Thus,

$$w_1 = 2^{1/6}e^{(10° + 0\cdot60°)i} = 2^{1/6}e^{(10°)i}$$
$$w_2 = 2^{1/6}e^{(10° + 1\cdot60°)i} = 2^{1/6}e^{(70°)i}$$
$$w_3 = 2^{1/6}e^{(10° + 2\cdot60°)i} = 2^{1/6}e^{(130°)i}$$
$$w_4 = 2^{1/6}e^{(10° + 3\cdot60°)i} = 2^{1/6}e^{(190°)i}$$
$$w_5 = 2^{1/6}e^{(10° + 4\cdot60°)i} = 2^{1/6}e^{(250°)i}$$
$$w_6 = 2^{1/6}e^{(10° + 5\cdot60°)i} = 2^{1/6}e^{(310°)i}$$

The six roots are easily graphed in the complex plane after the first root is located. The roots are equally spaced around a circle with radius $2^{1/6}$ at an angular increment of 60° from one root to the next, as shown in Figure 1.   ■

Matched Problem 2   Find the three distinct cube roots of $1 + i$, and leave the answers in polar form. Plot the roots in a complex plane.   ■

EXAMPLE 3

Solving a Quadratic Equation

Solve $x^2 + 25i = 0$. Write final answers in rectangular form.

Solution $x^2 + 25i = 0$
$$x^2 = -25i$$

Any solution x is therefore a square root of $-25i$. We write $-25i$ in polar form and then use the nth-root theorem:

$$-25i = 25e^{(270°)i}$$

The two square roots of $-25i$ are given by

$$25^{1/2}e^{(270°/2 + k360°/2)i}, \qquad k = 0, 1$$

Therefore,

$$w_1 = 5e^{(135°)i} = 5(\cos 135° + i \sin 135°) = -\frac{5}{\sqrt{2}} + \frac{5}{\sqrt{2}} i$$

$$w_2 = 5e^{(315°)i} = 5(\cos 315° + i \sin 315°) = \frac{5}{\sqrt{2}} - \frac{5}{\sqrt{2}} i$$

Note that the solutions w_1 and w_2 are exactly opposite points on the circle of radius 5, with center the origin.   ■

Matched Problem 3   Solve $x^2 - 36i = 0$. Write final answers in rectangular form.   ■

Solving a Cubic Equation

Solve $x^3 - 1 = 0$. Write the final answers in rectangular form and plot them in a complex plane.

Solution
$$x^3 - 1 = 0$$
$$x^3 = 1$$

Therefore, x is a cube root of 1, and there are three cube roots. First, we write 1 in polar form and then use the nth-root theorem:

$$1 = 1e^{(0°)i}$$

All three cube roots of 1 are given by

$$1^{1/3}e^{(0°/3 + k360°/3)} \qquad k = 0, 1, 2$$

So,

$$w_1 = e^{(0°)i} = \cos 0° + i \sin 0° = 1$$

$$w_2 = e^{(120°)i} = \cos 120° + i \sin 120° = -\frac{1}{2} + \frac{\sqrt{3}}{2} i$$

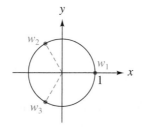

FIGURE 2

$$w_3 = e^{(240°)i} = \cos 240° + i \sin 240° = -\frac{1}{2} - \frac{\sqrt{3}}{2} i$$

The three roots are graphed in Figure 2.   ■

Matched Problem 4   Solve $x^3 + 1 = 0$. Write the final answers in rectangular form and plot them in a complex plane.   ■

We have only touched on a subject that has far-reaching consequences. The theory of functions of a complex variable provides a powerful tool for engineers, scientists, and mathematicians.

Answers to Matched Problems

1. $8e^{(90°)i}$; $8i$

2. $w_1 = 2^{1/6}e^{(15°)i}$, $w_2 = 2^{1/6}e^{(135°)i}$, $w_3 = 2^{1/6}e^{(255°)i}$

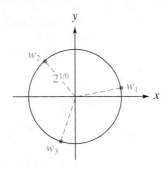

3. $w_1 = 3\sqrt{2} + 3i\sqrt{2}$, $w_2 = -3\sqrt{2} - 3i\sqrt{2}$

4. $w_1 = \dfrac{1}{2} + \dfrac{\sqrt{3}}{2}i$, $w_2 = -1$, $w_3 = \dfrac{1}{2} - \dfrac{\sqrt{3}}{2}i$

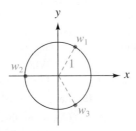

EXERCISE 7.4

A *In Problems 1–8, find the value of each expression using De Moivre's theorem. Leave your answer in polar form.*

1. $(3e^{(15°)i})^3$

2. $(2e^{(30°)i})^8$

3. $(\sqrt{2}e^{(45°)i})^{10}$

4. $(\sqrt{2}e^{(60°)i})^8$

5. $(2e^{(\pi/5)i})^4$

6. $(10e^{(\pi/3)i})^2$

7. $(5e^{(11\pi/6)i})^2$

8. $(4e^{(7\pi/4)i})^3$

B *In Problems 9–14, find the value of each expression using De Moivre's theorem, and write the result in exact rectangular form.*

9. $(-1 + i)^4$

10. $(-\sqrt{3} - i)^4$

11. $(-\sqrt{3} + i)^5$

12. $(1 - i)^8$

13. $\left(-\dfrac{1}{2} - \dfrac{\sqrt{3}}{2}i\right)^3$

14. $\left(-\dfrac{1}{2} + \dfrac{\sqrt{3}}{2}i\right)^3$

In Problems 15–20, find all nth roots of z for n and z as given. Leave answers in polar form.

15. $z = 4e^{(30°)i}$; $n = 2$

16. $z = 16e^{(60°)i}$; $n = 2$

17. $z = 8e^{(90°)i}$; $n = 3$

18. $z = 27e^{(120°)i}$; $n = 3$

19. $z = -1 + i$; $n = 5$

20. $z = 1 - i$; $n = 5$

In Problems 21–26, find all nth roots of z for n and z as given. Write answers in polar form and plot in a complex plane.

21. $z = -8$; $n = 3$

22. $z = -16$; $n = 4$

23. $z = 1; n = 4$ **24.** $z = 8; n = 3$

25. $z = -i; n = 5$ **26.** $z = i; n = 6$

27. (A) Show that $-\sqrt{2} + i\sqrt{2}$ is a root of $x^4 + 16 = 0$. How many other roots does the equation have?

(B) The root $-\sqrt{2} + i\sqrt{2}$ is located on a circle of radius 2 in the complex plane, as shown in the figure. Without using the nth-root theorem, locate all other roots on the figure, and explain geometrically how you found their locations.

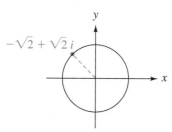

(C) Verify that each complex number found in part (B) is a root of $x^4 + 16 = 0$. Show your steps.

28. (A) Show that 2 is a root of $x^3 - 8 = 0$. How many other roots does the equation have?

(B) The root 2 is located on a circle of radius 2 in the complex plane, as shown in the figure. Without using the nth-root theorem, locate all other roots on the figure and explain geometrically how you found their locations.

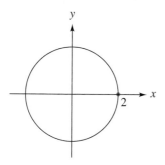

(C) Verify that each complex number found in part (B) is a root of $x^3 - 8 = 0$. Show your steps.

In Problems 29–36, solve each equation for all roots. Write the final answers in exact rectangular form.

29. $x^2 - i = 0$ **30.** $x^2 + i = 0$

31. $x^2 + 4i = 0$ **32.** $x^2 - 16i = 0$

33. $x^3 + 27 = 0$ **34.** $x^3 - 27 = 0$

35. $x^3 - 64 = 0$ **36.** $x^3 + 64 = 0$

C **37.** Show that
$$r^{1/n}e^{(\theta/n + k360°/n)i}$$
is the same number for $k = 0$ and $k = n$.

38. Show that
$$\left(r^{1/n}e^{(\theta/n + k360°/n)i}\right)^n = re^{i\theta}$$
for any natural number n and any integer k.

In Problems 39–42, solve each equation for all roots. Write final answers in rectangular form, $a + bi$, where a and b are computed to three decimal places.

39. $x^5 - 1 = 0$ **40.** $x^4 + 1 = 0$

41. $x^3 + 5 = 0$ **42.** $x^5 - 6 = 0$

43. Let $z = e^{(60°)i}$.

(A) Show that z^k is a sixth root of 1 for k an integer, $1 \le k \le 6$.

(B) Which of the 6 sixth roots of 1 from part (A) are square roots of 1?

(C) Which of the 6 sixth roots of 1 from part (A) are cube roots of 1?

44. Let $z = e^{(45°)i}$.

(A) Show that z^k is an eighth root of 1 for k an integer, $1 \le k \le 8$.

(B) Which of the 8 eighth roots of 1 from part (A) are square roots of 1?

(C) Which of the 8 eighth roots of 1 from part (A) are fourth roots of 1?

45. Let $z = e^{(36°)i}$.

(A) Show that z^k is a tenth root of 1 for k an integer, $1 \le k \le 10$.

(B) Which of the 10 tenth roots of 1 from part (A) are square roots of 1?

(C) Which of the 10 tenth roots of 1 from part (A) are fifth roots of 1?

46. Let $z = e^{(30°)i}$

(A) Show that z^k is a twelfth root of 1 for k an integer, $1 \le k \le 12$.

(B) Which of the 12 twelfth roots of 1 from part (A) are square roots of 1?

(C) Which of the 12 twelfth roots of 1 from part (A) are cube roots of 1?

(D) Which of the 12 twelfth roots of 1 from part (A) are fourth roots of 1?

(E) Which of the 12 twelfth roots of 1 from part (A) are sixth roots of 1?

CHAPTER 7 GROUP ACTIVITY

Orbits of Planets

I. CONICS AND ECCENTRICITY

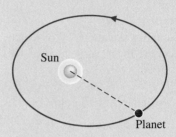

Plane curves formed by intersecting a plane with a right circular cone of two nappes (visualize an hour glass) are called **conic sections**, or **conics**. Circles, ellipses, parabolas, and hyperbolas are conics. Conics can also be defined in terms of *eccentricity e,* an approach that is of great use in space science. We start with a fixed point *F,* called the **focus,** and a fixed line not containing the focus, called the **directrix** (Fig. 1). For *e* > 0, a conic is defined as the set of points *P* having the property that the distance from *P* to the focus *F,* divided by the distance from *P* to the directrix *d,* is the constant *e.* An ellipse, a parabola, or a hyperbola will result by choosing the constant *e* appropriately.

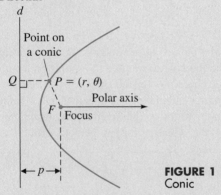

FIGURE 1
Conic

Problem 1 **Polar Equation of a Conic**
Using the eccentricity definition of a conic above, show that the **polar equation of a conic** is given by

$$r = \frac{ep}{1 - e\cos\theta} \tag{1}$$

where *p* is the distance between the focus *F* and the directrix *d,* the pole of the polar axis is at *F,* and the polar axis is perpendicular to *d* and is pointing away from *d* (see Fig. 1). ∎

Problem 2 **Exploration with a Graphing Calculator**
Explore and discuss the effect of varying the positive values of the eccentricity *e* and the distance *p* on the type and shape of conic produced. Based on your exploration, complete Table 1.

(Continued)

TABLE 1

Eccentricity e	Type of conic
$0 < e < 1$	
$e = 1$	
$e > 1$	

Problem 3   Maximum and Minimum Distances from Focus

Show that if $0 < e < 1$, then the points on the conic of equation (1) for which $\theta = 0$ and $\theta = \pi$ are farthest from and closest to the focus, respectively. ■

Problem 4   Eccentricity

Suppose that $(M, 0)$ and (m, π) are the polar coordinates of two points on the conic of equation (1). Show that

$$e = \frac{M - m}{M + m} \quad \text{and} \quad p = \frac{2Mm}{M - m}$$ ■

II. PLANETARY ORBITS

All the planets in our solar system have elliptical orbits around the sun. If we place the sun at the pole in a polar coordinate system, we will be able to get polar equations for the orbits of the planets, which can then be graphed on a graphing calculator. The data in Table 2 are taken from the *World Almanac* and rounded to three significant digits. Table 2 gives us enough information to find the polar equation for any planet's orbit.

TABLE 2
The Planets

Planet	Maximum distance from sun (million miles)	Minimum distance from sun (million miles)
Mercury	43.4	28.6
Venus	67.7	66.8
Earth	94.6	91.4
Mars	155	129
Jupiter	507	461
Saturn	938	838
Uranus	1,860	1,670
Neptune	2,820	2,760

Problem 5   Polar Equation for Venus's Orbit

Show that Venus's orbit is given approximately by

$$r = \frac{6.72 \times 10^7}{1 - 0.00669 \cos \theta}$$

Graph the equation on a graphing calculator.   ■

Problem 6   Polar Equation for Earth's Orbit

Show that the orbit of the earth is given approximately by

$$r = \frac{9.29 \times 10^7}{1 - 0.0172 \cos \theta}$$

Graph the equation on a graphing calculator.   ■

Problem 7   Polar Equation for Saturn's Orbit

Show that Saturn's orbit is given approximately by

$$r = \frac{8.85 \times 10^8}{1 - 0.0563 \cos \theta}$$

Graph the equation on a graphing calculator.   ■

CHAPTER 7 REVIEW

7.1 POLAR AND RECTANGULAR COORDINATES

Figure 1 illustrates a **polar coordinate** system superimposed on a rectangular coordinate system. The fixed point O is called the **pole**, or **origin**, and the horizontal half-line is the **polar axis**. The **polar–rectangular relationships** are used to transform coordinates and equations from one system to the other.

FIGURE 1
Polar–rectangular relationships

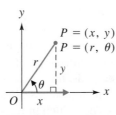

$$r^2 = x^2 + y^2$$

$$\sin \theta = \frac{y}{r} \qquad y = r \sin \theta$$

$$\cos \theta = \frac{x}{r} \qquad x = r \cos \theta$$

$$\tan \theta = \frac{y}{x}$$

[*Note:* The signs of x and y determine the quadrant for θ. The angle θ is usually chosen so that $-180° < \theta \le 180°$ or $-\pi < \theta \le \pi$.]

7.2 SKETCHING POLAR EQUATIONS

Polar graphs can be obtained by **point-by-point** plotting of points on the graph, which are usually computed with a calculator. **Rapid polar sketching techniques** use the uniform variation of $\sin \theta$ and $\cos \theta$ to quickly produce rough sketches of a polar graph. Table 1 illustrates some standard polar curves.

TABLE 1   Some Standard Polar Curves

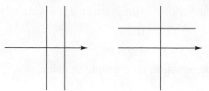 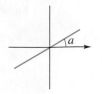

(a) Vertical line:
$r = \dfrac{a}{\cos \theta}$

(b) Horizontal line:
$r = \dfrac{a}{\sin \theta}$

(c) Radial line:
$\theta = a$

(d) Circle:
$r = a$

(e) Circle:
$r = a \cos \theta$

(f) Circle:
$r = a \sin \theta$

(g) Cardioid:
$r = a + a \cos \theta$

(h) Cardioid:
$r = a + a \sin \theta$

(i) Three-leaved
rose:
$r = a \cos 3\theta$

(j) Four-leaved
rose:
$r = a \cos 2\theta$

(k) Lemniscate:
$r^2 = a^2 \cos 2\theta$

(l) Archimedes'
spiral: $r = a\theta$

7.3
THE COMPLEX PLANE

If x and y are real numbers and i is the imaginary unit, then the **complex number** $x + yi$ is associated with the ordered pair (x, y). The plane is called the **complex plane**, the x axis is the **real axis**, the y axis is the **imaginary axis,** and $x + yi$ is the **rectangular form** of the number, as illustrated in Figure 2 on the next page. Complex numbers can also be written in **polar form**, or **trigonometric form,** as shown in Figure 3 on the next page.

From Figure 3 we have the following polar–rectangular equations for complex numbers:

$$x = r \cos \theta \qquad\qquad r = \sqrt{x^2 + y^2}$$

$$y = r \sin \theta \qquad\qquad \tan \theta = \frac{y}{x}, x \neq 0$$

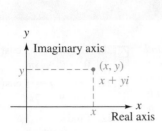

FIGURE 2
Complex plane

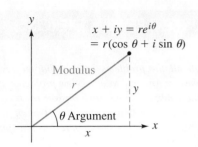

FIGURE 3
Rectangular, polar, and trigonometric forms

The number r is called the **modulus,** or **absolute value,** of z and is denoted by **mod** z or $|z|$. The modulus of z is the distance from z to the origin. The number θ is called the **argument** or **angle,** of z, denoted **arg** z. The argument of z is the angle that the line joining z to the origin makes with the positive real axis (we usually choose the argument θ so that $-\pi < \theta \leq \pi$ or $-180° < \theta \leq 180°$).

Products and **quotients** of complex numbers in polar form are found as follows: If

$$z_1 = r_1 e^{i\theta_1} \quad \text{and} \quad z_2 = r_2 e^{i\theta_2}$$

then

1. $z_1 z_2 = r_1 e^{i\theta_1} r_2 e^{i\theta_2} = r_1 r_2 e^{i(\theta_1 + \theta_2)}$

2. $\dfrac{z_1}{z_2} = \dfrac{r_1 e^{i\theta_1}}{r_2 e^{i\theta_2}} = \dfrac{r_1}{r_2} e^{i(\theta_1 - \theta_2)}$

**7.4
DE MOIVRE'S
THEOREM AND THE
nTH-ROOT THEOREM**

De Moivre's theorem and the related nth-root theorem make the process of finding natural number powers and all the nth roots of a complex number relatively easy. **De Moivre's theorem** is stated as follows: If $z = x + iy = re^{i\theta}$ and n is a natural number, then

$$z^n = (x + iy)^n = (re^{i\theta})^n = r^n e^{n\theta i}$$

From De Moivre's theorem, we can derive the **nth-root theorem:** For n a positive integer greater than 1 and θ in degrees,

$$r^{1/n} e^{(\theta/n + k360°/n)i} \qquad k = 0, 1, \ldots, n - 1$$

are the n distinct nth roots of $re^{i\theta}$, and there are no others.

CHAPTER 7 REVIEW EXERCISE

Work through all the problems in this chapter review and check the answers. Answers to all review problems appear in the back of the book; following each answer is an italic number that indicates the section in which that type of problem is discussed. Where weaknesses show up, review the appropriate sections in the text.

A *In Problems 1–4, plot in a polar coordinate system.*

1. $A = (5, 210°); B = (-7, 180°); C = (-5, -45°)$

2. $r = 5 \sin \theta$ 3. $r = 4 + 4 \cos \theta$

4. $r = 8$

Verify the graphs of Problems 2–4 on a graphing calculator.

5. Change $(2\sqrt{2}, \pi/4)$ to rectangular coordinates.

6. Change $(-\sqrt{3}, 1)$ to polar coordinates $(r \geq 0, 0 \leq \theta < 2\pi)$.

7. Graph $-3 - 2i$ in a rectangular coordinate system.

8. Plot $z = 5(\cos 60° + i \sin 60°) = 5e^{(60°)i}$ in a polar coordinate system.

9. A point in a polar coordinate system has coordinates $(-8, 30°)$. Find all other polar coordinates for the point, $-360° < \theta \leq 360°$, and verbally describe how the coordinates are associated with the point.

10. Find $z_1 z_2$ and z_1/z_2 for $z_1 = 9e^{(42°)i}$ and $z_2 = 3e^{(37°)i}$. Leave answers in polar form.

11. Find $[2e^{(10°)i}]^4$ using De Moivre's theorem. Leave answer in polar form.

12. Find a polar equation for the horizontal line $y = 7$.

B *Plot Problems 13–16 in a polar coordinate system.*

13. $A = (-5, \pi/4); B = (5, -\pi/3); C = (-8, 4\pi/3)$

14. $r = 8 \sin 3\theta$

15. $r = 4 \sin 2\theta$

16. $\theta = \pi/6$

Verify the graphs of Problems 14 and 15 on a graphing calculator.

17. Change $8x - y^2 = x^2$ to polar form.

18. Change $r(3 \cos \theta - 2 \sin \theta) = -2$ to rectangular form.

19. Change $r = -3 \cos \theta$ to rectangular form.

20. Graph $r = 5 \cos(\theta/5)$ on a graphing calculator for $0 \leq \theta \leq 5\pi$.

21. Graph $r = 5 \cos(\theta/7)$ on a graphing calculator for $0 \leq \theta \leq 7\pi$.

22. Find $z_1 z_2$ and z_1/z_2 for $z_1 = 32e^{(3\pi/4)i}$ and $z_2 = 8e^{(\pi/2)i}$. Leave answers in polar form.

23. Find $(6e^{(5\pi/6)i})^3$ using De Moivre's theorem. Leave answer in polar form.

24. Convert $-\sqrt{3} - i$ to polar form $(r \geq 0, -180° < \theta \leq 180°)$.

25. Convert $3\sqrt{2}e^{(3\pi/4)i}$ to exact rectangular form.

26. Convert the factors in $(2 + 2i\sqrt{3})(-\sqrt{2} + i\sqrt{2})$ to polar form and evaluate. Leave answer in polar form.

27. Find $(-\sqrt{2} + i\sqrt{2})/(2 + 2i\sqrt{3})$ by first converting the numerator and denominator to polar form. Leave answer in polar form.

28. Find $(-1 - i)^4$ using De Moivre's theorem and write the result in exact rectangular form.

29. Solve $x^2 + 9i = 0$. Write answers in exact rectangular form.

30. Solve $x^3 - 64 = 0$. Write answers in exact rectangular form and plot them in the complex plane.

31. Find all cube roots of $-4\sqrt{3} - 4i$. Write answers in polar form using degrees.

32. Find all cube roots of i. Write answers in exact polar form using radians.

33. Show that $2e^{(30°)i}$ is a square root of $2 + 2i\sqrt{3}$.

34. (A) A cube root of a complex number is shown in the figure. Geometrically locate all other cube roots of the number and explain how you located them.

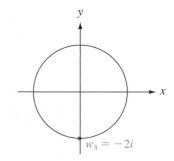

(B) Determine geometrically (without using the nth-root theorem) the two cube roots you graphed in part (A). Give answers in exact rectangular form.

(C) Cube each cube root from part (B).

35. Graph $r = 5(\sin \theta)^{2n}$ in a different viewing window for $n = 1, 2$, and 3. How many leaves do you expect the graph will have for arbitrary n?

36. Graph $r = 2/(1 - e \sin \theta)$ in separate viewing windows for the following values of e. Identify each curve as a hyperbola, ellipse, or parabola.

(A) $e = 1.6$ (B) $e = 1$ (C) $e = 0.4$

37. Change $r(\sin \theta - 2) = 3$ to rectangular form.

38. Find all solutions of $x^3 - 12 = 0$. Write answers in rectangular form $a + bi$, where a and b are computed to three decimal places.

C 39. Show that
$$[r^{1/3}e^{(\theta/3 + k120°)i}]^3 = re^{i\theta} \quad \text{for } k = 0, 1, 2$$

40. (A) Graph $r = 10 \sin \theta$ and $r = -10 \cos \theta$, $0 \le \theta \le \pi$, in the same viewing window. Use TRACE to determine which intersection point has coordinates that satisfy both equations simultaneously.

(B) Solve the equations simultaneously to verify the results in part (B).

(C) Explain why the pole is not a simultaneous solution, even though the two curves intersect at the pole.

41. Describe the location in the complex plane of all solutions of $x^{360} - 1 = 0$. How many solutions are there? How many are real?

CUMULATIVE REVIEW EXERCISE CHAPTERS 1–7

Work through all the problems in this cumulative review and check the answers. Answers to all review problems appear in the back of the book; following each answer is an italic number that indicates the section in which that type of problem is discussed. Where weaknesses show up, review the appropriate sections in the text. Review problems flagged with a star ☆ are from optional sections.

Where applicable, quantities in the problems refer to a triangle labeled as shown in the figure.

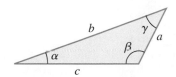

A 1. Which angle has the larger measure: $\alpha = 2\pi/7$ or $\beta = 51°25'40''$? Explain how you got your answer.

2. The sides of a right triangle are 1.27 cm and 4.65 cm. Find the hypotenuse and the acute angles in degree measure.

3. Find the values of $\sec \theta$ and $\tan \theta$ if the terminal side of θ contains $(7, -24)$.

4. Verify the following identities. Try not to look at a table of identities.

(A) $\cot x \sec x \sin x = 1$

(B) $\tan \theta + \cot \theta = \sec \theta \csc \theta$

Evaluate Problems 5–8 exactly as real numbers.

5. $\sin \dfrac{11\pi}{6}$ **6.** $\tan \dfrac{-5\pi}{3}$

7. $\cos^{-1}(-0.5)$ **8.** $\csc^{-1}\sqrt{2}$

Evaluate Problems 9–12 as real numbers to four significant digits using a calculator.

9. $\sin 43°22'$ **10.** $\cot \dfrac{2\pi}{5}$

11. $\sin^{-1} 0.8$ **12.** $\sec^{-1} 4.5$

☆ **13.** Write $\sin 3t + \sin t$ as a product.

14. Explain what is meant by an angle of radian measure 2.5.

In Problems 15–17 solve each triangle, given the indicated measures of angles and sides.

15. $\beta = 110°, \quad \gamma = 42°, \quad b = 68$ m

16. $\beta = 34°, \quad a = 16$ in., $\quad c = 24$ in.

17. $a = 18$ ft, $b = 23$ ft, $c = 32$ ft

☆ **18.** Find the area of the triangle in Problem 16.

19. A point in a polar coordinate system has coordinates $(-7, 30°)$. Find all other polar coordinates for the point, $-180° < \theta \le 180°$, and verbally describe how the coordinates are associated with the point.

20. Find the magnitudes of the horizontal and vertical components of the vector **v** located in a coordinate system as indicated in the figure.

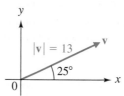

Figure for 20

21. Two vectors **u** and **v** are located in a coordinate system as indicated in the figure. Find the magnitude and the direction (relative to the x axis) of $\mathbf{u} + \mathbf{v}$ if $|\mathbf{u}| = 6.4$ and $|\mathbf{v}| = 3.9$.

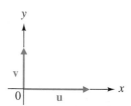

Figure for 21

22. Given $A = (4, -2)$ and $B = (-3, 7)$, represent the geometric vector AB as an algebraic vector and find its magnitude.

☆ **23.** Find the angle (in decimal degrees, to one decimal place) between $\mathbf{u} = 2\mathbf{i} - 7\mathbf{j}$ and $\mathbf{v} = 3\mathbf{i} + 8\mathbf{j}$.

24. Plot in a polar coordinate system: $A = (6, 240°)$; $B = (-4, 225°)$; $C = (9, -45°)$; $D = (-7, -60°)$.

25. Plot in a polar coordinate system: $r = 5 + 5 \sin \theta$.

26. Change $(3\sqrt{2}, 3\pi/4)$ to exact rectangular coordinates.

27. Change $(-2\sqrt{3}, 2)$ to exact polar coordinates.

28. Change $z = 2 - 2i$ to exact polar form using radians.

29. Change $z = 3e^{(3\pi/2)i}$ to exact rectangular form.

30. Find $z_1 z_2$ and z_1/z_2 for $z_1 = 3e^{(50°)i}$ and $z_2 = 5e^{(15°)i}$. Leave answers in polar form.

31. Find $(3e^{(25°)i})^4$ using De Moivre's theorem. Leave answer in polar form.

B **32.** Find an equation of the form $y = k + A \cos Bx$ that produces the graph shown in the following graphing calculator display:

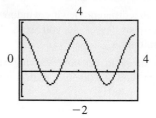

33. Find the exact values of $\csc \theta$ and $\cos \theta$ if $\cot \theta = 4$ and $\sin \theta < 0$.

34. Graph $y = 1 + 2 \sin(2x + \pi)$ for $-\pi \le x \le 2\pi$. State the amplitude, period, frequency, and phase shift.

Verify the identities in Problems 35–37.

35. $\dfrac{\cos x}{1 - \sin x} + \dfrac{\cos x}{1 + \sin x} = 2 \sec x$

36. $\tan \dfrac{\theta}{2} = \dfrac{1}{\csc \theta + \cot \theta}$

37. $\dfrac{\cos x - \sin x}{\cos x + \sin x} = \sec 2x - \tan 2x$

38. Find the exact values of $\tan(x/2)$ and $\sin 2x$, given $\cos x = \frac{24}{25}$ and $\sin x > 0$.

39. Use a graphing calculator to test whether each equation is an identity. If the equation appears to be an identity, verify it. If the equation does not appear to be an identity, find a value of x for which both sides are defined but are not equal.

(A) $\dfrac{\cos^2 x}{(\cos x - 1)^2} = \dfrac{1 + \cos x}{1 - \cos x}$

(B) $\dfrac{\sin^2 x}{(\cos x - 1)^2} = \dfrac{1 + \cos x}{1 - \cos x}$

40. Find the exact value of $\sec(\sin^{-1} \frac{3}{4})$.

41. (A) A cube root of a complex number is shown in the figure on the next page. Geometrically locate all other cube roots of the number and explain how you located them.

(B) Determine geometrically (without using the nth-root theorem) the two cube roots you graphed in part (A). Give answers in exact rectangular form.

(C) Cube each cube root from part (B).

y

w₂ = −2

Figure for 41

42. Find exactly all real solutions for

$$\sin 2x + \sin x = 0$$

43. Find all real solutions (to four significant digits) for

$$2 \cos 2x = 5 \sin x - 4$$

44. Solve the triangle with $b = 17.4$ cm and $\alpha = 49°30'$ for each of the following values of a:

(A) $a = 11.5$ cm   (B) $a = 14.7$ cm

(C) $a = 21.1$ cm

45. Given the vector diagram below, find $|\mathbf{u} + \mathbf{v}|$ and θ.

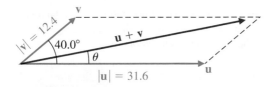

46. Find $3\mathbf{u} - 4\mathbf{v}$ for

(A) $\mathbf{u} = \langle 1, -2 \rangle$; $\mathbf{v} = \langle 0, 3 \rangle$

(B) $\mathbf{u} = 2\mathbf{i} + 3\mathbf{j}$; $\mathbf{v} = -\mathbf{i} + 5\mathbf{j}$

47. Find a unit vector $\mathbf{u}$ with the same direction as $\mathbf{v} = \langle 7, -24 \rangle$.

48. Express $\mathbf{v}$ in terms of $\mathbf{i}$ and $\mathbf{j}$ unit vectors if $\mathbf{v} = \overrightarrow{AB}$, with $A = (-3, 2)$ and $B = (-1, 5)$.

☆ **49.** Determine which vector pairs are orthogonal using properties of the dot product.

(A) $\mathbf{u} = \langle 4, 0 \rangle$; $\mathbf{v} = \langle 0, -5 \rangle$

(B) $\mathbf{u} = \langle 3, 2 \rangle$; $\mathbf{v} = \langle -3, 4 \rangle$

(C) $\mathbf{u} = \mathbf{i} - 2\mathbf{j}$; $\mathbf{v} = 6\mathbf{i} + 3\mathbf{j}$

50. Plot $r = 8 \cos 2\theta$ in a polar coordinate system.

51. Change $x^2 = 6y$ to polar form.

52. Change $r = 4 \sin \theta$ to rectangular form.

53. Convert the factors in $(3 + 3i)(-1 + i\sqrt{3})$ to polar form using degrees and evaluate. Leave answer in polar form.

54. Convert the numerator and denominator in $(-1 + i\sqrt{3})/(3 + 3i)$ to polar form using degrees and evaluate. Leave answer in polar form.

55. Use De Moivre's theorem to evaluate $(1 - i)^6$, and write the result in exact rectangular form.

56. Find all cube roots of $8i$. Write answers in exact rectangular form and plot them in a complex plane.

In Problems 57–59, graph each function in a graphing calculator over the indicated interval.

57. $y = -2 \cos^{-1}(2x - 1), 0 \le x \le 1$

58. $y = 2 \sin^{-1} 2x, -\frac{1}{2} \le x \le \frac{1}{2}$

59. $y = \tan^{-1}(2x + 5), -5 \le x \le 0$

In Problems 60–62, use a graphing calculator to approximate all solutions over the indicated intervals. Compute solutions to three decimal places.

60. $\tan x = 5, -\pi \le x \le \pi$

61. $\cos x = \sqrt[3]{x}$, all real x

62. $3 \sin 2x \cos 3x = 2, 0 \le x \le 2\pi$

63. Graph $r = 5 \sin(\theta/4)$ in a graphing calculator for $0 \le \theta \le 8\pi$.

64. Graph $r = 2/[1 - e \sin(\theta + 0.6)]$ for the indicated values of e, each in a different viewing window. Identify each curve as a hyperbola, ellipse, or parabola.

(A) $e = 0.7$ (B) $e = 1$ (C) $e = 1.5$

C 65. In the circle shown, find the exact radian measure of θ and the coordinates of B to four significant digits if the arc length s is 8 units.

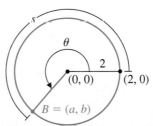

Figure for 65 and 66

66. Refer to the figure. Find θ and s to four significant digits if $(a, b) = (-1.6, -1.2)$.

67. Graph $y = 2 \sec(\pi x + \pi/4)$ for $-1 \le x \le 3$.

68. Express $\sec(2 \tan^{-1} x)$ as an algebraic expression in x free of trigonometric and inverse trigonometric functions.

69. Verify the identity: $\tan 3x = \dfrac{3 - \tan^2 x}{\cot x - 3 \tan x}$.

70. Change $r(\cos \theta + 1) = 1$ to rectangular form.

71. (A) Graph $r = -8 \sin \theta$ and $r = -8 \cos \theta$, $0 \le \theta \le \pi$, in the same viewing window.

(B) Explain why the pole is not a simultaneous solution, even though the two curves intersect at the pole.

72. Find all roots of $x^3 - 4 = 0$. Write answers in the form $a + bi$, where a and b are computed to three decimal places.

73. De Moivre's theorem can be used to derive multiple angle identities. For example, consider

$$\cos 2\theta + i \sin 2\theta = (\cos \theta + i \sin \theta)^2$$
$$= \cos^2 \theta - \sin^2 \theta + 2i \sin \theta \cos \theta$$

Equating the real and imaginary parts of the left side with the real and imaginary parts of the right side yields two familiar identities:

$$\cos 2\theta = \cos^2 \theta - \sin^2 \theta \qquad \sin 2\theta = 2 \sin \theta \cos \theta$$

(A) Apply De Moivre's theorem to $(\cos \theta + i \sin \theta)^3$ to derive identities for $\cos 3\theta$ and $\sin 3\theta$.

(B) Use identities from Chapter 4 to verify the identities obtained in part (A).

In Problems 74–76, graph $f(x)$ in a graphing calculator. From the graph, find a function of the form $g(x) = K + A \cdot t(Bx)$, where $t(x)$ is one of the six trigonometric functions, that produces the same graph. Then verify the identity $f(x) = g(x)$.

74. $f(x) = 2 \cos^2 x - 4 \sin^2 x$

75. $f(x) = \dfrac{6 \sin^2 x - 2}{2 \cos^2 x - 1}$

76. $f(x) = \dfrac{\sin x + \cos x - 1}{1 - \cos x}$

Applications

77. **Surveying** To determine the length CB of the lake in the figure, a surveyor makes the following measurements: $AC = 520$ ft, $\angle BCA = 52°$, and $\angle CAB = 77°$. Find the approximate length of the lake.

78. **Surveying** Refer to Problem 77. Find the approximate length of a similar lake using the following measurements: $AC = 430$ ft, $AB = 580$ ft, $\angle CAB = 64°$.

79. **Tree Height** A tree casts a 35 ft shadow when the angle of elevation of the sun is 54°.

(A) Find the height of the tree if its shadow is cast on level ground.

(B) Find the height of the tree if its shadow is cast straight down a hillside that slopes 11° relative to the horizontal.

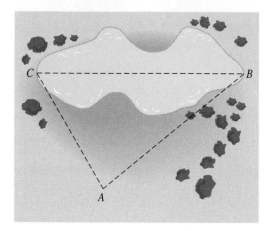

Figure for 77 and 78

80. **Navigation**   Two tracking stations located 4 mi apart on a straight coastline sight a ship (see the figure). How far is the ship from each station?

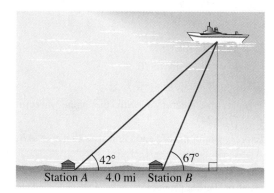

Figure for 80

81. **Electrical Circuits**   The voltage E in an electrical circuit is given by an equation of the form $E = 110 \cos Bt$.

(A) If the frequency is 70 Hz, what is the period? What is the value of B?

(B) If the period is 0.0125 sec, what is the frequency? What is the value of B?

(C) If $B = 100\pi$, what is the period? What is the frequency?

☆ **82. Water Waves**   A wave with an amplitude of exactly 2 ft and a period of exactly 4 sec has an equation of the form $y = A \sin Bt$ at a fixed position. How high is the wave from trough to crest? What is its wavelength (in feet)? How fast is it traveling (in feet per second)? Compute answers to the nearest foot.

Precalculus   *Problems 83–86 refer to the region OCBA inside the first-quadrant portion of a unit circle centered at the origin, as shown in the figure.* *

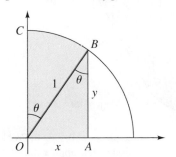

Figure for 83–86

83. Express the area of *OCBA* in terms of θ.

84. Express the area of *OCBA* in terms of x.

85. If the area of $OCBA = 0.5$, use the result of Problem 83 and a graphing calculator to find θ to three decimal places.

86. If the area of $OCBA = 0.4$, use the result of Problem 84 and a graphing calculator to find x to three decimal places.

87. Solar Energy †   A truncated conical solar collector is aimed directly at the sun, as shown in part (a) of the

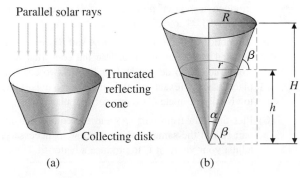

Parallel solar rays

Truncated reflecting cone

Collecting disk

(a)   (b)

Figure for 87

figure. An analysis of the amount of solar energy absorbed by the collecting disk requires certain equations relating the quantities shown in part (b) of the figure. Find an equation that expresses:

(A) R in terms of h, H, r, and α

(B) β in terms of h, H, r, and R

88. Navigation   An airplane can cruise at 265 mph in still air. If a steady wind of 81.5 mph is blowing from the east, what compass heading should the pilot fly in order for the true course of the plane to be north (0°)? Compute the ground speed for this course.

89. Static Equilibrium   A weight suspended from a cable deflects the cable relative to the horizontal as indicated in the figure.

(A) If w is the weight of the object, T_L is the tension in the left side of the cable, and T_R is the tension in the right side, show that

$$T_L = \frac{w \cos \alpha}{\sin(\alpha + \beta)} \quad \text{and} \quad T_R = \frac{w \cos \beta}{\sin(\alpha + \beta)}$$

(B) If $\alpha = \beta$, show that $T_L = T_R = \frac{1}{2} w \csc \alpha$.

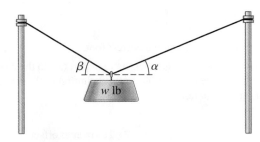

w lb

Figure for 89

90. Railroad Construction   North American railroads do not express specifications for circular track in terms of the radius. The common practice is to define the degree of a track curve as the degree measure of the central angle θ subtended by a 100 ft chord of the circular arc of track (see the figure on the next page).

(A) Find the radius of a 10° track curve to the nearest foot.

(B) If the radius of a track curve is 2,000 ft, find the degree of the track curve to the nearest 0.1°.

---

* See "On the Derivatives of Trigonometric Functions" by M. R. Spiegle in the *American Mathematical Monthly*, Vol. 63 (1956).

† Based on the article "The Solar Concentrating Properties of a Conical Reflector" by Don Leake in *The UMAP Journal*, Vol. 8, No. 4 (1987).

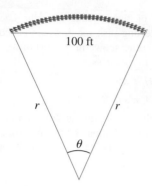

Figure for 90

91. Railroad Construction Actually using a compass to lay out circular arcs for railroad track is impractical. Instead, track layers work with the distance that curved track is offset from straight track.

(A) Suppose a 50 ft length of rail is bent into a circular arc, resulting in a horizontal offset of 1 ft (see the figure). Show that the radius r of this track curve satisfies the equation

$$r \cos^{-1}\left(\frac{r-1}{r}\right) = 50$$

and approximate r to the nearest foot.

(B) Find the degree of the track curve in part (A) to the nearest 0.1°. (See Problem 90 for the definition of the degree of a track curve.)

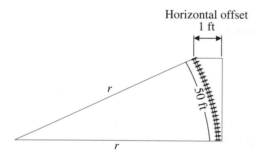

Figure for 91

92. Engineering A circular log of radius r is cut lengthwise into three pieces by two parallel cuts equidistant from the center of the log (see the figure). Express the cross-sectional area of the center piece in terms of r and θ.

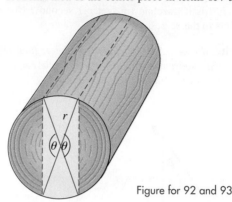

Figure for 92 and 93

93. Engineering If all three pieces of the log in Problem 92 have the same cross-sectional area, approximate θ to four decimal places.

94. Modeling Twilight Duration Periods of twilight occur just before sunrise and just after sunset. The length of these periods varies with the time of year and latitudinal position. At the equator, twilight lasts about an hour all year long. Near the poles, there will be times during the year when twilight lasts for 24 hours and other times when it does not occur at all. Table 1 gives the duration of twilight on the 11th day of each month for 1 year at 20°N latitude.

(A) Convert the data in Table 1 from hours and minutes to two-place decimal hours. Enter the data for a 2 year period in your graphing calculator and produce a scatter plot in the following viewing window: $1 \le x \le 24, 1 \le y \le 5$.

(B) A function of the form $y = k + A \sin(Bx + C)$ can be used to model the data. Use the converted data from Table 1 to determine k and A (to two decimal places), and B (exact value). Use the graph in part (A) to visually estimate C (to one decimal place).

(C) Plot the data from part (A) and the equation from part (B) in the same viewing window. If necessary, adjust your value of C to produce a better fit.

TABLE 1												
x (months)	1	2	3	4	5	6	7	8	9	10	11	12
y (twilight duration, hr:min)	1:37	1:49	2:21	2:59	3:33	4:07	4:03	3:30	2:48	2:13	1:48	1:34

Comments On Numbers

APPENDIX

A

A.1 Real Numbers

A.2 Complex Numbers

A.3 Significant Digits

A.1 Real Numbers

- The Real Number System
- The Real Number Line
- Basic Real Number Properties

This appendix provides a brief review of the real number system and some of its basic properties. The real number system and its properties are fundamental to the study of mathematics.

■ The Real Number System

The real number system is the number system you have used most of your life. Informally, a **real number** is any number that has a decimal representation. Table 1 describes the set of real numbers and some of its important subsets. Figure 1 illustrates how these various sets of numbers are related to each other.

TABLE 1
The Set of Real Numbers

Name	Description	Examples
Natural numbers	Counting numbers (also called positive integers)	$1, 2, 3, \ldots$
Integers	Natural numbers, their negatives, and zero	$\ldots, -2, -1, 0, 1, 2, \ldots$
Rational numbers	Numbers that can be represented as a/b, where a and b are integers, $b \neq 0$; decimal representations are repeating or terminating	$-7, 0, 1, 36, \frac{2}{3}, 3.8,$ $-0.66\overline{6},^* 6.38\overline{38}$
Irrational numbers	Numbers that can be represented as nonrepeating and nonterminating decimal numbers	$\sqrt{3}, \sqrt[3]{7}, \pi, 3.27393...$
Real numbers	Rational numbers and irrational numbers	

* The bar over the 6 means that 6 repeats infinitely, $-0.666666666.\ldots$. Similarly, $\overline{38}$ means that 38 is repeated infinitely.

The set of **integers** contains all the **natural numbers** and something else—their negatives and zero. The set of **rational numbers** contains all the integers and

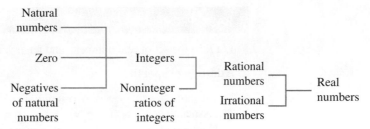

FIGURE 1
The real number system

something else—noninteger ratios of integers. And the set of **real numbers** contains all the rational numbers and something else—the **irrational numbers.**

■ The Real Number Line

A one-to-one correspondence exists between the set of real numbers and the set of points on a line. That is, each real number corresponds to exactly one point, and each point corresponds to exactly one real number. A line with a real number associated with each point, and vice versa, as in Figure 2, is called a **real number line,** or simply a **real line.** Each number associated with a point is called the **coordinate** of the point. The point with coordinate 0 is called the **origin.** The arrow on the right end of the real line in Figure 2 indicates a positive direction. The coordinates of all points to the right of the origin are called **positive real numbers,** and those to the left of the origin are called **negative real numbers.**

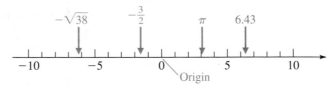

FIGURE 2
A real number line

■ Basic Real Number Properties

In the box on the next page, we list some basic properties of the real number system that enable us to convert algebraic expressions into equivalent forms. Do not let the names of these properties intimidate you. Most of the ideas presented are quite simple. In fact, you have been using many of these properties in arithmetic for a long time. We summarize these properties here for convenient reference because they represent some of the basic rules of the "game of algebra," and you cannot play the game very well unless you know the rules.

BASIC PROPERTIES OF THE SET OF REAL NUMBERS

Let R be the set of real numbers, and let x, y, and z be arbitrary elements of R.

Addition properties

Closure	$x + y$ is a unique element in R.
Associative	$(x + y) + z = x + (y + z)$
Commutative	$x + y = y + x$
Identity	0 is the additive identity; that is, for all x in R, $0 + x = x + 0 = x$, and 0 is the only element in R with this property.
Inverse	For each x in R, $-x$ is its unique additive inverse; that is, $x + (-x) = (-x) + x = 0$, and $-x$ is the only element in R relative to x with this property.

Multiplication properties

Closure	xy is a unique element in R.
Associative	$(xy)z = x(yz)$
Commutative	$xy = yx$
Identity	1 is the multiplicative identity; that is, for all x in R, $(1)x = x(1) = x$, and 1 is the only element in R with this property.
Inverse	For each x in R, $x \neq 0$, $1/x$ is its unique multiplicative inverse; that is, $x(1/x) = (1/x)x = 1$, and $1/x$ is the only element in R relative to x with this property.

Combined property

Distributive	$x(y + z) = xy + xz$
	$(x + y)z = xz + yz$

EXERCISE A.1

1. Give an example of a negative integer, an integer that is neither positive nor negative, and a positive integer.

2. Give an example of a negative rational number, a rational number that is neither positive nor negative, and a positive rational number.

3. Give an example of a rational number that is not an integer.

4. Give an example of an integer that is not a natural number.

5. Indicate which of the following are true:

   (A) All natural numbers are integers.

   (B) All real numbers are irrational.

   (C) All rational numbers are real numbers.

6. Indicate which of the following are true:

   (A) All integers are natural numbers.

   (B) All rational numbers are real numbers.

   (C) All natural numbers are rational numbers.

In Problems 7 and 8, express each number in decimal form to the capacity of your calculator. Observe the repeating decimal representation of the rational numbers and the apparent nonrepeating decimal representation of the irrational numbers. Indicate whether each number is rational or irrational.

7. (A) $\frac{4}{11}$ (B) $\frac{7}{9}$ (C) $\sqrt{7}$ (D) $\frac{13}{8}$

8. (A) $\frac{19}{6}$ (B) $\sqrt{23}$ (C) $\frac{9}{16}$ (D) $\frac{31}{111}$

9. Each of the following real numbers lies between two successive integers on a real number line. Indicate which two.

(A) $\frac{26}{9}$ (B) $-\frac{19}{5}$ (C) $-\sqrt{23}$

10. Each of the following real numbers lies between two successive integers on a real number line. Indicate which two.

(A) $\frac{13}{4}$ (B) $-\frac{5}{3}$ (C) $-\sqrt{8}$

In Problems 11–24, replace each question mark with an appropriate expression that will illustrate the use of the indicated real number property.

11. Commutative property (+): $3 + y = ?$

12. Commutative property (+): $u + v = ?$

13. Associative property ($\cdot$): $3(2x) = ?$

14. Associative property ($\cdot$): $7(4z) = ?$

15. Commutative property ($\cdot$): $x7 = ?$

16. Commutative property ($\cdot$): $yx = ?$

17. Associative property (+): $5 + (7 + x) = ?$

18. Associative property (+): $9 + (2 + m) = ?$

19. Identity property (+): $3m + 0 = ?$

20. Identity property (+): $0 + (2x + 3) = ?$

21. Identity property ($\cdot$): $1(u + v) = ?$

22. Identity property ($\cdot$): $1(xy) = ?$

23. Distributive property: $2x + 3x = ?$

24. Distributive property: $7(x + y) = ?$

25. Indicate whether each is true or false, and for each false statement find real number replacements for a and b to illustrate that it is false. For all real numbers a and b:

(A) $a + b = b + a$ (B) $a - b = b - a$

(C) $ab = ba$ (D) $a/b = b/a$

26. Indicate whether each is true or false, and for each false statement find real number replacements for a, b, and c to illustrate that it is false. For all real numbers a, b, and c:

(A) $(a + b) + c = a + (b + c)$

(B) $(a - b) - c = a - (b - c)$

(C) $a(bc) = (ab)c$

(D) $(a/b)/c = a/(b/c)$

A.2 Complex Numbers

The Pythagoreans (500–275 BC) found that the simple equation

$$x^2 = 2 \tag{1}$$

had no rational number solutions. (A **rational number** is any number that can be expressed as P/Q, where P and Q are integers and $Q \neq 0$.) If equation (1) were to have a solution, then a new kind of number had to be invented—the **irrational number.**

The irrational numbers $\sqrt{2}$ and $-\sqrt{2}$ are both solutions to equation (1). The invention of these irrational numbers evolved over a period of about 2,000 years, and it was not until the 19th century that they were finally put on a rigorous foundation. The rational numbers and irrational numbers together constitute the real number system.

Is there any need to extend the real number system still further? Yes, if we want the simple equation

$$x^2 = -1$$

to have a solution. Since $x^2 \geq 0$ for any real number x, this equation has no real number solutions. Once again we are forced to invent a new kind of number, a number that has the possibility of being negative when it is squared. These new numbers are called *complex numbers*. The complex numbers, like the irrational numbers, evolved over a long period of time, dating mainly back to Girolamo Cardano (1501–1576). But it was not until the 19th century that they were firmly established as numbers in their own right. A **complex number** is defined to be a number of the form

$$a + bi$$

where a and b are real numbers, and i is called the **imaginary unit.** Thus,

$$5 + 3i \qquad \tfrac{1}{3} - 6i \qquad \sqrt{7} + \tfrac{1}{4}i \qquad 0 + 6i \qquad \tfrac{1}{2} + 0i \qquad 0 + 0i$$

are all complex numbers. Particular kinds of complex numbers are given special names, as follows:

$a + 0i = a$	Real number
$0 + bi = bi$	(Pure) imaginary number
$0 + 0i = 0$	Zero
$a - bi$	Conjugate of $a + bi$
a	Real part of $a + bi$
bi	Imaginary part of $a + bi$

Just as every integer is a rational number, every real number is a complex number.

To use complex numbers we must know how to add, subtract, multiply, and divide them. We start by defining equality, addition, and multiplication.

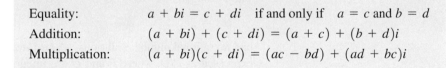

BASIC DEFINITIONS FOR COMPLEX NUMBERS

Equality: $a + bi = c + di$ if and only if $a = c$ and $b = d$

Addition: $(a + bi) + (c + di) = (a + c) + (b + d)i$

Multiplication: $(a + bi)(c + di) = (ac - bd) + (ad + bc)i$

These definitions, particularly the one for multiplication, may seem a little strange to you. But it turns out that if we want many of the same basic properties that hold for real numbers to hold for complex numbers, and if we also want negative real numbers to have square roots, then we must define addition and multiplication as shown. Let us use the definition of multiplication to see what happens to i when it is squared:

$$
\begin{aligned}
i^2 &= (0 + 1i)(0 + 1i) \\
&= (0 \cdot 0 - 1 \cdot 1) + (0 \cdot 1 + 1 \cdot 0)i \\
&= -1 + 0i \\
&= -1
\end{aligned}
$$

So

$$i^2 = -1$$

Fortunately, you do not have to memorize the definitions of addition and multiplication. We can show that the complex numbers, under these definitions, are closed, associative, and commutative, and multiplication distributes over addition. As a consequence, we can manipulate complex numbers as if they were algebraic expressions in the variable i, with the exception that i^2 **is to be replaced with** -1. Example 1 illustrates the mechanics of carrying out addition, subtraction, multiplication, and division.

 **EXAMPLE 1**

Performing Operations with Complex Numbers

Write each of the following in the form $a + bi$:
(A) $(2 + 3i) + (3 - i)$ (B) $(2 + 3i) - (3 - i)$
(C) $(2 + 3i)(3 - i)$ (D) $(2 + 3i)/(3 - i)$

Solution   We treat these as we would ordinary binomials in elementary algebra, with one exception: Whenever i^2 turns up, we replace it with -1.

(A) $(2 + 3i) + (3 - i) = 2 + 3i + 3 - i$
$$= 2 + 3 + 3i - i$$
$$= 5 + 2i$$

(B) $(2 + 3i) - (3 - i) = 2 + 3i - 3 + i$
$$= 2 - 3 + 3i + i$$
$$= -1 + 4i$$

(C) $(2 + 3i)(3 - i) = 6 + 7i - 3i^2$
$$= 6 + 7i - 3(-1)$$
$$= 6 + 7i + 3$$
$$= 9 + 7i$$

(D) To eliminate i from the denominator, we multiply the numerator and denominator by the complex conjugate of $3 - i$, namely, $3 + i$. [Recall from elementary algebra that $(a - b)(a + b) = a^2 - b^2$.]

$$\frac{2 + 3i}{3 - i} \cdot \frac{3 + i}{3 + i} = \frac{6 + 11i + 3i^2}{9 - i^2}$$
$$= \frac{6 + 11i - 3}{9 + 1}$$
$$= \frac{3 + 11i}{10} = \frac{3}{10} + \frac{11}{10}i$$ ∎

Matched Problem 1   Write each of the following in the form $a + bi$:

(A) $(3 - 2i) + (2 + i)$ (B) $(3 - 2i) - (2 - i)$
(C) $(3 - 2i)(2 - i)$ (D) $(3 - 2i)/(2 - i)$ ∎

At this time, your experience with complex numbers has likely been limited to solutions of equations, particularly quadratic equations. Recall that if the expression $b^2 - 4ac$ is negative in

$$x = \frac{-b \pm \sqrt{b^2 - 4ac}}{2a}$$

then the solutions to the quadratic equation $ax^2 + bx + c = 0$ are complex numbers that are not real. This is easy to verify, since a square root of a negative number can be written in the form

$$\sqrt{-k} = i\sqrt{k} \quad \text{for } k > 0$$

To check this last equation, we square $i\sqrt{k}$ to see if we get $-k$:

$$(i\sqrt{k})^2 = i^2(\sqrt{k})^2 = -k$$

[*Note:* We write $i\sqrt{k}$ instead of $\sqrt{k}\, i$ so that i will not mistakenly be included under the radical.]

EXAMPLE 2

Converting Square Roots of Negative Numbers to Complex Form

Write in the form $a + bi$:

(A) $\sqrt{-4}$

(B) $4 + \sqrt{-4}$

(C) $\dfrac{-3 - \sqrt{-7}}{2}$

(D) $\dfrac{1}{1 - \sqrt{-9}}$

Solution

(A) $\sqrt{-4} = i\sqrt{4} = 2i$

(B) $4 + \sqrt{-4} = 4 + i\sqrt{4} = 4 + 2i$

(C) $\dfrac{-3 - \sqrt{-7}}{2} = \dfrac{-3 - i\sqrt{7}}{2} = -\dfrac{3}{2} - \dfrac{\sqrt{7}}{2}\, i$

(D) $\dfrac{1}{1 - \sqrt{-9}} = \dfrac{1}{1 - 3i} = \dfrac{1}{1 - 3i} \cdot \dfrac{1 + 3i}{1 + 3i} = \dfrac{1 + 3i}{1 - 9i^2}$

$$= \dfrac{1 + 3i}{10} = \dfrac{1}{10} + \dfrac{3}{10}\, i$$

Matched Problem 2   Write in the form $a + bi$:

(A) $\sqrt{-16}$

(B) $5 + \sqrt{-16}$

(C) $\dfrac{-5 - \sqrt{-2}}{2}$

(D) $\dfrac{1}{3 - \sqrt{-4}}$

Answers to Matched Problems

1. (A) $5 - i$ (B) $1 - i$ (C) $4 - 7i$ (D) $\frac{8}{5} - \frac{1}{5}i$

2. (A) $4i$ (B) $5 + 4i$ (C) $-\frac{5}{2} - (\sqrt{2}/2)i$ (D) $\frac{3}{13} + \frac{2}{13}i$

EXERCISE A.2

In Problems 1–18, perform the indicated operations and write each answer in the standard form a + bi.

1. $(3 - 2i) + (4 + 7i)$ **2.** $(4 + 6i) + (2 - 3i)$

3. $(3 - 2i) - (4 + 7i)$ **4.** $(4 + 6i) - (2 - 3i)$

5. $(6i)(3i)$ **6.** $(5i)(4i)$

7. $2i(3 - 4i)$ **8.** $4i(2 - 3i)$

9. $(3 - 4i)(1 - 2i)$ **10.** $(5 - i)(2 - 3i)$

11. $(3 + 5i)(3 - 5i)$ **12.** $(7 - 3i)(7 + 3i)$

13. $\dfrac{1}{2 + i}$ **14.** $\dfrac{1}{3 - i}$

15. $\dfrac{2 - i}{3 + 2i}$ **16.** $\dfrac{3 + i}{2 - 3i}$

17. $\dfrac{-1 + 2i}{4 + 3i}$ **18.** $\dfrac{-2 - i}{3 - 4i}$

In Problems 19–26, convert square roots of negative numbers to complex forms, perform the indicated operations, and express answers in the standard form a + bi.

19. $(3 + \sqrt{-4}) + (2 - \sqrt{-16})$

20. $(2 + \sqrt{-9}) + (3 - \sqrt{-25})$

21. $(5 - \sqrt{-1}) - (2 - \sqrt{-36})$

22. $(2 + \sqrt{-9}) - (3 - \sqrt{-25})$

23. $(-3 - \sqrt{-1})(-2 + \sqrt{-49})$

24. $(3 - \sqrt{-9})(-2 - \sqrt{-1})$

25. $\dfrac{5 - \sqrt{-1}}{2 + \sqrt{-4}}$ **26.** $\dfrac{-2 + \sqrt{-16}}{3 - \sqrt{-25}}$

In Problems 27–30, evaluate:

27. $(1 - i)^2 - 2(1 - i) + 2$

28. $(1 + i)^2 - 2(1 + i) + 2$

29. $\left(-\dfrac{1}{2} + \dfrac{\sqrt{3}}{2}i\right)^3$ **30.** $\left(-\dfrac{1}{2} - \dfrac{\sqrt{3}}{2}i\right)^3$

A.3 Significant Digits

- Scientific Notation
- Significant Digits
- Calculation Accuracy

Many calculations in the real world deal with figures that are only approximate. After a series of calculations with approximate measurements, what can be said about the accuracy of the final answer? It seems reasonable to assume that a final answer cannot be any more accurate than the least accurate figure used in the calculation. This is an important point, since calculators tend to give the impression that greater accuracy is achieved than is warranted. In this section we introduce *scientific notation* and we use this concept in a discussion of *significant digits*. We will then be able to set up conventions for indicating the accuracy of the results of certain calculations involving approximate quantities. We will be guided by these conventions throughout the text when writing a final answer to a problem.

■ Scientific Notation

Work in science and engineering often involves the use of very large numbers. For example, the distance that light travels in 1 yr is called a **light-year.** This distance is approximately

9,440,000,000,000 km

Very small numbers are also used. For example, the mass of a water molecule is approximately

0.000 000 000 000 000 000 000 03 g

It is generally troublesome to write and work with numbers of this type in standard decimal form. In fact, these two numbers cannot even be entered into most calculators as they are written. Fortunately, it is possible to represent any decimal form as the product of a number between 1 and 10 and an integer power of 10, that is, in the form

$a \times 10^n$ $1 \leq a < 10$, n **an integer,** a **in decimal form**

A number expressed in this form is said to be in **scientific notation.**

EXAMPLE 1

Using Scientific Notation

Each number is written in scientific notation.

$$4 = 4 \times 10^0 \qquad\qquad 0.36 = 3.6 \times 10^{-1}$$
$$63 = 6.3 \times 10 \qquad\qquad 0.0702 = 7.02 \times 10^{-2}$$
$$805 = 8.05 \times 10^2 \qquad\qquad 0.005\,32 = 5.32 \times 10^{-3}$$
$$3{,}143 = 3.143 \times 10^3 \qquad\qquad 0.000\,67 = 6.7 \times 10^{-4}$$
$$7{,}320{,}000 = 7.32 \times 10^6 \qquad 0.000\,000\,54 = 5.4 \times 10^{-7}$$ ∎

Can you discover a rule that relates the number of decimal places a decimal point is moved to the power of 10 used?

$$7{,}320{,}000 = 7.320\ 000. \times 10^6 = 7.32 \times 10^6$$

6 places left

Positive exponent

$$0.000\ 000\ 54 = 0.000\ 000\ 5.4 \times 10^{-7} = 5.4 \times 10^{-7}$$

7 places right

Negative exponent

Matched Problem 1   Write in scientific notation:

(A) 450   (B) 360,000   (C) 0.0372   (D) 0.000 001 43   ∎

Most calculators display very large and small numbers in scientific notation. Some common methods for displaying scientific notation on a calculator are shown in Table 1.

TABLE 1
Scientific Notation in Calculators

Number entered	Typical scientific calculator display	Typical graphing calculator display
$5.427\,493 \times 10^{-17}$	$5.427493 - 17$	$5.427493E - 17$
$2.359\,779 \times 10^{12}$	$2.359779\ 12$	$2.359779E12$

Note that the numbers in the first column of Table 1 can be entered in a scientific or graphing calculator exactly as they are written (Fig. 1), regardless of the manner in which the calculator displays scientific notation.

FIGURE 1

■ Significant Digits

Suppose we wish to compute the area of a rectangle with the dimensions shown in Figure 1. As often happens with approximations, we have one dimension to one-decimal-place accuracy and the other to two-decimal-place accuracy.

Using a calculator and the formula for the area of a rectangle, $A = ab$, we have

$$A = (11.4)(6.27) = 71.478$$

How many decimal places are justified in this calculation? We will answer this question later in this section. First, we must introduce the idea of *significant digits*.

Whenever we write a measurement such as 11.4 mm, we assume that the measurement is accurate to the last digit written. Thus, 11.4 mm indicates that the measurement was made to the nearest tenth of a millimeter—that is, the actual length is between 11.35 mm and 11.45 mm. In general, the digits in a number that indicate the accuracy of the number are called **significant digits.** If (going from left to right) the first digit and the last digit of a number are not 0, then all the digits are significant. So, the measurements 11.4 and 6.27 in Figure 1 have three significant digits, the number 100.8 has four significant digits, and the number 10,102 has five significant digits.

If the last digit of a number is 0, then the number of significant digits may not be clear. Suppose we are given a length of 23.0 cm. Then we assume the

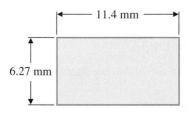

FIGURE 1

measurement has been taken to the nearest tenth and say that the number has three significant digits. However, suppose we are told that the distance between two cities is 3,700 mi. Is the stated distance accurate to the nearest hundred, ten, or unit? That is, does the stated distance have two, three, or four significant digits? We cannot really tell. In order to resolve this ambiguity, we give a precise definition of significant digits using scientific notation.

SIGNIFICANT DIGITS

If a number x is written in scientific notation as

$$x = a \times 10^n \qquad 1 \le a < 10, n \text{ an integer}$$

then the number of significant digits in x is the number of digits in a.

3.7×10^3 has two significant digits

3.70×10^3 has three significant digits

3.700×10^3 has four significant digits

All three of these measurements have the same decimal representation, 3,700, but each represents a different accuracy.

EXAMPLE 2   Determining the Number of Significant Digits

Indicate the number of significant digits in each of the following numbers:

(A) 9.1003×10^{-3} \qquad (B) 1.080×10
(C) 5.92×10^{22} \qquad (D) 7.9000×10^{-13}

Solution   In all cases the number of significant digits is the number of digits in the number to the left of the multiplication sign (as stated in the definition).

(A) Five \qquad (B) Four \qquad (C) Three \qquad (D) Five   ■

Matched Problem 2   Indicate the number of significant digits in each of the following numbers:

(A) 4.39×10^{12} \qquad (B) 1.020×10^{-7}
(C) 2.3905×10^{-1} \qquad (D) 3.00×10 ■

The definition of significant digits tells us how to write a number so that the number of significant digits is clear, but it does not tell us how to interpret the accuracy of a number that is not written in scientific notation. We will use the following convention for numbers that are written as decimal fractions.

SIGNIFICANT DIGITS IN DECIMAL FRACTIONS

The number of significant digits in **a number with no decimal point** is found by counting the digits from left to right, starting with the first digit and ending with the last *nonzero* digit.

The number of significant digits in **a number containing a decimal point** is found by counting the digits from left to right, starting with the first *nonzero* digit and ending with the last digit (which may be 0).

Applying this convention to the number 3,700, we conclude that this number (as written) has two significant digits. If we want to indicate that it has three or four significant digits, we must use scientific notation. The significant digits in the following numbers are underlined.

<u>34,00</u>7   <u>92</u>0,000   <u>25.300</u>   0.00<u>63</u>   0.000 <u>430</u>

■ Calculation Accuracy

When performing calculations, we want an answer that is as accurate as the numbers used in the calculation warrant, but no more. In calculations involving multiplication, division, powers, and roots, we adopt the following accuracy convention (which is justified in courses in numerical analysis).

ACCURACY OF CALCULATED VALUES

The number of significant digits in a calculation involving multiplication, division, powers, and/or roots is the same as the number of significant digits in the number in the calculation with the smallest number of significant digits.

Applying this convention to the calculation of the area A of the rectangle in Figure 1, we have

$$A = (11.4)(6.27)$$ *Both numbers have three significant digits.*
$$\boxed{= 71.478}$$ *Calculator computation*
$$= 71.5$$ *The computed area is accurate only to three significant digits.*

 **EXAMPLE 3**   **Determining the Accuracy of Computed Values**

Perform the indicated operations on the given approximate numbers. Then use the accuracy conventions stated above to round each answer to the appropriate accuracy.

(A) $\dfrac{(204)(34.0)}{120}$ (B) $\dfrac{(2.50 \times 10^5)(3.007 \times 10^7)}{2.4 \times 10^6}$

Solution   (A) $\dfrac{(204)(34.0)}{120}$ *120 has the least number of significant digits (two).*

$\boxed{= 57.8}$ *Calculator computation*

$= 58$ *Answer must have the same number of significant digits as the number with the least number of significant digits (two).*

(B) $\dfrac{(2.50 \times 10^5)(3.007 \times 10^7)}{2.4 \times 10^6}$ *2.4×10^6 has the least number of significant digits (two).*

$\boxed{= 3.132\,291 \ldots \times 10^6}$ *Calculator computation*

$= 3.1 \times 10^6$ *Answer must have the same number of significant digits as the number with the least number of significant digits (two).* ∎

Matched Problem 3   Perform the indicated operations on the given approximate numbers. Then use the rounding conventions stated previously to write each answer with the appropriate accuracy.

(A) $\dfrac{2.30}{(0.0341)(2.674)}$ (B) $\dfrac{1.235 \times 10^8}{(3.07 \times 10^{-3})(1.20 \times 10^4)}$ ∎

We complete this section with two important observations:

1. Many formulas have constants that represent exact quantities. Such quantities are assumed to have infinitely many significant digits. For example, the formula for the circumference of a circle, $C = 2\pi r$, has two constants that are exact, 2 and π. Consequently, the final answer will have as many significant digits as in the measurement r (radius).

2. How do we round a number if the last nonzero digit is 5? For example, the following product should have only two significant digits. Do we change the 2 to a 3 or leave it alone?

$(1.3)(2.5) = 3.25$ *Calculator result: Should be rounded to two significant digits.*

We will round to 3.2 following the conventions given in the box.

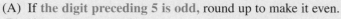

ROUNDING NUMBERS WHEN LAST NONZERO DIGIT IS 5

(A) If **the digit preceding 5 is odd,** round up to make it even.
(B) If **the digit preceding 5 is even,** do not change it.

We have adopted these conventions in order to prevent the accumulation of rounding errors with numbers having the last nonzero digit 5. The idea is to round up 50% of the time. (There are a number of other ways to accomplish this—flipping a coin, for example.)

EXAMPLE 4

Rounding Numbers When Last Nonzero Digit is 5

Round each number to three significant digits.
(A) 3.1495   (B) 0.004 135
(C) 32,450   (D) $4.314\ 764\ 09 \times 10^{12}$

Solution   (A) 3.15   (B) 0.00414   (C) 32,400   (D) 4.31×10^{12} ■

Matched Problem 4   Round each number to three significant digits.

(A) 43.0690   (B) 48.05

(C) 48.15   (D) $8.017\ 632 \times 10^{-3}$ ■

Answers to Matched Problems

1. (A) 4.5×10^2 (B) 3.6×10^5 (C) 3.72×10^{-2} (D) 1.43×10^{-6}
2. (A) Three   (B) Four   (C) Five   (D) Three
3. (A) 25.2   (B) 3.35×10^6
4. (A) 43.1   (B) 48.0   (C) 48.2   (D) 8.02×10^{-3}

EXERCISE A.3

A *In Problems 1–12, write in scientific notation.*

1. 640
2. 384
3. 5,460,000,000
4. 38,400,000
5. 0.73
6. 0.00493
7. 0.000 000 32
8. 0.0836
9. 0.000 049 1
10. 435,640
11. 67,000,000,000
12. 0.000 000 043 2

In Problems 13–20, write as a decimal fraction.

13. 5.6×10^4
14. 3.65×10^6
15. 9.7×10^{-3}
16. 6.39×10^{-6}
17. 4.61×10^{12}
18. 3.280×10^9
19. 1.08×10^{-1}
20. 3.004×10^{-4}

In Problems 21–32, indicate the number of significant digits in each number.

21. 12.3
22. 123
23. 12.300
24. 0.00123
25. 0.01230
26. 12.30
27. 6.7×10^{-1}
28. 3.56×10^{-4}
29. 6.700×10^{-1}
30. 3.560×10^{-4}
31. 7.090×10^5
32. 6.0050×10^7

B *In Problems 33–38, round each to three significant digits.*

33. 635,431
34. 4,089,100
35. 86.85
36. 7.075
37. 0.004 652 3
38. 0.000 380 0

In Problems 39–44, write in scientific notation, rounding to two significant digits.

39. 734

40. 908

41. 0.040

42. 700

43. 0.000 435

44. 635.46813

In Problems 45–50. indicate how many significant digits should be in the final answer.

45. $(32.8)(0.2035)$

46. $(0.00230)(25.67)$

47. $\dfrac{(7.21)(360)}{1{,}200}$

48. $\dfrac{(0.0350)(621)}{8{,}543}$

49. $\dfrac{(5.03 \times 10^{-3})(6 \times 10^{4})}{8.0}$

50. $\dfrac{3.27(1.8 \times 10^{7})}{2.90 \times 10}$

In Problems 51–56, use a calculator and express each answer with the appropriate accuracy.

51. $\dfrac{6.07}{0.5057}$

52. $(53{,}100)(0.2467)$

53. $(6.14 \times 10^{9})(3.154 \times 10^{-1})$

54. $\dfrac{7.151 \times 10^{6}}{9.1 \times 10^{-1}}$

55. $\dfrac{6{,}730}{(2.30)(0.0551)}$

56. $\dfrac{63{,}100}{(0.0620)(2{,}920)}$

In Problems 57–62, the indicated constants are exact. Compute the answer to an accuracy appropriate for the given approximate values of the variables.

57. **Circumference of a Circle** $C = 2\pi r$; $r = 25.31$ cm

58. **Area of a Circle** $A = \pi r^2$; $r = 2.5$ in.

59. **Area of a Triangle** $A = \frac{1}{2}bh$; $b = 22.4$ ft, $h = 8.6$ ft

60. **Area of an Ellipse** $A = \pi ab$; $a = 0.45$ cm, $b = 1.35$ cm

61. **Surface Area of a Sphere** $S = 4\pi r^2$; $r = 1.5$ mm

62. **Volume of a Sphere** $S = \frac{4}{3}\pi r^3$; $r = 1.8$ in.

C *In Problems 63–66, the indicated constants are exact. Compute the value of the indicated variable to an accuracy appropriate for the given approximate values of the other variables in the formula.*

63. **Volume of a Rectangular Parallelepiped (a Box)** $V = lwh$; $V = 24.2$ cm^3, $l = 3.25$ cm, $w = 4.50$ cm, $h = ?$

64. **Volume of a Right Circular Cylinder** $V = \pi r^2 h$; $V = 1{,}250$ ft^3, $h = 6.4$ ft, $r = ?$

65. **Volume of a Right Circular Cone** $V = \frac{1}{3}\pi r^2 h$; $V = 1{,}200$ in.^{3}, $h = 6.55$ in., $r = ?$

66. **Volume of a Pyramid** $V = \frac{1}{3}Ah$; $V = 6{,}000$ m^3, $A = 1{,}100$ m, $h = ?$

Functions and Inverse Functions

B

B.1   Functions

B.2   Graphs and Transformations

B.3   Inverse Functions

497

B.1   Functions

- Definition of a Function—Rule Form
- Definition of a Function—Set Form
- Function Notation
- Function Classification

Seeking correspondences between various types of phenomena is undoubtedly one of the most important aspects of science. A physicist attempts to find a correspondence between the current in an electrical circuit and time; a chemist looks for a correspondence between the speed of a chemical reaction and the concentration of a given substance; an economist tries to determine a correspondence between the price of an object and the demand for the object; and so on. The list could go on and on.

Establishing and working with correspondences among various types of phenomena—whether through tables, graphs, or equations—is so fundamental to pure and applied science that it has become necessary to describe this activity in the precise language of mathematics.

■   Definition of a Function–Rule Form

What do all the examples cited above have in common? Each describes the matching of elements from one set with the elements in a second set. Consider Charts 1–3, which list values for the cube, square, and square root, respectively.

CHART 1	
Number	Cube
$-2 \longrightarrow -8$	
$-1 \longrightarrow -1$	
$0 \longrightarrow 0$	
$1 \longrightarrow 1$	
$2 \longrightarrow 8$	

CHART 2	
Number	Square
-2	
-1	4
0	1
1	0
2	

CHART 3	
Number	Square root
$0 \longrightarrow 0$	
1	1
	-1
4	2
	-2
9	3
	-3

Charts 1 and 2 define functions, but Chart 3 does not. Why? The definition of *function,* given in the box on the next page, will explain.

FUNCTION—RULE FORM

A **function** is a rule that produces a correspondence between two sets of elements such that to each element in the first set there corresponds *one and only one* element in the second set.

The first set is called the **domain**. The set of all corresponding elements in the second set is called the **range**.

A variable representing an arbitrary element from the domain is called an **independent variable**. A variable representing an arbitrary element from the range is called a **dependent variable**.

Charts 1 and 2 define functions, since to each domain value there corresponds exactly one range value. For example, the cube of -2 is -8 and no other number. On the other hand, Chart 3 does not specify a function, since to at least one domain value there is more than one corresponding range value. For example, to the domain value 4 correspond -2 and 2, both square roots of 4.

Some equations in two variables define functions. If in an equation in two variables, say x and y, there corresponds exactly one range value y for each domain value x (x is independent and y is dependent), then the correspondence established by the equation is a function.

EXAMPLE 1

Determining Whether an Equation Defines a Function

Determine which of the following equations define functions with independent variable x and domain all real numbers.

(A) $y - x = 1$ (B) $y^2 - x^2 = 1$

Solution (A) Solving for the dependent variable y, we have

$$y - x = 1 \tag{1}$$
$$y - 1 + x$$

Since $1 + x$ is a real number for each real number x, equation (1) assigns exactly one value of the dependent variable, $y = 1 + x$, to each value of the independent variable x. So, equation (1) defines a function.

(B) Solving for the dependent variable y, we have

$$y^2 - x^2 = 1$$
$$y^2 = 1 + x^2 \tag{2}$$
$$y = \pm \sqrt{1 + x^2}$$

Since $1 + x^2$ is always a positive real number and since each positive real number has two real square roots, each value of the independent variable x corresponds to two values of the dependent variable, $y = -\sqrt{1 + x^2}$ and $y = \sqrt{1 + x^2}$. So, equation (2) does not define a function.   ■

Matched Problem 1   Determine which of the following equations define functions with independent variable x and domain all real numbers.

(A) $y = x + 3$ (B) $y^2 = x^2 + 3$ ■

It is very easy to determine whether an equation defines a function by examining the graph of the equation. The two equations considered in Example 1 are graphed in Figure 1.

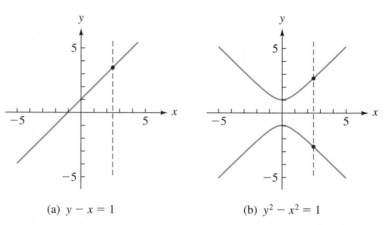

(a) $y - x = 1$ (b) $y^2 - x^2 = 1$

FIGURE 1
Graphs of equations and the vertical-line test

In Figure 1a, any vertical line intersects the graph in exactly one point, illustrating graphically the fact that each value of the independent variable x corresponds to exactly one value of the dependent variable y. On the other hand, Figure 1b shows that there are vertical lines that intersect this graph in two points. This indicates that there are values of the independent variable x that correspond to two different values of the dependent variable y. These observations form the basis for the **vertical-line test** stated in the following box.

VERTICAL-LINE TEST

An equation defines a function if and only if each vertical line in the rectangular coordinate system passes through at most one point on the graph of the function.

■ Definition of a Function–Set Form

Since elements in the range of a function are paired with elements in the domain by some rule or process, this correspondence (pairing) can be illustrated by using **ordered pairs** of elements, where the first component represents a domain element and the second component represents a corresponding range element. We can write functions 1 and 2 specified in Charts 1 and 2 as sets of pairs as follows:

Function 1: $\{(-2, -8), (-1, -1), (0, 0)\ (1, 1), (2, 8)\}$

Function 2: $\{(-2, 4), (-1, 1), (0, 0)\ (1, 1), (2, 4)\}$

In both cases, notice that no two ordered pairs have the same first component and different second components. On the other hand, if we list the set S of ordered pairs determined by Chart 3, we have

$$S = \{(0, 0), (1, 1), (1, -1), (4, 2), (4, -2), (9, 3), (9, -3)\}$$

In this case, there are ordered pairs with the same first component and different second components. For example, $(1, 1)$ and $(1, -1)$ both belong to the set S. Once again, we see that Chart 3 does not specify a function.

This discussion suggests an alternative but equivalent way of defining functions that produces additional insight into this concept.

FUNCTION—SET FORM

A **function** is a set of ordered pairs with the property that no two ordered pairs have the same first component and different second components. The set of all first components in a function is called the **domain** of the function and the set of all second components is called the **range.**

EXAMPLE 2

Functions Defined as Sets of Ordered Pairs

Given the sets:

$$H = \{(1, 1), (2, 1), (3, 2), (3, 4)\}$$
$$G = \{(2, 4), (3, -1), (4, 4)\}$$

(A) Which set specifies a function?

(B) Give the domain and range of the function.

Solution

(A) Set H does not specify a function, since $(3, 2)$ and $(3, 4)$ both have the same first component. The domain value 3 corresponds to more than one range value. Set G does specify a function, since each domain value corresponds to exactly one range value.

(B) Domain of $G = \{2, 3, 4\}$; Range of $G = \{-1, 4\}$ ■

Matched Problem 2   Repeat Example 2 for the following sets:

$$M = \{(3, 4), (5, 4), (6, -1)\}$$
$$N = \{(-1, 2), (0, 4), (1, 2), (0, 1)\}$$ ■

■ Function Notation

If x represents an element in the domain of a function f, then we will often use the symbol $f(x)$ in place of y to designate the number in the range of f to which x is paired. It is important not to be confused by this new symbol and think of it as a product of f and x. The symbol is read "f of x" or "the value of f at x." The correct use of this function symbol should be mastered early. For example, if

$$f(x) = 2x + 3$$

then

$$f(5) = 2(5) + 3 = 13$$

That is, the function f assigns the range value 13 to the domain value 5. Thus, (5, 13) belongs to f. Can you find another ordered pair that belongs to f?

The correspondence between domain values and range values is usually illustrated in one of the two ways shown in Figure 2.

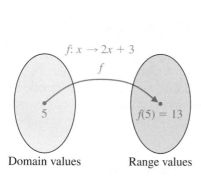

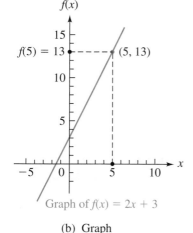

(a) Map

(b) Graph

FIGURE 2
Function notation

EXAMPLE 3

Evaluating Functions

If $f(x) = (x/2) - 1$ and $g(x) = 1 - x^2$, find:

(A) $f(4)$ (B) $g(-3)$ (C) $f(2) - g(0)$

(D) $f(2 + h)$ (E) $[f(3 + h) - f(3)]/h$ (F) $f(g(3))$

Solution (A) $f(4) = \dfrac{4}{2} - 1 = 2 - 1 = 1$

(B) $g(-3) = 1 - (-3)^2 = 1 - 9 = -8$

(C) $f(2) - g(0) = \left(\dfrac{2}{2} - 1\right) - (1 - 0^2) = 0 - 1 = -1$

(D) $f(2 + h) = \dfrac{2 + h}{2} - 1 = \dfrac{2 + h - 2}{2} = \dfrac{h}{2}$

(E) $\dfrac{f(3 + h) - f(3)}{h} = \dfrac{\left(\dfrac{3 + h}{2} - 1\right) - \left(\dfrac{3}{2} - 1\right)}{h}$

$$= \dfrac{\dfrac{1 + h}{2} - \dfrac{1}{2}}{h} = \dfrac{1}{2}$$

(F) $f(g(3)) = f(1 - 3^2) = f(-8) = \dfrac{-8}{2} - 1 = -5$ ■

Matched Problem 3 If $f(x) = 2x - 3$ and $g(x) = x^2 - 2$, find:

(A) $f(3)$ (B) $g(-2)$ (C) $f(0) + g(1)$

(D) $f(3 + h)$ (E) $[f(2 + h) - f(2)]/h$ (F) $f(g(2))$ ■

EXAMPLE 4

Finding the Range of a Function

If the function f defined by $f(x) = x^2 - x$ has domain $X = \{-2, -1, 0, 1, 2\}$, what is the range Y of f?

Solution
$$f(-2) = (-2)^2 - (-2) = 6$$
$$f(-1) = (-1)^2 - (-1) = 2$$
$$f(0) = 0^2 - 0 = 0$$
$$f(1) = 1^2 - 1 = 0$$
$$f(2) = 2^2 - 2 = 2$$

The range of f is $Y = \{0, 2, 6\}$. ■

Matched Problem 4 Repeat Example 4 for $g(x) = x^2 - 4$, $X = \{-2, -1, 0, 1, 2\}$. ■

■ **Function Classification**

Functions are classified in special categories for more efficient study. You have already had some experience with many *algebraic functions* (defined by means of the algebraic operations addition, subtraction, multiplication, division, powers, and roots), *exponential functions,* and *logarithmic functions.* In this book, we add two more classes of functions to this list, namely, the *trigonometric functions* and the *inverse trigonometric functions.* These five classes of functions are called the **elementary functions.** They are related to each other as follows:

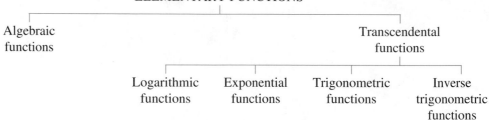

ELEMENTARY FUNCTIONS

Algebraic functions

Transcendental functions

Logarithmic functions

Exponential functions

Trigonometric functions

Inverse trigonometric functions

Answers to Matched Problems

1. (A) Defines a function   (B) Does not define a function
2. (A) M is a function   (B) Domain $= \{3, 5, 6\}$, Range $= \{-1, 4\}$
3. (A) 3   (B) 2   (C) -4 (D) $3 + 2h$ (E) 2   (F) 1
4. $Y = \{-4, -3, 0\}$

EXERCISE B.1

In Problems 1–6, for $f(x) = 4x - 1$*, evaluate.*

1. $f(1)$
2. $f(2)$
3. $f(-1)$
4. $f(-2)$
5. $f(0)$
6. $f(5)$

In Problems 7–12, for $g(x) = x - x^2$*, evaluate.*

7. $g(1)$
8. $g(3)$
9. $g(5)$
10. $g(4)$
11. $g(-2)$
12. $g(-3)$

In Problems 13–24, for $f(x) = 1 - 2x$ *and* $g(x) = 4 - x^2$*, evaluate.*

13. $f(0) + g(0)$
14. $g(0) - f(0)$
15. $\dfrac{f(3)}{g(1)}$
16. $[g(-2)][f(-1)]$
17. $2f(-1)$
18. $\frac{1}{5}g(-3)$
19. $f(2 + h)$
20. $g(2 + h)$
21. $\dfrac{f(2 + h) - f(2)}{h}$
22. $\dfrac{g(2 + h) - g(2)}{h}$
23. $g[f(2)]$
24. $f[g(2)]$

In Problems 25–32, does the equation specify a function, given that x is the independent variable?

25. $x^2 + y^2 = 25$
26. $y = 3x + 1$
27. $2x - 3y = 6$
28. $y = x^2 - 2$
29. $y^2 = x$
30. $y = x^2$
31. $y = |x|$
32. $|y| = x$

33. If the function f, defined by $f(x) = x^2 - x + 1$, has domain $X = \{-2, -1, 0, 1, 2\}$, find the range Y of f.

34. If the function g, defined by $g(x) = 1 + x - x^2$, has domain $X = \{-2, -1, 0, 1, 2\}$, find the range Y of g.

35. Indicate which set specifies a function and write down its domain and range.

$$F = \{(-2, 1), (-1, 1), (0, 0)\}$$
$$G = \{(-4, 3), (0, 3), (-4, 0)\}$$

36. Repeat Problem 35 for the following sets:

$$H = \{(-1, 3), (2, 3), (-1, -1)\}$$
$$L = \{(-1, 1), (0, 1), (1, 1)\}$$

Applications

Precalculus Problems 37–39 pertain to the following relationship: The distance d (in meters) that an object falls in a vacuum in t seconds is given by

$$d = s(t) = 4.88t^2$$

37. Find $s(0)$, $s(1)$, $s(2)$, and $s(3)$ to two decimal places.

38. The expression $[s(2 + h) - s(2)]/h$ represents the average speed of the falling object over the time interval from $t = 2$ to $t = 2 + h$. Use a calculator to compute each of the following to four significant digits. Then guess the speed of a free-falling object at the end of 2 sec.

(A) $\dfrac{s(3) - s(2)}{1}$ (B) $\dfrac{s(2.1) - s(2)}{0.1}$

(C) $\dfrac{s(2.01) - s(2)}{0.01}$ (D) $\dfrac{s(2.001) - s(2)}{0.001}$

(E) $\dfrac{s(2.0001) - s(2)}{0.0001}$

39. Find $[s(2 + h) - s(2)]/h$ and simplify. What happens as h gets closer and closer to 0? Interpret physically.

B.2 Graphs and Transformations

- Basic Functions
- Vertical and Horizontal Shifts
- Reflections
- Stretching and Shrinking

The functions

$$g(x) = x^2 - 2 \qquad h(x) = (x - 2)^2 \qquad k(x) = -2x^2$$

can be expressed in terms of the function $f(x) = x^2$ as follows:

$$g(x) = f(x) - 2 \qquad h(x) = f(x - 2) \qquad k(x) = -2f(x)$$

The graphs of functions g, h, and k are closely related to the graph of function f. Before reviewing relationships like these, we identify some basic functions and summarize their properties.

▇ Basic Functions

Figure 1 on the next page shows six basic functions that you will encounter frequently. You should know the definition, domain, and range of each and be able to recognize their graphs.

▇ Vertical and Horizontal Shifts

If performing an operation on a given function forms a new function, then the graph of the new function is called a **transformation** of the graph of the original function. For example, graphs of both $y = f(x) + k$ and $y = f(x + h)$ are transformations of the graph of $y = f(x)$.

FIGURE 1
Some basic functions
and their graphs

NOTE: Letters used to
designate the func-
tions in Figure 1 may
vary from context to
context; R is the set of
real numbers.

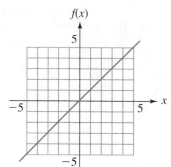

(A) **Identity function**
$f(x) = x$
Domain: R
Range: R

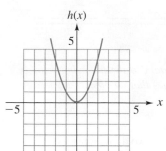

(B) **Square function**
$h(x) = x^2$
Domain: R
Range: $[0, \infty)$

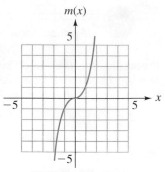

(C) **Cube function**
$m(x) = x^3$
Domain: R
Range: R

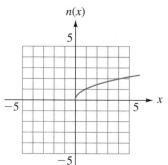

(D) **Square root function**
$n(x) = \sqrt{x}$
Domain: $[0, \infty)$
Range: $[0, \infty)$

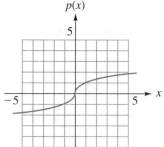

(E) **Cube root function**
$p(x) = \sqrt[3]{x}$
Domain: R
Range: R

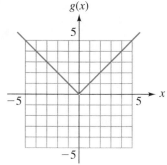

(F) **Absolute value function**
$g(x) = |x|$
Domain: R
Range: $[0, \infty)$

 **EXAMPLE 1**   **Vertical and Horizontal Shifts**

(A) Graph $f(x) = x^2$, $g(x) = x^2 + 1$, and $h(x) = x^2 - 4$ simultaneously in the same coordinate system. How are the graphs of g and h related to the graph of f?

(B) Graph $f(x) = x^2$, $G(x) = (x - 3)^2$, and $H(x) = (x + 2)^2$ simultaneously in the same coordinate system. How are the graphs of G and H related to the graph of f?

Solution   (A) The graph of g is the same as the graph of f shifted upward 1 unit, and the graph of h is the same as the graph of f shifted downward 4 units (see Fig. 2 on the next page).

(B) The graph of G is the same as the graph of f shifted 3 units to the right, and the graph of H is the same as the graph of f shifted 2 units to the left (see Fig. 3 on the next page).

FIGURE 2
Vertical shifts

FIGURE 3
Horizontal shifts

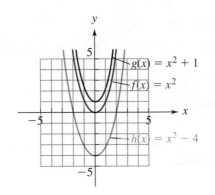

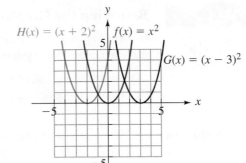

Matched Problem 1

(A) Graph $f(x) = |x|$, $g(x) = |x| - 4$, and $h(x) = |x| + 2$ simultaneously in the same coordinate system. How are the graphs of g and h related to the graph of f?

(B) Graph $f(x) = |x|$, $G(x) = |x + 1|$, and $H(x) = |x - 3|$ simultaneously in the same coordinate system. How are the graphs of G and H related to the graph of f? ■

Comparing the graphs of $y = f(x) + k$ with the graph of $y = f(x)$ (Fig. 2), we see that the graph of $y = f(x) + k$ can be obtained from the graph of $y = f(x)$ by **vertically shifting (translating)** the graph of the latter upward k units if k is positive, and downward $|k|$ units if k is negative. Comparing the graphs of $y = f(x + h)$ with the graph of $y = f(x)$ (Fig. 3), we see that the graph of $y = f(x + h)$ can be obtained from the graph of $y = f(x)$ by **horizontally shifting (translating)** the graph of the latter h units to the left if h is positive, and $|h|$ units to the right if h is negative.

■ Reflections

A complete definition of reflection through a line requires concepts from Euclidean geometry that will not be discussed in this text. Here we are interested in reflecting the graph of $y = f(x)$ through the x axis or the y axis. Figure 4 illustrates these reflections for a single point (a, b).

FIGURE 4
Reflections through the coordinate axes

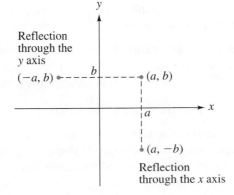

EXAMPLE 2

Reflecting the Graph of a Function

Graph $f(x) = (x - 1)^2$, $g(x) = -f(x)$, and $h(x) = f(-x)$ simultaneously in the same coordinate system.

(A) How is the graph of g related to the graph of f?
(B) How is the graph of h related to the graph of f?

Solution   (A) The graph of g is the reflection through the x axis of the graph of f (see Fig. 5).
(B) We simplify h before graphing:

$$h(x) = f(-x) = (-x - 1)^2$$
$$= [(-1)(x + 1)]^2$$
$$= (-1)^2 (x + 1)^2$$
$$= (x + 1)^2$$

The graph of h is the reflection through the y axis of the graph of f (see Fig. 5).

FIGURE 5
Reflections

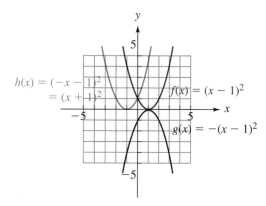

Matched Problem 2   Graph $f(x) = |x + 2|$, $g(x) = -f(x)$, and $h(x) = f(-x)$ simultaneously in the same coordinate system.

(A) How is the graph of g related to the graph of f?
(B) How is the graph of h related to the graph of f?

Examining the graphs in Figure 5, we see that the graph of $y = -f(x)$ is the **reflection through the x axis** of the graph of $y = f(x)$ and the graph of $y = f(-x)$ is the **reflection through the y axis** of the graph of $y = f(x)$.

■ Stretching and Shrinking

Horizontal shifts, vertical shifts, and reflections are called **rigid transformations** because they do not change the shape of a graph, only its location. Now we consider some **nonrigid transformations** that stretch or shrink a graph.

Stretching or Shrinking a Graph

(A) Graph $f(x) = x^3$, $g(x) = 2x^3$, and $h(x) = \frac{1}{2}x^3$ simultaneously in the same coordinate system. How are these graphs related?

(B) Graph $f(x) = x^3$, $G(x) = (2x)^3$, and $H(x) = (\frac{1}{2}x)^3$ simultaneously in the same coordinate system. How are these graphs related?

Solution (A) The graph of g can be obtained from the graph of f by multiplying each y value by 2. This stretches the graph of f vertically by a factor of 2 (Fig. 6). The graph of h can be obtained from the graph of f by multiplying each y value by $\frac{1}{2}$. This shrinks the graph of f vertically by a factor of $\frac{1}{2}$ (Fig. 6).

FIGURE 6
Vertical stretching and shrinking

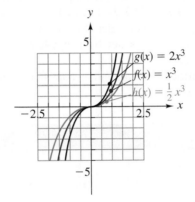

(B) The graph of G can be obtained from the graph of f by multiplying each x value by $\frac{1}{2}$. This shrinks the graph of f horizontally by a factor of $\frac{1}{2}$ (Fig. 7). The graph of H can be obtained from the graph of f by multiplying each x value by 2. This stretches the graph of f horizontally by a factor of 2 (Fig. 7).

FIGURE 7
Horizontal stretching and shrinking

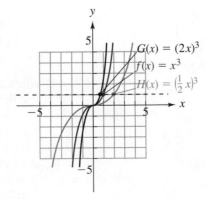

Matched Problem 3   (A) Graph $f(x) = \sqrt{x}$, $g(x) = 2\sqrt{x}$, and $h(x) = \frac{1}{2}\sqrt{x}$ simultaneously in the same coordinate system. How are these graphs related?

(B) Graph $f(x) = \sqrt{x}$, $G(x) = \sqrt{2x}$, and $H(x) = \sqrt{\frac{1}{2}x}$ simultaneously in the same coordinate system. How are these graphs related?   ∎

Plotting a few points, as we did in Figures 6 and 7, will help you differentiate between stretches and shrinks. In general, the graph of $y = af(x)$ can be obtained from the graph of $y = f(x)$ by multiplying each y value by a. This **vertically stretches** the graph of f by a factor of a if $a > 1$ and **vertically shrinks** the graph of f by a factor of a if $0 < a < 1$. The graph of $y = f(ax)$ can be obtained from the graph of $y = f(x)$ by multiplying each x value by $1/a$. This **horizontally stretches** the graph of f by a factor of $1/a$ if $0 < a < 1$ and **horizontally shrinks** the graph of f by a factor of $1/a$ if $a > 1$. The terms **expansion** and **compression** are also used to describe stretching and shrinking, respectively.

Refer to Example 3. Notice that $G(x) = (2x)^3 = 8x^3$, so shrinking a graph horizontally can be equivalent to stretching it vertically. This happens only for a few functions. It does not happen when these transformations are applied to trigonometric functions.

The various transformations considered previously are summarized next for easy reference:

GRAPH TRANSFORMATIONS		
Vertical Shift $y = f(x) + k$	$k > 0$	Shifts graph of f up k units.
	$k < 0$	Shifts graph of f down $|k|$ units.
Horizontal Shift $y = f(x + h)$	$h > 0$	Shifts graph of f left h units.
	$h < 0$	Shifts graph of f right $|h|$ units.
Reflections		
$y = -f(x)$		Reflects graph of f through the x axis.
$y = f(-x)$		Reflects graph of f through the y axis.
Stretches and Shrinks		
$y = af(x)$	$a > 1$	Stretches graph of f vertically by multiplying each y value by a.
	$0 < a < 1$	Shrinks graph of f vertically by multiplying each y value by a.
$y = f(ax)$	$a > 1$	Shrinks graph of f horizontally by multiplying each x value by $1/a$.
	$0 < a < 1$	Stretches graph of f horizontally by multiplying each x value by $1/a$.

EXAMPLE 4

Combining Transformations

Use a sequence of transformations to describe the relationship between the graph of $f(x) = x^2$ and the graph of $g(x) = 2(x - 1)^2 - 3$. Illustrate the sequence graphically, starting with the graph of f and ending with the graph of g.

Solution

Here is one sequence that will describe the relationship between the graph of f and the graph of g:

1. Shift the graph of x^2 right 1 unit.

2. Stretch the graph of $(x - 1)^2$ vertically 2 units.

3. Shift the graph of $2(x - 1)^2$ down 3 units.

This sequence of transformations can be expressed more compactly as

$$x^2 \xrightarrow{(1)} (x - 1)^2 \xrightarrow{(2)} 2(x - 1)^2 \xrightarrow{(3)} 2(x - 1)^2 - 3$$

The graphs are shown in Figure 8.

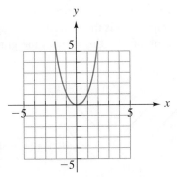

(a) $y = x^2$

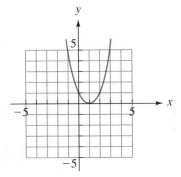

(b) $y = (x - 1)^2$

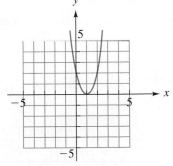

(c) $y = 2(x - 1)^2$

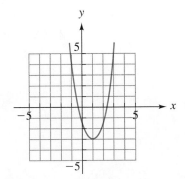

(d) $y = 2(x - 1)^2 - 3 = g(x)$

FIGURE 8
A sequence of transformations

Matched Problem 4   Repeat Example 4 for $g(x) = -3(x + 2)^2 + 4$.   ■

Refer to Example 4. Sequences of transformations that change one graph into another are not always unique. Here is a second sequence that also provides a solution to Example 4:

$$x^2 \rightarrow 2x^2 \rightarrow 2x^2 - 3 \rightarrow 2(x - 1)^2 - 3$$

Answers to Matched Problems

1. (A) The graph of g is the same as the graph of f shifted downward 4 units, and the graph of h is the same as the graph of f shifted upward 2 units.

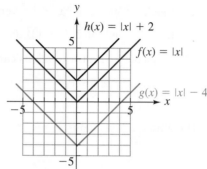

(B) The graph of G is the same as the graph of f shifted 1 unit to the left, and the graph of H is the same as the graph of f shifted 3 units to the right.

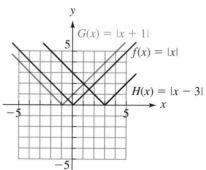

2. (A) The graph of g is the reflection through the x axis of the graph of f.

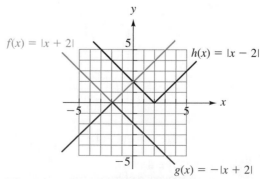

(B) The graph of h is the reflection through the y axis of the graph of f.

3. (A) The graph of g can be obtained from the graph of f by multiplying each y value by 2. This stretches the graph of f vertically. The graph of h can be obtained from the graph of f by multiplying each y value by $\frac{1}{2}$. This shrinks the graph of f vertically.

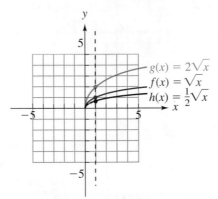

(B) The graph of G can be obtained from the graph of f by multiplying each x value by $\frac{1}{2}$. This shrinks the graph of f horizontally. The graph of H can be obtained from the graph of f by multiplying each x value by 2. This stretches the graph of f horizontally.

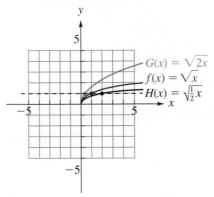

4. One sequence of transformations that will transform the graph of f into the graph of g is: Shift left 2 units, stretch vertically by a factor of 3, reflect through the x axis, and shift up 4 units.

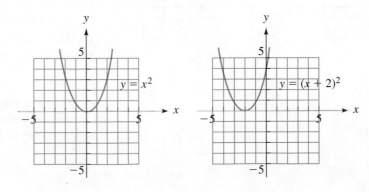

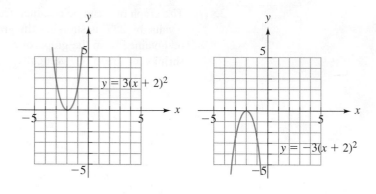

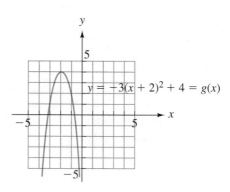

EXERCISE B.2

A *Graph each of the functions in Problems 1–16 using the graphs of functions f and g that follow.*

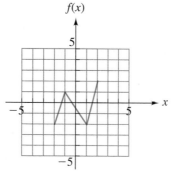

1. $y = f(x) - 2$ **2.** $y = g(x) + 1$

3. $y = f(x - 2)$ **4.** $y = g(x + 1)$

5. $y = g(x + 3)$ **6.** $y = f(x - 3)$

7. $y = g(x) + 3$ **8.** $y = f(x) - 3$

9. $y = -f(x)$ **10.** $y = -g(x)$

11. $y = f(-x)$ **12.** $y = g(-x)$

13. $y = 0.5\, g(x)$ **14.** $y = 2\, f(x)$

15. $y = g(0.5x)$ **16.** $y = f(2x)$

B *In Problems 17–24, indicate how the graph of each function is related to the graph of one of the six basic functions in Figure 1. Sketch a graph of each function.*

17. $g(x) = -|x - 3|$ **18.** $h(x) = -|x + 5|$

19. $f(x) = (x + 3)^2 + 3$ **20.** $m(x) = -(x - 2)^2 - 4$

21. $f(x) = 5 - \sqrt{x}$ **22.** $g(x) = -4 + \sqrt[3]{x}$

23. $h(x) = -\sqrt[3]{2x}$ **24.** $m(x) = -\sqrt{\frac{1}{2}x}$

Each graph in Problems 25–32 is the result of applying a sequence of transformations to the graph of one of the six basic functions in Figure 1 on page 506. Identify the basic function and describe the transformation verbally. Write an equation for the given graph.

25.

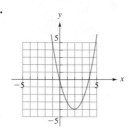

26.

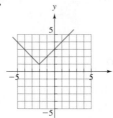

27.

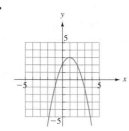

28.

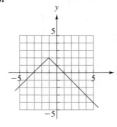

29.

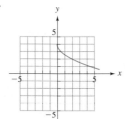

30.

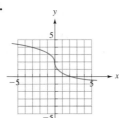

31.

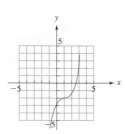

32.

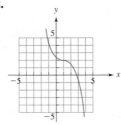

In Problems 33–38, the graph of the function g is formed by applying the indicated sequence of transformations to the given function f. Find an equation for the function g and graph g using $-5 \leq x \leq 5$ *and* $-5 \leq y \leq 5$.

33. The graph of $f(x) = \sqrt{x}$ is shifted 3 units to the left, stretched vertically by a factor of 2, and shifted 4 units down.

34. The graph of $f(x) = |x|$ is stretched vertically by a factor of 3, shifted 2 units to the left, and shifted 5 units down.

35. The graph of $f(x) = |x|$ is shrunk vertically by a factor of $\frac{1}{2}$, reflected through the x axis, shifted 1 unit to the right, and shifted 4 units up.

36. The graph of $f(x) = |x|$ is shifted 2 units to the right, reflected through the y axis, shrunk vertically by a factor of $\frac{3}{4}$, and shifted 3 units down.

37. The graph of $f(x) = x^3$ is reflected through the y axis, shifted 2 units to the left, and shifted down 1 unit.

38. The graph of $f(x) = x^2$ is reflected through the x axis, shifted 2 units to the right, reflected through the y axis, and shifted 4 units up.

Each graph in Problems 39–44 is the result of applying a sequence of transformations to the graph of one of the six basic functions in Figure 1 on page 506. Identify the basic function and describe the transformation verbally. Write an equation for the given graph.

39.

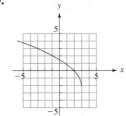

40.

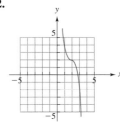

41.

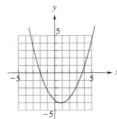

42.

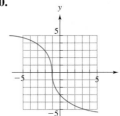

43.

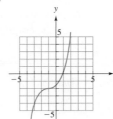

44.

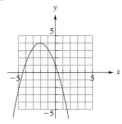

B.3   Inverse Functions

- One-to-One Functions
- Inverse Functions
- Geometric Relationship

In this section we develop techniques for determining whether the *inverse function* exists, some general properties of inverse functions, and methods for finding the rule of correspondence that defines the inverse function.

An inverse function is formed by reversing the correspondence in a given function. However, the reverse correspondence for a given function may or may not specify a function. As we will see, only functions that are *one-to-one* have inverse functions.

Many important functions are formed as inverses of existing functions. Logarithmic functions, for example, are inverses of exponential functions, and inverse trigonometric functions are inverses of trigonometric functions with restricted domains.

■ One-to-One Functions

A **one-to-one correspondence** exists between two sets if each element of the first set corresponds to exactly one element of the second set and each element of the second set corresponds to exactly one element of the first set.

ONE-TO-ONE FUNCTION

A function f is **one-to-one** if each element in the range corresponds to exactly one element from the domain. (Since f is a function, we already know that each element in the domain corresponds to exactly one element in the range.)

Thus, if a function f is one-to-one, then there exists a one-to-one correspondence between the domain elements and the range elements of f. To illustrate these concepts, consider the two functions f and g given in the following charts:

Function f	
Domain	Range
0 ———→ 0	
1 ———→ 1	
2 ———→ 8	

Function g	
Domain	Range
0 ———→ 0	
−2 ⟍	
2 ⟋ ———→ 4	

Function f is one-to-one, since each range element corresponds to exactly one domain element. Function g is not one-to-one, since the range element 4 corresponds to two domain elements, -2 and 2.

Geometrically, if a horizontal line intersects the graph of the function in two or more points, then the function is not one-to-one (see Fig. 1a). However, if each horizontal line intersects the graph of the function in at most one point, then the function is one-to-one (see Fig. 1b).

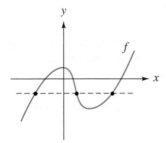 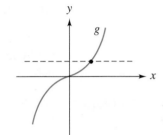

(a) Function f is not one-to-one   (b) Function g is one-to-one

FIGURE 1
Graphs of functions and the
horizontal-line test

These observations form the basis for the **horizontal-line test.**

HORIZONTAL-LINE TEST

A function is one-to-one if and only if each horizontal line intersects the graph of the function in at most one point.

EXAMPLE 1

Determining Whether a Function Is One-to-One

Graph each function and determine which is one-to-one by the horizontal-line test.

(A) $f(x) = x^2$ (B) $g(x) = x^2, \ x \geq 0$

Solution   (A) We graph the function f (Fig. 2 on the next page) and note that it fails the horizontal-line test. Therefore, function f is not one-to-one.
(B) We graph the function g (which is function f with a restricted domain) and find that g passes the horizontal-line test (Fig. 3 on the next page). Therefore, function g is one-to-one.

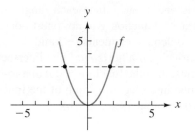

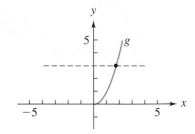

FIGURE 2 **FIGURE 3** ■

Example 1 illustrates an important point: A function may not be one-to-one, but by restricting the domain it can be made one-to-one. This is exactly what is done to the trigonometric functions to form the inverse trigonometric functions.

Matched Problem 1 Graph each function and determine which is one-to-one by the horizontal-line test.

(A) $f(x) = (x - 1)^2$ (B) $g(x) = (x - 1)^2$, $x \geq 1$ ■

■ Inverse Functions

As we mentioned at the beginning of this section, we are interested in forming new functions by reversing the correspondence in a given function—that is, by reversing all the ordered pairs in the given function. The concept of a one-to-one function plays a critical role in this process. If we reverse all the ordered pairs in a function that is not one-to-one, the resulting set does not define a function. For example, reversing the ordered pairs in the function.

$$g = \{(-2, 4), (0, 0), (2, 4)\} \qquad \textit{Function g is not one-to-one.}$$

produces the set

$$h = \{(4, -2), (0, 0), (4, 2)\} \qquad \textit{Set h is not a function.}$$

which does not define a function. However, reversing all the ordered pairs in a one-to-one function f always produces a new function, called the *inverse function,* and denoted by the symbol f^{-1}.* For example, reversing the ordered pairs in the function

$$f = \{(0, 2), (1, 3), (2, 4)\} \qquad \textit{Function f is one-to-one.}$$

produces the inverse function

$$f^{-1} = \{(2, 0), (3, 1), (4, 2)\} \qquad \textit{Set } f^{-1} \textit{ is a function.}$$

Furthermore, we see that

$$\text{Domain of } f = \{0, 1, 2\} = \text{Range of } f^{-1}$$
$$\text{Range of } f = \{2, 3, 4\} = \text{Domain of } f^{-1}$$

---

* *Note:* f^{-1} is a special symbol used to represent the inverse of the function f. It does *not* mean $1/f$.

In other words, reversing all the ordered pairs also reverses the domain and range. This discussion is summarized in the following definition:

INVERSE FUNCTION

The **inverse** of a one-to-one function f, denoted by f^{-1}, is the function formed by reversing all the ordered pairs in the function f. Symbolically,

$$f^{-1} = \{(b, a) | (a, b) \text{ is an element of } f\}$$

Domain of f^{-1} = Range of f

Range of f^{-1} = Domain of f

Immediate consequences of the definition of an inverse function are the function–inverse function identities:

FUNCTION–INVERSE FUNCTION IDENTITIES

If f is a one-to-one function, then f^{-1} exists and

$$f^{-1}(f(x)) = x \quad \text{for all } x \text{ in the domain of } f$$

and

$$f(f^{-1}(x)) = x \text{ for all } x \text{ in the domain of } f^{-1}$$

These identities are illustrated schematically in Figure 4.

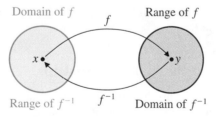

if the function f is one-to-one and if f maps x into y, then f^{-1} maps y back into x.

FIGURE 4
Function–inverse function identities

EXAMPLE 2

Finding the Inverse of a Function

For the one-to-one function $f(x) = 2x - 3$, find $f^{-1}(x)$, and check by showing that $f(f^{-1}(x)) = x$ and $f^{-1}(f(x)) = x$.

Solution   For both f and f^{-1} we keep x as the independent variable and y as the dependent variable.

Step 1 Replace $f(x)$ with y:

$$f: y = 2x - 3 \qquad \text{x is independent.}$$

Step 2 Interchange the variables x and y to form f^{-1}:

$$f^{-1}: x = 2y - 3 \qquad \text{x is independent.}$$

Step 3 Solve the equation for f^{-1} for y in terms of x:

$$x = 2y - 3$$
$$y = \frac{x + 3}{2}$$

Step 4 Replace y with $f^{-1}(x)$:

$$f^{-1}(x) = \frac{x + 3}{2}$$

Step 5 Check by showing that $f(f^{-1}(x)) = x$ and $f^{-1}(f(x)) = x$.

$$f(f^{-1}(x)) = 2(f^{-1}(x)) - 3 \qquad\qquad f^{-1}(f(x)) = \frac{f(x) + 3}{2}$$

$$= 2\left(\frac{x + 3}{2}\right) - 3 \qquad\qquad = \frac{(2x - 3) + 3}{2}$$

$$= x + 3 - 3 = x \qquad\qquad = \frac{2x}{2} = x \qquad ∎$$

Matched Problem 2 For the one-to-one function $g(x) = 3x + 2$, find $g^{-1}(x)$, and check by showing that $g(g^{-1}(x)) = x$ and $g^{-1}(g(x)) = x$. ∎

■ Geometric Relationship

We conclude this discussion of inverse functions by observing an important relationship between the graph of a one-to-one function and its inverse. As an example, let f be the one-to-one function

$$f(x) = 2x - 2 \tag{1}$$

Its inverse is

$$f^{-1}(x) = \frac{x + 2}{2} \tag{2}$$

Any ordered pair of numbers that satisfies (1), when reversed in order, will satisfy (2). For example, $(4, 6)$ satisfies (1) and $(6, 4)$ satisfies (2). (Check this.) The graphs of f and f^{-1} are given in Figure 5.

FIGURE 5
Symmetry property of the graphs
of a function and its inverse

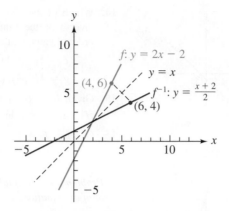

Notice that we sketched the line $y = x$ in Figure 5 to show that if we fold the paper along this line, then the graphs of f and f^{-1} will match. Actually, we can graph f^{-1} by drawing f with wet ink and folding the paper along $y = x$ before the ink dries; f will print f^{-1}. [To prove this, we need to show that the line $y = x$ is the perpendicular bisector of the line segment joining (a, b) to (b, a).]

Knowing that the graphs of f and f^{-1} are symmetric relative to the line $y = x$ makes it easy to graph f^{-1} if f is known, and vice versa.

Answers to Matched Problems

1. (A) f is not one-to-one

(B) g is one-to-one

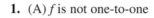

2. $g^{-1}(x) = \dfrac{x - 2}{3}$

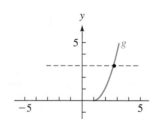

EXERCISE B.3

In Problems 1–12, indicate which functions are one-to-one.

1.

Domain	Range
−2	−4
−1	−2
0	0
1	2
2	4

2.

Domain	Range
−2	
−1	−3
0	0
1	
2	5

3.

Domain	Range
0	
1	
2	9
3	
4	

4.

Domain	Range
0	5
1	3
2	1
3	2
4	4

5.

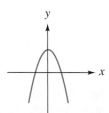

6.

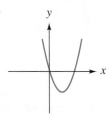

7.

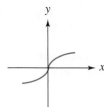

8.

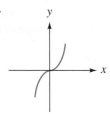

9.

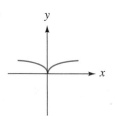

10.

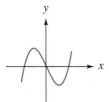

11.

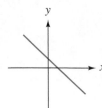

12.

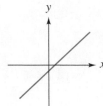

13. Which of the given functions is one-to-one? Write the inverse of the function that is one-to-one as a set of ordered pairs, and indicate its domain and range.

$$f = \{(-1, 0), (1, 1), (2, 0)\}$$
$$g = \{(-2, -8), (1, 1), (2, 8)\}$$

14. Which of the given functions is one-to-one? Write the inverse of the function that is one-to-one as a set of ordered pairs, and indicate its domain and range.

$$f = \{(-2, 4), (0, 0), (2, 4)\}$$
$$g = \{(9, 3), (4, 2), (1, 1)\}$$

15. Given $H = \{(-1, 0.5), (0, 1), (1, 2), (2, 4)\}$, graph H, H^{-1}, and $y = x$ in the same coordinate system.

16. Given $F = \{(-5, 0), (-2, 1), (0, 2), (1, 4), (2, 7)\}$, graph F, F^{-1}, and $y = x$ in the same coordinate system.

In Problems 17–20, find the inverse for each function in the form of an equation.

17. $f(x) = 2x - 7$

18. $g(x) = \dfrac{x}{2} + 1$

19. $h(x) = \dfrac{x + 3}{3}$

20. $f(x) = \dfrac{x - 2}{3}$

In Problems 21 and 22, graph the indicated function, its inverse, and $y = x$ in the same coordinate system.

21. f and f^{-1} in Problem 17

22. g and g^{-1} in Problem 18

23. For $f(x) = 2x - 7$, find $f^{-1}(x)$ and $f^{-1}(3)$.

24. For $g(x) = (x/2) + 1$, find $g^{-1}(x)$ and $g^{-1}(-3)$.

25. For $h(x) = (x/3) + 1$, find $h^{-1}(x)$ and $h^{-1}(2)$.

26. For $m(x) = 3x + 2$, find $m^{-1}(x)$ and $m^{-1}(5)$.

27. Find $f(f^{-1}(4))$ for Problem 23.

28. Find $g^{-1}(g(2))$ for Problem 24.

29. Find $h^{-1}(h(x))$ for Problem 25.

30. Find $m(m^{-1}(x))$ for Problem 26.

31. Find $h(h^{-1}(x))$ for Problem 25.

32. Find $m^{-1}(m(x))$ for Problem 26.

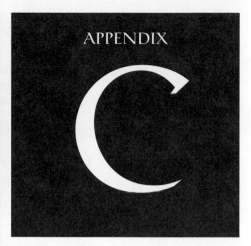

Plane Geometry: Some Useful Facts

C.1 Lines and Angles

C.2 Triangles

C.3 Quadrilaterals

C.4 Circles

The following four sections include a brief list of plane geometry facts that are of particular use in studying trigonometry. They are grouped together for convenient reference.

C.1   Lines and Angles

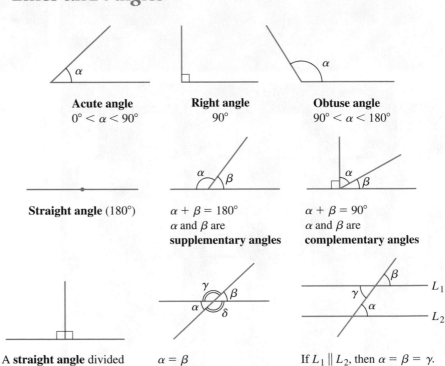

Acute angle
$0° < \alpha < 90°$

Right angle
$90°$

Obtuse angle
$90° < \alpha < 180°$

Straight angle $(180°)$

$\alpha + \beta = 180°$
α and β are
supplementary angles

$\alpha + \beta = 90°$
α and β are
complementary angles

A **straight angle** divided
into equal parts forms
two **right angles**.

$\alpha = \beta$
$\gamma = \delta$
$\alpha + \delta = \beta + \gamma = 180°$
$\alpha + \delta + \beta + \gamma = 360°$

If $L_1 \parallel L_2$, then $\alpha = \beta = \gamma$.
If $\alpha = \beta = \gamma$, then $L_1 \parallel L_2$.

C.2 Triangles

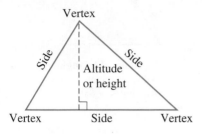

Vertex
Side
Side
Altitude or height
Vertex
Side
Vertex

Triangle

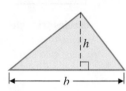

h
b

Area: $A = \frac{1}{2}bh$

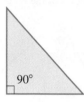

90°

Right triangle
One right angle

(a) **Acute triangle**
All acute angles

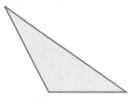

(b) **Obtuse triangle**
One obtuse angle

Oblique triangles
No right angles

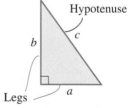

Hypotenuse
b
c
Legs
a

$c^2 = a^2 + b^2$
Pythagorean theorem

β
α

$\alpha + \beta = 90°$

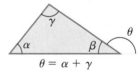

γ
α
β
θ

$\theta = \alpha + \gamma$

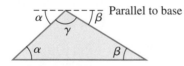

Parallel to base
α
β
γ
α
β

$\alpha + \beta + \gamma = 180°$
The sum of angle measures of all angles in a triangle is 180°.

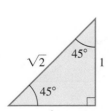

$\sqrt{2}$
45°
1
45°
1

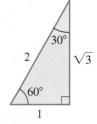

30°
2
$\sqrt{3}$
60°
1

Special triangles

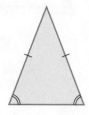

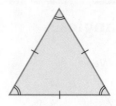

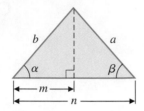

Isosceles triangle
At least two equal sides
At least two equal angles

Equilateral triangle
All sides equal
All angles equal

If $a = b$, then $\alpha = \beta$ and $m = n/2$.
If $\alpha = \beta$, then $a = b$ and $m = n/2$.

Similar triangles
Two triangles are similar if two angles of one
triangle are equal to two angles of the other.

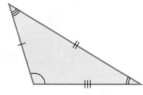

$$\alpha = \alpha', \beta = \beta', \gamma = \gamma'$$

$$\frac{a}{a'} = \frac{b}{b'} = \frac{c}{c'}$$

Euclid's Theorem

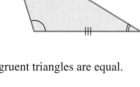

Congruent triangles
Corresponding parts of congruent triangles are equal.

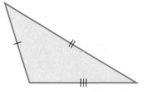

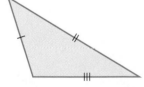

SSS
If three sides of one triangle are equal to the corresponding
sides of another triangle, the two triangles are congruent.

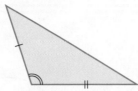 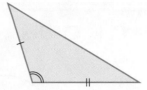

SAS

If two sides and the included angle of one triangle are equal to the corresponding parts of another triangle, the two triangles are congruent.

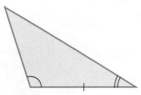 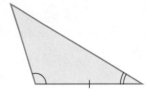

ASA

If two angles and the included side of one triangle are equal to the corresponding parts of another triangle, the two triangles are congruent.

C.3 Quadrilaterals

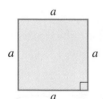 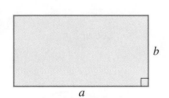 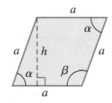 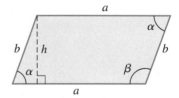

Square
$A = a^2$
$P = 4a$

Rectangle
$A = ab$
$P = 2a + 2b$

Rhombus
Opposite sides are parallel.
$A = ah$
$P = 4a$
$\alpha + \beta = 180°$

Parallelogram
Opposite sides are parallel.
$A = ah$
$P = 2a + 2b$
$\alpha + \beta = 180°$

C.4 Circles

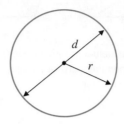

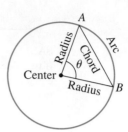

$r = $ Radius
$d = $ Diameter
$d = 2r$
$A = \pi r^2$ (Area)
$C = 2\pi r = \pi d$ (Circumference)
$C/d = \pi$

$\widehat{AB} = $ Arc AB
$AB = $ Chord AB
$\angle \theta$ subtends AB
$\angle \theta$ subtends $\widehat{AB}$
$\widehat{AB}$ subtends $\angle \theta$
AB subtends $\angle \theta$

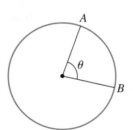

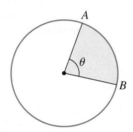

$$\frac{\widehat{AB}}{C} = \frac{\theta}{360°}$$

($C = $ Circumference)

Sector

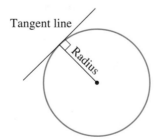

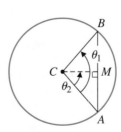

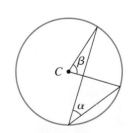

A radius of a circle is $\perp$ to a tangent line at the point of tangency.

If $CM \perp AB$, then

$$AM = \frac{1}{2}AB$$

$$\theta_2 = \frac{1}{2}\theta_1$$

$$\beta = 2\alpha$$

SELECTED ANSWERS

CHAPTER 1

Exercise 1.1

1. $180°, 60°, 135°, 210°$ **3.** $\dfrac{1}{8}, \dfrac{5}{12}, \dfrac{3}{4}$ **5.** Acute
7. Right **9.** Obtuse **11.** None **15.** $25.365°$
17. $11.135°$ **19.** $195.469°$ **21.** $15°7'30''$
23. $79°12'4''$ **25.** $159°38'20''$ **27.** $47°33'41''$
29. $\alpha > \beta$ **31.** $\alpha = \beta$ **33.** $\alpha < \beta$ **35.** $110°18'4''$
37. $22°22'31''$ **39.** 0 **41.** 100 cm **43.** 450 km
45. $250{,}000$ mi **47.** $1{,}440$ cm **49.** 262 cm^2
51. $71.0°$ **53.** 5.58 mm **55.** 11.5 mm **57.** 680 mi
59. 553 mi **61.** 590 nautical mi **63.** 480 nautical mi
65. (A) 70 ft
  (B) The arc length of a circular sector is very close to
  the chord length if the central angle of the sector is
  small and the radius of the sector is large, which is
  the case in this problem. The arc length s is easily
  found using the formula $\dfrac{s}{2\pi r} = \dfrac{\theta}{360}$.
67. $0.056°$ **69.** $758{,}000$ miles **71.** 87 miles

Exercise 1.2

1. The measures of the third angle in each triangle are the
same, since the sum of the measures of the three angles in
any triangle is $180°$. (Subtract the sum of the measures of
the two given angles in each triangle 3. 4. 8 from $180°$.)
5. $b' = 21$ **7.** $c = 90$ **9.** $c' = 2.9$
11. Two similar triangles can have equal sides only if they
are congruent, that is, if the two triangles coincide when
one is moved on top of the other.
13. $a = 24$ in., $c = 56$ in. **15.** $b = 50$ m, $c = 55$ m
17. $a = 1.2 \times 10^9$ yd, $c = 2.7 \times 10^9$ yd
19. $a = 3.6 \times 10^{-5}$ mm, $b = 7.6 \times 10^{-5}$ mm
21. $c \approx 108$ ft **23.** 19.5 ft **25.** 63 ft **27.** 28 ft
29. 34 ft **31.** 1.1 km

33. (A) Triangles *PAC, FBC, ACP'*, and *ABF'* are all right
  triangles. Angles *APC* and *BFC* are equal, and
  angles *CP'A* and *BF'A* are equal—alternate interior
  angles of parallel lines cut by a transversal are
  equal (see Appendix C.1). Thus, triangles *PAC* and
  FBC are similar, and triangles *ACP'* and *ABF'* are
  similar.
  (B) Start with the proportions $AC/PA = BC/BF$ and
  $AC/CP' = AB/BF'$.
  (C) Add the two equations in part (B) together and
  divide both sides of the result by $(h + h')$.
  (D) 50.847 mm
35. 150 ft **37.** $x = 16$, $y = 20$

Exercise 1.3

1. a/c **3.** b/a **5.** c/a **7.** $\sin\theta$ **9.** $\tan\theta$ **11.** $\csc\theta$
13. 0.569 **15.** 0.572 **17.** 1.350 **19.** 0.115 **21.** 0.431
23. 1.012 **25.** $60°$ **27.** $54.34°$ **29.** $26°50'$ **31.** $70°2'$
33. The triangle is uniquely determined. Angle α can be
found using $\tan\alpha = a/b$; angle $\beta = 90° - \alpha$. The
hypotenuse c can be found using the Pythagorean
theorem or by using $\sin\alpha = b/c$.
35. The triangle is not uniquely determined. In fact, there are
infinitely many triangles of different sizes with the same
acute angles—all are similar to each other.
37. $90° - \theta = 31°20'$, $a = 7.80$ mm, $b = 12.8$ mm
39. $90° - \theta = 6.3°$, $a = 0.354$ km, $c = 3.23$ km
41. $90° - \theta = 18.5°$, $a = 4.28$ in., $c = 13.5$ in.
43. $\theta = 28°30'$, $90° - \theta = 61°30'$, $a = 118$ ft
45. $\theta = 50.7°$, $90° - \theta = 39.3°$, $c = 171$ mi
47. The calculator was accidentally set in radian mode.
Changing the mode to degree,
  $a = 235\sin(14.1) = 57.2$ m and
  $b = 235\cos(14.1) = 228$ m.
53. $90° - \theta = 6°48'$, $b = 199.8$ mi, $c = 201.2$ mi
55. $\theta = 37°6'$, $90° - \theta = 52°54'$, $c = 70.27$ cm
57. $\theta = 44.11°$, $90° - \theta = 45.89°$, $a = 36.17$ cm

63. (A) $\sin \theta = \dfrac{AD}{OD} = \dfrac{AD}{1} = AD$

(B) $\tan \theta = \dfrac{AD}{OA} = \dfrac{DC}{OD} = \dfrac{DC}{1} = DC \ (\angle OED = \theta)$

(C) $\csc \theta = \dfrac{OE}{OD} = \dfrac{OE}{1} = OE$

65. (A) $\sin \theta$ approaches 1
(B) $\tan \theta$ increases without bound
(C) $\csc \theta$ approaches 1

67. (A) $\cos \theta$ approaches 1
(B) $\cot \theta$ increases without bound
(C) $\sec \theta$ approaches 1

69. $6\sqrt{3}\,\text{ft}^2$ **71.** 101.4 in.^2

Exercise 1.4

1. 7.0 m   **3.** 45 ft   **5.** 25°   **7.** 23 ft   **9.** 78 m
11. 33 m   **13.** 211 m   **15.** 1.1 km
17. 29,400 m, or 29.4 km   **19.** 22°
21. (A) 5.1 ft   (B) 2.6 ft
23. 134 ft   **25.** 8.4 ft, 21 ft
27. (A) $\cos \alpha = \dfrac{r}{r + h}$ (B) $r = \dfrac{h \cos \alpha}{1 - \cos \alpha}$
(C) 3,960 mi
29. 8.4 mi   **31.** 19,600 mi
33. (A) The lifeguard can run faster than he can swim, so
he should run along the beach first before entering
the water. [Parts (D) and (E) suggest how far the
lifeguard should run before swimming to get to the
swimmer in the least time.]
(B) $T = \dfrac{d - c \cot \theta}{p} + \dfrac{c \csc \theta}{q}$
(C) $T = 119.97$ sec
(D) T decreases, then increases as θ goes from 55°
to 85°; T has a minimum value of 116.66 sec
when $\theta = 70°$.
(E) $d = 352$ m
35. (A) Since the angle θ determines the length of pipe to
be laid on land and in the water, it appears that the
total cost of the pipeline depends on θ.
(B) $C = 160,000 \sec \theta + 20,000(10 - 4 \tan \theta)$
(C) \$344,200
(D) As θ increases from 15° to 45°, C decreases and
then increases; C has a minimum value of
\$338,600 when $\theta = 30°$.
(E) 4.62 mi in the water and 7.69 mi on land
37. 0.97 km   **39.** 120 ft   **41.** 386 ft apart; 484 ft high

43. 25 ft   **45.** $g = 32.0 \text{ ft/sec}^2$ **47.** 3.5 m   **49.** 9.8 yd
51. 5.69 cm   **53.** $\sin \theta = \dfrac{3}{5}$ **55.** 7.0 ft

Chapter 1 Review Exercise

1. 7,794″   [1.1]   **2.** 45°   [1.1]   **3.** 16,000   [1.2]
4. 36.33°   [1.1]
5. An angle of degree measure 1 is an angle formed by
rotating the terminal side of the angle $\frac{1}{360}$ of a complete
revolution in a counterclockwise direction.   [1.1]
6. All three are similar because all three have equal
angles.   [1.2]
7. No. Similar triangles have equal angles.   [1.2]
8. The sum of all the angles in a triangle is 180°. Two
obtuse angles would add up to more than 180°.   [1.2]
9. 720 ft   [1.2]
10. (A) b/c (B) c/a (C) b/a (D) c/b (E) a/c
(F) a/b [1.3]
11. $90° - \theta = 54.8°$, $a = 16.5 \text{ cm}$, $b = 11.6 \text{ cm}$ [1.3]
12. 144°   [1.1]   **13.** 4.19 in.   [1.1]
14. 27.25° is larger.   [1.1]
15. (A) 67.709°   (B) 129°19′1″   [1.1]
16. (A) 65°41′52″   (B) 327°48′27″   [1.1]
17. $7.1 \times 10^{-6} \text{ mm}$ [1.2]
18. (A) $\cos \theta$ (B) $\tan \theta$ (C) $\sin \theta$ (D) $\sec \theta$
(E) $\csc \theta$ (F) $\cot \theta$ [1.3]
19. Two right triangles having an acute angle of one
equal to an acute angle of the other are similar, and
corresponding sides of similar triangles are
proportional.   [1.3]
20. $90° - \theta = 27°40′$, $b = 7.63 \times 10^{-8} \text{ m}$,
$c = 8.61 \times 10^{-8} \text{ m}$ [1.3]
21. (A) 68.17°   (B) 68°10′   (C) 3°12′16″   [1.3]
22. Window (a) is in radian mode; window (b) is in degree
mode.   [1.3]
23. $\theta = 40.3°$, $90° - \theta = 49.7°$, $c = 20.6 \text{ mm}$ [1.3]
24. $\theta = 40°20′$, $90° - \theta = 49°40′$ [1.3]
25. 8.7 ft   [1.2]   **26.** 940 ft   [1.1]   **27.** 107 ft^2 [1.1]
28. $\theta = 66°17′$, $a = 93.56 \text{ km}$, $b = 213.0 \text{ km}$ [1.3]
29. $\theta = 60.28°$, $90° - \theta = 29.72°$, $b = 4,241 \text{ m}$ [1.3]
30. 1.0853   [1.3]   **31.** 8.8 ft   [1.2]   **32.** 6 ft   [1.2]
33. 5.6 ft   [1.4]   **34.** 71 ft   [1.4]
35. 24.5 ft; 24.1 ft   [1.4]   **36.** 2.3°; 0.07 or 7%   [1.4]
37. 955 mi   [1.1]   **38.** 46°   [1.4]   **39.** 830 m   [1.4]
40. 1,240 m   [1.4]   **41.** 760 mph   [1.4]
42. (A) $\beta = 90° - \alpha$ (B) $r = h \tan \alpha$
(C) $H - h = (R - r) \cot \alpha$ [1.4]

43. (A) In both cases the ladder must get longer.
  (B) $L = 5 \csc \theta + 4 \sec \theta$.
  (C)

θ	25°	35°	45°	55°	65°	75°	85°
L	16.24	13.60	12.73	13.08	14.98	20.63	50.91

  (D) L decreases and then increases; L has a minimum value of 12.73 when $\theta = 45°$.
  (E) Make up another table for values of θ close to 45° and on either side of 45°. *[1.4]*

44. $AC = 93$ ft, $BC = 142$ ft *[1.4]* **45.** 28,768 sq ft *[1.3]*

CHAPTER 2
Exercise 2.1

3. $\pi/6, \pi/3, \pi/2, 2\pi/3, 5\pi/6, \pi, 7\pi/6, 4\pi/3, 3\pi/2,$
  $5\pi/3, 11\pi/6, 2\pi$

7.
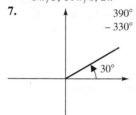
390°
−330°
30°

9.
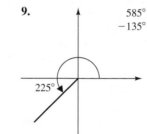
585°
−135°
225°

11.
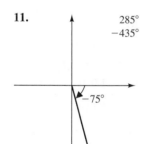
285°
−435°
−75°

13.
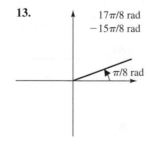
17π/8 rad
−15π/8 rad
π/8 rad

15.
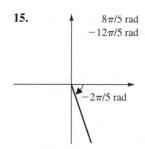
8π/5 rad
−12π/5 rad
−2π/5 rad

17.
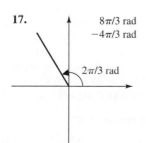
8π/3 rad
−4π/3 rad
2π/3 rad

21. $\pi/60$ rad ≈ 0.05236 rad   **23.** $11\pi/18$ rad ≈ 1.920 rad
25. $\pi/5$ rad ≈ 0.6283 rad   **27.** $(360/\pi)° \approx 114.6°$
29. $(108/\pi)° \approx 34.38°$ **31.** $(360/7)° \approx 51.43°$
33. (A) 28.648°   (B) 80.214°   (C) 355.234°
  (D) −262.988°
35. (A) 0.436 rad   (B) 2.862 rad   (C) 11.310 rad
  (D) −3.869 rad
37. (A) 0.4 rad; 22.9°   (B) 12 rad; 68.8°
  (C) 2.5 rad; 143.2°   (D) 4 rad; 229.2°
39. (A) 9.45 m   (B) 5.13 m
  (C) 4.57 m   (D) 15.27 m
41. Yes. Since $\theta_d = (180/\pi)\theta_r$, if θ_r is doubled, then θ_d also must be doubled (multiply both sides of the conversion equation by 2).
43. Since $\theta_r = s/r$, if s is held constant while r (the denominator) is doubled, then θ_r will be cut in half.
45. (A) 144 cm^2 (B) 31.4 cm^2 (C) 46.8 cm^2
  (D) 131.9 cm^2
47. Since $s = r\theta$ (θ in radian measure), $s = 1 \cdot m = m$.
49. II   **51.** III   **53.** II   **55.** III   **57.** III
59. 0.9795 rad   **61.** 26.5738°   **63.** 1.530 rad
65. 488.5714°   **67.** 1/3 rad   **69.** 62°
71. 8.9 km   **73.** 53 cm
75. $7\pi/12$ rad ≈ 1.83 rad   **77.** 12 cm   **79.** 19°
81. 256 cm   **83.** 1.4×10^6 km   **85.** 175 ft
87. 0.2 m; 7.9 in.   **89.** $\pi/26$ rad ≈ 0.12 rad
91. 6,800 mi   **93.** 31 ft
95. (A) Since one revolution corresponds to 2π radians, n revolutions corresponds to $2\pi n$ radians.
  (B) No. Radian measure is independent of the size of the circle used; so, it is independent of the size of the wheel used.
  (C) 31.42 rad; 22.62 rad
97. 6.5 revolutions; 40.8 rad   **99.** 859°   **101.** 384 in.

Exercise 2.2

5. 8.4 cm/min   **7.** 240 ft/hr   **9.** 7.8 rad/hr
11. 2.14 rad/min   **13.** 168 rad/hr
15. 6.29 rad/min   **19.** 75 m/sec
21. 4.65 rad/hr   **23.** 223.5 rad/sec or 35.6 rps
25. $\pi/4,380$ rad/hr; 66,700 mph
27. (A) 0.634 rad/hr   (B) 28,100 mph
29. 6,900 mph   **31.** 1.61 hr
33. (A) 1 rps corresponds to an angular velocity of 2π rad/sec, so that at the end of t sec, $\theta = 2\pi t$.
  (B) Using right triangle trigonometry, $a = 15 \tan \theta$. Then, substituting $\theta = 2\pi t$, $a = 15 \tan 2\pi t$.

(C) The speed of the light spot on the wall increases as t increases from 0.00 to 0.24. When $t = 0.25$, $a = 15\tan(\pi/2)$, which is not defined. The light has made one-quarter turn, and the spot is no longer on the wall.

t (sec)	0.00	0.04	0.08	0.12	0.16	0.20	0.24
a (ft)	0.00	3.85	8.25	14.09	23.64	46.17	238.42

35.   36.1 mph

Exercise 2.3

1. $\sin x = 4/5, \cos x = 3/5, \tan x = 4/3, \cot x = 3/4,$
$\sec x = 5/3, \csc x = 5/4$

3. $\sin x = 1/\sqrt{2}, \cos x = 1/\sqrt{2}, \tan x = 1, \cot x = 1,$
$\sec x = \sqrt{2}, \csc x = \sqrt{2}$

5. $\sin x = \sqrt{3}/2, \cos x = -1/2, \tan x = -\sqrt{3},$
$\cot x = -1/\sqrt{3}, \sec x = -2, \csc x = 2/\sqrt{3}$

7. $\sin x = 3/5, \cos x = 4/5, \tan x = 3/4, \cot x = 4/3,$
$\sec x = 5/4, \csc x = 5/3$

9. $\sin x = -24/25, \cos x = -7/25, \tan x = 24/7,$
$\cot x = 7/24, \sec x = -25/7, \csc x = -25/24$

11. $\sin x = 5/13, \cos x = -12/13, \tan x = -5/12,$
$\cot x = -12/5, \sec x = -13/12, \csc x = 13/5$

13. $\cos \theta = 4/5, \tan \theta = 3/4, \cot \theta = 4/3, \sec \theta = 5/4,$
$\csc \theta = 5/3$

15. $\sin \theta = -12/13, \tan \theta = -12/5, \cot \theta = -5/12,$
$\sec \theta = 13/5, \csc \theta = -13/12$

17. $\sin x = -3/\sqrt{13}, \cos x = -2/\sqrt{13}, \cot x = 2/3,$
$\sec x = -\sqrt{13}/2, \csc x = -\sqrt{13}/3$

19. $\sin x = 2/\sqrt{5}, \cos x = 1/\sqrt{5}, \tan x = 2, \sec x = \sqrt{5},$
$\csc x = \sqrt{5}/2$

21. $\sin x = -1/\sqrt{2}, \cos x = 1/\sqrt{2}, \tan x = -1,$
$\cot x = -1, \csc x = -\sqrt{2}$

23. $\sin x = -24/25, \cos x = -7/25, \tan x = 24/7,$
$\cot x = 7/24, \sec x = -25/7$

25. No. For all those values of x for which both are defined, $\cos x = 1/(\sec x)$; so, either both are positive or both are negative.

27. 0.8829   **29.** −0.2910   **31.** −1.530
33. −0.8829   **35.** −0.9004   **37.** 1.100

39. −0.9749   **41.** 1.851   **43.** −2.475

51. $\sin x = \frac{1}{2}, \cos x = \sqrt{3}/2, \tan x = 1/\sqrt{3}, \csc x = 2,$
$\sec x = 2/\sqrt{3}, \cot x = \sqrt{3}$

53. $\sin x = -\sqrt{3}/2, \cos x = \frac{1}{2}, \tan x = -\sqrt{3},$
$\csc x = -2/\sqrt{3}, \sec x = 2, \cot x = -1/\sqrt{3}$

55. I, IV   **57.** I, III   **59.**   I, IV   **61.** III, IV

63. II, IV   **65.** III, IV

67. $\cos x = -\sqrt{3}/2, \tan x = -1/\sqrt{3}, \cot x = -\sqrt{3},$
$\sec x = -2/\sqrt{3}, \csc x = 2$

69. $\sin x = -\sqrt{5}/3, \cos x = 2/3, \tan x = -\sqrt{5}/2,$
$\cot x = -2/\sqrt{5}, \csc x = -3/\sqrt{5}$

71. $\sin x = -1/2, \cos x = \sqrt{3}/2, \tan x = -1/\sqrt{3},$
$\sec x = 2/\sqrt{3}, \csc x = -2$

73. Yes, $\cos \alpha = \cos \beta$, since the terminal sides of these angles will coincide and the same point $P = (a, b) \neq (0, 0)$ can be chosen on the terminal sides; $\cos \alpha = a/r = \cos \beta$, where $r = \sqrt{a^2 + b^2} \neq 0$.

75. Use the reciprocal identity $\cot x = 1/(\tan x)$;
$\cot x = 1/(-2.18504) = -0.45766$.

77. 0.03179   **79.** 2.225   **81.** −14.82
83. −2.372   **85.** 0.2243   **87.** −0.08626
89. 1.212   **91.** 0.8838   **93.** −1.587

95. Tangent and secant; since $\tan x = b/a$ and $\sec x = r/a$, neither is defined when the terminal side of the angle lies along the positive or negative vertical axis, because a will be 0 (division by 0 is not defined).

97. (A) $x = 1.2$ rad
(B) $(a, b) = (5 \cos 1.2, 5 \sin 1.2) = (1.81, 4.66)$

99. (A) $x = 2$ rad
(B) $(a, b) = (1 \cos 2, 1 \sin 2) = (-0.416, 0.909)$

101. 3.22 units

103. $\sin x = -1/\sqrt{10}, \cos x = 3/\sqrt{10}, \cot x = -3,$
$\sec x = \sqrt{10}/3, \csc x = -\sqrt{10}$

105. $k, 0.94k, 0.77k, 0.50k, 0.17k$

107. Summer solstice: $E = 0.97k$; winter solstice:
$E = 0.45k$

109. (A)

n	6	10	100	1,000	10,000
A_n	2.59808	2.93893	3.13953	3.14157	3.14159

(B) The area of the circle is $A = \pi r^2 = \pi(1)^2 = \pi$, and A_n seems to approach π, the area of the circle, as n increases.

(C) No. An n-sided polygon is always a polygon, no matter the size of n, but the inscribed polygon can be made as close to the circle as you like by taking n sufficiently large.

113. $I = 24$ amperes
115. (A) $2.01; -0.14$ (B) $y = -0.75x + 3.74$

Exercise 2.4

9. 29.3°   **11.** 25.1°   **13.** 24.4°
15. The ball will appear higher than the real ball. The light path from the eye of a person standing on shore to the

real ball will bend downward at the water surface
because of refraction. But the ball will appear to lie on
the straight continuation of the light path from the eye
below the water surface.

17. $n_2 = 1.5$ **21.** 40 km/hr   **23.** 84°

25. 2.74×10^{10} cm/sec   **27.** −6°   **29.** 192 in.

Exercise 2.5

5. $\alpha = \theta = 60°$

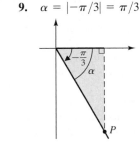

7. $\alpha = |-60°| = 60°$ **9.** $\alpha = |-\pi/3| = \pi/3$

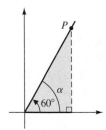

11. $\alpha = \pi - 3\pi/4 = \pi/4$

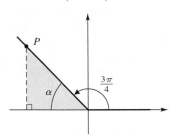

13. $\alpha = 210° - 180° = 30°$

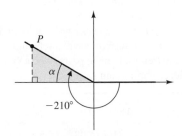

15. $\alpha = 5\pi/4 - \pi = \pi/4$

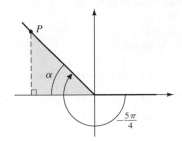

17. 0   **19.** 0   **21.** $\frac{1}{2}$ **23.** 1

25. $2/\sqrt{3}$,   **27.** −1

29. Not defined   **31.** $-1/\sqrt{2}$

33. $-\sqrt{3}$ **35.** $-1/\sqrt{2}$ **37.** $-\sqrt{3}$

39. Not defined   **41.** $-\sqrt{3}/2$ **43.** 1   **45.** $-\sqrt{2}$

47. The tangent function is not defined at $\theta = \pi/2$ and
$\theta = 3\pi/2$, because $\tan \theta = b/a$ and $a = 0$ for any
point on the vertical axis.

49. The cosecant function is not defined at $\theta = 0$, $\theta = \pi$,
and $\theta = 2\pi$, because $\csc \theta = r/b$ and $b = 0$ for any
point on the horizontal axis.

51. $\sin(-45°) = -1/\sqrt{2}$ **53.** $\tan(-\pi/3) = -\sqrt{3}$

55. (A) 30° (B) $\pi/6$ **57.** A) 120° (B) $2\pi/3$

59. (A) 120° (B) $2\pi/3$ **63.** All 0.9525

65. (A) Both 1.6   (B) Both −1.5 (C) Both 0.96

67. (A) Both −0.14 (B) Both 0.23 (C) Both 0.99

69. (A) 1.0   (B) 1.0   (C) 1.0

71. 1   **73.** csc x **75.** csc x **77.** cos x

79. 240°, 300°   **81.** $5\pi/6, 11\pi/6$ **83.** $3\pi/4$

85. $\pi/6$

87. (A) $x = 14, y = 7\sqrt{3}$

    (B) $x = 4/\sqrt{2}, y = 4/\sqrt{2}$

    (C) $x = 10/\sqrt{3}, y = 5/\sqrt{3}$

89. $s = \cos^{-1} 0.58064516 = 0.951 = \sin^{-1} 0.81415674$

91. (A) Identity (4)   (B) Identity (9)   (C) Identity (2)

93. 2π **95.** 1

99. $s_1 - 1, s_2 = 1.540302, s_3 = 1.570792, s_4 = 1.570796,$
$s_5 = 1.570796; \pi/2 \approx 1.570796$

Chapter 2 Review Exercise

1. (A) $\pi/3$ (B) $\pi/4$ (C) $\pi/2$ *[2.1]*

2. (A) 30° (B) 90° (C) 45° *[2.1]*

3. A central angle of radian measure 2 is an angle subtended
by an arc of twice the length of the radius. *[2.1]*

4. An angle of radian measure 1.5 is larger, since the corresponding degree measure of the angle would be approximately 85.94°. [2.1]
5. (A) 874.3° (B) −6.793 rad [2.1]
6. 185 ft/min [2.2] 7. 80 rad/hr [2.2]
8. $\sin\theta = \frac{3}{5}$; $\tan\theta = -\frac{3}{4}$ [2.3]
9. No, since $\csc x = 1/\sin x$, when one is positive so is the other. [2.3]
10. (A) 0.7355 (B) 1.085 [2.3]
11. (A) 0.9171 (B) 0.9099 [2.3]
12. (A) 0.9394 (B) 5.177 [2.3]
13. (A) $\alpha = 60°$

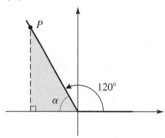

(B) $\alpha = \dfrac{\pi}{4}$ [2.5]

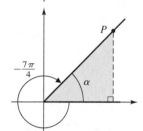

14. (A) $\sqrt{3}/2$ (B) $1/\sqrt{2}$ (C) 0 [2.5]
15. (A) (1, 0) (B) (−1, 0) (C) (0, 1) (D) (0, 1) (E) (−1, 0) (F) (0, −1) [2.3]
16. (A) 0 to 1 (B) 1 to 0 (C) 0 to −1 (D) −1 to 0 (E) 0 to 1 (F) 1 to 0 [2.3]
17. −11π/6 and 13π/6: When the terminal side of the angle is rotated any multiple of a complete revolution (2π rad) in either direction, the resulting angle will be coterminal with the original. In this case, for the restricted interval, this happens for π/6 ± 2π. [2.1]
18. 42° [2.1] 19. 6 cm [2.1] 20. 53π/45 [2.1]
21. 15° [2.1]
22. (A) −3.72 (B) 264.71° [2.1]
23. Yes, since $\theta_d = (180°/\pi\text{ rad})\theta_r$, if θ_r is tripled, then θ_d will also be tripled. [2.1]
24. No. For example, if $\alpha = \pi/6$ and $\beta = 5\pi/6$, α and β are not coterminal, but $\sin(\pi/6) = \sin(5\pi/6)$. [2.3]
25. Use the reciprocal identity:
$\csc x = 1/\sin x = 1/0.8594 = 1.1636.$ [2.3]

26. 0; not defined; 0; not defined [2.5]
27. (A) I (B) IV [2.3] 28. −0.992 [2.3]
29. −4.34 [2.3] 30. −1.30 [2.3]
31. 0.683 [2.3] 32. 0.284 [2.3]
33. −0.864 [2.3] 34. −√3/2 [2.5]
35. −1 [2.5] 36. −1 [2.5]
37. 0 [2.5] 38. √3/2 [2.5]
39. −2 [2.5] 40. −1 [2.5]
41. Not defined [2.5] 42. $\frac{1}{2}$ [2.5]
43. $\tan(-60°) = -\sqrt{3}$ [2.5]
44. 0.40724 [2.3] 45. −0.33884 [2.3]
46. 0.64692 [2.3] 47. 0.49639 [2.3]
48. $\cos\theta = \frac{3}{5}$; $\tan\theta = -\frac{4}{3}$ [2.3] 49. 7π/6 [2.5]
50. $\cos\theta = \sqrt{21}/5$, $\sec\theta = 5/\sqrt{21}$, $\csc\theta = -\frac{5}{2}$, $\tan\theta = -2/\sqrt{21}$, $\cot\theta = -\sqrt{21}/2$ [2.3]
51. 135°, 315° [2.5] 52. 5π/6, 7π/6 [2.5]
53. (A) 20.3 cm (B) 4.71 cm [2.1]
54. (A) 618 ft² (B) 1,880 ft² [2.1]
55. 215.6 mi [2.1] 56. 0.422 rad/sec [2.2]
57. A radial line from the axis of rotation sweeps out an angle at the rate of 12π rad/sec. [2.2]
58. All 0.754 [2.3]
59. (A) Both −0.871 (B) Both −1.40 (C) Both −3.38 [2.3]
60. (D) [2.5] 61. $\cos x$ [2.5] 62. $\cos x$ [2.5]
63. Since $P = (a, b)$ is moving clockwise, $x = -29.37$. By the definition of the trigonometric functions, $P = (\cos(-29.37), \sin(-29.37)) = (-0.4575, 0.8892)$. P lies in quadrant II, since a is negative and b is positive. [2.3]
64. The radian measure of a central angle θ subtended by an arc of length s is $\theta = s/r$, where r is the radius of the circle. In this case, $\theta = 1.3/1 = 1.3$ rad. [2.1]
65. Since $\cot x = 1/(\tan x)$ and $\csc x = 1/(\sin x)$, and $\sin(k\pi) = \tan(k\pi) = 0$ for all integers k, $\cot x$ and $\csc x$ are not defined for these values. [2.3]
66. $x = 4\pi/3$ [2.5] 67. $s = 0.8905$ unit [2.5]
68. 57 m [2.1] 69. 5.74 units [2.3]
70. 200 rad, $100/\pi \approx 31.8$ revolutions [2.1]
71. 7.5 rev, 15 rev [2.1] 72. 62 rad/sec [2.2]
73. 16,400 mph [2.2] 74. −17.6 amperes [2.3]
75. (A) $L = 10\csc\theta + 2\sec\theta$
(B) As θ decreases to 0 rad, L increases without bound; as θ increases to $\pi/2$, L increases without bound. Between these extremes, there appears to be a value of θ that produces a minimum L.
(C)

θ rad	0.70	0.80	0.90	1.00
L ft	18.14	16.81	15.98	15.59
θ rad	1.10	1.20	1.30	
L ft	15.63	16.25	17.85	

(D) $L = 15.59$ ft for $\theta = 1.00$ rad   [2.3]

76. $\beta = 23.3°$ [2.4]   **77.** $\alpha = 41.1°$ [2.4]

78. 11 mph [2.4]

CHAPTER 3

Exercise 3.1

1. $2\pi, 2\pi, \pi$

7. (A) 1 unit (B) Indefinitely far (C) Indefinitely far

9. $-3\pi/2, -\pi/2, \pi/2, 3\pi/2$ **11.** $-2\pi, -\pi, 0, \pi, 2\pi$

13. No x intercepts

15. (A) None (B) $-2\pi, -\pi, 0, \pi, 2\pi$
(C) $-3\pi/2, -\pi/2, \pi/2, 3\pi/2$

17.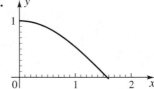

19.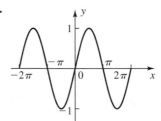

21.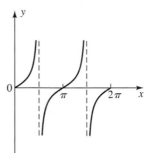

23.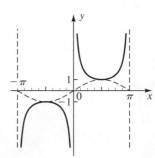

25. (A)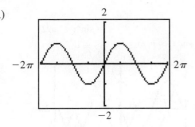

(B)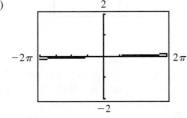

(C) The mode setting is crucial. Degree mode will make the graph totally different.

27. (A)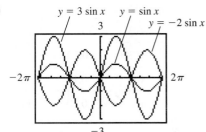

$y = 3 \sin x$ $y = \sin x$ $y = -2 \sin x$

(B) No (C) 2 units; 1 unit; 3 units

(D) The deviation of the graph from the x axis is changed by changing A. The deviation appears to be $|A|$.

29. (A)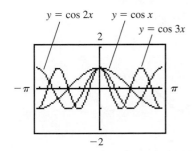

$y = \cos 2x$ $y = \cos x$ $y = \cos 3x$

(B) 1; 2; 3 (C) n

31. (A)

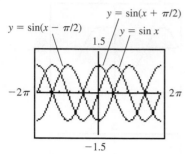

$y = \sin(x + \pi/2)$
$y = \sin(x - \pi/2)$
$y = \sin x$

(B) The graph of $y = \sin x$ is shifted $|C|$ units to the right if $C < 0$ and $|C|$ units to the left if $C > 0$.

33. For each case, the number is not in the domain of the function and an error message of some type will appear.

35. (A) The graphs are almost indistinguishable when x is close to the origin.

(B)

x	-0.3	-0.2	-0.1		
$\tan x$	-0.309	-0.203	-0.100		
x	0.0	0.1	0.2	0.3	
$\tan x$	0.000	0.100	0.203	0.309	

(C) It is not valid to replace $\tan x$ with x for small x if x is in degrees, as is clear from the graph.

37. For a given value of T, the y value on the unit circle and the corresponding y value on the sine curve are the same. This is a graphing calculator illustration of how the sine function is defined as a trigonometric function. See Figure 3 in this section.

39. $3\pi/2$ **41.** $3\pi/2$

43.

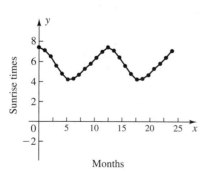

Months

45. $f(x) = [a(x) - 5.8]/1.6$

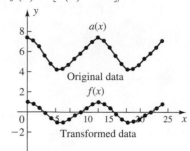

$a(x)$
Original data
$f(x)$
Transformed data

Exercise 3.2

5.

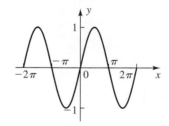

7. Amplitude $= 2$; period $= 2\pi$

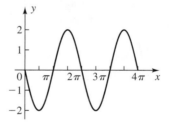

9. Amplitude $= \frac{1}{2}$; period $= 2\pi$

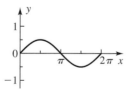

11. Amplitude $= 1$; period $= 1$

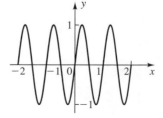

13. Amplitude = 1; period = 8π

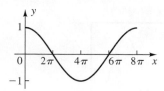

15. Amplitude = 2; period = $\pi/2$

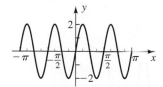

17. Amplitude = $\frac{1}{3}$; period = 1

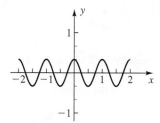

19. Amplitude = $\frac{1}{4}$; period = 4π

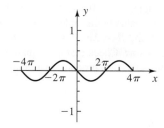

21. $y = 3\sin(\pi t/2)$ or $y = -3\sin(\pi t/2)$
23. $y = 9\cos(10\pi t)$
25. Since $P = 2\pi/B$, P approaches 0 as B increases without bound.
27.

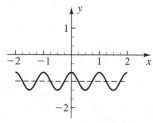

29.

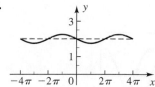

31. $y = 5\sin 2x$ **33.** $y = -4\sin(\pi x/2)$
35. $y = 8\cos(x/4)$ **37.** $y = -\cos(\pi x/3)$
39. $y = 0.5\sin 2x$

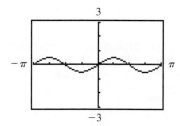

41. $y = 1 + \cos 2x$

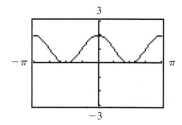

43. $y = 2\cos 4x$

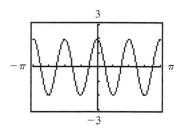

45. (A) $C = 0$ and $-\pi/2$:

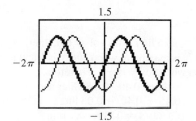

$C = 0$ and $\pi/2$:

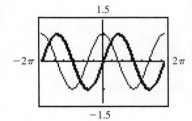

(D) Max vol = 0.85 liter; Min vol = 0.05 liter

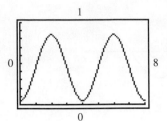

(B) If $C < 0$, then the graph of $y = \sin x$ is shifted $|C|$ units to the right. If $C > 0$, then the graph of $y = \sin x$ is shifted C units to the left.

47. $p = \pi$ **49.** $y = 2 \cos \dfrac{2\pi x}{3}$

51. 0.05 sec **53.** 0.002 sec

55. Amplitude = 110; period = $\frac{1}{60}$ sec; frequency = 60 Hz

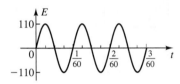

57. $E = 12 \cos 80\pi t$ **59.** No; they are 24.7 kg

61. (A) 2,220 kg

(B) $y = 0.2 \sin 2\pi t$

(C)

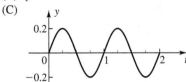

63. (A) Max vol = 0.85 liter; Min vol = 0.05 liter; $0.40 \cos \frac{\pi t}{2}$ is maximum when $\cos \frac{\pi t}{2}$ is 1 and is minimum when $\cos \frac{\pi t}{2}$ is -1. Therefore, max vol = $0.45 + 0.40 = 0.85$ liter and min vol = $0.45 - 0.40 = 0.05$ liter.

(B) 4.00 sec (C) 60/4 = 15 breaths/min

65.

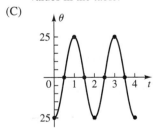

67. (A) The data for θ repeat every 2 sec, so the period is $P = 2$ sec. The angle θ deviates from 0 by $25°$ in each direction, so the amplitude is $|A| = 25°$.

(B) $\theta = A \sin Bt$ is not suitable because, for example, when $t = 0$, $A \sin Bt = 0$ no matter what the choice of A and B. $\theta = A \cos Bt$ appears suitable because, for example, if $t = 0$ and $A = -25$, then we can get the first value in the table—a good start. Choose $A = -25$ and $B = 2\pi/P = \pi$, which yields $\theta = -25 \cos \pi t$. Check that this equation produces (or comes close to producing) all the values in the table.

(C)

69. (A)

t (sec)	0	3	6	9	12	15	18	21	24
h (ft)	28	13	−2	13	28	13	−2	13	28

(B)

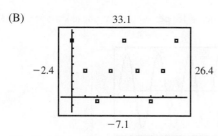

(C) Period = 12 sec. Because the maximum value of $h = k + A \cos Bt$ occurs when $t = 0$ (assuming A is positive), and this corresponds to a maximum value in Table 3 when $t = 0$.

(D) $h = 13 + 15 \cos(\pi t/6)$

(E)

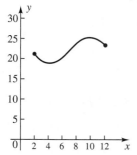

71. $t = 23.25 - 3.75 \sin\left(\dfrac{\pi x}{6}\right)$

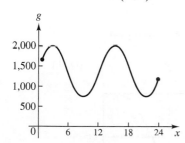

73. $g = 1{,}339 + 624 \sin\left(\dfrac{\pi x}{6}\right)$

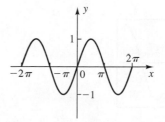

Exercise 3.3

1. The graph with no phase shift is moved 2 units to the left.

5. Phase shift $= -\pi/2$

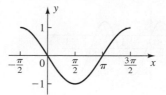

7. Phase shift $= \pi/4$

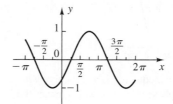

9. Amplitude $= 4$; period $= 2$; phase shift $= -\frac{1}{4}$

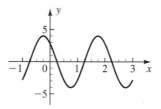

11. Amplitude $= 2$; period $= \pi$; phase shift $= -\pi/2$

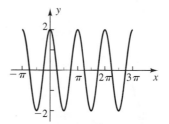

13. Both have the same graph; thus, $\cos(x - \pi/2) = \sin x$ for all x.

15.

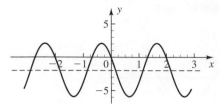

17.

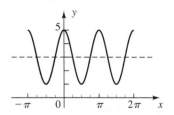

19. (B): The graph of the equation is a sine curve with a period of 2 and a phase shift of $\frac{1}{2}$, which means the sine curve is shifted $\frac{1}{2}$ unit to the right.

21. (A): The graph of the equation is a cosine curve with a period of π and a phase shift of $-\pi/4$, which means the cosine curve is shifted $-\pi/4$ unit to the left.

23. $y = 5 \sin(\pi x/2 - \pi/2)$

25. $y = -2 \cos(x/2 + \pi/4)$

27. Amplitude = 2; period = $2\pi/3$; phase shift = $\pi/6$

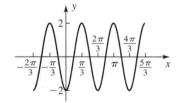

29.

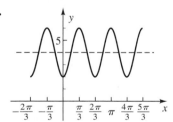

31. Amplitude = 2.3; period = 3; phase shift = 2

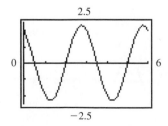

33. Amplitude = 18; period = 0.5; phase shift = -0.137

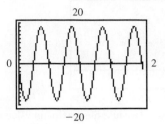

35. x intercept: -1.047; $y = 2 \sin(x + 1.047)$

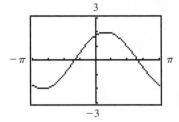

37. x intercept: 0.785; $y = 2 \sin(x - 0.785)$

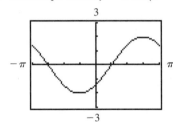

39. x intercept: -0.644; $y = 5 \sin(2x + 1.288)$

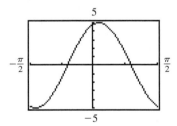

41. x intercept: 1.682; $y = 3 \sin(x/2 - 0.841)$

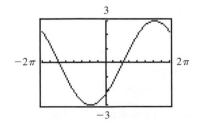

43. Amplitude = 5 m; period = 12 sec;
phase shift = −3 sec

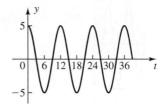

45. Amplitude = 30; period = $\frac{1}{60}$; frequency = 60 Hz;
phase shift = $\frac{1}{120}$

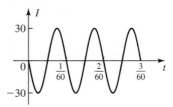

47. (A) $y = 517 - 406 \cos\left(\dfrac{\pi x}{6} - \dfrac{\pi}{3}\right)$

49. (A)

TABLE 1
*Distance d, t Seconds After the Second Hand
Points to 12*

t (sec)	0	5	10	15	20	25	30
d (in.)	0.0	3.0	5.2	6.0	5.2	3.0	0.0
t (sec)	35	40	45	50	55	60	
d (in.)	−3.0	−5.2	−6.0	−5.2	−3.0	0.0	

TABLE 2
*Distance d, t Seconds After the Second Hand
Points to 9*

t (sec)	0	5	10	15	20	25	
d (in.)	−6.0	−5.2	−3.0	0.0	3.0	5.2	
t (sec)	30	35	40	45	50	55	60
d (in.)	6.0	5.2	3.0	0.0	−3.0	−5.2	−6.0

(B) From the table values, and reasoning geometrically
from the clock, we see that relation (2) is 15 sec
out of phase with relation (1).

(C) Both have period $P = 60$ sec and amplitude
$|A| = 6$.

(D) (1) $d = 6 \sin\left(\dfrac{\pi t}{30}\right)$; (2) $d = 6 \sin\left(\dfrac{\pi t}{30} - \dfrac{\pi}{2}\right)$

(E) $d = 6 \sin\left(\dfrac{\pi t}{30} + \dfrac{\pi}{2}\right)$; phase shift = −15 sec

(F)

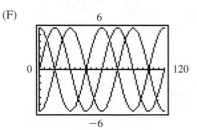

51. (A)

(B) $|A| = (\text{Max } y - \text{Min } y)/2 = (85 - 50)/2 = 17.5$;
period = 12 months, therefore, $B = 2\pi/P = \pi/6$;
$k = |A| + \text{Min } y = 17.5 + 50 = 67.5$. From the
scatter plot it appears that if we use $A = 17.5$, then
the phase shift will be about 3.0, which can be
adjusted for a better visual fit later, if necessary.
Thus, using Phase shift $= -C/B$,
$C = -(\pi/6)(3.0) = -1.6$, and
$y = 67.5 + 17.5 \sin(\pi t/6 - 1.6)$. Graphing this
equation in the same viewing window as the scatter
plot, we see that adjusting C to -2.1 produces a
slightly better visual fit. Thus, with this adjustment,
the equation and graph are

$$y = 67.5 + 17.5 \sin(\pi t/6 - 2.1)$$

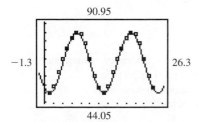

(C) $y = 68.7 + 17.1 \sin(0.5t - 2.1)$

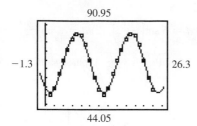

(D) The regression equation differs slightly in k, A, and B, but not in C. Both appear to fit the data very well.

55. $y = \sin 2\pi x/23$

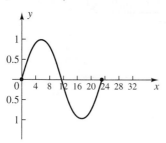

57. $y = \sin 2\pi x/33$

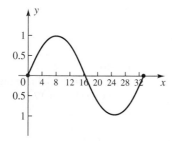

59.

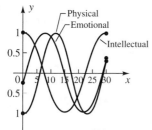

Exercise 3.4

1. (A) Amplitude = 10 amperes; frequency = 60 Hz; phase shift = 1/240 sec
   (B) The maximum current is the amplitude, which is 10 amperes.

(C) Six periods

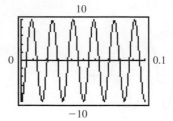

3. $I = 20 \cos 60\pi t$ **5.** 30 ft; 1,311 ft; 82 ft/sec

7.

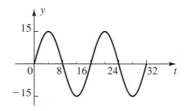

9. (A) $y = 2 \sin\dfrac{\pi}{75} r$ (B) $T = 393$ sec

11. (A) The equation models the vertical motion of the wave at the fixed point $r = 1{,}024$ ft from the source relative to time in seconds.
   (B) The appropriate choice is period, since period is defined in terms of time and wavelength is defined in terms of distance. Period $T = 10$ sec.
   (C)

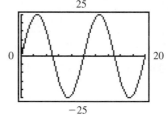

13. Period $= 10^{-8}$ sec; $\lambda = 3$ m
15. $B = 2\pi \times 10^{18}$
17. Period $= 10^{-6}$ sec; frequency $= 10^{6}$ Hz; no
19. (A) Simple harmonic

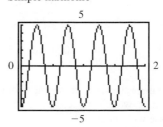

(B) Damped harmonic

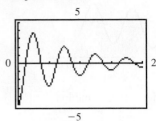

21. (A) High: 49 ft after 2.97 sec; low: 24 ft at 0 sec
(B) Five times   (C) 32 sec

23. Men: $m(x) - 36.7 - \sin(\pi x/12)$
Women: $w(x) = 35.65 - 2.45 \sin(\pi x/12)$

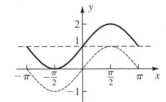

Exercise 3.5

3.

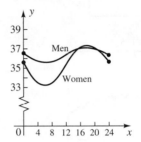

5.

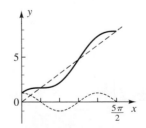

7.

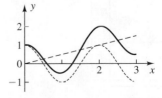

9.

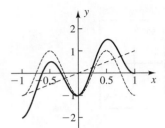

11.

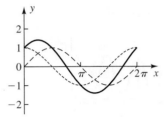

13.

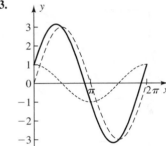

15.

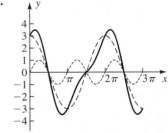

17.

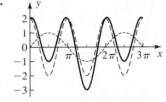

19.

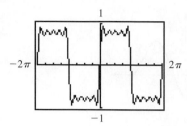

21. (A)

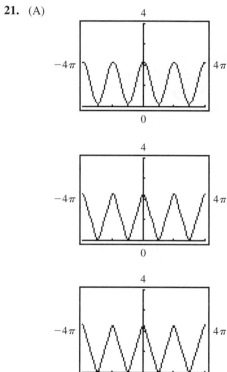

(B) Sawtooth wave form

(C)

23. $V = 0.45 - 0.35 \cos(\pi t/2), 0 \le t \le 8$

25. (A)

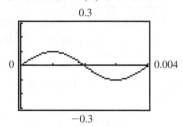

(B) \$11.5 million   (C) \$4 million

27. (A)

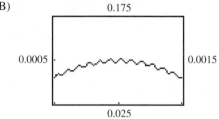

(B)

29.

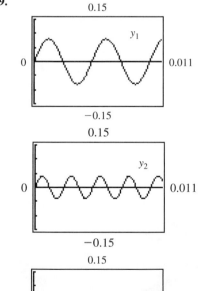

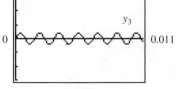

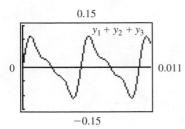

0.15

$y_1 + y_2 + y_3$

0

0.011

−0.15

The period of the combined tone is the same as the period of the fundamental tone.

31. (A) Constructive

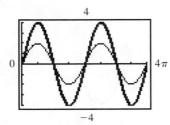

4

0

4π

−4

(B) Constructive

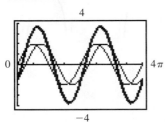

4

0

4π

−4

(C) Destructive

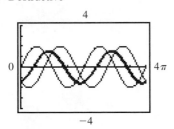

4

0

4π

−4

(D) Destructive

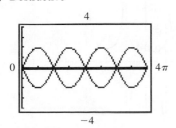

4

0

4π

−4

33. (A) $y_2 = 65 \sin(400\pi t - \pi)$

(B) y_2 added to y_1 produces a sound wave of 0 amplitude—no noise.

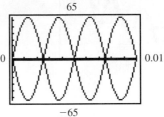

65

0

0.01

−65

35. (A)

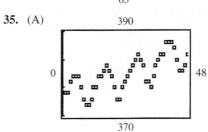

390

0

48

370

(B) $f(x) = 0.167x + 375$

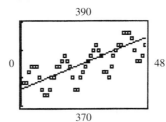

390

0

48

370

(C)

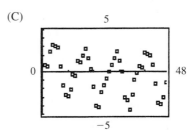

5

0

48

−5

(D) $s(x) = 2.39 \sin(0.516x - 0.419) - 0.305$

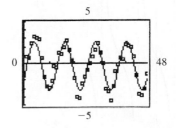

5

0

48

−5

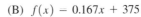

(E)

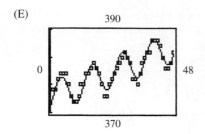

Exercise 3.6

5. *A* makes the graph steeper if $|A| > 1$ and less steep if $|A| < 1$. If $A < 0$, the graph is turned upside down.

7.

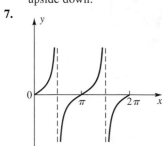

9.

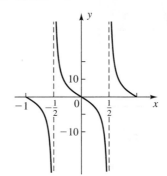

11. Period $= \pi/2$

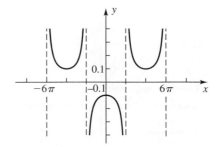

13. Period $= 2\pi$

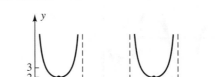

15. Period $= 4\pi$

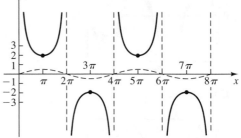

17. Period $= 1$

19. Period $= 8\pi$

21. Period $= \pi/2$; Phase shift $= \pi/2$

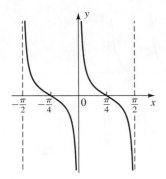

23. Period $= 2$; phase shift $= \frac{1}{2}$

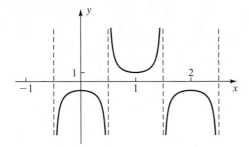

25. Period $= \pi$; phase shift $= \pi/4$

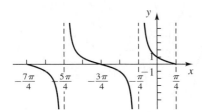

27. $y = \tan \dfrac{x}{2}$

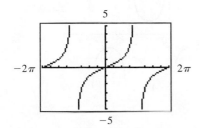

29. $y = 2 \csc 2x$

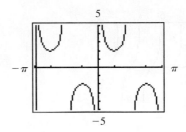

31. Period $= 1$; phase shift $= 1$

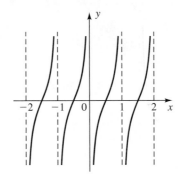

33. Period $= 2$; phase shift $= \frac{1}{2}$

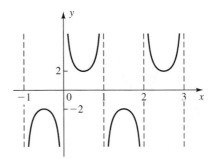

35. $y = \sec 2x$

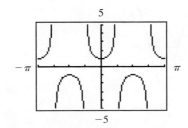

37. $y = \cot 3x$

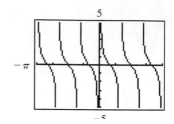

39. (A) $a = 15 \tan 2\pi t$
     (B)

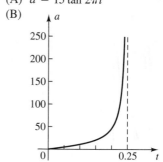

     (C) a increases without bound.

Chapter 3 Review Exercise

1.

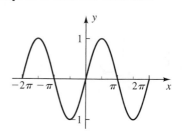

[3.1]

2.

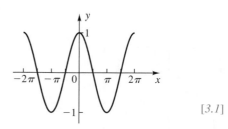

[3.1]

3.

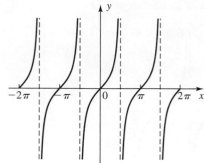

[3.1]

4.

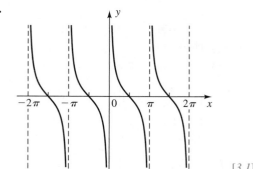

[3.1]

5.

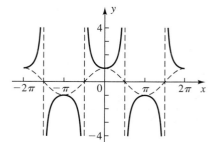

[3.1]

6.

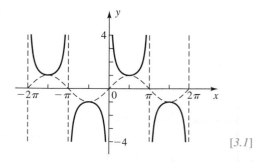

[3.1]

7.

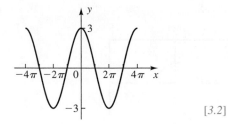

[3.2]

8.

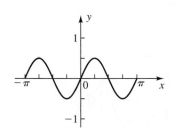

[3.2]

9.

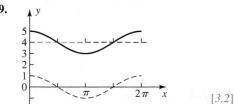

[3.2]

10. All six trigonometric functions are periodic. This is the key property shared by all.   [3.1]

11. If B increases, the period decreases; if B decreases, the period increases.   [3.2]

12. If C increases, the graph is moved to the left; if C decreases, the graph is moved to the right.   [3.3]

13. (A) cosecant   (B) cotangent   (C) sine   [3.1]

14. Period $= 2/3$; amplitude $= 3$ [3.2]

15. Period $= 4\pi$; amplitude $= 1/4$; phase shift $= 4\pi$ [3.3]

16. Period $= 3$ [3.6]

17. Period $= \pi$; phase shift $= -7$ [3.6]

18. Period $= 1$; phase shift $= 2$ [3.6]

19. Period $= 2\pi/5$ [3.6]

20.

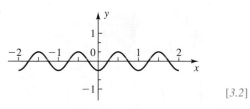

[3.2]

21.

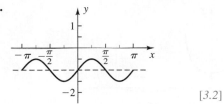

[3.2]

22.

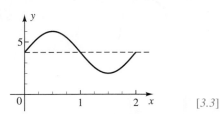

[3.3]

23.

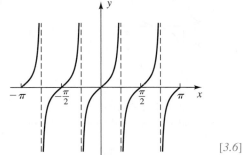

[3.6]

24.

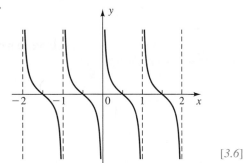

[3.6]

25.

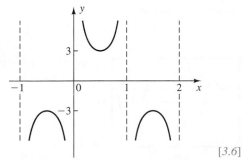

[3.6]

26.

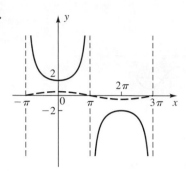

[3.6]

27.

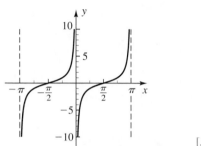

[3.6]

28. (A) y_2 is y_1 with half the period.
  (B) y_2 is y_1 reflected across the x axis with twice the amplitude.
  (C) $y_1 = \sin \pi x$; $y_2 = \sin 2\pi x$
  (D) $y_1 = \cos \pi x$; $y_2 = -2 \cos \pi x$ [3.2]
29. $y = 3 + 4 \sin(\pi x/2)$ [3.2]
30. $y = 65 \sin 200\pi t$ [3.2]
31. (2) $y_2 = y_1 + y_3$: For each value of x, y_2 is the sum of the ordinate values for y_1 and y_3.   [3.5]

32.

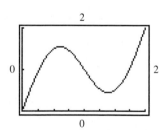

[3.5]

33.

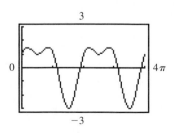

[3.5]

34. $y = \frac{1}{2} + \frac{1}{2} \cos 2x$

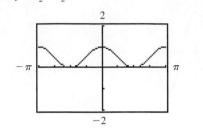

[3.2]

35.

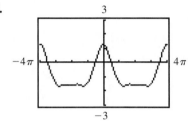

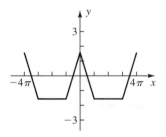

[3.5]

36. (A) $y = \sec x$

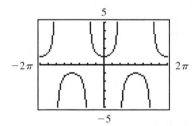

(B) $y = \csc x$

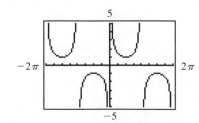

(C) $y = \cot x$

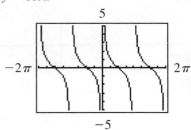

(D) $y = \tan x$

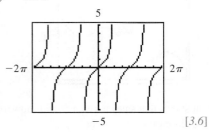

[3.6]

37. A horizontal shift of $\pi/2$ to the left and a reflection across the x axis   [3.1]

38.

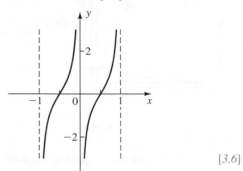

[3.6]

39.

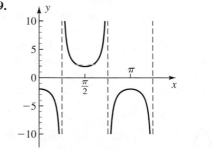

[3.6]

40. $y = 2 \sin 2\pi x$ [3.2]

41. $y = -\sin(\pi x - \pi/4)$ [3.3]

42. x intercept: -0.464; $y = 2 \sin(2x + 0.928)$

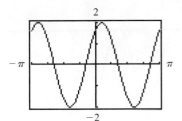

[3.3]

43. (A)

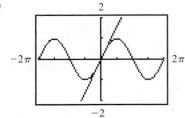

(B)

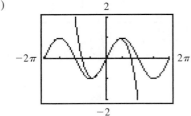

(C)

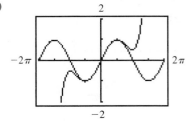

(D) As more terms of the series are used, the resulting approximation of sin x improves over a wider interval.   [3.1]

44. $p = \pi$ [3.1, 3.2]

45. $y = -4 \cos 16\pi t$; $y = A \sin Bt$ cannot be used to model the motion, because when $t = 0$, y cannot equal -4 for any values of A and B.   [3.2]

46. $P = 1 + \cos(\pi n/26), 0 \le n \le 104$; $y = k + A \sin Bn$ will not work, because the curve starts at $(0, 2)$ and oscillates 1 unit above and below the line $P = 1$; a sine curve would have to start at $(0, 1)$.   [3.2]

47. (A) 179 kg (B) $y = 0.6 \sin(2.5\pi t)$

(C)

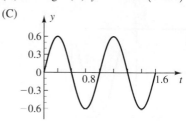

[3.2]

48. (A) Period $= \frac{1}{280}$ sec; $B = 560\pi$

(B) Frequency $= 400$ Hz; $B = 800\pi$

(C) Period $= \frac{1}{350}$ sec; frequency $= 350$ Hz   [3.2]

49. $E = 18 \cos 60\pi t$ [3.2]

50. Amplitude $= 6$ m; period $= 20$ sec; phase shift $= 5$ sec

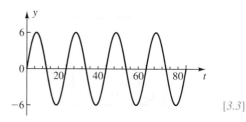

[3.3]

51. Wave height $= 24$ ft; wavelength $= 184$ ft; speed $= 31$ ft/sec   [3.4]

52. Period $= 10^{-15}$ sec; wavelength $= 3 \times 10^{-7}$ m   [3.4]

53. (A) Simple harmonic

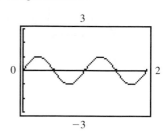

(B) Resonance

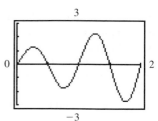

(C) Damped harmonic

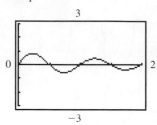

[3.4]

54. (A)

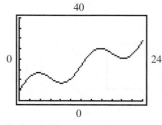

(B) It shows that sales have an overall upward trend with seasonal variations.   [3.5]

55. (A) $h = 1{,}000 \tan \theta$

(B)

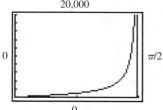

(C) As θ approaches $\pi/2$, h increases without bound.   [3.1]

56. (A) The data for θ repeat every second, so the period is $P = 1$ sec. The angle θ deviates from 0 by 36° in each direction, so the amplitude is $|A| = 36°$.

(B) $\theta = A \cos Bt$ is not suitable, because, for example, for $t = 0$, $A \cos Bt = A$, which would be 36° and not 0°. $\theta = A \sin Bt$ appears suitable, because, for example, if $t = 0$, then $\theta = 0°$, a good start. Choose $A = 36$ and $B = 2\pi/P = 2\pi$, which yields $\theta = 36 \sin 2\pi t$.

(C)

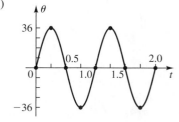

[3.2]

57. (A)

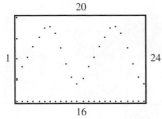

(B) $y = 18.22 + 1.37 \sin\left(\dfrac{\pi x}{6} - 1.7\right)$

(C)

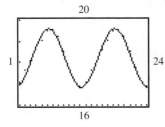

[3.3]

Cumulative Review Exercise Chapters 1–3

1. 43.53°   [1.1]   **2.** 1.55 rad   [2.1]
3. 2.13 rad   [2.1]   **4.** 157.56°   [2.1]
5. −83°5′   [2.1]
6. $\sin\theta = 24/25$; $\cos\theta = 7/25$; $\tan\theta = 24/7$;
  $\cot\theta = 7/24$; $\sec\theta = 25/7$; $\csc\theta = 25/24$ [2.3]
7. Third leg = 6.9 in.; acute angles are 58° and 32°   [1.3]
8. (A) 0.9537   (B) −0.2566   (C) −0.8978   [2.3]
9. (A) −2.538   (B) 2.328   (C) 1.035   [2.3]
10. (A) $\alpha = \dfrac{\pi}{6}$ (B) $\alpha = 45°$

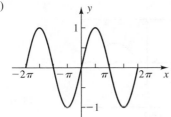

[2.5]

11. (A)

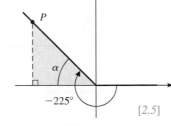

(B)

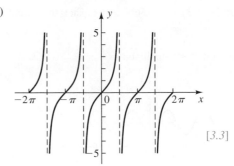

[3.3]

(C)

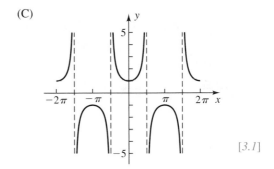

[3.1]

12. The central angle of a circle subtended by an arc that is 1.5 times the length of the radius of the circle.   [2.1]
13. Yes. For example, for any x such that $\pi/2 < x < \pi$, $\cos x$ is negative and $\csc x = 1/(\sin x)$ is positive.   [2.3]
14. No. The sum of all three angles in any triangle is 180°. An obtuse angle is one that has a measure between 90° and 180°, so a triangle with more than one obtuse angle would have two angles whose measures add up to more than 180°, contradicting the first statement.   [1.3]
15. 43°30′   [1.3]   **16.** 20.94 cm   [1.1, 2.1]
17. 31.4 cm/min   [2.2]   **18.** 8   [1.2]
19. $4\pi/15$ [2.1]
20. (A) Degree mode   (B) Radian mode   [2.3]
21. Yes. Each is a different measure of the same angle, so if one is doubled, the other must be doubled. This can be seen from the conversion formula $\theta_r = (\pi/180)\theta_d$.   [2.1]
22. Use the identity $\cot x = 1/(\tan x)$. Thus, $\cot x = 1/0.5453 = 1.8339$.   [2.5]
23. (A) $-1/\sqrt{2}$ (B) $-\sqrt{3}/2$ (C) $\sqrt{3}$
  (D) Not defined   [2.5]
24. $\sin\theta = \sqrt{5}/3$, $\tan\theta = -\sqrt{5}/2$, $\sec\theta = -\frac{3}{2}$,
  $\csc\theta = 3/\sqrt{5}$, $\cot\theta = -2/\sqrt{5}$ [2.3]

25. $\sin(-\pi/3) = -0.8660$ is not exact; the exact form is $-\sqrt{3}/2$. [2.5]

26. Period $= \pi$; amplitude $= \frac{1}{2}$

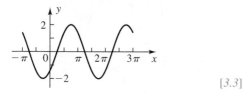

[3.2]

27. Period $= 2\pi$; amplitude $= 2$; phase shift $= \pi/4$

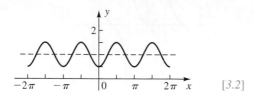

[3.3]

28. Period $= \pi/4$

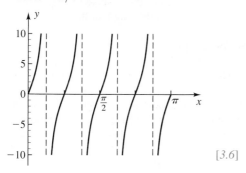

[3.6]

29. Period $= 4\pi$

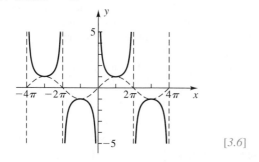

[3.6]

30. Period $= 2$

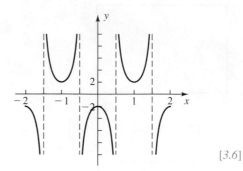

[3.6]

31. Period $= 1$; Phase shift $= -\frac{1}{2}$

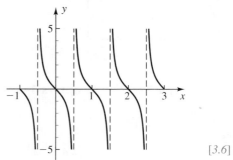

[3.6]

32. $y = -2\sin(2\pi x)$ [3.2]
33. $y = 1 + 2\cos \pi x$ [3.2]
34. $\sec x$ [2.6]
35. $210°, 330°$ [2.5]
36. 44.1 in.; $57.8°$; $32.2°$ [1.3]
37.

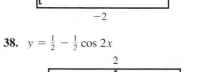

[3.5]

38. $y = \frac{1}{2} - \frac{1}{2}\cos 2x$

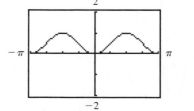

[3.2]

39.

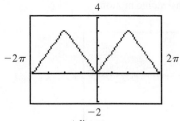

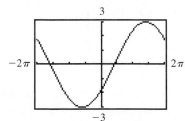

[3.5]

40. $\sin \theta = \dfrac{a}{\sqrt{1 + a^2}}$, $\cos \theta = \dfrac{1}{\sqrt{1 + a^2}}$, $\cot \theta = \dfrac{1}{a}$,

$\csc \theta = \dfrac{\sqrt{1 + a^2}}{a}$, $\sec \theta = \sqrt{1 + a^2}$ [2.3, 2.5]

41. 18.37 units   [2.3]

42. Since the point moves clockwise, x is negative, and the coordinates of the point are:
$P = (\cos(-53.077), \sin(-53.077))$
$= (-0.9460, -0.3241)$. The quadrant in which P lies is determined by the signs of the coordinates. In this case, P lies in quadrant III, because both coordinates are negative.   [2.3]

43. $s = \sin^{-1} 0.8149 = 0.9526$ unit, or
$s = \cos^{-1} 0.5796 = 0.9526$ unit   [2.3]

44. $y = 4 \sin(\pi x - \pi/3)$ [3.3]

45. x intercept: 1.287; $y = 3 \sin\left(\dfrac{x}{2} - 0.6435\right)$

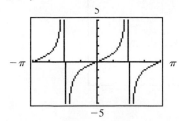

[3.3]

46. (A) $y = \tan x$

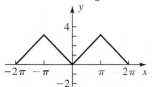

(B) $y = \sec x$

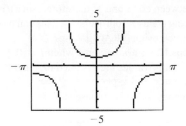

(C) $y = \csc x$

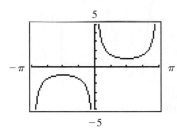

(D)

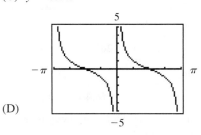

[3.6]

47. 773 mi   [1.1, 2.1]   **48.** 45 cm³   [1.2]
49. 41°; 36 ft   [1.4]   **50.** $\frac{4}{3}$ ft   [1.2, 1.4]
51. 1.7°; 5%   [1.4]
52. Building: 51 m high; street: 17 m wide   [1.4]
53. (A) 1.7 mi (B) 1.3 mi   [1.4]
54. (A) 942 rad/min (B) 64,088 in./min   [2.2]
55. 27.8°   [2.4]
56. 84°   [2.4]
57. (C) $\cos x$ approaches 1 as x approaches 0, and $(\sin x)/x$ is between $\cos x$ and 1; therefore, $(\sin x)/x$ must approach 1 as x approaches 0.   [2.1, 2.3]
58. (A)

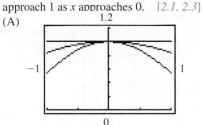

(B) $y_1 < y_2 < y_3$

(C) cos x approaches 1 as x approaches 0, and $(\sin x)/x$ is between cos x and 1; therefore, $(\sin x)/x$ must approach 1 as x approaches 0. This can also be observed using [TRACE]. [2.1, 2.3]

59. $y = 3.6 \cos 12\pi t$; no, because when $t = 0$, y will be 0 no matter what values are assigned to A and B. [3.2]

60. (A) Amplitude $= 12$; period $= \frac{1}{30}$ sec; frequency $= 30$ Hz; phase shift $= \frac{1}{60}$ sec

(B)
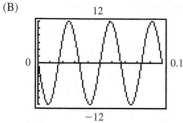
[3.2]

61. Period $= 2 \times 10^{-13}$ sec [3.4]

62. (A) $L = 100 \csc \theta + 70 \sec \theta, 0 < \theta < \pi/2$

(C)

TABLE 1

θ (rad)	0.50	0.60	0.70	0.80
L (ft)	288.3	261.9	246.7	239.9
θ (rad)	0.90	1.00	1.10	
L (ft)	240.3	248.4	266.5	

From the table, the shortest walkway is 239.9 ft when $\theta = 0.80$ rad.

(D) The minimum $L = 239.16$ ft when $\theta = 0.84$ rad.

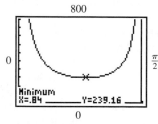
[2.3, 3.2]

63. (A) $d = 50 \tan \theta, 0 < \theta < \pi/2$ (B) $\theta = 40\pi t$
(C) $d = 50 \tan 40\pi t$
(D) d increases without bound as t approaches $\frac{1}{80}$ min.

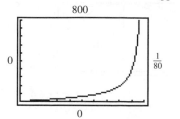
[2.2]

64. (A) Damped harmonic motion
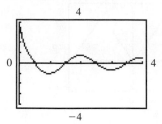

(B) Simple harmonic motion
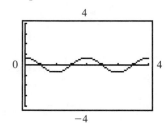

(C) Resonance
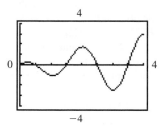
[3.4]

65. (A)
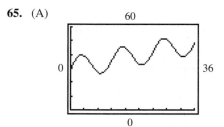

(B) The sales trend is up but with the expected seasonal variations. [3.5]

66. (A)
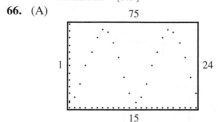

(B) $y = 45 + 26 \sin\left(\dfrac{\pi x}{6} - 2.2\right)$

(C)

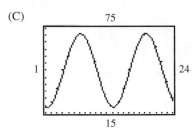

[3.3]

CHAPTER 4
Exercise 4.1

1. Reciprocal identities: $\csc x = 1/(\sin x)$, $\sec x = 1/(\cos x)$, $\cot x = 1/(\tan x)$. Identities for negatives: $\sin(-x) = -\sin x$, $\cos(-x) = \cos x$, $\tan(-x) = -\tan x$

3. (1) is a conditional equation, and (2) is an identity. (1) is true for only one value of x and is false for all other values of x. (2) is true for all values of x for which both sides are defined.

5. $\tan x = 2$, $\cot x = 1/2$, $\sec x = \sqrt{5}$, $\csc x = \sqrt{5}/2$

7. $\sin x = -3/\sqrt{10}$, $\tan x = -3$, $\cot x = -1/3$, $\sec x = \sqrt{10}$

9. $\sin x = -1/4$, $\cos x = -\sqrt{15}/4$, $\cot x = \sqrt{15}$, $\csc x = -4$

11. 1   13. $\sec x$ 15. $\tan x$ 17. $\sec\theta$

19. $\cot^2\beta$ 21. 2

23. Not necessarily. For example, $\sin x = 0$ for infinitely many values ($x = k\pi$, k any integer), but the equation is not an identity. The left side is not equal to the right side for values other than $x = k\pi$, k an integer; for example, when $x = \pi/2$, $\sin x = 1 \ne 0$.

25. $\cos x = -\sqrt{21}/5$, $\tan x = -2/\sqrt{21}$, $\cot x = -\sqrt{21}/2$, $\sec x = -5/\sqrt{21}$, $\csc x = 5/2$

27. $\sin x = 1/\sqrt{5}$, $\cos x = -2/\sqrt{5}$, $\cot x = -2$, $\sec x = -\sqrt{5}/2$, $\csc x = \sqrt{5}$

29. $\sin x = \sqrt{15}/4$, $\cos x = 1/4$, $\tan x = \sqrt{15}$, $\cot x = 1/\sqrt{15}$, $\csc x = 4/\sqrt{15}$

35. (A) -0.4350 (B) $1 - 0.1892 = 0.8108$

37. $-\cot y$ 39. $\csc x$ 41. 0

43. $\csc^2 x$ 45. $\cot w$ 47. No

49. Yes   51. Yes   53. No

55. Yes   57. No

59. (A) 1   (B) 1

61. I, II   63. II, III   65. All

67. I, IV   69. $a \cos x$ 71. $a \sec x$

73. (A) $\dfrac{x^2}{25} + \dfrac{y^2}{16} = 1$

(B) Elliptical

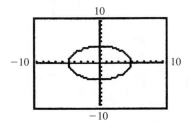

75. (A) $\dfrac{y^2}{9} - \dfrac{x^2}{25} = 1$

(B) Hyperbolic

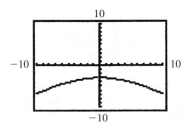

Exercise 4.2

1. $\sin x$ 3. $\cos x$ 5. $\sec x$

7. $\csc x$ 9. $\cot x$ 37. Yes

39. No   41. No   43. No

45. No   47. Yes

81. (A) No. Use TRACE and move from one curve to the other, comparing y values for different values of x. You will see that, though close, they are not exactly the same.

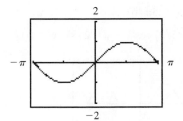

(B) Outside the interval $[-\pi, \pi]$ the graphs differ widely.

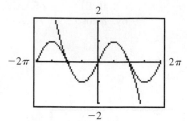

83. An identity   **85.** Not an identity
87. Not an identity   **89.** An identity

Exercise 4.3

13. $\frac{1}{2}(\sqrt{3}\sin x + \cos x)$

15. $\frac{1}{2}(\cos x - \sqrt{3}\sin x)$

17. $\dfrac{\tan x + 1}{1 - \tan x}$ **19.** $\dfrac{\sqrt{3} + \tan x}{1 - \sqrt{3}\tan x}$

21. $\dfrac{\sqrt{3} - 1}{2\sqrt{2}}$ **23.** $\dfrac{1}{\sqrt{2}}$

25. $\dfrac{1}{2}$ **27.** 1   **29.** $-\dfrac{1}{4} + \dfrac{\sqrt{14}}{6}$

31. $-\dfrac{21}{\sqrt{442}}$ **33.** $\dfrac{1 + \sqrt{120}}{\sqrt{15} - \sqrt{8}}$

35. No   **37.** Yes   **39.** No   **41.** Yes

57. Find values of x and y for which both sides are defined and the left side is not equal to the right side. Evaluate both sides for $x = 2$ and $y = 1$, for example.

59. Graph $y_1 = \sin(x - 2)$ and $y_2 = \sin x - \sin 2$ in the same viewing window, and observe that the graphs are not the same:

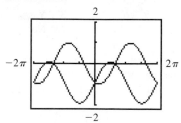

61. $y_1 = \cos(x + 5\pi/6)$,

$y_2 = (-\sqrt{3}/2)\cos x - (1/2)\sin x$

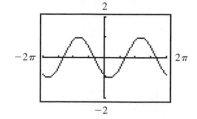

63. $y_1 = \tan(x - \pi/4),\quad y_2 = \dfrac{\tan x - 1}{1 + \tan x}$

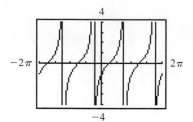

65. $y_1 = \sin 3x \cos x - \cos 3x \sin x,\quad y_2 = \sin 2x$

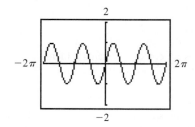

67. $y_1 = \sin(\pi x/4)\cos(3\pi x/4) + \cos(\pi x/4)\sin(3\pi x/4)$
$y_2 = \sin \pi x$

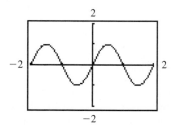

73. $\theta = 0.322$ rad   **75.** (C) 3,940 ft

Exercise 4.4

1. $\cos x$ **3.** $\cos x$ **5.** $\sin x$

7. $\dfrac{\sqrt{2 + \sqrt{3}}}{2}$ **9.** $2 - \sqrt{3}$

11.

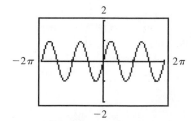

13.

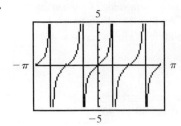

33. No   **35.** Yes   **37.** No
39. Yes   **41.** No   **43.** No

45. $\sin 2x = -\dfrac{336}{625}, \cos 2x = \dfrac{527}{625}, \tan 2x = -\dfrac{336}{527}$

47. $\sin 2x = -\dfrac{840}{1,369}, \cos 2x = -\dfrac{1,081}{1,369}, \tan 2x = \dfrac{840}{1,081}$

49. $\sin(x/2) = \sqrt{\dfrac{3}{8}}, \cos(x/2) = \sqrt{\dfrac{5}{8}}$

51. $\sin(x/2) = \sqrt{\dfrac{2}{3}}, \cos(x/2) = \sqrt{\dfrac{1}{3}}$

53. $\sin(x/2) = -\sqrt{\dfrac{2}{5}}, \cos(x/2) = \sqrt{\dfrac{3}{5}}$

55. (A) Since θ is a first-quadrant angle and $\sec 2\theta$ is negative for 2θ in the second quadrant and not for 2θ in the first, 2θ is a second-quadrant angle.

    (B) Construct a reference triangle for 2θ in the second quadrant with $a = -4$ and $r = 5$. Use the Pythagorean theorem to find $b = 3$. Thus, $\sin 2\theta = \frac{3}{5}$ and $\cos 2\theta = -\frac{4}{5}$.

    (C) The double-angle identities $\cos 2\theta = 1 - 2\sin^2 \theta$ and $\cos 2\theta = 2\cos^2 \theta - 1$

    (D) Use the identities in part (C) in the form $\sin \theta = \sqrt{(1 - \cos 2\theta)/2}$ and $\cos \theta = \sqrt{(1 + \cos 2\theta)/2}$. The positive radicals are used because θ is in quadrant I.

    (E) $\sin \theta = 3/\sqrt{10}; \cos \theta = 1/\sqrt{10}$

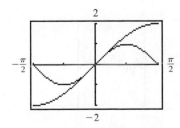

57. (A) Approximation improves

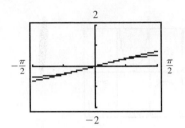

    (B) Approximation improves

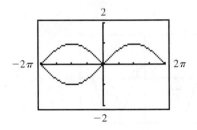

59. $0 \le x \le 2\pi$

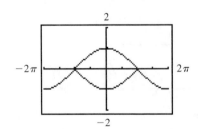

61. $-2\pi \le x \le -\pi, \pi \le x \le 2\pi$

63. $\sin x = \dfrac{5}{\sqrt{146}}, \cos x = \dfrac{11}{\sqrt{146}}, \tan x = \dfrac{5}{11}$

65. $\sin x = \dfrac{7}{\sqrt{53}}, \cos x = \dfrac{2}{\sqrt{53}}, \tan x = \dfrac{7}{2}$

67. $\sin x = -\dfrac{4}{\sqrt{65}}, \cos x = \dfrac{7}{\sqrt{65}}, \tan x = -\dfrac{4}{7}$

75. $g(x) = \cot \dfrac{x}{2}$

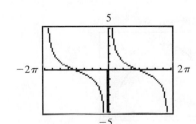

77. $g(x) = \csc 2x$

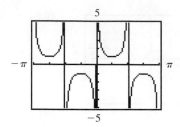

79. $g(x) = 2\cos x - 1$

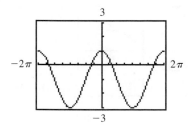

81.

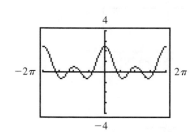

83. (A) $d = \dfrac{v_0^2 \sin 2\theta}{32}$

(B) d is maximum when $\sin 2\theta$ is maximum, and $\sin 2\theta$ is maximum when $2\theta = 90°$—that is, when $\theta = 45°$.

(C) As θ increases from $0°$ to $90°$, d increases to a maximum of 312.5 ft when $\theta = 45°$, then decreases.

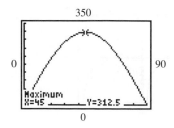

85. (B) $\sin 2\theta$ has a maximum value of 1 when $2\theta = 90°$, or $\theta = 45°$. When $\theta = 45°$, $V = 64$ ft³.

(C) Max $V = 64.0$ ft³ occurs when $\theta = 45°$.

TABLE 2							
θ (deg)	30	35	40	45	50	55	60
V (ft³)	55.4	60.1	63.0	64.0	63.0	60.1	55.4

(D) Max $V = 64$ ft³ when $\theta = 45°$.

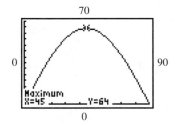

87. $x = 2\sqrt{3} \approx 3.464$ cm; $\theta = 30.000°$ **89.** $\cos \theta = \frac{4}{5}$

Exercise 4.5

1. $\frac{1}{2}\cos 5w + \frac{1}{2}\cos 3w$ **3.** $\frac{1}{2}\sin 3u - \frac{1}{2}\sin u$

5. $\frac{1}{2}\sin 7B - \frac{1}{2}\sin 3B$ **7.** $\frac{1}{2}\cos m - \frac{1}{2}\cos 7m$

9. $2\cos 4\theta \cos \theta$ **11.** $2\cos 4u \sin 2u$

13. $2\sin 4B \cos B$ **15.** $2\sin 3w \sin 2w$

17. $(2 - \sqrt{3})/4$ **19.** $\frac{1}{4}$ **21.** $1/\sqrt{2}$ **23.** $1/\sqrt{2}$

27. Let $x = u + v$ and $y = u - v$, and solve the resulting system for u and v in terms of x and y. Then substitute the results into the first identity.

37. No   **39.** Yes

41. Yes   **43.** No

45. $y_2 = \frac{1}{2}(\cos 8x + \cos 2x)$

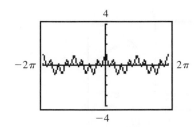

47. $y_2 = \frac{1}{2}(\sin 2.4x - \sin 1.4x)$

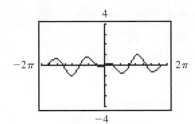

49. $y_2 = 2\cos 2x \cos x$

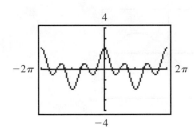

51. $y_2 = 2\cos 1.3x \sin 0.8x$

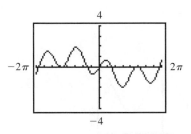

55. (A)

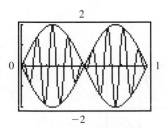

(B) $y_1 = \sin(18\pi x) - \sin(14\pi x)$
   Graph same as part (A).

57. (A)

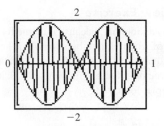

(B) $y_1 = \cos(22\pi x) - \cos(26\pi x)$
   Graph same as part (A).

59. $y = 2k \sin 517\pi t \cos 5\pi t$; $f_b = 5$ Hz

61. (A)

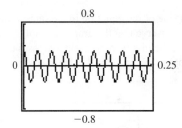

(B)

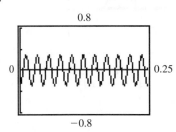

(C)

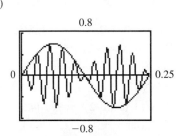

(D) $y_1 = 0.6 \sin 80\pi t \sin 8\pi t$

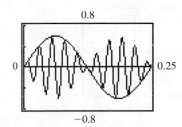

Chapter 4 Review Exercise

1. Equation (1) is an identity, because it is true for all replacements of x by real numbers for which both sides are defined. Equation (2) is a conditional equation, because it is true only for $x = -2$ and $x = 3$; it is not true, for example, for $x = 0$.   *[4.1]*

13. $\frac{1}{2} = \frac{1}{2}$ *[4.4]*

14. $1/\sqrt{2} = 1/\sqrt{2}$ *[4.4]*

15. $\frac{1}{2}\cos 3t - \frac{1}{2}\cos 13t$ *[4.5]*

16. $2\sin 3w \cos 2w$ *[4.5]*

21. The equation is not an identity. The equation is not true for $x = \pi/2$, for example, and both sides are defined for $x = \pi/2$.   *[4.1]*

22. Graph each side of the equation in the same viewing window and observe that the graphs are not the same, except where the graph of $y_1 = \sin x$ crosses the x axis. (Note that the graph of $y_2 = 0$ is the x axis.)   *[4.1]*

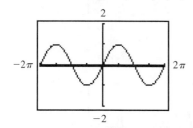

23. No   *[4.1]*   **24.** No   *[4.1]*   **25.** Yes   *[4.3]*
26. No   *[4.3]*   **27.** No   *[4.3]*   **28.** Yes   *[4.3]*
29. No   *[4.4]*   **30.** Yes   *[4.5]*

46. $\frac{1}{2} - \frac{\sqrt{3}}{4}$ *[4.5]*   **47.** $-\sqrt{6}/2$ *[4.5]*

48. $\sec x = -\frac{3}{2}$, $\sin x = \sqrt{5}/3$, $\csc x = 3/\sqrt{5}$, $\tan x = -\sqrt{5}/2$, $\cot x = -2/\sqrt{5}$ *[4.1]*

49. $\sin 2x = \frac{24}{25}$, $\cos 2x = -\frac{7}{25}$, $\tan 2x = -\frac{24}{7}$ *[4.1, 4.4]*

50. $\sin(x/2) = -3/\sqrt{13}$, $\cos(x/2) = 2/\sqrt{13}$, $\tan(x/2) = -\frac{3}{2}$ *[4.2]*

51. $\tan(x + \pi/4) = (\tan x + 1)/(1 - \tan x)$

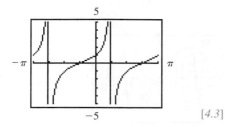

[4.3]

52. $y_2 = \cos(1.8x)$

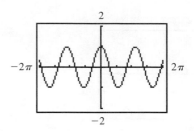

[4.3]

53. Identities for $-2\pi \le x \le 0$

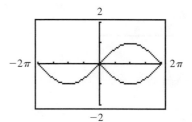

[4.4]

54. (A) Not an identity; for example, both sides are defined for $x = 0$ but are not equal.
(B) An identity   *[4.2]*

55. $\sin x = -5/\sqrt{26}$, $\cos x = 1/\sqrt{26}$, $\tan x = -5$ *[4.4]*

62. $g(x) = 4 + 2\cos x$

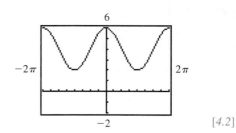

[4.2]

63. $g(x) = \tan 2x$

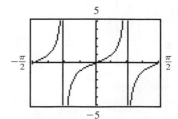

[4.4]

64. $g(x) = 2 - \cos 2x$

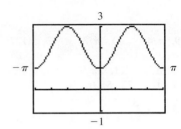

[4.4]

65. $g(x) = -2 + \sec 2x$

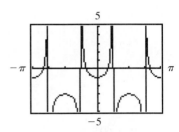

[4.4]

66. $g(x) = 2 + \cot(x/2)$

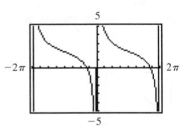

[4.4]

67. $-2\pi \le x < -\pi, 0 \le x < \pi$

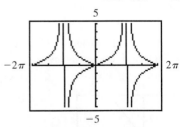

[4.4]

68. $-\pi < x \le 0, \pi < x \le 2\pi$

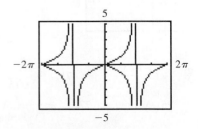

[4.4]

69.

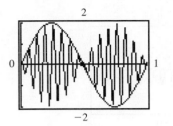

[4.5]

70. $y_1 = \sin 32\pi x - \sin 28\pi x$

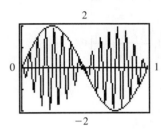

[4.5]

71. $a \tan x$ [4.1]   **72.** $57.5°$ [4.3]
73. $x = \frac{445}{39} \approx 11.410; \theta \approx 32.005°$ [4.4]
74. $\theta = 0.464$ rad   [4.3]
75. (B) L steadily increases.
(C)

TABLE 1				
θ (deg)	30	35	40	45
L (ft)	250.7	252.6	254.6	256.6
θ (deg)	50	55	60	
L (ft)	258.7	260.8	263.1	

(D) Min $L = 250.7$ ft; Max $L = 263.1$ ft

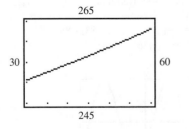

[4.4]

76. $y = 0.6 \sin 130\pi t \sin 10\pi t$; beat frequency $= 10$ Hz
[4.5]

77. (A)

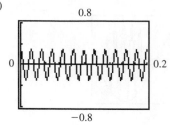

(B)

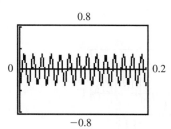

(C)

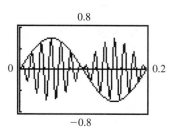

(D)

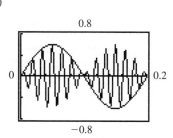

[*4.5*]

78. $y = 10 \sin(3t + 3.79)$; amplitude = 10 cm;
period = $2\pi/3$ sec; frequency = $3/(2\pi)$ Hz;
phase shift = -1.26 sec   [*Chap. 4 Group Activity*]

79. t intercepts: $-1.26, -0.21, 0.83, 1.88$;
phase shift = -1.26

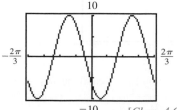

[*Chap. 4 Group Activity*]

CHAPTER 5

Exercise 5.1

5. 0   **7.** $\pi/6$ **9.** $\pi/4$ **11.** $\pi/3$ **13.** 3.127
15. 1.329   **17.** Not defined
19. $x = \sin 37 = 0.601815$ **21.** $2\pi/3$
23. $-\pi/4$ **25.** $-\pi/3$ **27.** $5\pi/6$
29. -0.6 **31.** $\sqrt{2}/2$ **33.** $2/\sqrt{5}$
35. $1/\sqrt{5}$ **37.** -1.328 **39.** 1.001   **41.** -0.1416
43.

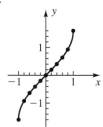

45. 120°   **47.** $-45°$ **49.** $-60°$
51. Not defined   **53.** 60°   **55.** 71.80°
57. $-54.16°$ **59.** $-86.69°$
61. 0.3; does not illustrate a cosine–inverse cosine identity,
because $\cos^{-1}(\cos x) = x$ only if $0 \le x \le \pi$.
63. (A)

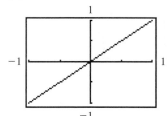

(B) The domain of $\sin^{-1}$ is restricted to $-1 \le x \le 1$;
no graph will appear for other values of x.

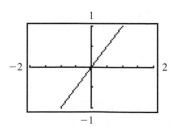

65. $-\dfrac{1}{2}$ **67.** $-\dfrac{24}{25}$ **69.** $\dfrac{2\sqrt{2} - 2}{3\sqrt{5}}$ **71.** $-\dfrac{7}{\sqrt{15}}$
73. $-\sqrt{\dfrac{5 - \sqrt{5}}{10}}$ **75.** $\sqrt{1 - x^2}$ **77.** $\dfrac{x}{\sqrt{1 - x^2}}$

81. (A) $\cos^{-1} x$ has domain $-1 \le x \le 1$; therefore, $\cos^{-1}(2x - 3)$ has domain $-1 \le 2x - 3 \le 1$, or $1 \le x \le 2$.

(B) The graph appears only for the domain values $1 \le x \le 2$.

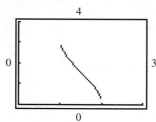

83. (A) $h^{-1}(x) = 1 + \sin^{-1}[(x - 3)/5]$

(B) $\sin^{-1} x$ has domain $-1 \le x \le 1$; therefore, $1 + \sin^{-1}[(x - 3)/5]$ has domain $-1 \le (x - 3)/5 \le 1$, or $-2 \le x \le 8$.

85. (A)

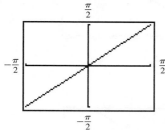

(B) The domain for $\sin x$ is $(-\infty, \infty)$ and the range is $[-1, 1]$, which is the domain for $\sin^{-1} x$. So, $y = \sin^{-1}(\sin x)$ has a graph over the interval $(-\infty, \infty)$, but $\sin^{-1}(\sin x) = x$ only on the restricted domain of $\sin x$, $[-\pi/2, \pi/2]$.

91. $30°$

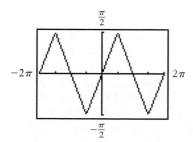

93. (A) $\theta = 2 \sin^{-1}(1/M)$, $M > 1$

(B) $72°$; $52°$

95. (A) It appears that θ increases and then decreases.

(C) From the table, Max $\theta = 9.21°$ when $x - 25$ yd.

TABLE 1

x (yd)	10	15	20	25
θ (deg)	6.44	8.19	9.01	9.21
x (yd)	30	35		
θ (deg)	9.04	8.68		

(D) The angle θ increases rapidly until a maximum is reached and then declines more slowly. Max $\theta = 9.21°$ when $x = 24.68$ yd.

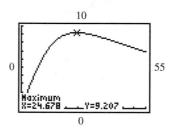

97. (B) 248 ft^3

(C) 2.6 ft

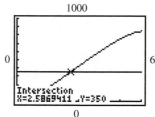

99. $y = \sin^2 x$

101. (A)

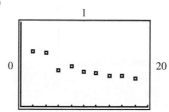

101. (B)

y	0.657	0.642	0.437	0.481	0.425	0.399	0.374	0.370	0.341
$\sin^{-1}\sqrt{y}$	0.945	0.929	0.722	0.766	0.710	0.684	0.658	0.654	0.624

101. (C)

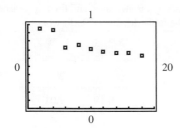

(D) Regression line: $y = -0.0193x + 0.937$

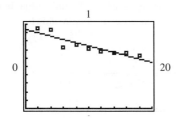

(E)

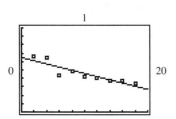

103. (A)

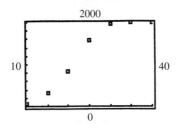

(B)

(C)

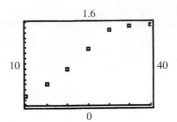

(D) Regression line: $y = 0.05009x - 0.2678$

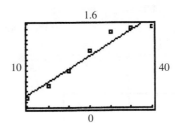

(E)

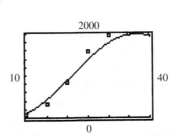

Exercise 5.2

5. $\pi/6$ **7.** $\pi/2$ **9.** $\pi/4$ **11.** 1 **13.** $\frac{4}{3}$
15. $3\pi/4$ **17.** $2\pi/3$ **19.** $-\pi/6$ **21.** Not defined
23. $\frac{4}{5}$ **25.** $-\frac{4}{3}$ **27.** $-\frac{1}{2}$ **29.** 33.4 **31.** -4
33. 1.398 **35.** 1.536 **37.** Not defined **39.** 2.875
41. 1.637 **43.** 120° **45.** 135° **47.** $-60°$
49. Not defined **51.** $-45°$ **53.** 165°
55. 71.80° **57.** $-54.16°$ **59.** 93.31°

y	69	325	836	1,571	1,965	1,993	1,999
$\sin^{-1}\sqrt{y/2{,}000}$	0.1868	0.4149	0.7030	1.089	1.438	1.512	1.548

61. $-\frac{8}{19}$ **63.** $\frac{24}{7}$ **65.** $\dfrac{1}{\sqrt{x^2+1}}$

67. $\dfrac{|x|}{\sqrt{x^2-1}}$ **69.** $\dfrac{2x}{x^2+1}$

73.

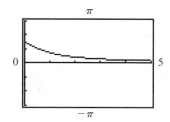

77. Identity
79. Not an identity
81. Identity
83. Not an identity

Exercise 5.3

5. $\pi/6 + 2k\pi, 11\pi/6 + 2k\pi$, k any integer
7. $\pi/4 + k\pi$, k any integer
9. $3\pi/2 + 2k\pi$, k any integer
11. $2\pi/3, 4\pi/3$
13. $2\pi/3 + 2k\pi, 4\pi/3 + 2k\pi$, k any integer
15. $45°, 135°$
17. $45° + k(360°), 135° + k(360°)$, k any integer
19. No solution
21. $104.9314°$
23. $1.1593, 5.1239$
25. $0.6696 + 2k\pi, 2.4720 + 2k\pi$, k any integer
27. $\pi/2, 3\pi/2$

29. $90° + k(180°), 45° + k(180°)$, k any integer
31. $\pi/2$
33. $180°$
35. $90°, 270°$
37. $\pi/6 + 2k\pi, 5\pi/6 + 2k\pi, 3\pi/2 + 2k\pi$, k any integer
39. $\pi/3 + 2k\pi, \pi/2 + 2k\pi, 3\pi/2 + 2k\pi, 5\pi/3 + 2k\pi$, k any integer
41. $k(720°), 180° + k(720°), 540° + k(720°)$, k any integer
43. $60°, 300°, 540°$
45. $\pi/2 + 4k\pi, 7\pi/2 + 4k\pi$, k any integer
47. $104.5°$
49. $0.9987, 5.284$
51. $0.9987 + 2k\pi, -0.9987 + 2k\pi$, k any integer
53. $0.6579 + 2k\pi, 5.625 + 2k\pi$, k any integer
55. $1.946 + 2k\pi, 4.338 + 2k\pi$, k any integer
57. $\cos^{-1}(-0.7334)$ has exactly one value, 2.3941; the equation $\cos x = -0.7334$ has infinitely many solutions, which are found by adding $2\pi k$, k any integer, to each solution in one period of $\cos x$.
59. $0, \pi/2$ **61.** 0
63. $\pi/3, \pi, 5\pi/3$, and $7\pi/3$ sec, or 1.05, 3.14, 5.24, and 7.33 sec
65. 0.002613 sec   **67.** $3.5, 8.5, 15.5, 20.5$, and 27.5 sec
69. $33.21°$ **71.** $64.1°$
73. $(r, \theta) = (1, 30°), (1, 150°)$
75. $\theta = 45°$
77. $k = 12.17, A = -2.833, B = \pi/6$; 2.9 months, 9.1 months
79. $y = 2.818 \sin(0.5108x - 1.605) + 12.14$; 3.0 months, 9.4 months

Exercise 5.4

3. 0.4502 **5.** 0.6167 **7.** $(0.7854, 3.9270)$
9. $[0.7391, \infty)$ **11.** $0.9987, 5.2845$
13. $0.9987 + 2k\pi, 5.2845 + 2k\pi$, k any integer
15. $(-1.5099, 1.8281)$
17. $[0.4204, 1.2346], [2.9752, \infty)$
19. Because $\sin^2 x - 2 \sin x + 1 = (\sin x - 1)^2$, and the latter is greater than or equal to 0 for all real x.
21. $0.006104, 0.006137$ **23.** $[-2\pi, 2\pi]$
25. $-2.331, 0, 2.331$
27. $(1.364, 2.566) \cup (3.069, \pi]$
29. $[0.2974, 1.6073] \cup [2.7097, 3.1416]$
31. 1.78 rad   **33.** After 72 weeks   **35.** 35.64 ft^2
37. (A) $L = 12.4575$ mm   (B) $y = 2.6495$ mm
39. (A) $32.7°$ or $57.3°$
(B) No. The equation $600 = (130^2/16) \sin \alpha \cos \alpha$ has no solution.
41. $y = 58.1 + 24.5 \sin(\pi x/6 + 4.1)$; $5.1 < x < 9.2$

43. $y = 57.5 + 25.5 \sin(0.478x - 1.83)$;
    $x < 3.2$ or $x > 11$

Chapter 5 Review Exercise

1. $-\pi/4$ [5.1]   **2.** $\pi/3$ [5.1]   **3.** 0   [5.1]
4. $-\pi/6$ [5.1]   **5.** Not defined   [5.1]
6. $3\pi/4$ [5.1]   **7.** $-\pi/3$ [5.2]   **8.** $\pi/2$ [5.2]
9. $2\pi/3$ [5.2]   **10.** Not defined   [5.2]
11. 0.6813   [5.1]   **12.** 2.898   [5.1]
13. 1.528   [5.1]   **14.** 2.526   [5.2]
15. Not defined   [5.1]   **16.** 83.16°   [5.1]
17. $-5.80°$ [5.1]
18. $4\pi/3 + 2k\pi, 5\pi/3 + 2k\pi, k$ any integer   [5.3]
19. $5\pi/3 + k\pi, k$ any integer   [5.3]
20. $\pi/4 + 2k\pi, 7\pi/4 + 2k\pi, k$ any integer   [5.3]
21. No solution   [5.3]   **22.** $\pi/6, 11\pi/6$ [5.3]
23. $0°, 30°, 150°, 180°$ [5.3]
24. $\pi/6, 5\pi/6, 7\pi/6, 11\pi/6$ [5.3]
25. $120°, 180°, 240°$ [5.3]   **26.** $\pi/16, 3\pi/16$ [5.3]
27. $-120°$ [5.3]   **28.** $x = 0.906308$ [5.1]
29. 0.315   [5.1]   **30.** -1.5 [5.1]   **31.** $-\frac{3}{5}$ [5.1]
32. $-2/\sqrt{5}$ [5.1]   **33.** $\sqrt{10}/3$ [5.2]
34. $2\sqrt{6}/5$ [5.2]   **35.** -1.192 [5.1]
36. Not defined   [5.1]   **37.** -0.9975 [5.1]
38. 1.095   [5.1]   **39.** 0.9894   [5.2]
40. 1.001   [5.2]
41. Figure (a) is in degree mode and illustrates a sine–inverse
    sine identity, since 2 is in the domain for this identity,
    $-90° \leq \theta \leq 90°$. Figure (b) is in radian mode and does
    not illustrate this identity, since 2 is not in the domain for
    the identity, $-\pi/2 \leq x \leq \pi/2$.   [5.1]
42. $90°, 270°$ [5.3]
43. $\pi/12, 5\pi/12$ [5.3]
44. $-\pi$ [5.3]
45. $k(360°), 180° + k(360°), 30° + k(360°),$
    $150° + k(360°), k$ any integer   [5.3]
46. $2k\pi, \pi + 2k\pi, \pi/6 + 2k\pi, -\pi/6 + 2k\pi, k$ any
    integer   [5.3]
47. $120°, 240°$ [5.3]   **48.** $7\pi/12, 11\pi/12$ [5.3]
49. $0.7878 + 2k\pi, 2.354 + 2k\pi, k$ any integer   [5.3]
50. $-1.343 + k\pi, k$ any integer   [5.3]
51. $0.6259 + 2k\pi, 2.516 + 2k\pi, k$ any integer   [5.3]
52. $1.178 + k\pi, -0.3927 + k\pi, k$ any integer   [5.3]
53. 0.253, 2.889   [5.4]   **54.** $-0.245, 2.897$ [5.4]
55. ± 1.047 [5.3]   **56.** ± 0.824 [5.4]
57. 0   [5.4]   **58.** 4.227, 5.197   [5.4]
59. 0.604, 2.797, 7.246, 8.203   [5.4]
60. 0.228, 1.008   [5.4]

61. The domain for y_1 is the domain for $\sin^{-1} x$,
    $-1 \leq x \leq 1$.

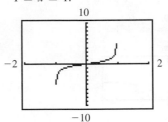

[5.1]

62. (A) $\sin^{-1} x$ has domain $-1 \leq x \leq 1$; therefore,
        $\sin^{-1}[(x - 2)/2]$ has domain
        $-1 \leq (x - 2)/2 \leq 1$, or $0 \leq x \leq 4$.
    (B) The graph appears only for the domain values
        $0 \leq x \leq 4$.

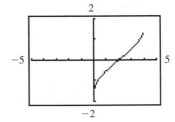

[5.1]

63. No, $\tan^{-1} 23.255$ represents only one number and one
    solution. The equation has infinitely many solutions,
    which are given by $x = \tan^{-1} 23.255 + k\pi, k$ any
    integer.   [5.3]
64. $0, \pi/2$ [5.4]   **65.** $0, \pi/3, 2\pi/3$ [5.4]
66. 0.375, 2.77   [5.4]   **67.** $-\frac{24}{25}$ [5.1]
68. $\frac{24}{25}$ [5.1]   **69.** $\dfrac{x}{\sqrt{1 - x^2}}$ [5.1]
70. $\dfrac{1}{\sqrt{x^2 + 1}}$ [5.1]
71. (A)

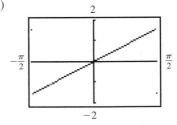

   (B) The domain for $\tan x$ is the set of all real numbers,
        except $x = \pi/2 + k\pi, k$ an integer. The range is
        R, which is the domain for $\tan^{-1} x$. So,
        $y = \tan^{-1}(\tan x)$ has a graph for all real x, except
        $x = \pi/2 + k\pi, k$ an integer. But

$\tan^{-1}(\tan x) = x$ only on the restricted domain of $\tan x$, $-\pi/2 < x < \pi/2$.   [5.1]

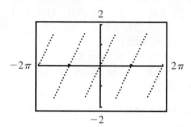

72. 0.00024 sec   [5.3]

73. 0.001936 sec   [5.3]

74. (A) $\theta = 2\arctan(500/x)$ (B) 45.2°   [5.1]

75. 31.2°   [5.1]

76. (C) From Table 1, Max $\theta = 2.73°$ when $x = 15$ ft.

TABLE 1

x (ft)	0	5	10	15
θ (deg)	0.00	1.58	2.47	2.73
x (ft)	20	25	30	
θ (deg)	2.65	2.46	2.25	

(D) As x increases from 0 ft to 30 ft, θ increases rapidly at first to a maximum of about 2.73° at 15 ft, then decreases more slowly.

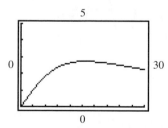

(E) From the graph, Max $\theta = 2.73°$ when $x = 15.73$ ft.

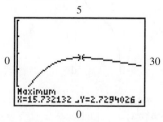

(F) $\theta = 2.5°$ when $x = 10.28$ ft or when $x = 24.08$ ft

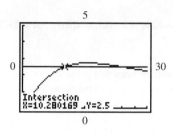

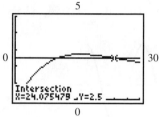

[5.1, 5.4]

77. (B)

TABLE 2

d (ft)	8	10	12	14	16
A (ft²)	704	701	697	690	682
d (ft)	18	20			
A (ft²)	670	654			

(C) $d = 17.2$ ft

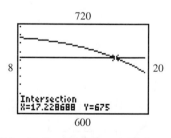

[5.1, 5.4]

78. 2.82 ft   [5.4]   **79.** 5.2 sec   [5.4]

80. (A)

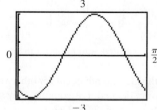

(B) 0.74 sec, 1.16 sec
(C) 0.37 sec, 1.52 sec   [5.4]

Cumulative Review Exercise Chapters 1–5

1. 241.22°   [2.1]
2. 8.83 rad   [1.1, 2.1]
3. An angle of radian measure 0.5 is the central angle of a circle subtended by an arc with measure half that of the radius of the circle.   [2.1]
4. 54°, 36°, 6.1 m   [1.3]
5. $\cos\theta = -\frac{5}{13}$; $\cot\theta = \frac{5}{12}$ [2.3]
6. (A) 0.3786   (B) 15.09   (C) −0.5196   [2.3]
7. (A)

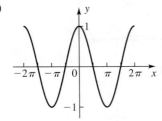

(B)

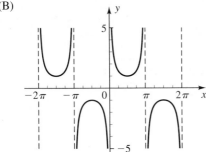

(C)

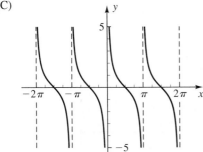

[3.1, 3.2, 3.6]

8. No, because $\sin\theta = 1/\csc\theta$, and either both are positive or both are negative.   [2.3]
13. No, because both sides are not equal for other values of x for which they are defined; for example, they are not equal for $x = 0$ or $x = \pi/2$.   [4.1]
14. $1/\sqrt{2}$ [2.5]   15. $\sqrt{3}$ [2.5]

16. Undefined   [2.5]   17. 0   [5.1]
18. $5\pi/6$ [5.1]   19. Undefined   [5.1]
20. $5\pi/6$ [5.2]   21. 0.0505   [1.3, 5.1]
22. 2.379   [1.3, 5.1]   23. −1.460   [1.3, 5.1]
24. 0.3218   [5.2]   25. 1.176   [5.2]
26. $x = \tan(-1) = -1.557408$ [5.1]
27. 30°, 150°   [5.3]   28. $-\pi/6$ [5.3]
29. $-\pi$ [5.3]   30. $\frac{1}{2}(\sin 10u + \sin 4u)$ [4.5]
31. $-2\sin 3w \sin 2w$ [4.5]
32. Yes; using the formula $\theta_d = \dfrac{180°}{\pi}\,\theta_r$, if θ_r is halved, then θ_d will also be halved.   [2.1]
33. 92°27′43″   [1.1]   34. 144°   [1.1]
35. $x = 6$; $\theta = 33.7°$ [1.3]   36. $2\pi/5$ [2.1]
37. $\sin\theta = -1/\sqrt{5}$, $\cos\theta = -2/\sqrt{5}$, $\cot\theta = 2$, $\sec\theta = -\sqrt{5}/2$, $\csc\theta = -\sqrt{5}$ [2.3]
38. Period $= 4\pi$; amplitude $= 2$ [3.2]

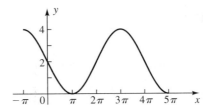

39. Period $= \pi$; amplitude $= 3$; phase shift $= \pi/2$ [3.3]

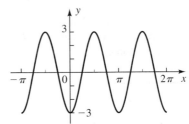

40. Period $= 1$; phase shift $= \frac{1}{4}$ [3.6]

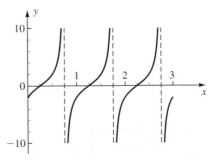

41. $y = 1 + 3\cos(\pi x)$ [3.2]
42. $y = -1 + 2\sin(\pi x/2)$ [3.2]
43. Hypotenuse: 46.0 cm; angles: 25°0′, 65°0′   [1.3]

49. (A) −0.4969 (B) 0.7531   [4.1]

50. $y = \sin(3x - x) = \sin 2x$ [4.3]

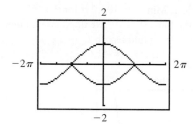

51. $\sin(x/2) = 4/\sqrt{17}$; $\cos 2x = \frac{161}{289}$ [2.3, 4.4]

52. $[-2\pi, \pi] \cup [\pi, 2\pi]$ [4.4]

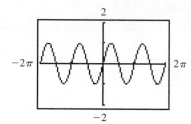

53. (A) Not an identity. Both sides are defined at $x = \pi/4$, but are not equal. (B) An identity.   [4.2]

54. (A) $\sqrt{21}/5$ [5.1] (B) $\sqrt{6}$ [5.1]
(C) 3   [5.1]   (D) $\sqrt{15}$ [5.2]

55. 60°   [1.3, 5.1] **56.** 64.01°   [1.3, 5.1]

57. $7\pi/6, 3\pi/2, 11\pi/6$ [5.3]

58. $90° + k(180°)$, k any integer   [5.3]

59. $\pi/3 + k\pi, 2\pi/3 + k\pi$, k any integer   [5.3]

60. $3.745 + 2k\pi, 5.679 + 2k\pi$, k any integer   [5.3]

61. $1.130 + 2k\pi, 5.153 + 2k\pi$, k any integer   [5.3]

62. $1.823 + 2k\pi, 4.460 + 2k\pi$, k any integer   [5.3]

63. Figure (a) does not, because −3 is not in the restricted domain for the cosine–inverse cosine identity. Figure (b) does, because 2.51 is in the restricted domain for the cosine–inverse cosine identity.   [5.1]

64. −1.893, 1.249   [5.4] **65.** 0.642   [5.4]

66. 0, 3.895, 5.121, 9.834, 12.566   [5.4]

67. The coordinates are (cos 28.703, sin 28.703) = (−0.9095, −0.4157). The point P is in the third quadrant, since both coordinates are negative.   [2.3]

68. $s = \cos^{-1} 0.5313 = 1.0107$ or
$s = \sin^{-1} 0.8472 = 1.0107$ [2.3]

69. $\sin \theta = -\sqrt{1 - a^2}$, $\tan \theta = \dfrac{-\sqrt{1 - a^2}}{a}$,
$\cot \theta = \dfrac{-a}{\sqrt{1 - a^2}}$, $\sec \theta = \dfrac{1}{a}$, $\csc \theta = \dfrac{-1}{\sqrt{1 - a^2}}$
[2.3, 2.5]

71. $\frac{7}{9}$ [4.4, 5.1] **72.** $\dfrac{\sqrt{1 - x^2} - x^2}{\sqrt{1 + x^2}}$ [5.1]

73. $\pi/2 + 2k\pi, \pi + 2k\pi$, k any integer   [5.4]

74. 0.6662, 2.475   [5.4]

75. $g(x) = 3 - \cos x$ [4.2]

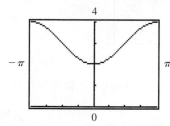

76. $g(x) = 4 + 2 \cos 2x$ [4.4]

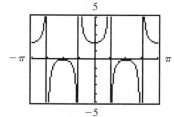

77. $g(x) = 1 + \sec 2x$ [4.4]

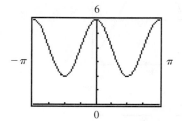

78. $g(x) = 3 - \cot(x/2)$ [4.4]

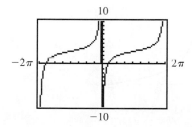

79. [0.694, 2.000]   [5.4]   **80.** 1,429 m   [1.4]

81. $a \cot x$ [4.1]   **82.** $x = 2\sqrt{3}$; $\theta = 30°$ [4.4]

83. 0.443 sec   [5.3]

84. 0.529 sec, 0.985 sec   [5.4]

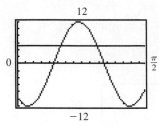

85. $\theta = 0.650$ rad   [1.4, 4.3]

86. (A) Amplitude: 60; period: $\frac{1}{45}$ sec; frequency 45 Hz; phase shift: $\frac{1}{180}$

(B)

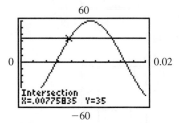

(C) $t = 0.0078$ sec   [3.3]

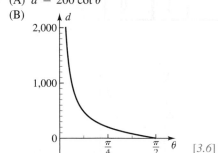

87. (A) $d = 200 \cot \theta$

(B)

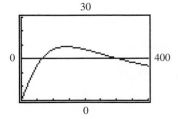

[3.6]

88. (B) 12.5°   [5.1]

89. 64.9 ft or 308.3 ft   [5.1]

90. 49.5π rad   [2.1]

91. $11,550\pi \approx 36,300$ cm/min   [2.2]

92. (A) L will decrease, then increase.

(C)

TABLE 1			
θ (deg)	35	40	45	50
L (ft)	117.7	110.4	106.1	104.2
θ (deg)	55	60	65	
L (ft)	104.6	107.7	114.3	

(D) From Table 1, Min $L = 104.2$ ft for $\theta = 50°$. The length of the longest log that will go around the corner is the minimum length L.

(E) Min $L = 104.0$ ft for $\theta = 51.6°$ [1.4, 2.3]

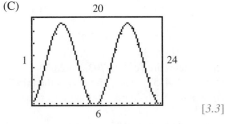

93. (A)

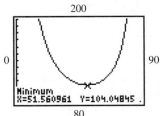

(B) $y = 12.435 + 6.835 \sin(\pi x/6 - 1.6)$

(C)

[3.3]

CHAPTER 6

Exercise 6.1

5. Yes   **7.** No   **9.** No

11. $\alpha = 101°$, $b = 64$ cm, $c = 55$ cm

13. $\alpha = 98.0°$, $b = 4.32$ mm, $c = 7.62$ mm

15. $\gamma = 22.9°$, $a = 27.3$ cm, $c = 12.6$ cm
17. $\alpha = 67°20'$, $a = 55.1$ km, $c = 58.8$ km
19. One triangle; case (d), where α is acute and $a \geq b$ $(a = 5, b = 4)$
21. Zero triangles; case (a), where α is acute and $0 < a < h$ $(a = 3, h = 4)$
23. Two triangles; case (c), where α is acute and $h < a < b$ $(h = 3\sqrt{2}, a = 5, b = 6)$
25. One triangle; case (f), where α is obtuse and $a > b$ $(a = 3, b = 2)$
27. One triangle; case (b), where α is acute and $a = 4 = h$
29. Zero triangles; case (e), where α is obtuse and $0 < a \leq b$ $(a = 2, b = 3)$
31. $\beta = 27°$, $\gamma = 19°$, $c = 17$ cm
33. Triangle I: $\beta = 81°$, $\gamma = 30°$, $c = 46$ ft
Triangle II: $\beta' = 99°$, $\gamma' = 12°$, $c' = 19$ ft
35. Triangle I: $\beta = 28°$, $\gamma = 131°$, $c = 9.9$ in.
Triangle II: $\beta' = 152°$, $\gamma' = 7°$, $c' = 1.6$ in.
37. $\gamma = 63°$, $b = 66$ m, $c = 64$ m
39. $\alpha = 90°$, $\gamma = 60°$, $c = 47$ cm
41. $\alpha = 39°$, $\gamma = 84°$, $c = 74$ ft
43. No solution
45. $\gamma = 18°10'$, $\alpha = 91.0$ cm, $b = 62.3$ cm
47. Triangle I: $\alpha = 57°50'$, $\gamma = 100°$, $c = 65.5$ mm
Triangle II: $\alpha' = 122°10'$, $\gamma' = 35°40'$, $c' = 38.8$ mm
49. $26.63 \approx 26.66$
53. $k = a \sin \beta = 66.8 \sin 46.8° \approx 48.7$
55. $AC = 12.8$ mi
57. 8.37 mi from A; 4.50 mi from B
59. 230 m   **61.** 103 mi   **63.** 8,180 ft   **65.** 159 ft
67. $\beta = 131.8°$, $c = 5.99$ m
69. $r = 7.76$ mm, $s = 13.4$ mm
71. 396 mi   **73.** 46°30'   **75.** 8.0 cm

Exercise 6.2

3. Law of cosines   **5.** Law of sines
7. Law of cosines   **9.** Law of sines
11. A triangle can have at most one obtuse angle. Since β is acute, then, if the triangle has an obtuse angle, it must be the angle opposite the longer of the two sides, a and c. Thus, γ, the angle opposite the shorter of the two sides, c, must be acute.
13. $a = 6.00$ mm, $\beta = 65°0'$, $\gamma = 64°20'$
15. $c = 26.2$ m, $\alpha = 33.3°$, $\beta = 12.7°$
17. If the triangle has an obtuse angle, then it must be the angle opposite the longest side—in this case, α.
19. $\alpha = 62.8°$, $\beta = 36.3°$, $\gamma = 80.9°$
21. $\alpha = 29°12'$, $\beta = 63°26'$, $\gamma = 87°22'$
23. $\alpha = 30.3°$, $a = 46.2$ ft, $c = 27.2$ ft
25. $b = 19.3$ m, $\alpha = 40.6°$, $\gamma = 72.9°$

27. No solution
29. $\alpha = 84.1°$, $\beta = 29.7°$, $\gamma = 66.2°$
31. No solution
33. $\beta = 28.2°$, $a = 984$ m, $c = 1,310$ m
35. No solution
37. Triangle 1: $\beta = 65.3°$, $\gamma = 68.0°$, $c = 23.1$ yd;
Triangle 2: $\beta' = 114.7°$, $\gamma' = 18.6°$, $c' = 7.93$ yd
39. An infinite number of similar triangles have those angles.
41. $b^2 = a^2 + c^2 - 2ac \cos 90° = a^2 + c^2 - 0 = a^2 + c^2$
43. $-0.872 \approx -0.873$
47. $\cos^{-1}(0.75) \approx 41.4°$
49. $CB = 613$ m   **51.** 113.3°   **53.** 180 km
55. 9.6 hr   **57.** 29 cm
59. (A) 5.87 mi   (B) 16.1°
61. $AC = 23.5$ ft; $AB = 18.9$ ft   **63.** 24,800 mi
65. 64°   **67.** 150 m

Exercise 6.3

3. 102 m^2 **5.** $15/4$ m^2 **7.** 12 cm^2
9. 11.6 in.^{2} **11.** $40,900$ ft^2 **13.** 45.4 cm^2
15. 129 m^2 **17.** 2.0 ft^2 **19.** 480 in.^{2}
21. 0.18 cm^2 **23.** True   **25.** False
27. $A_1 = A_3 = \dfrac{ab}{2} \sin(180° - \theta) = \dfrac{ab}{2} \sin \theta = A_2 = A_4$
29. 84 ft^2 **31.** $2,300$ m^2
33. $12,100   **35.** $47,000

Exercise 6.4

3. $|\mathbf{u} + \mathbf{v}| = \sqrt{241}$

5. $|\mathbf{u} + \mathbf{v}| = 54$ km/hr; $\theta = 36°$
7. $|\mathbf{u} + \mathbf{v}| = 101$ lb; $\theta = 77°$
9. $|\mathbf{u} + \mathbf{v}| = 32.3$ knots; $\theta = 61.1°$
11. $|\mathbf{H}| = 87$ lb; $|\mathbf{V}| = 37$ lb
13. $|\mathbf{H}| = 23.2$ km/hr; $|\mathbf{V}| = 324$ km/hr
15. No. The magnitude of a geometric vector is the length (a nonnegative quantity) of the directed line segment.
17. $|\mathbf{u} + \mathbf{v}| = 190$ lb, $\alpha = 18°$
19. $|\mathbf{u} + \mathbf{v}| = 699$ mi/hr, $\alpha = 7.4°$
21. Same   **23.** Opposite   **25.** Neither   **27.** Same
29. Since the zero vector has an arbitrary direction, it can be parallel to any vector.
33. Magnitude = 5.0 km/hr; direction = 143°
35. 250 mi/hr; 350°
37. Magnitude = 2,500 lb; direction = 13° (relative to $\mathbf{F}_1$)

39. 32°
41. 2,500 sin 15° = 650 lb; 2,500 cos 15° = 2,400 lb
43. |**H**| = 40 lb; |**V**| = 33 lb   **45.** To the left

Exercise 6.5

5.

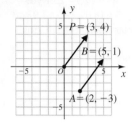

7.

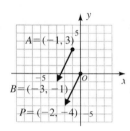

9. ⟨4, 2⟩   **11.** ⟨4, −4⟩   **13.** 5   **15.** √29
17. Two geometric vectors are equal if and only if they have the same magnitude and direction.
19. (A) ⟨−2, 6⟩ (B) ⟨4, 2⟩ (C) ⟨11, 2⟩ (D) ⟨6, 18⟩
21. (A) ⟨1, −6⟩ (B) ⟨3, 0⟩ (C) ⟨7, 3⟩ (D) ⟨3, −6⟩
23. **u** = ⟨1/√2, −1√2⟩
25. **u** = ⟨−$\frac{4}{5}$, $\frac{3}{5}$⟩   **27.** **v** = −10**i** + 7**j**   **29.** **v** = −12**i**
31. **v** = 10**i** − 14**j**   **33.** 9.1**i** + 7.9**j**   **35.** 2.3**i** − 10.8**j**
37. 10**i** − 10**j**   **39.** −**i** − 2**j**   **41.** 16**i** − 20**j**
43. Any one of the force vectors must have the same magnitude as the resultant of the other two force vectors and be directed opposite to the resultant of the other two.
45. ⟨15/√13, 10/√13⟩
47. ⟨$\frac{8}{5}$, −$\frac{6}{5}$⟩
61. Left side: 676 lb; Right side: 677 lb
63. (B) 300 lb
65. For *AB*, a compression of 9,000 lb; for *CB*, a tension of 7,500 lb
67. θ = 18.6°; Line tension = 10.1 lb

Exercise 6.6

3. 10   **5.** 21   **7.** 0   **9.** 31
11. 45.0°   **13.** 1.2°   **15.** 165.7°

17. Negative, because cos θ is negative for θ an obtuse angle.
19. Orthogonal   **21.** Not orthogonal
23. No   **25.** Yes
27. 12.0   **29.** 0
31. −0.0673   **39.** ⟨3, 0⟩
41. −$\frac{36}{13}$**i** − $\frac{24}{13}$**j**   **47.** False
49. True   **51.** 4,900 ft-lb
53. 800 ft-lb; more work is done
55. 85 ft-lb   **57.** 9 ft-lb
59. 20 ft-lb

Chapter 6 Review Exercise

1. To use the law of sines, we need to have an angle and a side opposite the angle given, which is not the case here.   [6.1]
2. The two forces are oppositely directed; that is, the angle between the two forces is 180°.   [6.4]
3. The forces are acting in the same direction; that is, the angle between the two forces is 0°.   [6.4]
4. (A) One triangle, since h = 8 sin 30° = 4 = a
     (B) Two triangles, since h = 7 sin 30° = 3.5 and
         h < a < b
     (C) Zero triangles, since h = 8 sin 30° = 4 and
         a < h   [6.1]
5. β = 22°, a = 90 cm, c = 110 cm   [6.1]
6. γ = 82°, a = 21 m, c = 22 m   [6.1]
7. a = 21 in., β = 53°, γ = 78°   [6.2]
8. β = 49°, γ = 69°, c = 15 cm   [6.1]
9. 224 in.²   [6.3]   **10.** 79 cm²   [6.3]
11. 9.4 at 32°   [6.4]
12. |**H**| = 9.8; |**V**| = 6.9; **v** = 9.8**i** + 6.9**j**   [6.4]
13. ⟨2, −5⟩   [6.5]   **14.** 13   [6.5]
15. −8   [6.6]   **16.** 4   [6.6]
17. 36.9°   [6.6]   **18.** 123.7°   [6.6]
19. (A) Neither (B) Parallel (C) Orthogonal   [6.6]
20. No triangle is possible, since sin β cannot exceed 1.   [6.1]
21. a = 97.7 m, β = 72.8°, γ = 42.2°   [6.2]
22. β = 43°30′, γ = 101°10′, c = 22.4 in.   [6.1]
23. β = 136°30′, γ = 8°10′, c = 3.24 in.   [6.1]
24. α = 50°, β = 59°, γ = 71°   [6.2]
25. 3,380 m²   [6.3]   **26.** 980 mm²   [6.3]
27. The two vectors have the same magnitude, but different directions.   [6.4]
28. They are equal if and only if a = c and b = d.   [6.5]
29. |**u** + **v**| = 23.3; θ = 15.9°   [6.4]
30. (A) ⟨2, −3⟩ (B) ⟨6, 3⟩ (C) ⟨16, 6⟩ (D) ⟨15, 8⟩
     [6.5]
31. (A) 5**i** − 4**j** (B) **i** + 2**j** (C) 5**i** + 3**j** (D) 5**j** [6.5]

32. $\mathbf{u} = \langle -\frac{8}{17}, \frac{15}{17} \rangle$ [6.5]

33. (A) $\mathbf{v} = -5\mathbf{i} + 7\mathbf{j}$ (B) $\mathbf{v} = -3\mathbf{j}$ (C) $\mathbf{v} = -4\mathbf{i} - \mathbf{j}$
[6.5]

34. (A) Orthogonal   (B) Not orthogonal   [6.6]

35. (A) $17/\sqrt{10} = 5.38$ (B) $-7/\sqrt{10} = -2.21$ [6.6]

36. Any one of the force vectors must have the same
magnitude as the resultant of the other two force vectors
and must be oppositely directed to the resultant of the
other two.   [6.5]

37. $-2.68 \approx -2.70$ (checks)   [6.1]

38. $k = 12.7 \sin 52.3°$ [6.1]

45. 5.9 cm, 17 cm   [6.2]   **46.** 252 m   [6.1]

47. 25 cm   [6.1]   **48.** 540 ft   [6.2]

49. 220,000 ft^2 [6.3]   **50.** 4.00 km   [6.2]

51. 660 mi   [6.2]   **52.** 325 mi   [6.1]

53. 260 km/hr at 57°   [6.4]   **54.** 1.5 hr   [6.2]

55. Magnitude = 489 lb; Direction = 13.4°   [6.4]

56. 10,800 ft-lb   [6.6]

57. Compression = 150 lb; Tension = 300 lb   [6.5]

58. (A) $|\mathbf{u}| = |\mathbf{w}| \cos\theta$; $|\mathbf{v}| = |\mathbf{w}| \sin\theta$
   (B) $|\mathbf{u}| = 40$ lb; $|\mathbf{v}| = 124$ lb   (C) $\theta = 80°$ [6.5]

59. 56 ft-lb   [6.6]

60. $\theta = 15.4°$; Tension = 19.5 lb   [6.5]

CHAPTER 7

Exercise 7.1

1.

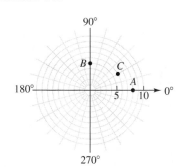

3.

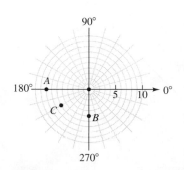

5.

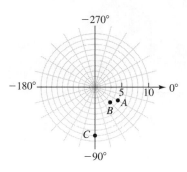

7.

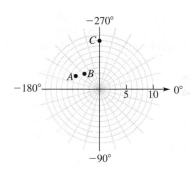

9.

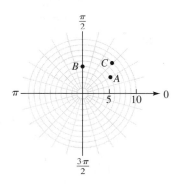

11.

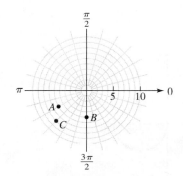

13.

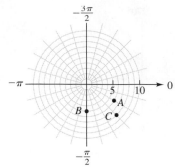

3.

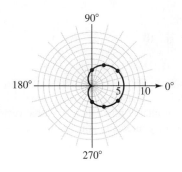

19. $(0, -2)$ **21.** $(5, 5\sqrt{3})$ **23.** $(3, -3)$

25. $(6, \pi/2)$ **27.** $(4\sqrt{2}, 3\pi/4)$ **29.** $(2\sqrt{3}, -2\pi/3)$

31. $(-6, -210°)$: The polar axis is rotated 210° clockwise (negative direction) and the point is located 6 units from the pole along the negative polar axis.
$(-6, 150°)$: The polar axis is rotated 150° counterclockwise (positive direction) and the point is located 6 units from the pole along the negative polar axis.
$(6, 330°)$: The polar axis is rotated 330° counterclockwise (positive direction) and the point is located 6 units along the positive polar axis.

33. $(3.900, 65.374°)$ **35.** $(10.045, -6.299°)$

37. $(8.992, -110.962°)$ **39.** $(2.314, -1.426)$

41. $(4.249, -0.241)$ **43.** $(0.051, 0.902)$

45. $r = 6 \cos \theta$ **47.** $r(2 \cos \theta + 3 \sin \theta) = 5$

49. $r^2 = 9$, or $r = \pm 3$ **51.** $r^2 = \dfrac{1}{\sin 2\theta}$

5.

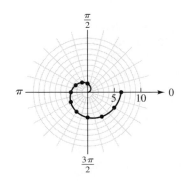

53. $r^2 = \dfrac{4}{4 - 5 \sin^2 \theta}$ **55.** $r = \dfrac{3}{\cos \theta}$

57. $r = -\dfrac{7}{\sin \theta}$ **59.** $2x + y = 4$

61. $x^2 + y^2 = 8x$ **63.** $x^2 + y^2 = 2x + 3y$

65. $x^2 - y^2 = 4$ **67.** $(x^2 + y^2)^2 = 3x^2 - 3y^2$

69. $x^2 + y^2 = 16$ **71.** $y = \dfrac{1}{\sqrt{3}} x$

7.

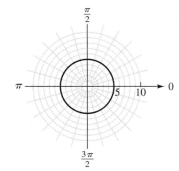

73. $y - 3 = -2\sqrt{x^2 + y^2}$, or $(y - 3)^2 = 4(x^2 + y^2)$

77. 1.615 units

Exercise 7.2

1.

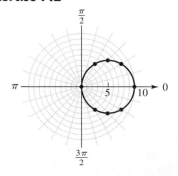

9.

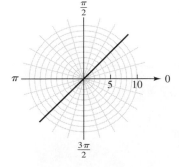

11.

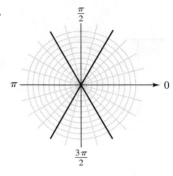

13.

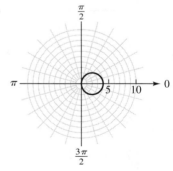

15.

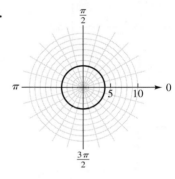

17.

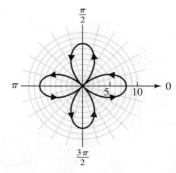

19.

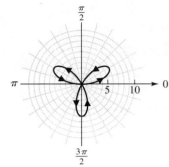

21.

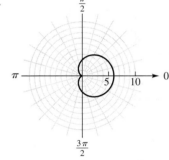

23.

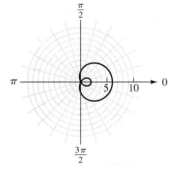

25.

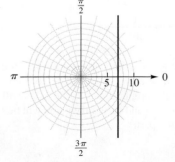

27.

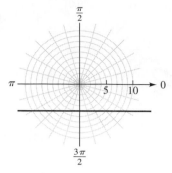

29. (A) $r = 5 + 5 \cos \theta$

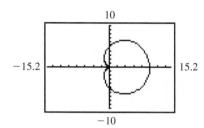

(B) $r = 5 + 4 \cos \theta$

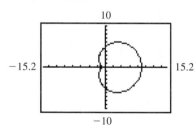

(C) $r = 4 + 5 \cos \theta$

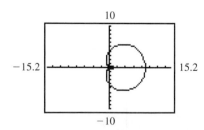

(D) If $a = b$, the graph will touch but not pass through the origin. If $a > b$, the graph will not touch or pass through the origin. If $a < b$, the graph will go through the origin and part of the graph will be inside the other part.

31. (A) $r = 9 \cos \theta$

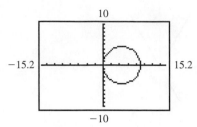

$r = 9 \cos 3\theta$

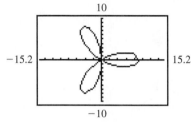

$r = 9 \cos 5\theta$

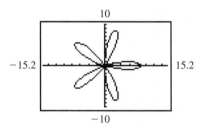

(B) 7 leaves   (C) n leaves

33. (A) $r = 9 \cos 2\theta$

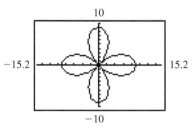

$r = 9 \cos 4\theta$

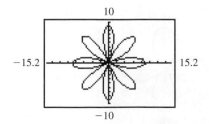

$r = 9 \cos 6\theta$

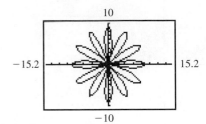

(B) 16 leaves   (C) $2n$ leaves

35. (A) $r = 9 \cos(\theta/2)$

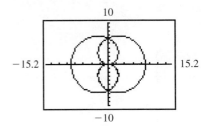

(B) $r = 9 \cos(\theta/4)$

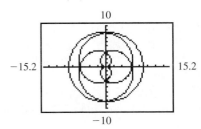

(C) n times

37.

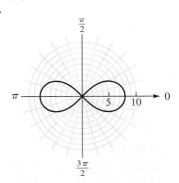

39. For each n, there are n large petals and n small petals.
For n odd, the small petals are within the large petals;
for n even, the small petals are between the large petals.

41.

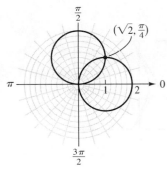

$(r, \theta) = (\sqrt{2}, \pi/4)$ [Note that $(0, 0)$ is not a solution to
the system even though the graphs cross at the origin.]

43.

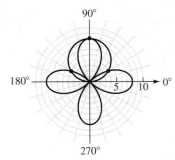

$(r, \theta) = (4, 30°), (4, 150°), (-8, 270°)$
[Note that $(0, 0)$ is not a solution to the system even
though the graphs cross at the origin.]

45. (a), (d), (e), (g), (i), (j), (k)

49. (A) 9.5 knots   (B) 12.0 knots
(C) 13.5 knots   (D) 12.0 knots

51. (A) Ellipse

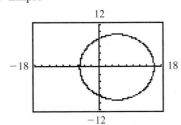

(B) Parabola

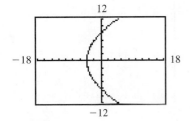

(C) Hyperbola

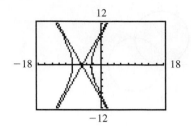

53. (A) Perihelion $= 2.85 \times 10^7$ mi;
Aphelion $= 4.33 \times 10^7$ mi

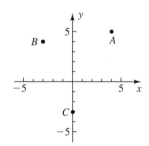

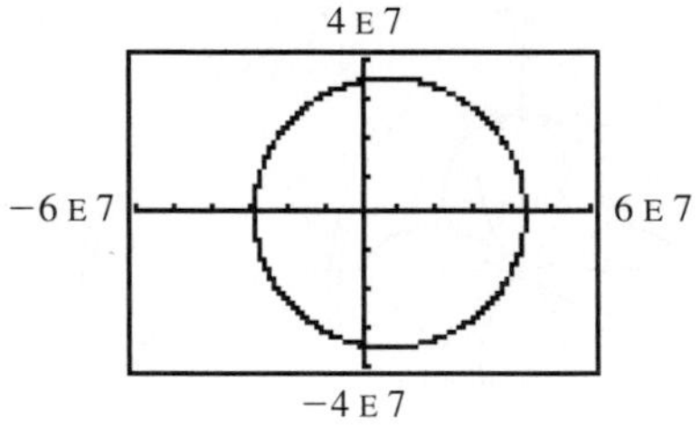

(B) Faster at perihelion

Exercise 7.3

1.

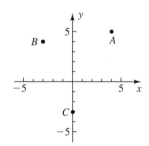

3.

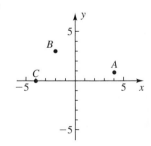

5.

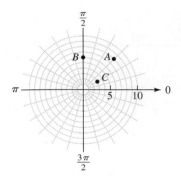

7.

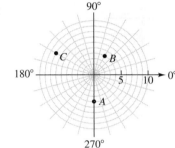

9. (A) $2e^{(-\pi/6)i}$ (B) $2\sqrt{2}e^{(3\pi/4)i}$ (C) $7.81e^{(-0.69)i}$
11. (A) $21e^{(\pi/4)i}$ (B) $12e^{(-\pi/6)i}$ (C) $6.32e^{(0.83)i}$
13. (A) $2e^{(120°)i}$ (B) $3e^{(-90°)i}$ (C) $8.06e^{(-150.26°)i}$
15. (A) $5e^{(30°)i}$ (B) $4e^{(-45°)i}$ (C) $2.83e^{(54°)i}$
17. (A) $\sqrt{3} + i$ (B) $-1 - i$ (C) $5.06 - 2.64i$
19. (A) $-i\sqrt{3}$ (B) $-1 + i$ (C) $-2.20 - 6.46i$
21. $32e^{(37°)i}$; $0.5e^{(13°)i}$
23. $12e^{(-12°)i}$; $3e^{(228°)i}$ or $3e^{(-132°)i}$
25. $7.37e^{2.18i}$; $2.58e^{0.36i}$ **27.** $20e^{(\pi/12)i}$; $0.8e^{(7\pi/12)i}$
29. $4i$; $4e^{(90°)i}$ **31.** $2i$; $2e^{(90°)i}$
33. $-2 + 2i$; $2\sqrt{2}e^{(135°)i}$ **39.** z^n appears to be $r^n e^{n\theta i}$
45. $re^{-i\theta}$
47. (A) $(8 + 0i) + (3\sqrt{3} + 3i) = (8 + 3\sqrt{3}) + 3i$
(B) $13.5(\cos 12.8° + i \sin 12.8°)$
(C) 13.5 lb at an angle of $12.8°$

Exercise 7.4

1. $27e^{(45°)i}$ **3.** $32e^{(450°)i} = 32e^{(90°)i}$
5. $16e^{(4\pi/5)i}$ **7.** $25e^{(11\pi/3)i}$ **9.** -4
11. $16\sqrt{3} + 16i$ **13.** 1 **15.** $2e^{(15°)i}, 2e^{(195°)i}$
17. $2e^{(30°)i}, 2e^{(150°)i}, 2e^{(270°)i}$
19. $2^{1/10}e^{(27°)i}, 2^{1/10}e^{(99°)i}, 2^{1/10}e^{(171°)i}, 2^{1/10}e^{(243°)i}$,
$2^{1/10}e^{(315°)i}$

21. $w_1 = 2e^{(60°)i}, w_2 = 2e^{(180°)i}, w_3 = 2e^{(300°)i}$

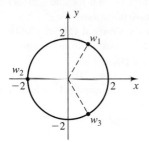

23. $w_1 = 1e^{(0°)i}, w_2 = 1e^{(90°)i}, w_3 = 1e^{(180°)i},$
$w_4 = 1e^{(270°)i}$

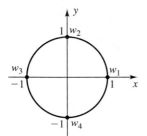

25. $w_1 = 1e^{(-18°)i}, w_2 = 1e^{(54°)i}, w_3 = 1e^{(126°)i},$
$w_4 = 1e^{(198°)i}, w_5 = 1e^{(270°)i}$

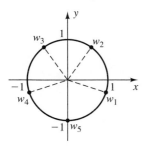

27. (A) $(-\sqrt{2} + i\sqrt{2})^4 + 16 = -16 + 16 = 0$; three
 (B) The four roots are equally spaced around the circle. Since there are four roots, the angle between successive roots on the circle is $360°/4 = 90°$.

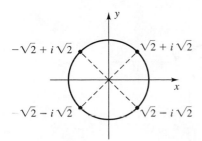

(C) $(\sqrt{2} + i\sqrt{2})^4 + 16 = -16 + 16 = 0,$
$(-\sqrt{2} - i\sqrt{2})^4 + 16 = -16 + 16 = 0,$
$(\sqrt{2} - i\sqrt{2})^4 + 16 = -16 + 16 = 0$

29. $\dfrac{\sqrt{2}}{2} + \dfrac{\sqrt{2}}{2}i, -\dfrac{\sqrt{2}}{2} - \dfrac{\sqrt{2}}{2}i$

31. $\sqrt{2} - i\sqrt{2}, -\sqrt{2} + i\sqrt{2}$

33. $\dfrac{3}{2} + \dfrac{3\sqrt{3}}{2}i, -3, \dfrac{3}{2} - \dfrac{3\sqrt{3}}{2}i$

35. $4, -2 + 2i\sqrt{3}, -2 - 2i\sqrt{3}$

39. $1, 0.309 + 0.951i, -0.809 + 0.588i,$
 $-0.809 - 0.588i, 0.309 - 0.951i$

41. $0.855 + 1.481i, -1.710, 0.855 - 1.481i$

43. (B) z^3, z^6
 (C) z^2, z^4, z^6

45. (B) z^5, z^{10}
 (C) $z^2, z^4, z^6, z^8, z^{10}$

Chapter 7 Review Exercise

1.

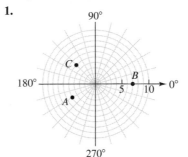

$[7.1]$

2.

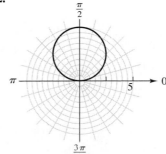

$[7.2]$

3.

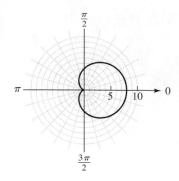

[7.2]

4.

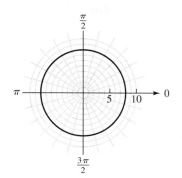

[7.2]

5. $(2, 2)$ [7.1]   **6.** $(2, 5\pi/6)$ [7.1]

7.

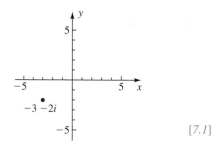

[7.1]

8.

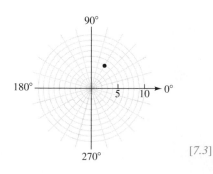

[7.3]

9. $(-8, -330°)$: The polar axis is rotated 330° clockwise (negative direction) and the point is located 8 units from the pole along the negative polar axis.

$(8, -150°)$: The polar axis is rotated 150° clockwise (negative direction) and the point is located 8 units from the pole along the positive polar axis.

$(8, 210°)$: The polar axis is rotated 210° counterclockwise (positive direction) and the point is located 6 units along the positive polar axis. [7.1]

10. $z_1 z_2 = 27e^{(79°)i}$; $z_1/z_2 = 3e^{(5°)i}$ [7.3]

11. $16e^{(40°)i}$ [7.4]

12. $r = \dfrac{7}{\sin \theta}$ [7.1]

13.

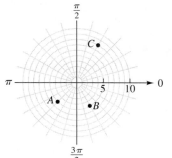

[7.1]

14.

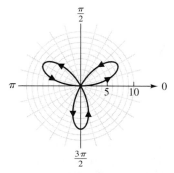

[7.2]

15.

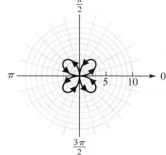

[7.2]

16.

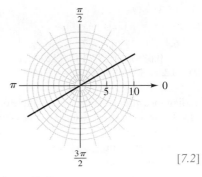

[7.2]

17. $r^2 = 8r \cos \theta, r = 8 \cos \theta$ [7.1]

18. $3x - 2y = -2$ [7.1]

19. $x^2 + y^2 = -3x$ [7.1]

20.

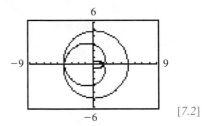

[7.2]

21.

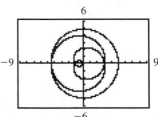

[7.2]

22. $256e^{(5\pi/4)i}; 4e^{(\pi/4)i}$ [7.3] **23.** $216e^{(5\pi/2)i}$ [7.4]

24. $2e^{(-150°)i}$ [7.3] **25.** $-3 + 3i$ [7.3]

26. $8e^{(195°)i}$ [7.3] **27.** $0.5e^{(75°)i}$ [7.3]

28. -4 [7.4]

29. $\dfrac{3\sqrt{2}}{2} - \dfrac{3\sqrt{2}}{2}i, -\dfrac{3\sqrt{2}}{2} + \dfrac{3\sqrt{2}}{2}i$ [7.4]

30. $w_1 = 4, w_2 = -2 + 2i\sqrt{3}, w_3 = -2 - 2i\sqrt{3}$

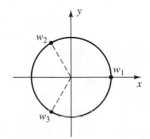

[7.4]

31. $2e^{(70°)i}, 2e^{(190°)i}, 2e^{(310°)i}$ [7.4]

32. $1e^{(\pi/6)i}, 1e^{(5\pi/6)i}, 1e^{(3\pi/2)i}$ [7.4]

33. $(2e^{(30°)i})^2 = 4e^{(60°)i} = 4(\cos 60° + i \sin 60°) = 2 + 2i\sqrt{3}$ [7.4]

34. (A) There are a total of three cube roots, and they are spaced equally around a circle of radius 2.

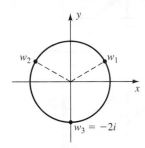

(B) $w_1 = \sqrt{3} + i, w_2 = -\sqrt{3} + i$

(C) The cube of each cube root is $8i;$ that is, each root is a cube root of $8i$. [7.4]

35. $n = 1$

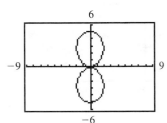

$n = 2$

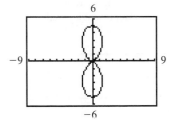

$n = 3$

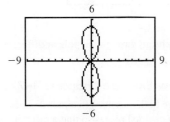

The graph always has two leaves. [7.2]

36. (A) Hyperbola

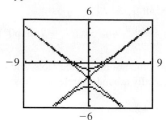

(B) Parabola

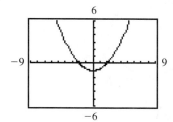

(C) Ellipse

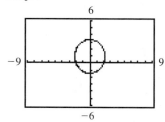

[7.3]

37. $y - 3 = -2\sqrt{x^2 + y^2}$, or
$(y - 3)^2 = 4(x^2 + y^2)$ [7.1]
38. $2.289, -1.145 + 1.983i, -1.145 - 1.983i$ [7.4]
40. (A) The coordinates of P represent a simultaneous
solution.

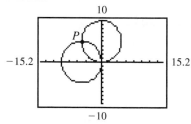

(B) $r = 5\sqrt{2}, \theta = 3\pi/4$
(C) The two graphs go through the pole at different
values of θ.   [7.2]

Cumulative Review Exercise Chapters 1-7

1. α is larger. Change both measures to either degrees or
radians to enough decimal places so that a difference
becomes apparent.   [2.1]

2. 4.82 cm, 74.7°, 15.3°   [1.3]
3. $\sec \theta = \frac{25}{7}, \tan \theta = -\frac{24}{7}$ [2.3]
5. -0.5 [2.5]   **6.** $\sqrt{3}$ [2.5]   **7.** $2\pi/3$ [5.1]
8. $\pi/4$ [5.2]   **9.** 0.6867   [2.3] **10.** 0.3249   [2.3]
11. 0.9273   [5.1]   **12.** 1.347   [5.2]
13. $2 \sin 2t \cos t$ [4.5]
14. An angle of radian measure 2.5 is the central angle of a
circle subtended by an arc with measure 2.5 times that
of the radius of the circle.   [2.1]
15. $\alpha = 28°, a = 34$ m, $c = 48$ m   [6.1]
16. $\alpha = 40°, \gamma = 106°, b = 14$ in.   [6.2]
17. $\alpha = 33°, \beta = 45°, \gamma = 102°$ [6.2]
18. 110 in.^{2} [6.3]
19. $(7, -150°)$: Rotate the polar axis 150° clockwise
(negative direction) and go 7 units along the positive
polar axis.   [7.1]
20. $|\mathbf{H}| = 12; |\mathbf{V}| = 5.5$ [6.4]   **21.** 7.5 at 31°   [6.4]
22. $\langle -7, 9 \rangle; \sqrt{130}$ [6.5]   **23.** 143.5°   [6.6]

24.

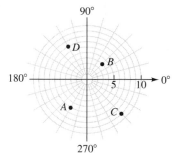

[7.1]

25.

[7.2]

26. $(-3, 3)$ [7.1]   **27.** $(4, 5\pi/6)$ [7.1]
28. $2\sqrt{2}e^{(-\pi/4)i}$ [7.3]   **29.** $-3i$ [7.3]
30. $z_1 z_2 = 15e^{(65°)i}; z_1/z_2 = 0.6e^{(35°)i}$ [7.3]
31. $81e^{(100°)i}$ [7.4]   **32.** $y = 1 + 2 \cos \pi x$ [3.2]
33. $\csc \theta = -\sqrt{17}, \cos \theta = -4/\sqrt{17}$ [2.3]

34. Amplitude $= 2$; period $= \pi$; frequency $= 1/\pi$; phase shift $= -\pi/2$

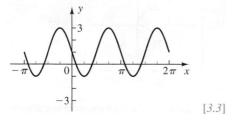

[3.3]

38. $\tan(x/2) = \frac{1}{7}$, $\sin 2x = \frac{336}{625}$ [2.3, 4.4]

39. (A) Not an identity; both sides are defined at $x = \pi/2$ but are not equal.

(B) An identity [4.2]

40. $4/\sqrt{7}$ [5.1]

41. (A) There are a total of three cube roots, and they are spaced equally around a circle of radius 2.

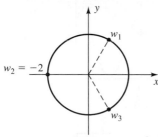

(B) $w_1 = 1 + i\sqrt{3}$, $w_3 = 1 - i\sqrt{3}$

(C) The cube of each cube root is -8. [7.4]

42. $k\pi, 2\pi/3 + 2k\pi, 4\pi/3 + 2k\pi, k$ any integer [5.4]

43. $0.8481 + 2k\pi, 2.294 + 2k\pi, k$ any integer [5.4]

44. (A) No triangle exists.

(B) $\beta = 64°10'$, $\gamma = 66°20'$, $c = 17.7$ cm
$\beta' = 115°50'$, $\gamma' = 14°40'$, $c' = 4.89$ cm

(C) $\beta = 38°50'$, $\gamma = 91°40'$, $c = 27.7$ cm; [6.1]

45. $|\mathbf{u} + \mathbf{v}| = 41.9$; $\theta = 11.0°$ [6.4]

46. (A) $\langle 3, -18 \rangle$ (B) $10\mathbf{i} - 11\mathbf{j}$ [6.5]

47. $\mathbf{u} = \langle 0.28, -0.96 \rangle$ [6.5] **48.** $2\mathbf{i} + 3\mathbf{j}$ [6.5]

49. (A) Orthogonal (B) Not orthogonal

(C) Orthogonal [6.6]

50.

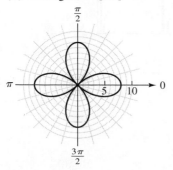

[7.2]

51. $r = 6 \tan \theta \sec \theta$ [7.1] **52.** $x^2 + y^2 = 4y$ [7.1]

53. $6\sqrt{2}\, e^{(165°)i}$ [7.3] **54.** $(\sqrt{2}/3)e^{(75°)i}$ [7.3]

55. $8i$ [7.4]

56. $w_1 = \sqrt{3} + i$, $w_2 = -\sqrt{3} + i$, $w_3 = -2i$

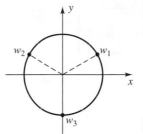

[7.4]

57.

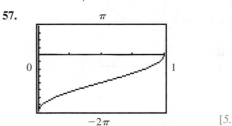

[5.1]

58.

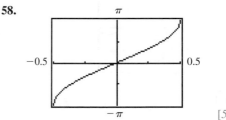

[5.1]

59.

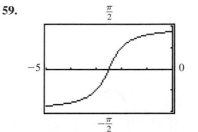

[5.1]

60. $-1.768, 1.373$ [5.3] **61.** 0.582 [5.4]

62. $3.909, 4.313, 5.111, 5.516$ [5.4]

63.

64. (A) Ellipse

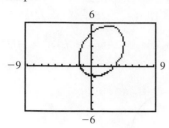

(B) Parabola

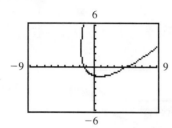

(C) Hyperbola

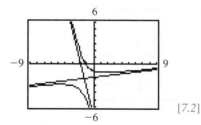

[7.2]

65. $\theta = 4$ rad; $(a, b) = (-1.307, -1.514)$ [2.3]
66. $\theta = 3.785$; $s = 7.570$ [2.3]
67.

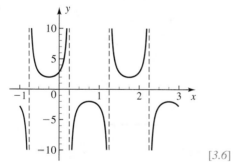

[3.6]

68. $\dfrac{1 + x^2}{1 - x^2}$ [4.4, 5.1]

70. $x - 1 = -\sqrt{x^2 + y^2}$, or $(x - 1)^2 = x^2 + y^2$ [7.1]

71. (A)

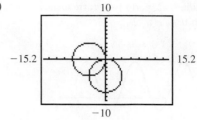

(B) The two graphs go through the pole at different values of θ. [7.2]
72. $1.587, -0.794 + 1.375i, -0.794 - 1.375i$ [7.4]
73. (A) $\cos 3\theta = \cos^3 \theta - 3 \cos \theta \sin^2 \theta$,
  $\sin 3\theta = 3 \cos^2 \theta \sin \theta - \sin^3 \theta$ [7.4]
74. $g(x) = -1 + 3 \cos 2x$

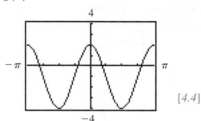

[4.4]

75. $g(x) = \sec 2x - 3$

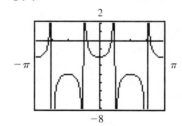

[4.4]

76. $g(x) = -1 + \cot(x/2)$

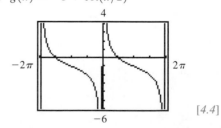

[4.4]

77. 650 ft [6.1] **78.** 550 ft [6.2]
79. (A) 48 ft   (B) 41 ft [1.4, 6.1]
80. Station *A:* 8.7 mi; Station *B:* 6.3 mi [1.4, 6.1]

81. (A) Period $= \frac{1}{70}$ sec; $B = 140\pi$

(B) Frequency $= 80$ Hz; $B = 160\pi$

(C) Period $= \frac{1}{50}$ sec; frequency $= 50$ Hz   *[3.2]*

82. Wave height $= 4$ ft; wavelength $= 82$ ft; speed $= 20$ ft/sec   *[3.4]*

83. Area $= \frac{1}{2}\theta + \frac{1}{2}\sin\theta\cos\theta$ *[2.1, 2.3]*

84. Area $= \frac{1}{2}\sin^{-1}x + \frac{1}{2}x\sqrt{1 - x^2}$ *[2.3, 5.1]*

85. $\theta = 0.553$ *[5.4]*   **86.** $x = 0.412$ *[5.4]*

87. (A) $R = r + (H - h)\tan\alpha$

(B) $\beta = \tan^{-1}\left(\dfrac{H - h}{R - r}\right)$ *[1.4]*

88. $18°$; 252 mph   *[6.4]*

90. (A) 574 ft   (B) $2.9°$ *[6.1, 6.2]*

91. (A) $1{,}250$ ft   (B) $4.6°$ *[1.4, 2.1, 6.2]*

92. $A = r^2(\pi - \theta + \sin\theta)$ *[5.4]*

93. 2.6053 rad   *[5.4]*

94. (A)

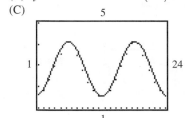

(B) $y = 2.845 + 1.275\sin(\pi x/6 - 1.8)$

(C)

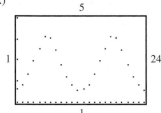

[3.3]

APPENDIX A

Exercise A.1

1. $-3, 0, 5$ (answers vary)   **3.** $\frac{2}{3}$ (answers vary)

5. (A) True   (B) False   (C) True

7. (A) $0.36\overline{36}$, rational   (B) $0.77\overline{7}$, rational

(C) $2.64575131\ldots$, irrational

(D) 1.62500, rational

9. (A) 2 and 3 (B) -4 and -3

(C) -5 and -4

11. $y + 3$ **13.** $(3 \cdot 2)x$ **15.** $7x$

17. $(5 + 7) + x$ **19.** $3m$ **21.** $u + v$

23. $(2 + 3)x$

25. (A) True   (B) False; $4 - 2 \ne 2 - 4$

(C) True   (D) False; $8/2 \ne 2/8$

Exercise A.2

1. $7 + 5i$ **3.** $-1 - 9i$ **5.** -18 **7.** $8 + 6i$

9. $-5 - 10i$ **11.** 34 **13.** $\frac{2}{5} - \frac{1}{5}i$

15. $\frac{4}{13} - \frac{7}{13}i$ **17.** $\frac{2}{25} + \frac{11}{25}i$ **19.** $5 - 2i$

21. $3 + 5i$ **23.** $13 - 19i$ **25.** $1 - \frac{3}{2}i$

27. 0 **29.** 1

Exercise A.3

1. 6.4×10^2 **3.** 5.46×10^9 **5.** 7.3×10^{-1}

7. 3.2×10^{-7} **9.** 4.91×10^{-5} **11.** 6.7×10^{10}

13. $56{,}000$ **15.** 0.0097 **17.** $4{,}610{,}000{,}000{,}000$

19. 0.108 **21.** Three   **23.** Five

25. Four   **27.** Two   **29.** Four

31. Four   **33.** $635{,}000$ **35.** 86.8

37. 0.00465 **39.** 7.3×10^2 **41.** 4.0×10^{-2}

43. 4.4×10^{-4} **45.** Three   **47.** Two

49. One   **51.** 12.0 **53.** 1.94×10^9

55. $53{,}100$ **57.** 159.0 cm   **59.** 96 ft^2

61. 28 mm^2 **63.** 1.65 cm   **65.** 13 in.

APPENDIX B

Exercise B.1

1. 3 **3.** -5 **5.** -1 **7.** 0 **9.** -20

11. -6 **13.** 5 **15.** $-\frac{5}{3}$ **17.** 6

19. $-3 - 2h$ **21.** -2

23. $g(f(2)) = g(-3) = -5$ **25.** Not a function

27. Function   **29.** Not a function

31. Function   **33.** $Y = \{1, 3, 7\}$

35. $F; X = \{-2, -1, 0\}, Y = \{0, 1\}$

37. 0 m; 4.88 m; 19.52 m; 43.92 m

39. $19.52 + 4.88h$; the ratio tends to 19.52 m/sec, the speed at $t = 2$

Exercise B.2

1.

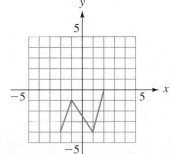

3.

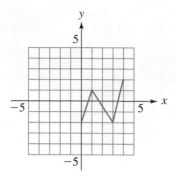

5.

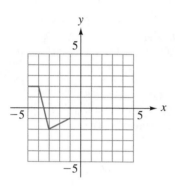

7.

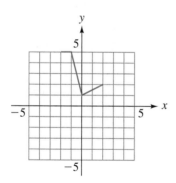

9.

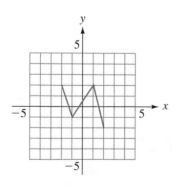

11.

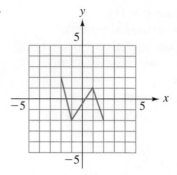

13.

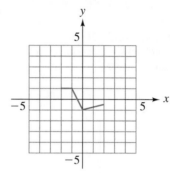

15.

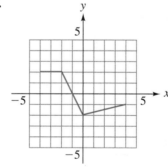

17. The graph of $g(x) = -|x - 3|$ is the graph of $y = |x|$ shifted right 3 units and reflected through the x axis.

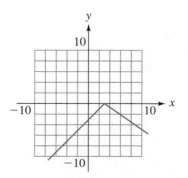

19. The graph of $f(x) = (x + 3)^2 + 3$ is the graph of $y = x^2$ shifted left 3 units and up 3 units.

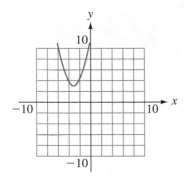

21. The graph of $f(x) = 5 - \sqrt{x}$ is the graph of $y = \sqrt{x}$ reflected through the x axis and shifted up 5 units.

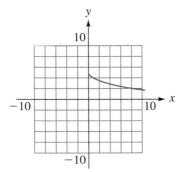

23. The graph of $h(x) = -\sqrt[3]{2x}$ is the graph of $y = \sqrt[3]{x}$ shrunk horizontally by a factor of $\frac{1}{2}$ and reflected through the x axis.

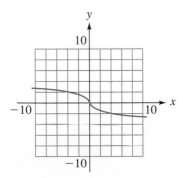

25. The basic function $y = x^2$ is shifted right 2 units and down 4 units to produce the graph of $y = (x - 2)^2 - 4$.

27. The basic function $y = x^2$ is shifted right 1 unit, reflected through the x axis, and shifted up 3 units to produce the graph of $y = 3 - (x - 1)^2$.

29. The basic function $y = \sqrt{x}$ is shrunk horizontally by a factor of $\frac{1}{2}$, reflected through the x axis, and shifted up 4 units to produce the graph of $y = 4 - \sqrt{2x}$.

31. The basic function $y = x^3$ is shifted right 1 unit, shrunk vertically by a factor of $\frac{1}{2}$, and shifted 2 units down to produce the graph of $y = \frac{1}{2}(x - 1)^3 - 2$.

33. $g(x) = 2\sqrt{x + 3} - 4$

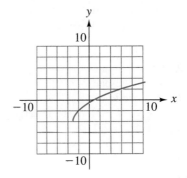

35. $g(x) = -\dfrac{1}{2}|x - 1| + 4$

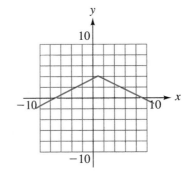

37. $g(x) = -(x + 2)^3 - 1$

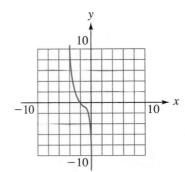

39. The graph of $y = \sqrt{x}$ is reflected through the y axis, shifted right 3 units, stretched vertically by a factor of 2, and shifted down 2 units to produce the graph of $y = 2\sqrt{3 - x} - 2$.

41. The graph of $y = x^2$ is shifted 1 unit to the right, shrunk vertically by a factor of $\frac{1}{2}$, and shifted down 4 units to produce the graph of $y = \frac{1}{2}(x - 1)^2 - 4$.

43. The graph of $y = x^3$ is shifted left 1 unit, shrunk vertically by a factor of $\frac{1}{4}$, and shifted down 2 units to produce the graph of $y = \frac{1}{4}(x + 1)^3 - 2$.

Exercise B.3

1. One-to-one   **3.** Not one-to-one
5. One-to-one   **7.** Not one-to-one
9. One-to-one   **11.** Not one-to-one
13. g is one-to-one; $g^{-1} = \{(-8, -2), (1, 1), (8, 2)\}$;
Domain: $\{-8, 1, 8\}$; Range: $\{-2, 1, 2\}$

15.

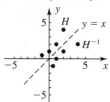

17. $f^{-1}(x) = \dfrac{x + 7}{2}$ **19.** $h^{-1}(x) = 3x - 3$

21.

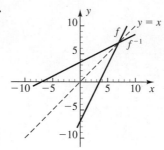

23. $f^{-1}(x) = \dfrac{x + 7}{2}; f^{-1}(3) = 5$

25. $h^{-1}(x) = 3x - 3; h^{-1}(2) = 3$

27. 4   **29.** x **31.** x

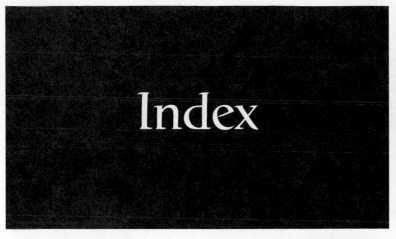

Index

A
Absolute value, 456
Accuracy (of calculations), 14, 28–29
Acoustics:
    frequency determinations, 155
    musical instruments, 194–196, 200
    noise control, 200–201
    sonic booms, 93–94, 227, 305–306
    sound frequency, 149–150
    sound waves, 221–222
    static, 200
Actual velocity, 388
Acute angles, 4, 524
Acute triangles, 352
Addition of ordinates, 190–192
Addition properties:
    of real numbers, 484
    of vectors, 403
Additive identity (of vectors), 403
Additive inverse (of vectors), 403
Adjacent side, 23
Advertising, 343
Aeronautics. *See also specific head-*
        ings, e.g.: Airplanes
    aircraft design, 365–366
    altitude determinations, 37
    distance determinations, 37, 38, 49
    navigation, 347, 373, 376, 429, 479
    rocket flight, 222
    sonic booms, 93–94
    spacecraft orbits, 236–237
Airplanes:
    air-to-air refueling, 34–35
    design of, 365–366
    ground speed of, 395
    navigation by, 347, 373, 376, 429,
        479
Algebraic functions, 503–504
Algebraic vector(s), 397–407

addition of, 399–401
conversion to/from geometric vec-
        tors, 397–399
defined, 398
determining magnitude of, 399
properties of, 403–404
scalar multiplication of, 400–401
unit vectors, 401–403
Alternating current, 84–85, 122,
        175–176, 187, 343
AM (amplitude modulation), 180
Ambiguous case (of SSA triangles),
        358
America's Cup, 452
Amplitude, 144, 175
    comparing, 142
    defined, 141
    and secant/cosecant functions, 208
    and tangent/cotangent functions, 203
Amplitude modulation (AM), 180
Anatomy, of the eye, 336–337, 366
Anchorage, Alaska, 349
Angle(s), 2–3
    acute, 4, 524
    and arcs, 6–7
    central, 6, 53–54
    complementary, 4, 24
    of complex numbers, 456
    coterminal, 3, 56–57
    critical, 91, 97
    degree measure of, 3–6
    of depression, 34
    of elevation, 34
    of incidence, 86, 91
    of inclination, 85
    obtuse, 4, 524
    plane geometry of, 524
    positive vs. negative, 3
    quadrantal, 99–100

radian measure of, 52–55
reference, 104–105
of reflection, 86
of refraction, 97
right, 4, 524
in standard position, 55–58
straight, 4, 524
subtended, 6
supplementary, 4, 524
    between two vectors, 412–414
Angular velocity, 66, 68–69
Aphelion, 453
Apparent diameter, 13
Apparent velocity, 388
Arbitrary triangles, 383
Arc(s):
    and angles, 6–7
    intercepted, 9
    subtended, 6
Archimedes, 8
Archimedes' spiral, 441
Architecture:
    area determinations in, 336
    height determinations in, 344
    length determinations in, 284–285
    roof overhang in, 38
Arc length (of circle), 58–59
Arcsin, 27
Arcsin transformation, 308
Area(s):
    of cones, 479
    of sector of a circle, 59–60
    of triangle, 378–383
        arbitrary triangles, 383
        given base and height, 378–379
        given three sides, 380–383
        given two sides and included
            angle, 379–380
Arguments, 456

Arms, motion of, 157
Associative property (of vectors), 403
Associative (term), 386
Astronomy:
  apparent diameters, 13
  Earth's angular velocity, 68
  Earth's orbit, 64–65, 71
  Earth-Venus distance, 366
  Hubble space telescope, 69
  Jupiter's orbit, 71
  lunar craters, 20
  lunar geography, 38
  lunar mountains, 20
  lunar movement relative to Jupiter, 72
  lunar phases, 166–168, 334
  Mercury's orbit, 328
  Neptune's velocity, 71
  planetary orbits, 469–471
  planetary speed, 453
  sun-Mercury distance, 36, 452–453
  sun's diameter, 64
  sun's rotation, 71
  sun-Venus distance, 35
  total solar eclipse, 10
  velocity due to rotation, 71
Auto-racing, 392–393
Axis of rotation, 83

Balloon flight, 49
Barrow, Alaska, 201–202
Baseball, 337
Beacon, rotating, 72
Beat (music), 273–274
Bell X-1 rocket plane, 94
Berliner and Berliner (psychologists), 95–96
Berlin Marathon, 5
Bicycling, 65, 70, 72, 349
Billiards, 226
Bioengineering:
  limb movement, 157, 222
  range of motion, 63–64
Biology:
  biorhythms, 174
  bird populations, 308
  of the eye, 12, 336–337
  modeling body temperature, 189
  perception of light, 177–178
  predator-prey analysis, 213–215
Biomechanics, traction in, 396, 410, 430
Biorhythms, 174
Birds, 308
Blood pressure, 156
Boating, 92–93, 122

determining distance in, 348
lightning protection in, 38–39
navigation in, 343, 365, 376, 388–389, 395, 478
velocities in, 395
Body temperature, 189
Bow waves, 92–93, 98, 122
Breaking waves, 188
Breathing rate, 156
Bungee jumping, 189, 344
Buoy, mass of a, 151, 221
Business, seasonal sales cycles, 200, 222, 228
Business applications:
  advertising, 343
  seasonal sales cycles, 336

Calculator(s). See also Graphing calculator(s)
  evaluating trigonometric functions with, 78–80
  evaluating trigonometric ratios with, 25–28
  finding values involving inverse cosine with, 298
  finding values involving inverse sine function with, 294
  finding values involving inverse tangent function with, 301
Camera lenses, 20–21
Carbon dioxide, modeling atmospheric, 201
Cardano, Girolamo, 486
Cardioids, 443–444
Central angle(s):
  defined, 6
  radian measure of, 53–54
Cerenkov, Pavel A., 94–95
Chair lifts, 405–406
Chord, 9
Circle(s):
  arc length of, 58–59
  area of sector of, 58–60
  great, 306
  plane geometry of, 528
  sketching, in polar coordinates, 442
  unit, 73–78
Circuits, electrical, 155, 171, 222, 227, 348, 478
Circumference, of Earth, 7–8
Climate, 337
Closure property (of real numbers), 484
Coastal patrol, 364
Coastal piloting, 36–37, 365
Cofunctions, 25, 250

Cofunction identities, 249–250, 252
Columbus, Ohio, 328
Combined forms of trigonometric functions, graphing, 190–197
  Fourier series, 196–197
  with graphing calculator, 191–193
  by hand, 190–191
  sound waves, 193–196
Combined property (of real numbers), 484
Communications, fiber optics in, 91
Commutative property:
  of real numbers, 484
  of vectors, 403
Commutative (term), 386
Complementary angles, 4, 24, 524
Complementary relationships (in right triangles), 25
Complex numbers, 453–460, 485–488
  converting, from rectangular to polar form, 456–458
  defined, 486–487
  finding powers of, 463–464
  operations with, 487–488
  plotting, 454–455
  in polar form, 453, 454
  polar-rectangular equations for, 456
  products/quotients of, in polar form, 459–460
  in rectangular form, 453–454
  roots of, 465
  and square roots of negative numbers, 488
Complex plane, 453
Components (of vectors), 386, 391
Compression (in graphing), 510
Concorde, 94
Conditional equations, 230, 315
Conics, 469
Construction:
  area calculations for, 21
  cost determinations in, 385
  crane positioning, 225
  of fire lookouts, 377
  ladder placement, 37
  length determinations in, 268
  of ramps, 48
  retaining walls, 98
  volume determinations in, 268
Conversions:
  DD to DMS, 6
  between degrees and radians, 54–55
  DMS to DD, 5
  of expressions with fundamental identities, 233–234

geometric vectors to/from algebraic vectors, 386–387
polar coordinates to/from rectangular coordinates, 436–437
polar form to/from rectangular form, 456–458
Coordinate, 483
Coplanar concurrent forces, 404–405
Cordelia (Uranus's moon), 13
Cornea, 12, 336
Cos, *see* Cosine
Cosecant (csc), 23–25
  graphing:
    $y = A \csc(Bx + C)$, 207–210
    $y = \csc x$, 132–134
  inverse cosecant function, 309–310, 312–313
Cosine (cos), 23–25. *See also* Law of cosines
  cofunction identity for, 249
  double-angle identity for, 258–259
  equations involving, 317–318
  graphing:
    $y = \cos x$, 128–129
    $y = k + A \cos Bx$, 140–151
    $y = k + A \cos(Bx + C)$, 159–168
  half-angle identity for, 261
  inverse cosine function, 294–298
  as periodic function, 107–108
  product-sum identities for, 269–270
  properties of, 175
  sum and difference identities for, 247–249
  sum-product identities for, 270–271
Cotangent (cot), 23–25
  graphing:
    $y = A \cot(Bx + C)$, 202–206
    $y = \cot x$, 132
  inverse cotangent function, 309–313
Coterminal angles, 3, 56–58
Coyotes, populations of, 213–214
Cranes, 225
Crest, 181
Critical angle, 91, 97
Csc, *see* Cosecant
The Curious History of Trigonometry (Katz), 17*n.*
Current, electric, *see* Electric current

Damped harmonic motion, 183–186
Daylight, 349
Decimal degree (DD), 4–6
Degrees, conversion to/from radians, 54–55

Degree measure (of angles), 3–5
Degree-minute-second (DMS), 4–6
De Moivre, Abraham, 462–463
De Moivre's theorem, 463–464
Dependent variables, 499
Descartes, René, 52, 87
Descartes' law, 87
Diameter:
  apparent, 13
  of sun, 9
Diastolic pressure, 156
Difference identities, 252
  for cosine, 247–248
  for sine, 250
  for tangent, 250–251
  using, 252–255
Dimensionless numbers, 16*n.*
Directed line segments, 386
Directrix, 469
Displacement vector, 417
Distance, between two points, 248
Distributive property:
  of real numbers, 484
  of vectors, 403
Diving, 97
DMS, *see* Degree-minute-second
Dogsledding, 416
Domain:
  for cotangent/secant/cosecant, 310
  rule form definition, 499
  set form definition, 501
Dot product, 410–417
  and angle between vectors, 412–413
  defined, 410
  orthogonality determinations with, 414
  properties of, 411
  and vector projection, 414–417
  in work problems, 416–417
Double-angle identities, 258–260
Drive wheels and shafts, 65

Earth, 35–36
  approximating circumference of, 7–8
  axis of rotation, 83
  distance to Venus, 366
  polar equation of orbit, 471
  radius of, 39
  speed of orbit, 71
Earth science:
  carbon dioxide in atmosphere, 201
  climate, 337
  daylight duration, 349
  hours of daylight, 328
  lake temperatures, 199

pollution, 221
twilight duration, 480
Eccentricity (conic), 469, 470
EIA (Energy Information Administration), 158
Einstein, Albert, 175
Electrical circuits, 155, 171, 222, 227, 348, 478
Electrical wind generators, 67, 79–80
Electric current, 84, 122, 175–177, 187, 327, 343
Electromagnetic spectrum, 178
Electromagnetic waves, 177–181, 188–189, 222, 227
Electrons, 179
Electronic synthesizers, 184
Elements (Euclid), 13–14
Elementary functions, 504
Emergency vehicles, 228
Energy:
  particle, 94–95
  solar, 83–84
Energy consumption, 171–172
Energy Information Administration (EIA), 158
Engineering:
  bicycle chains, 349
  bicycle movement, 70
  determining compression, 430
  fire lookout, 377
  freeway design, 395
  fuel tanks, 307
  gear revolutions, 121
  gears, 226–227
  oil pumps, 63
  piston movement, 84
  pistons, in, 366
  pulley systems, 307–308
  race track design, 392–393
  railroad track, 479–480
  railroad tunnels, 344
  rotation in water, 157–158
  shaft rotation, 70
  shafts and drive wheels, 65
  tunnel design, 376–377, 429
  velocity of automobile wheels, 121
Envelope, 185
Epimetheus (Saturn's moon), 13
Equal vectors, 386, 398
Equation(s). *See also* Trigonometric equations
  conditional, 230, 315
  finding, for a graph, 148–149
  solving cubic, 466
  solving quadratic, 465–466

Eratosthenes, 7
Euclid, 13–14, 86
Euclid's theorem, 14–17, 526
Euler, Leonhard, 52, 460
Expansion (of graphs), 510
Extraneous solutions, 321*n.*
Eye, 12, 59, 336–337, 366

Factoring, 319–323
Faraday, Michael, 175
Fermat, Pierre, 86
Fermat's least-time principle, 86
Ferris wheels, 156–157, 327
Fiber optics, 91
Field (in physics), 175
Field of view, 64
Firefighting, 226, 364, 377
Floating objects, 150–151, 156, 221
FM (frequency modulation), 180
Focal length, 20
Focal point, 20
Focus (of conics), 469
Foot-pound (ft-lb), 416*n.*
Force:
    centrifugal, 392–393
    vectors of, 390–391
Forestry, 226, 364
Fourier, Joseph, 196
Fourier series, 184, 196–197
Four-leaved rose, 444–445
Frank, Il′ja, 94–95
Freeway design, 395
Frequency, 149–150, 175
Frequency modulation (FM), 180
Fuel tanks, 307
Function(s), 498–512
    algebraic, 503–504
    classification of, 503–504
    elementary, 504
    graphing, 505–512
    inverse, 518–521
    notation for, 502–503
    one-to-one, 516–518
    rule form definition of, 498–500
    set form definition of, 501
    vertical-line test of, 500
Fundamental identities, 109–110,
        230–234, 237–245
    and conversion of expressions to
        equivalent forms, 233–234
    and evaluation of trigonometric
        functions, 231–232
    for negatives, 231
    Pythagorean identities, 231
    quotient identities, 230

reciprocal identities, 230
testing, using a graphing calculator,
        243–245
verifying, 237–243
Fundamental tone, 184

Galileo Galilei, 42
Gears, 72, 121
Generators, alternating current,
        175–177, 187
Geography:
    depth of canyons, 17–18
    distances between cities, 12, 49, 225
    latitude length, 40
Geology, determining height in, 257
Geometric vector(s), 386–393,
        396–399
    addition of, 386–387
    and centrifugal force, 392–393
    conversion to/from algebraic vectors,
        386–387
    for force, 390–391
    resolution of, into components,
        391–392
    standard position for, 397
    sum of two nonparallel vectors, 386
    sum of two vectors **u** and **v,** 386
    for velocity, 388–389
Geostationary orbit, 377
Geostationary satellites, 71
Gnomon, 356
Golf, 97, 337
Graphing. *See also* Graphing of
        trigonometric functions;
        Sketching
    basic functions, 505, 506
    combined transformations, 511
    inverse of one-to-one functions,
        520–521
    reflections, 507–508
    standard position, sketching angles
        in, 56–57
    stretching/shrinking in, 508–510
    transformations of functions, 505–512
    translations in, 507
    vertical/horizontal shifts, 505–507
Graphing calculator(s):
    degree mode in, 25
    finding equation of simple harmonic
        with, 165–166
    graphing trigonometric functions
        with, 134–137, 191–193
    solving trigonometric equations
        with, 329–333
    solving trigonometric inequalities

with, 333–334
testing identities with, 243–245,
        252–253
Graphing of trigonometric functions,
        124–125
    combined forms, 190–197
    Fourier series, 196–197
    with graphing calculator, 191–193
    by hand, 190–191
    sound waves, 193–196
    with graphing calculator, 134–137,
        191–193
    $y = A \sec(Bx + C)$

        $[y = A \csc (Bx + C)]$,
        207–210
    $y = A \tan(Bx + C)$

        $[y = A \cot (Bx + C)]$,
        202–206
    $y = \cos x$, 128–129
    $y = \cot x$, 132
    $y = \csc x$, 132–134
    $y = k + A \sin Bx$,
        $[y = k + A \cos Bx]$, 140–151
    $y = k + A \sin(Bx + C)$
        $[y = k + A \cos(Bx + C)]$,
        159–168
    $\sec x$, 132–134
    $\sin x$, 125–127
    $\tan x$, 129–131
Gravitational constant, 42
Great circle, 306
Greeks, ancient, 2, 7, 14

Half-angle identities, 261–264
Half Dome, 257
Harmonics, 159–160, 165–166, 175
Harmonic motion, 183–186
    damped, 183–186
    simple, 159
Harmonic tone, 184
Heartbeat, 156
Helicopters, 343
Heron of Alexandria, 380
Heron's formula, 380–382
Hertz, Heinrich Rudolph, 149
Hipparchus, 2
Horizontal-line test, 517
Horizontal shifts, in graphing, 507
Horizontal shrinking, 510
Horizontal stretching, 510
Horizontal translation, 160
Hubble space telescope, 69

Huygens (European Space Agency), 13
Hypotenuse, 22–24
Hz (unit), 149

Identities, 230–282
    cofunction, 249–250, 252
    defined, 230
    double-angle, 258–260
    fundamental, 109–110, 230–234,
        237–245
    half-angle, 261–264
    of inverse functions, 519
        inverse cosecant, 313
        inverse cosine, 295
        inverse cotangent, 313
        inverse secant, 313
        inverse sine, 291
        inverse tangent, 300
    for negatives, 231
    product-sum, 269–270
    Pythagorean, 231
    quotient, 230
    reciprocal, 230
    solving trigonometric equations
        using, 320–324
    sum and difference, 251–252
        for cosine, 247–249
        for sine and tangent, 250–251
        using, 252–255
    sum-product, 270–274
    testing, with graphing calculator,
        243–245, 252–253
    transformation, 278–279
    verifying, 237–243, 254–255, 260,
        264
Identity property (of real numbers),
    484
Imaginary axis, 453
Imaginary part (of complex numbers),
    453
Imaginary unit, 486
Independent variables, 499
Index of refraction, 86–87, 89, 97, 98
Indirect measurement, 2
Initial point, 386
Initial side, 2
Inner product, 410
Integers, 56n., 482
Intercepted arc, 9
Inverse cosecant function, 309–310,
    312–313
    defined, 310
    finding values of, with calculator,
        313
    identities involving, 313

Inverse cosine function, 294–298
    defined, 295
    exact values of equations involving,
        296–297
    finding values of, with calculator,
        298
    identities involving, 295
Inverse cotangent function, 309–313
    defined, 310
    exact values of equations involving,
        311
    finding values of, with calculator,
        312–313
    identities involving, 313
Inverse functions, 518–521
    defined, 519
    determining, 519–520
    graphs of, 520–521
    identities, 519
Inverse property (of real numbers), 484
Inverse secant function, 309–313
    defined, 310
    exact values of equations involving,
        311–312
    identities involving, 313
Inverse sine function, 288–294
    defined, 290
    exact values of equations involving,
        292–293
    finding values of, with calculator,
        294
    identities involving, 291
Inverse tangent function, 298–302
    defined, 299
    equivalent algebraic expressions
        involving, 300
    exact values of equations involving,
        300–301
    finding values of, with calculator,
        301
    identities involving, 300
Inverse trigonometric functions, 288
Irrational numbers, 483, 485

Javelin throw, 267
Joule, 416n.
Juliet (Uranus's moon), 13
Jupiter, 71

Katz, Victor J., 17n.
Kepler, Johannes, 453

Lakes, 199
Lambert, Johann, 8
Latitude, 12, 40

Lawn mowing, 416–417
Law of cosines, 367–374
    derivation of, 367–368
    solving SAS case using, 369–370
    solving SSS case using, 371–373
Law of reflection, 86
Law of refraction, 87
Law of sines, 352–362
    derivation of, 353–354
    solving ASA and AAS cases using,
        355–357
    solving SSA cases using, 358–362
Legs (of humans), 157, 222
Legs (of right triangles), 22–23
Leibniz, Gottfried, 8
Lemniscate (polar curve), 448
Lenses, camera, 20–21
Lifeguarding, 40–41
Light:
    filters for, 327
    intensity of, 83
    refraction of, 86–91, 122, 258
    speed of, 114–115
Light cones, 95, 98
Lightning rods, 38–39
Light waves, 177–179, 227
Light-year, 489
Linear velocity, 66–69
Line segments, directed, 386
Liu Hui, 8
Longitude, 12
Lunar phases, modeling, 166–168
Lungs, 155

Mach, Ernst, 94
Mach number, 94
Magnitude, 386, 399
Marathon, 5
Mass, determinations for floating
    objects, 151, 156, 221
Mathematical induction, 463
Mauna Loa, Hawaii, 201
Maxwell, James, 175
Mean sidereal day, 68
Mean solar day, 68
Measurement, indirect, 2
Medicine:
    eye surgery, 336–337
    fiber optics in, 91
    stress tests, 48–49
    traction in, 430
    use of traction, 396, 410
Mercury, 36, 328, 452–453
Meter, 416n.
Miles, nautical, 12

Milwaukee, Wisconsin, 139–140, 228
Mining:
    length of air shafts, 34
    length of air vents in, 16–17
Minutes (angle measurement), 4
Mirror Lake, 257
Modulus (mod), 456
Mollweide's equation, 364
Moon, 10, 13, 20, 38, 59, 166–168, 334
Moving wave equation, 182
Multiplication identity (of vectors), 403
Multiplication properties (of real numbers), 484
Music:
    beats in, 273–274, 276, 285
    sound intensity, 72
    tones in, 279
Musical instruments:
    modeling sounds produced by, 194–196, 200
    tuning, 273

National Data Buoy Center (NDBC), 158
Natural gas:
    consumption of, 171–172
    modeling storage of, 158–159
Natural numbers, 482
Nautical miles, 12
Navigation:
    by airplanes, 347, 373, 376, 429, 479
    by boat, 343, 365, 376, 388–389, 395, 478
    coastal piloting, 36–37
    determining distances, 37–38, 361–362
    determining location in, 39–40
    by ships, 343
    and sun, 49
Navigational compass, 388
Negatives, identities for, 231
Negative angles, 3
Negative real numbers, 483
Neptune, 71
New Orleans, Louisiana, 328
Newton, Isaac, 175, 462
Newton-meter (unit), 416n.
Newton's second law of motion, 404
Newton (unit), 416n.
Nobel prize, 94
Noise control, 200–201
Nonrigid transformations, 508
Norm (of vectors), 399

nth-root theorem, 464–467
Nuclear particles, 70–71
Numbers:
    complex, 485–488
    dimensionless, 16n.
    irrational, 483, 485
    natural, 482
    rational, 482, 485
    real, 482–484
    square roots of negative, 488

Oblique triangles, 352
Obtuse angles, 4, 524
Obtuse triangles, 352
Oceanography:
    modeling sea temperatures, 158
    tidal current curve, 448
Offshore pipelines, 41
Oil pumping, 63
One-to-one correspondence, 516
"On the Derivatives of Trigonometric Functions," 479n.
Opposite direction (of vectors), 386
Opposite side, 23
Optics:
    modeling light waves, 86–91, 177–179
    refraction, 86–91, 122
    refraction of light, 227
    speed of light in water, 114–115
Optical illusions, 95–96
Orbiting Astronomical Observatory, 12
Ordered pairs, 501
Ordinates, addition of, 190–192
Origin, 432, 483
Orthogonal vectors, 412, 414
Overtones, 194

Paddle wheels, 157
Parallels of latitude, 40
Parallelograms, 527
Parallelogram rule, 386
Parallel vectors, 412
Parking lots, 39
Partial tones, 194
Particle energy, 94–95
Pendulums, 63
Perception, 95–96
Perihelion, 453
Period(s) (of periodic function), 108
    comparing, 143
    formula for, 144
    and frequency, 149
Periodic functions, 107–108
Perpendicular vectors, 412

Perspective illusions, 95–96
Phase shift, 160–161, 175
Photography:
    field of view, 12, 64
    lens focusing, 20–21
    polarizing filters, 327
Physics. See also Acoustics; Optics
    atomic particle movement, 70–71
    bow wave modeling, 92–93
    bow waves, modeling of, 92–93
    compression determinations, 406–407
    electric current modeling, 175–177, 187
    electromagnetic waves modeling, 177–181, 188–189, 222
    floating objects, 150–151
    harmonic motion, 183–186
    harmonic motion modeling, 183–186
    light refraction, 258
    mass determinations, 156
    particle energy modeling, 94–95
    spring—mass systems, 155, 189, 221, 222, 227, 228, 279, 285, 324–325, 327, 344, 348
    tension determinations, 405–407, 409, 479
    velocity studies in, 42
    water waves modeling, 181–183, 187–188, 222
    wave movement in, 171
    wind farms, 308
Physiology:
    blood pressure, 156
    of breathing, 155, 199
Pi, estimating, 113
Pianos, 274
Piloting, coastal, 36–37
Pipelines, offshore, 41
Pistons, 84, 366
Pittsburgh, Pennsylvania, 337
Pixels, 134
Plane geometry, 524–528
    angles, 524
    circles, 528
    quadrilaterals, 527
    triangles, 525–527
Planetary orbits, 469–471
Point(s), distance between two, 248
Point-by-point sketching, 441–442
Polar axis, 432
Polar coordinates, 432–438
    application example, 448–449
    calculator conversion of, 436–437
    complex numbers in, 453–460

converting, to/from rectangular coordinates, 434–439, 456–458
defined, 432
plotting points in, 433
sketching equations in, 440–445
standard polar curves, 447–448
Polar coordinate systems, 432–433
Polar equations:
of conics, 469
graphing, on calculators, 445–446
sketching, 440–445
Polar form (of complex numbers), 454
Polarizing light filters, 327
Pole, 432
Pollution, 221
Position vector, 397
Positive angles, 3
Positive real numbers, 483
Predator-prey relationships, 213–215
Principal values, 295
Products, of complex numbers, 459–460
Product-sum identities, 269–270
Propellers, 158
Psychology, of perception, 95–96, 98
Pulleys, 65, 121, 307–308
Pulse rate, 156
Pumps, 63
Pythagoreans, 485
Pythagorean identities, 231
Pythagorean theorem, 22–23, 73, 75n., 525

Quadrants, 62n.
Quadrantal angles, 99–100
Quadratic formula, 323–324
Quadrilaterals, 527
Quantities, scalar vs. vector, 385
Quotients, of complex numbers, 459–460
Quotient identities, 230

Rabbits, 213–214
Radians, conversion to/from degrees, 54–55
Radian measure, 52–54
Radio towers, 348
Radio waves, 180
Radius (of Earth), 39
Radius vector, 397
Railroads, 479–480
grades of, 226
tunnels for, 344
Range:
rule form definition, 499

set form definition, 501
of x and y, 134
Range of motion, 63–64
Rapid polar sketching, 443
Ratios, trigonometric, 23–28
Rational numbers, 482, 485
Rays, 2
Real axis, 453
Real numbers, 482–484
properties of, 483–484
and real lines, 483
system of, 482–483
Real number line, 455, 483
Real part (of complex numbers), 453
Reciprocals, 25n.
Reciprocal identities, 230
Reciprocal relationships, 25–26
Rectangles, 527
Rectangular coordinates, converting to/from polar coordinates, 434–439, 456–458
Rectangular coordinate system, 52
quadrants in, 62n.
standard position in, 55–56
Rectangular form (of complex numbers), 453
Reduced-gravity simulation, 430
Reference angles, 104–105
Reference triangles, 104–107
Reflected light, 89–91
Reflections, in graphing, 507–508
Refracted light, 86–88, 97, 114–115
Refraction, 227
Resonance, 183–185
Resultant force, 390–391
Resultant vectors, 386
Resultant velocity, 388
Retaining walls, 98
Retina, image of moon on, 59
Rhind Papyrus, 8
Rhombuses, 527
Right angles, 4, 524
Right triangles, 22–23, 525
applications of, 33–36
complementary relationships in, 25
reciprocal relationships in, 26
solving, 22, 28–31
trigonometric ratios for, 23–25
Rigid transformations, 508
Roads, planning, 268
Rockets, 222
Roof overhang, 38
Rotary motion, 156–157

Sailing, 462
Sales cycles, 200, 222, 228, 336
Same direction (of vectors), 386
San Antonio, Texas, 173
San Francisco Bay, California, 173–174, 223
San Francisco Light Station, 448–449
Satellites, 64, 70, 121–122, 236, 366, 377, 429
geostationary, 71
telescopes as, 12
Saturn, 13, 471
Saws, 226–227
Sawtooth waves, 197
Scalars, 400
Scalar component(s):
defined, 398
of u on v, 415
Scalar multiplication properties (of vectors), 403
Scalar product, 410
Scalar quantities, 385
Scale (of x and y), 134
Scales (weighing), 155
Search and rescue, 376
Seasons, 83–84
Secant (sec), 23–25
confunction identity for, 250
graphing:
$y = A \sec(Bx + C)$, 207–210
$y = \sec x$, 132–134
inverse secant function, 309–313
Seconds (angle measurement), 4
Sector (of a circle):
area of, 59–60
defined, 59
Semiperimeter, 380
Sensory perception, 95–96
Shafts, 65, 70
Shrinking (of graphs), 508–510
Side(s):
initial, 2
terminal, 2–3
Side adjacent, 23
Side opposite, 23
Significant digits, 489–495
in decimal fractions, 493
defined, 491, 492
determining, for calculated values, 493–494
and rounding, 494–495
and scientific notation, 489–491
for solving triangles, 353
Sign (of trigonometric functions), 80–81

Similar triangles, 13–18
Simple harmonic, 159–160
Simple harmonic motion, 159
Simple harmonics, 159, 165–166, 175
Sine (sin), 23–25. *See also* Law of
sines
  confunction identity for, 249
  double-angle identity for, 258
  equations involving, 316–317
  graphing:
    $y = k + A \sin Bx$, 140–151
    $y = k + A \sin(Bx + C)$,
      159–168
    $y = \sin x$, 125–127
  graphing horizontal translations of,
    159–161
  half-angle identity for, 261
  inverse sine function, 288–294
  as periodic function, 107–108
  product-sum identities for, 269–270
  properties of, 175
  sum and difference identities for,
    250
  sum-product identities for, 270–271
Sine waves, 193
Sinusoidal curve, 173
Sinusoidal regression, 173
Sketching. *See also* Graphing
  point-by-point, 441–442
  of polar equations, 440–445
  rapid polar, 443–445
Skiing, 405–406
Skylights, 284–285
Slingshot effect, 236–237
Snell, Willebrod, 87
Snell's law, 87
Soccer, 306
Solar eclipse, total, 10
Solar energy, 49–50, 83–84, 479
Solutions, extraneous, 321n.
Solving triangles, 352–353
  right, 22, 28–31
  strategy for oblique triangles, 371
  strategy for SAS cases, 369
  strategy for SSS cases, 371
Sonic booms, 93–94, 227, 305–306
Sound, 149–150, 155, 200. *See also*
  Acoustics
Sound barrier, 70
Sound cones, 93–94, 98
Sound waves, 93–94, 193–196,
  221–222
Spacecraft, 236–237, 306
Space flight, 12
  angular velocity, 70

communication, 377
determining distances during
  orbit, 39
linear velocity, 69
orbiting spacecraft, 306
reduced-gravity simulation, 430
re-entry of shuttles, 38
satellites, 71, 121–122, 366, 377,
  429
shuttle movement relative to Earth,
  71–72
Space shuttle, 71–72
Speigle, M. R., 479n.
Spirals, sketching, 441
Sports:
  auto-racing, 392–393
  baseball, 337
  bicycling, 65, 70, 72, 349
  billiards, 226
  bungee jumping, 344
  golfing, 97, 337
  javelin throwing, 267
  marathon, 5
  sailboat racing, 452
  soccer, 306
  tennis, 19
Spring—mass systems, 155, 189, 221,
  222, 227, 228, 279, 285,
  324–325, 327, 344, 348
Squares, 527
Square roots (of negative numbers),
  488
Square wave, 197
Standard position:
  angles in, 55–58
  for geometric vectors, 397
Standard vector, 397
Standard viewing window (graphing
  calculator), 134
Static, modeling sound and, 200
Static equilibrium, 404–407, 409, 479
Statue of Liberty, 343
Straight angles, 4, 524
Stress tests, 48–49
Stretching (of graphs), 509–510
Subtended (term), 6
Sum:
  of two nonparallel vectors, 386
  of two vectors **u** and **v,** 386
Sum identities, 251
  for cosine, 249
  for sine, 250
  for tangent, 251
  using, 252–254
Sum-product identities, 270–274

Sun, 13
  approximating diameter of, 9
  distance to Mercury, 452–453
  eclipse of, 10
  rotation of, 71
Sundials, 356–357
Sunrise times, modeling, 139–140, 173
Sunset times, modeling, 139–140, 223
Supersonic aircraft, 94
Supplementary angles, 4, 524
Surveying:
  determining angles, 371–372
  determining distance, 50, 364, 365,
    375–376, 378, 429, 478
  determining height, 37, 41–42, 48,
    64, 226, 257, 365
  determining land cost, 385
  positioning walkways, 227–228
Suspension systems, 183–185
Systolic pressure, 156

Tacoma Narrows Bridge, 185
Tail-to-tip rule, 386
Tamm, Igor, 94–95
Tangent (tan), 23–25
  confunction identity for, 250
  double-angle identity for, 259
  equations involving, 318
  graphing:
    $y = A \tan(Bx + C)$, 202–206
    $y = \tan x$, 129–131
  half-angle identity for, 261
  inverse tangent function, 298–302
  sum and difference identities for,
    250–251
Temperature, modeling, 158, 173, 228
Tergat, Paul, 5
Terminal point, 386
Terminal side, 2–3
Three-leaved rose, 448
Tides, 448
Tightrope walking, 409
Time, measuring, 65
Titan (Saturn's moon), 13
Torque, 404
Total solar eclipse, 10
Transformations (of functions), 505
Transformation identity, 278–279
Translation:
  horizontal, 160
  vector, 386
  vertical, 146–147
Treadmills, 48–49
Triangle(s):
  acute, 352, 525

congruent, 526–527
equilateral, 526
finding area of, 378–383
    arbitrary triangles, 383
    given base and height, 378–379
    given three sides, 380–383
    given two sides and included
        angle, 379–380
isosceles, 526
oblique, 352, 525
obtuse, 352, 525
plane geometry of, 525–527
reference, 104–107
right, *see* Right triangles
similar, 13–18, 526
solving, *see* Solving triangles
special, 525
sum of angle measures, 14
Trigonometric equations, 315–325
approximate solutions to, 322–324
cosine equation, 317–318
exact solutions to, 319–322
sine equation, 316–317
solving, algebraically, 316–319
solving, using identities, 320–324
solving, with factoring, 319–323
solving, with graphing calculator,
    329–333
spring—mass system example,
    324–325
tangent equation, 318
Trigonometric form (of complex num-
    bers), 456
Trigonometric functions:
calculator evaluation of, 78–80
defined, 73–74
evaluating, 75–77
exact values of, at special angles,
    99–104
fundamental identities involving,
    109–110
graphing, *see* Graphing of trigono-
    metric functions
inverse, *see* Inverse trigonometric
    functions
periodic, 107–108
and reference triangles, 104–107
sign properties of, 80–81
unit circle approach to, 73–77
Trigonometric inequalities, using
    graphing calculator to solve,
    333–334

Trigonometric ratios, 23–25, 52
calculator evaluation of, 25–28
complementary relationships with, 25
finding inverses of, 28
reciprocal relationships with, 26
for right triangles, 23–25
Trigonometry, spherical, 362
Troughs (of waves), 181
Tsu Ch'ung-chih, 8
Tsunami, 188
Tuning forks, 149, 184, 221–222
Tunnels, 376–377, 429
Twilight duration, 480

Ultraviolet waves, 181
Unit circle, 73–78
U.S. Air Force, 94
U.S. Coast and Geodetic Survey, 448
Unit vectors, 401–403
Uranus, 13

Values, principal, 295
Variables, independent vs. dependent,
    499
Vectors. *See also* Algebraic vector(s);
    Geometric vector(s)
addition properties of, 403
angle between two, 412–414
displacement, 417
dot product of, *see* Dot product
equal, 386, 398
force, 390–391
with opposite direction, 386
orthogonal, 412, 414
parallel, 412
perpendicular, 412
projection of, 414
resultant, 386
with same direction, 386
scalar component of, 414–415
scalar multiplication properties of,
    403
in static equilibrium problems,
    404–407
translations of, 386
unit, 401–403
velocity, 388–389
zero, 386, 398
Vector quantities, 385
Velocity:
actual, 388
angular, 66, 68–69

apparent, 388
linear, 66–69
resultant, 388
vectors for, 388–389
Velocity vector, 388–389
Venus, 35
distance to Earth, 366
polar equation of orbit, 471
Verification of identities, 237–243,
    254–255, 260, 264
Vertex, 3
Vertical-line tests (of functions), 500
Vertical shifts, in graphing, 507
Vertical shrinking, 510
Vertical stretching, 510
Vertical translation, 146–147
Viewing window (graphing calculator),
    134
Volume:
of a cone, 225
maximum interior, 268
Voyager I, 236
Voyager II, 236

Walkways, 227–228
Warbling, 273
Water temperatures, 199
Waves:
bow, 92–93, 122
breaking, 188
electromagnetic, 177–181, 188–189,
    222, 227
light, 86–91, 122, 177–179, 227
radio, 180
sawtooth, 197
sound, 93, 94, 221–222
square, 197
ultraviolet, 181
water, 92, 122, 171, 181–183,
    187–188, 222, 479
Wheels:
angular velocity of, 121
and shafts, 65
Wind farms, 308
Wind generators, 67, 79–80
Work, 416–417

Yeager, Chuck, 94
Yosemite National Park, 257

Zero vectors, 386, 398

GRAPHING TRIGONOMETRIC FUNCTIONS (3.1 – 3.3)

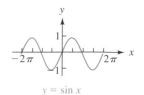

$y = \sin x$

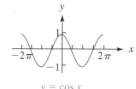

$y = \cos x$

$y = A \sin(Bx + C)$ $y = A \cos(Bx + C)$

$\text{Amplitude} = |A|$ $\text{Period} = \dfrac{2\pi}{B}$

$\text{Frequency} = \dfrac{B}{2\pi}$

$\text{Phase shift} = -\dfrac{C}{B} \begin{cases} \text{left} & \text{if} -C/B < 0 \\ \text{right} & \text{if} -C/B > 0 \end{cases}$

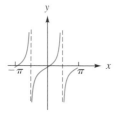

$y = \tan x$

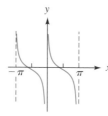

$y = \cot x$

$y = A \tan(Bx + C)$ $y = A \cot(Bx + C)$

$\text{Period} = \dfrac{\pi}{B}$

$\text{Phase shift} = -\dfrac{C}{B} \begin{cases} \text{left} & \text{if} -C/B < 0 \\ \text{right} & \text{if} -C/B > 0 \end{cases}$

LAW OF SINES (6.1)

$$\frac{\sin \alpha}{a} = \frac{\sin \beta}{b} = \frac{\sin \gamma}{c}$$

LAW OF COSINES (6.2)

$a^2 = b^2 + c^2 - 2bc \cos \alpha$

$b^2 = a^2 + c^2 - 2ac \cos \beta$

$c^2 = a^2 + b^2 - 2ab \cos \gamma$

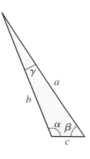

HERON'S FORMULA FOR AREA (6.3)

If the semiperimeter s is

$$s = \frac{a + b + c}{2}$$

then

$$A = \sqrt{s(s - a)(s - b)(s - c)}$$

INVERSE TRIGONOMETRIC FUNCTIONS (5.1, 5.2)

$y = \sin^{-1} x$ means $x = \sin y$

where $-\dfrac{\pi}{2} \leq y \leq \dfrac{\pi}{2}$ and $-1 \leq x \leq 1$

$y = \cos^{-1} x$ means $x = \cos y$

where $0 \leq y \leq \pi$ and $-1 \leq x \leq 1$

$y = \tan^{-1} x$ means $x = \tan y$

where $-\dfrac{\pi}{2} < y < \dfrac{\pi}{2}$

and x is any real number

$y = \cot^{-1} x$ means $x = \cot y$

where $0 < y < \pi$ and x is any real number

$y = \sec^{-1} x$ means $x = \sec y$

where $0 \leq y \leq \pi, y \neq \dfrac{\pi}{2}$,

and $x \leq -1$ or $x \geq 1$

$y = \csc^{-1} x$ means $x = \csc y$

where $-\dfrac{\pi}{2} \leq y \leq \dfrac{\pi}{2}, y \neq 0$,

and $x \leq -1$ or $x \geq 1$

The ranges for $\sec^{-1}$ and $\csc^{-1}$ are sometimes selected differently.

VECTORS (6.4 – 6.6)

Vector $\mathbf{v} = \vec{AB}$